Precision Technologies for Digital Agriculture

Precision Technologies for Digital Agriculture

Harnessing IoT, Big Data, Crop Modeling, and AI for Agricultural Production

Edited by

Sushil Kumar Himanshu

Agricultural Systems and Engineering, Faculty of Food, Agriculture and Natural Resources, Asian Institute of Technology, Pathum Thani, Thailand

Hemendra Kumar

College of Agriculture and Natural Resources, University of Maryland, College Park, MD, United States

Pankaj Kumar Gupta

National Institute of Urban Affairs (NIUA), Ministry of Urban and Housing Affairs, New Delhi, India

ELSEVIER

ACADEMIC PRESS

An imprint of Elsevier

Academic Press is an imprint of Elsevier
125 London Wall, London EC2Y 5AS, United Kingdom
50 Hampshire Street, 5th Floor, Cambridge, MA 02139, United States

For accessibility purposes, images in electronic versions of this book are accompanied by alt text descriptions independently provided by Elsevier, generated with the aid of AI. For more information, see https://www.elsevier.com/about/accessibility or Radarweg 29, 1043 NX Amsterdam, Netherlands.

Books and Journals published by Elsevier comply with applicable product safety requirements. For any product safety concerns or queries, please contact our authorised representative, Elsevier B.V., at productsafety@elsevier.com.

Publisher's note: Elsevier takes a neutral position with respect to territorial disputes or jurisdictional claims in its published content, including in maps and institutional affiliations.

Notices
Knowledge and best practice in this field are constantly changing. As new research and experience broaden our understanding, changes in research methods, professional practices, or medical treatment may become necessary.

Practitioners and researchers must always rely on their own experience and knowledge in evaluating and using any information, methods, compounds, or experiments described herein. In using such information or methods they should be mindful of their own safety and the safety of others, including parties for whom they have a professional responsibility.

To the fullest extent of the law, neither the Publisher nor the authors, contributors, or editors, assume any liability for any injury and/or damage to persons or property as a matter of products liability, negligence or otherwise, or from any use or operation of any methods, products, instructions, or ideas contained in the material herein.

ISBN: 978-0-443-40513-6

For information on all Academic Press publications visit our website at https://www.elsevier.com/books-and-journals

Publisher: Andre Wolff
Acquisitions Editor: Chiara Giglio
Editorial Project Manager: Sneha Roy
Production Project Manager: Fahmida Sultana
Cover Designer: Mark Rogers

Typeset by Integra Software Services

Contents

Contributors

Aliza Adhikari Agricultural Systems and Engineering, Faculty of Food, Agriculture and Natural Resources, Asian Institute of Technology, Pathum Thani, Thailand

Ridwan Ahmed Department of Agronomy, Bangladesh Agricultural University, Mymensingh, Bangladesh

Sheikh Faruk Ahmed Department of Crop Botany, Gazipur Agricultural University, Gazipur, Bangladesh

Ozgur B. Akan Center for neXt-generation Communications (CXC), Department of Electrical and Electronics Engineering, Koç University, Istanbul, Turkey; Internet of Everything (IoE) Group, Electrical Engineering Division, Department of Engineering, University of Cambridge, Cambridge, United Kingdom

Anamika Department of Agronomy, Gazipur Agricultural University, Gazipur, Bangladesh

Rohit Anand Department of Food, Agricultural, and Biological Engineering, The Ohio State University, Columbus, OH, United States

Thin Thiri Aung Agricultural Systems and Engineering, Faculty of Food, Agriculture and Natural Resources, Asian Institute of Technology, Pathum Thani, Thailand

Al Bashir Department of Biological and Agricultural Engineering, Texas A&M AgriLife Research, Texas A&M University, Dallas, TX, United States

Subhrasita Behera Department of Biological and Agricultural Engineering, Texas A&M University, College Station, TX, United States

Dattatray G. Bhalekar Department of Biological Systems Engineering, Center for Precision and Automated Agricultural Systems, Washington State University, Prosser, WA, United States

Prachi Bhatnagar Department of Biosciences and Bioengineering, Indian Institute of Technology (IIT), Roorkee, Uttarakhand, India

Fatih E. Bilgen Center for neXt-generation Communications (CXC), Department of Electrical and Electronics Engineering, Koç University, Istanbul, Turkey

Suriyan Cha-um National Center for Genetic Engineering and Biotechnology, National Science and Technology Development Agency, Pathum Thani, Thailand

Subir Kumar Chakraborty Central Institute of Agricultural Engineering, Bhopal, India

Narendra Singh Chandel Central Institute of Agricultural Engineering, Bhopal, India

Jeevan Kumar Cheruku Maulana Azad National Urdu University, Hyderabad, India

Pooja Chouhan Central Institute of Agricultural Engineering, Bhopal, India

Avishek Datta Agricultural Systems and Engineering, Faculty of Food, Agriculture and Natural Resources, Asian Institute of Technology, Pathum Thani, Thailand

Subhankar Debnath Agricultural Systems and Engineering, Faculty of Food, Agriculture and Natural Resources, Asian Institute of Technology, Pathum Thani, Thailand

Prasad Deshpande Department of Agronomy, Kansas State University, Manhattan, KS, United States

Wri Dutta Department of Agronomy, Kansas State University, Manhattan, KS, United States

Achraf El Allali College of Computing, University Mohammed VI Polytechnic, Benguerir, Morocco

Oumnia Ennaji College of Computing; College of Agriculture and Environmental Sciences, University Mohammed VI Polytechnic, Benguerir, Morocco

Chawanrat Euafua Faculty of Food, Agriculture, and Natural Resources, Asian Institute of Technology, Bangkok, Thailand

Shubham Gade Agricultural Systems and Engineering, Faculty of Food, Agriculture and Natural Resources, Asian Institute of Technology, Pathum Thani, Thailand

Shubham Anil Gade Agricultural Systems and Engineering, Faculty of Food, Agriculture and Natural Resources, Asian Institute of Technology, Pathum Thani, Thailand

Partho Ghose Department of Biological and Agricultural Engineering, Texas A&M AgriLife Research, Texas A&M University, Dallas, TX, United States

Keri Grant College of Agriculture and Natural Resources, University of Maryland, College Park, MD, United States

Md. Masudul Hassan Department of Agricultural Finance and Management, Sher-e-Bangla Agricultural University, Dhaka, Bangladesh

Sushil Kumar Himanshu Agricultural Systems and Engineering, Faculty of Food, Agriculture and Natural Resources, Asian Institute of Technology, Pathum Thani, Thailand

Dilip Jat Central Institute of Agricultural Engineering, Bhopal, India

Gaurav Jha Department of Agronomy, Kansas State University, Manhattan, KS, United States

Kanyanat Kasetsuntorn National Center for Genetic Engineering and Biotechnology, National Science and Technology Development Agency, Pathum Thani, Thailand

Vishal Katekar National Institute of Disaster Management (NIDM), Ministry of Home Affairs, Government of India, New Delhi, India

Abia Katimbo Department of Biological Systems Engineering, University of Nebraska-Lincoln, Lincoln, NE, United States

Prashant Kaushik Chaudhary Charan Singh Haryana Agricultural University, Hisar, Haryana, India

Ahmet B. Kilic Center for neXt-generation Communications (CXC), Department of Electrical and Electronics Engineering, Koç University, Istanbul, Turkey

Hemendra Kumar Digital and Precision Agriculture Lab; College of Agriculture and Natural Resources, University of Maryland, College Park, MD; Central Maryland Research and Education Center, University of Maryland Extension, Upper Marlboro, MD, United States

Navneet Kumar School of Biotechnology, Faculty of Applied Sciences and Biotechnology, Shoolini University; Microbiology Department, Shoolini Life Sciences Private Limited, Solan, Himachal Pradesh, India

Karishma Kumari Department of Agronomy, Horticulture and Plant Science, South Dakota State University, Brookings, SD, United States

Mallappa Jadiyappa Madolli Department of Agricultural Engineering, University of Agricultural Sciences, Bangalore, India; Agricultural Systems and Engineering, Faculty of Food, Agriculture and Natural Resources, Asian Institute of Technology, Pathum Thani, Thailand

Nerice Millet-Williams College of Agriculture and Natural Resources, University of Maryland, College Park, MD, United States

Sibiraj Murugesan School of Bioengineering and Food Technology, Faculty of Applied Sciences and Biotechnology, Shoolini University, Solan, Himachal Pradesh, India

Hope Njuki Nakabuye Department of Biological and Agricultural Engineering, Texas A&M University, College Station, TX, United States

Mike Ojo Department of Biological and Agricultural Engineering, Texas A&M AgriLife Research, Texas A&M University, Dallas, TX, United States

Abubakar Palli Digital and Precision Agriculture Lab; College of Agriculture and Natural Resources, University of Maryland, College Park, MD, United States

Rabindra Kumar Panda Centre for Climate Smart Agriculture, Siksha 'O' Anusandhan Deemed to be University, Bhubaneswar, Odisha, India

Dilwar Singh Parihar International Rice Research Institute, Varanasi, India

Roaf Ahmad Parray Division of Agricultural Engineering, ICAR-Indian Agricultural Research Institute, New Delhi, India

Sushma Perati Department of Biological and Agricultural Engineering, Texas A&M University, College Station, TX, United States

Piyanan Pipatsitee National Center for Genetic Engineering and Biotechnology, National Science and Technology Development Agency, Pathum Thani, Thailand

Md. Rashedur Rahman Department of Agronomy, Bangladesh Agricultural University, Mymensingh, Bangladesh

Md. Sadique Rahman Department of Agricultural Finance and Management, Sher-e-Bangla Agricultural University, Dhaka, Bangladesh

Aadarsh Rijal Agricultural Systems and Engineering, Faculty of Food, Agriculture and Natural Resources, Asian Institute of Technology, Pathum Thani, Thailand

Md. Rahat Bari Rupok Department of Agronomy, Gazipur Agricultural University, Gazipur, Bangladesh

Karan Saini Applied Phycology and Food Technology Laboratory, CSIR-Institute of Himalayan Bioresource Technology (IHBT), Palampur, Himachal Pradesh, India

Haley Sater College of Agriculture and Natural Resources, University of Maryland, College Park, MD, United States

Gaurav Saxena Department of Life Science, Faculty of Life Sciences, Mandsaur University, Mandsaur, Madhya Pradesh, India

Meenakshi Sharma Department of Chemistry, Ranchi University, Ranchi, Jharkhand; Joita Bioseed AI Pvt. Ltd., Kurukshetra, Haryana, India

Ronald Singh Digital and Precision Agriculture Lab, University of Maryland, College Park, MD, United States

Rajiv Kumar Srivastava Biological and Agricultural Engineering Department, Texas A&M University, College Station, TX, United States

Andrej Svyantek Department of Horticultural Sciences, Texas A&M University, College Station, TX, United States

Mriganka Kaustab Talukdar Agricultural Systems and Engineering, Faculty of Food, Agriculture and Natural Resources, Asian Institute of Technology, Pathum Thani, Thailand

Kamlesh Thakur School of Biotechnology, Faculty of Applied Sciences and Biotechnology, Shoolini University, Solan, Himachal Pradesh, India

Rujira Tisarum National Center for Genetic Engineering and Biotechnology, National Science and Technology Development Agency, Pathum Thani, Thailand

Abhishek Upadhyay Dr. Rajendra Prasad Central Agricultural University, Samastipur, Bihar, India

Uday Bhanu Prakash Vaddevolu Department of Biological and Agricultural Engineering, Texas A&M University, College Station, TX, United States

Leonardus Vergutz College of Agriculture and Environmental Sciences, University Mohammed VI Polytechnic, Benguerir, Morocco

Rishith Kumar Vogeti Department of Biological and Agricultural Engineering, Texas A&M University, College Station, TX, United States

Yibin Wang Department of Biological and Agricultural Engineering, Texas A&M AgriLife Research, Texas A&M University, Dallas, TX, United States

Nitiwat Wongchanla Agricultural Systems and Engineering, Faculty of Food, Agriculture and Natural Resources, Asian Institute of Technology, Pathum Thani, Thailand

Rahul Yadav National Accreditation Board for Testing and Calibration, Laboratories (NABL), Quality Council of India (QCI), Gurugram, Haryana, India

Jipanu Yamjabok National Center for Genetic Engineering and Biotechnology, National Science and Technology Development Agency, Pathum Thani, Thailand

Muhammad Yaseen Faculty of Food, Agriculture, and Natural Resources, Asian Institute of Technology, Bangkok, Thailand

Azlan Zahid Department of Biological and Agricultural Engineering, Texas A&M AgriLife Research, Texas A&M University, Dallas, TX, United States

Farhad Zulfiqar College of Agricultural and Marine Sciences, Sultan Qaboos University, Muscat, Oman

Section I

Smart precision technologies for digital agriculture

Smart agricultural technologies: Challenges and opportunities from today's fields to the future

1

Sushma Perati[a], Rishith Kumar Vogeti[a], Andrej Svyantek[b], Subhrasita Behera[a], Uday Bhanu Prakash Vaddevolu[a], Hope Njuki Nakabuye[a], Hemendra Kumar[c,d,e], and Abia Katimbo[f]

[a]Department of Biological and Agricultural Engineering, Texas A&M University, College Station, TX, United States

[b]Department of Horticultural Sciences, Texas A&M University, College Station, TX, United States

[c]Digital and Precision Agriculture Lab, University of Maryland, College Park, MD, United States

[d]College of Agriculture and Natural Resources, University of Maryland, College Park, MD, United States

[e]Central Maryland Research and Education Center, University of Maryland Extension, Upper Marlboro, MD, United States

[f]Department of Biological Systems Engineering, University of Nebraska-Lincoln, Lincoln, NE, United States

1 Introduction to digital agriculture

Digital agriculture (DA) refers to smart farming practices that leverage high-resolution digital information to support informed decision-making across the entire farm-to-table process. This includes large-scale, diverse data sources that enable data-driven decision support. The scope of digitalization extends beyond farming or livestock zones to cover the entire supply chain network (Shepherd et al., 2020). DA was defined by the United Nations as *"the use of new and advanced technologies, integrated into one system, to enable growers and other stakeholders within the agriculture value chain to improve food production"* (Puntel et al., 2022; United Nations, 2017).

This emerging transformation in the agriculture sector significantly reduces excess resource application by enabling real-time data for crop monitoring and management. In addition, it facilitates monitoring of soil moisture and water levels, which, in turn, supports controlled and targeted irrigation (Abdelmoneim et al., 2025a; Kumar et al., 2023a). It also reduces excess runoff and facilitates seeding, harvesting, and disease control which effectively reduce labor and other input costs (Papadopoulos et al., 2024a).

Precision Technologies for Digital Agriculture. https://doi.org/10.1016/B978-0-443-40513-6.00018-4

These technologies simplify farm operations while giving high yields and helping farmers better understand crop needs. They also improve decision-making by integrating agricultural machinery with software applications that log farm-specific datasets to assess soil conditions and zone-specific weather patterns (Clapp and Ruder, 2020). Based on this, precision agriculture helps mitigate adverse environmental impacts by optimizing resource use. These data-driven technologies, when combined with artificial intelligence (AI), enable early pest detection and provide insights into crop status for more efficient cultivation methods (Kariyanna and Sowjanya, 2024).

The digitalization of agriculture is especially critical for feeding a rising global population while using fewer and more precise agricultural inputs, which will lead to a sustainable path in an era of climate change, resource scarcity, and increasing pollution (Vilavan et al., 2024; Visser et al., 2021). With innovation and integration in smart systems, DA additionally enhances productivity and sustainability by lowering agricultural input use and energy consumption (Zhou and Yin, 2023). It also contributes to mitigating the emissions of methane (CH_4) and nitrous oxide (N_2O), two of the major man-made greenhouse gases (GHGs) from agriculture, which have significantly higher global warming potential than carbon dioxide (CO_2). Excess irrigation and improper fertilization are two of the key drivers of CH_4 and N_2O emissions (Chataut et al., 2023). For instance, a condition called a dead zone has been created in the Gulf of Mexico due to the low oxygen area (hypoxia), which is caused by the excess runoff of nutrients, primarily nitrogen and phosphorus, from agricultural fields into the Mississippi River Basin. These nutrients are carried by the Mississippi River and discharged into the Gulf of Mexico, causing excess growth of algae. When this algae dies and decomposes, an excess amount of dissolved oxygen is consumed, leading to low oxygen levels, which negatively impacts marine life in the Gulf of Mexico (National Oceanic and Atmospheric Administration, 2025). Precision technologies such as variable rate irrigation or zone-specific irrigation play a pivotal role in optimizing water and nutrient application, thereby reducing GHGs and enhancing resource efficiency (Balasundram et al., 2023). In addition, technologies such as vision-based intelligent systems, combined with AI and machine learning (ML) techniques, imitate visual recognition and cognitive abilities in humans to make informed decisions for agricultural management (Ghazal et al., 2024). Complementing these intelligent systems, the growing influence of big data further empowers these digital technologies by providing farmers with extensive information on the display screens of farm machinery, agricultural tools, apps, and mobile devices (Clapp and Ruder, 2020). One such application is variable rate technology (VRT), a precise agricultural technique used to adjust the application of farm inputs such as seed, fertilizer, irrigation water, and chemicals through farm machinery guided by map-based or sensor-based data (McFadden, 2023; AGRIVI, 2022). Overall, precision technologies in DA are changing farming practices by making them smarter, more efficient, and better for the environment. As these technologies continue to evolve, integrating them into everyday farming operations can lead to more sustainable, robust, and intelligent agricultural systems. The prime objective of this chapter is to investigate the current scenario, crucial technologies, and future directions of DA, with a focus on how data-driven innovations such as AI, ML, and precision farming tools contribute to sustainability, productivity, and climate resilience.

2 History and evolution of agricultural practices

The shift from traditional farming methods to data-driven technologies has greatly transformed agricultural practices. Instead of relying on manual labor, farmer intuition, and uniform input use, modern farming now employs mechanical equipment, sensor-based systems, and variable-rate application technologies. The growing population, climate change, and limited resources have driven agriculture toward more efficient resource use.

One of the major milestones in United States (US) agriculture was the invention of the first steam-powered tractor in 1890, followed by the development of gasoline-powered tractors that replaced animal-based tillage (Balasundram et al., 2023). These innovations marked the transition from manual labor to mechanized farming. Over time, nearly all farm operations, from soil preparation and seeding to postharvest handling, have evolved through the integration of modern machinery and precision tools. The following sections highlight how key farm activities have been transformed over time.

2.1 Precision tillage

Over time, tillage practices have evolved from digging with sharp sticks to using draft animal-pulled plows and harrows, and now to global positioning system (GPS)-guided tractors equipped with precision implements. These advanced tractors enable high-accuracy soil preparation, reduce overlap, minimize fuel use, and preserve soil structure (University of Minnesota Extension, 2022). The shift toward mechanized and precision tillage has been most prominent in countries such as the US, Canada, Australia, New Zealand, and South Africa, where larger farm sizes support advanced technologies (University of Minnesota Extension, 2022). In contrast, many developing nations still rely on handheld or animal-drawn tools due to small landholdings and limited resources, though such systems remain vital for local food security and livelihoods (University of Minnesota Extension, 2022).

As mechanization progressed, farmers moved from moldboard and shovel plows to multirow disc implements, which increased efficiency and coverage (DeJong-Hughes and Daigh, 2021). However, the Dust Bowl of the 1930s was caused by the increased migration to the Great Plains, which led to the conversion of large areas of prairie grasslands into agricultural land. Intensive farming, including extensive tillage and poor land management, reduced the deep-rooted natural grasses and vegetation. As a result, the protective topsoil was blown away by strong winds. This contributed to frequent severe droughts and extreme temperature shifts in the region (Alexander et al., 2018). This natural disaster exposed the damaging effects of excessive tillage, raising awareness of soil degradation and sparking the soil conservation movement (Lal et al., 2007) In addition, intensive tillage lowers soil organic carbon levels and contributes to increased GHG emissions (Srivastava et al., 2024).

By the 1970s, no-till farming gained popularity for minimizing soil disturbance and conserving soil health (Ghazal et al., 2024). Modern approaches, such as zone or strip tillage, continue this transformation, preparing seedbeds while applying precise amounts of fertilizer where crops need them most (AGRIVI, 2022). These methods conserve resources, protect soil organic matter, and improve input efficiency.

Despite advances in conservation tillage, such as multispecies cover cropping that enhances soil health by boosting microbial metabolic diversity and strengthening crop-microbe interactions (Chinthalapudi et al., 2023) and no-till farming for reducing soil disturbance and preserving long-term soil health (Thapa and Dura, 2024), poor management can still lead to soil erosion and long-term productivity losses (McFadden, 2023). To address this, precision tillage systems now integrate technologies such as Controller Area Network bus systems for real-time depth adjustment and variable-rate tillage, which adapts tillage intensity to soil type, compaction, and field conditions (Fox et al., 2018). Overall, the evolution of tillage reflects a shift toward precision-based, resource-efficient, and data-driven farming practices that balance productivity with sustainability.

2.2 Modern seeding technologies in agriculture

Seeding methods in agriculture have undergone a remarkable transformation over time. Traditionally, farmers relied on broadcast sowing, where seeds were scattered over tilled soil. While simple, this approach made weeding and harvesting difficult (Sean, 2023). Later, the dibbling technique emerged, placing seeds into holes made with sticks—a more accurate but labor-intensive method. With the rise of plowing, seeds began to be dropped into trenches formed by cultivators or plows, followed by harrowing to cover them. The introduction of seed drills marked a major breakthrough by placing seeds at consistent depths and spacing, resulting in better germination and higher yields (University of Minnesota Extension, 2022).

The adaptation of traditional tools such as horticultural dibblers, seeders, and transplanters into modern versions has gained new interest, with many companies offering innovative designs. Mechanized seed drills, typically pulled by tractors, use vacuum nozzles, rotating shafts, and drums to help with precise seed placement and to control seed density, thereby reducing seed wastage (Mukherjee et al., 2024). Building on this, GPS-operated seeding systems have improved navigation and enabled variable-rate seeding for greater efficiency. These precision technologies adjust seed density based on field variability and often include sensors for real-time monitoring of crop health. Recent innovations feature smart tools that assess soil moisture and temperature to determine optimal planting conditions. Remote sensing (RS) tools, including geographic information systems (GIS) and satellite mapping, further enhance accuracy by dividing fields into management zones based on soil and moisture characteristics (Kumar et al., 2022, 2023; Šarauskis et al., 2022).

Among the latest innovations, hopper seed drills are gaining popularity for their ability to perform variable-rate seeding with exceptional precision. These systems can monitor sowing patterns in real time, ensuring uniform seed distribution. Automated seeding machines, which integrate GPS, sensors, and data-driven control systems, are increasingly in demand for their ability to save time, reduce input use, and improve crop yields (DeJong-Hughes and Daigh, 2021).

2.3 Mechanization and automation in transplanting

Transplanting refers to the process of moving a seedling from one location to another, typically from the seed flat or nursery to the main growing field. Historically, this was done by hand, which was labor-intensive, time-consuming, and often resulted in uneven row spacing. To address these challenges, seedling transplanting machines were developed in the 1970s, offering a more efficient and uniform method for placing seedlings in the field (Taizy, 2020). The transplanter was a compact, rickshaw-pulled unit equipped with a gear-driven mechanism that operated the planting hooks (Nishio, 2019). Later, mechanical transplanters have been designed and developed to accommodate a wide range of soil types and cropping systems.

Today, manual, semi-automatic, and automatic transplanters are available in the market, made to meet the diverse needs of farms of varying sizes and farmer preferences. Semi-automatic transplanters require human assistance for certain tasks but are powered by tractors and operate through mechanical transmission systems (Sharma and Khar, 2024). These machines offer greater efficiency and consistency compared to fully manual transplanters. Automatic transplanters, typically mounted on tractors or other vehicles, are engineered to perform the entire transplanting operation autonomously, except for the task of restocking seedlings. Their operation requires only a driver to guide the vehicle through the field (Ishizaki et al., 2024). Currently, research is underway to develop autonomous transplanters that eliminate the need for a driver. However, such systems are not yet commercially available (UGA Extension, 2025).

2.4 Weed management

Early weed control methods relied mainly on manual labor, using hand tools such as knives, sweeps, spikes, blades, and harrows, along with basic tillage using plows. The invention of the grain drill by Jethro Tull revolutionized crop planting by enabling row-based sowing, which made interrow weeding much easier (American Society of Mechanical Engineers, 2012). To reduce manual effort, techniques such as flame weeding and the use of rod weeders were later developed. In flooded fields, particularly paddy fields, farmers adopted a biological approach by releasing fish into the water, where they consumed weeds and insects in an eco-friendly and cost-effective practice.

By the mid-20th century, herbicides such as 2,4-D (introduced in the 1940s) became popular for their effectiveness and ease of use (Ahmad et al., 2024). Application methods also advanced with the introduction of hand-powered and gasoline-powered sprayers. However, the environmental impact, fate, and off-target effects of synthetic herbicides contributed to an investigation of alternative weed control methods. Modern weed management now emphasizes cultural and preventive practices alongside or in place of conventional herbicides; these practices include crop rotation, mulching (plasticulture), cover cropping, and higher planting densities that suppress weed growth through canopy shading (Tarrant et al., 2024).

Mechanical control methods have progressed to include site-specific tillage and specialized implements such as rototillers, finger weeders, and brush weeders.

Flame weeding, once a traditional technique, has regained popularity among organic growers seeking nonchemical control (Upadhyay et al., 2024). Meanwhile, integrated weed management approaches increasingly use RS, GPS-based systems, and automated equipment like unmanned aerial and ground vehicles (Zhang et al., 2024). These technologies allow precise, site-specific weed detection and removal while reducing labor and chemical inputs (Mukherjee et al., 2024). Sprayers equipped with adaptive nozzles and flow sensors further enhance application efficiency.

Recent advances in digital imaging and data analytics now support high-resolution weed mapping using red, green, and blue (RGB), multispectral, and hyperspectral imagery. These techniques enable early-stage weed identification and targeted control, improving the accuracy and efficiency of field operations (Jahanbakht et al., 2025). In one study, (Silva et al., 2024) reported that a few different deep learning models were applied and tested for their performance in identifying and classifying weeds, bean, and soybean crops. Among them, the (small) YOLOv8s model achieved the highest accuracy of about 99.7% in detecting and segmenting weeds. Overall, weed management has transitioned from manual and chemical approaches to precision-driven, data-supported systems that promote sustainable, efficient, and productive farming.

2.5 Data-driven smart irrigation systems

Irrigation is the process of supplying water to crops, and it has been used by farmers for thousands of years to grow food in dry areas. Among the prior methods, surface and flood irrigation systems have historically relied on networks of canals, dikes, and gates, often configured as furrow or border setups to regulate water flow (Benin and Snezhko, 2021). In response to the seasonal overflows of the Nile River, ancient Egyptian farmers developed basin irrigation techniques, constructing earthen embankments to capture and utilize floodwaters effectively (Haug, 2017). Meanwhile, ancient Chinese civilizations introduced water wheels to lift river water into puddling fields, enhancing agricultural productivity (Angelakis et al., 2022). Later, the Indus Valley Civilization in India pioneered advanced irrigation infrastructure, including reservoirs, wells, and intricate canal systems, laying the foundation for sustainable water management in arid regions (Singh et al., 2020). Building on these early developments, step wells became popular in arid regions as a sustainable source of water, allowing communities to access ground water during dry seasons. Similarly, the Romans contributed significantly to irrigation by developing aqueducts, terracing methods, and underground clay pipes to transport water to distant farmlands (Deming, 2019).

During the medieval period, civilizations in the Middle East and Europe introduced the "Qanat" system, a network of subsurface tunnels designed to channel groundwater from aquifers to surface irrigation systems (Hassen et al., 2025). Later, in the Renaissance period, European societies enhanced agricultural productivity by introducing water mills and refining canal systems to improve water distribution and boost crop yields (DIG, 2025). However, despite these advancements, using poor-quality water for irrigation can lead to soil salinization, which, in turn, reduces soil fertility and contributes to declining crop yields (Angelakıs et al., 2020). In response to growing concerns over resource degradation driven by population growth, limited natural resources, and climate change, in the 19th and 20th centuries—a period often referred to as the beginning of the precision irrigation era—significant changes in irrigation practices were made. During this period, steam and motorized pumps were developed to extract water from wells and groundwater sources, significantly expanding irrigation capacity (Kent and Hamer, 2004). However, due to excessive withdrawal driven by high demand, resource conservation gained importance, leading to innovations such as sprinkler irrigation distributed as main lines, submainlines, and later line pipes, which emerged in the late 1800s as a more efficient alternative. This method sprays water over the crop canopy or field, mimicking rainfall, and is suitable for a wide range of crops on flat or gently sloped land (Hossain et al., 2024). Despite its advantages, it is prone to evaporation losses and wind interference. To irrigate large areas efficiently, the center pivot irrigation system was developed. It features a long boom that rotates around a central pivot point where the water source is located (Waller and Yitayew, 2016). This system is sustainable for large farms but involves high initial costs. It became especially popular in the Great Plains region of the US, enabling automatic irrigation across expansive farmland (Evett et al., 2020). However, due to increasing water scarcity issues during the mid-20th century, precision irrigation methods emerged. Modern drip irrigation, also known as drop-by-drop irrigation, was introduced in Israel and could save up to 70% of water consumption in rice cultivation (Surkes, 2021). It delivers water directly to the plant root zones or deep root zones using drippers or emitters, minimizing evaporation and runoff.

Modern pressurized irrigation systems are selected based on field size, crop type, soil conditions, and economic considerations. To further reduce water loss, subsurface drip irrigation was developed, where pipes are buried underground to deliver water directly to the root zone (Zheng et al., 2025). Today, drip irrigation continues to advance through integration with Internet of Things (IoT) technologies, such as sensors that monitor soil moisture and other parameters to determine optimal irrigation timing and volume (Pereira et al., 2023). Satellite-based data is also used for field mapping based on soil type and water-holding capacity, helping to decide irrigation frequency and quantity (Dari et al., 2025). Leveraging this data, techniques like variable-rate irrigation and deficit irrigation, which is an approach of supplying water to the crop below the full demand for optimal growth (Kumar et al., 2022a, 2022b, 2023a; Ray et al., 2023), are gaining popularity for their efficient and sustainable water use. In addition, the Crop Water Stress Index (CWSI) is increasingly used to guide irrigation decisions (Liao et al., 2024). Together, these innovations reflect a transformative shift toward data-driven irrigation strategies that not only conserve water but also enhance crop resilience and productivity, marking a critical step forward in sustainable agricultural practices.

In addition to irrigation, drainage techniques have also evolved from simple surface ditches and drains made with stones or branches of trees to the first tile drain technology in the US in 1838. Later, drainpipes replaced the open ditches or pipes. Clay and concrete drain tiles gained popularity in the mid-1940s, installed using gas-powered trenching tools (Yannopoulos et al., 2020). These excess water-routing techniques have evolved into modern methods, including drainpipes made of plastic.

Considerable changes in installation processes have been driven by the utilization of laser surveying to improve precision. In the DA era, the application of GPS and GIS for designing drainage systems has replaced laser grade control. Tile-drained systems have been used to control the flow of water from fields to streams through manual or automated systems (Kumar et al., 2023b). These advancements have made agricultural drainage systems more durable, easier to implement, and highly economical.

2.6 Precision fertilizer application and management

Until the early 1900s, fertilizers such as animal manure or dung were applied manually using methods such as broadcasting, band placement, side dressing, and top-dressing. Small tools like shovels were also used to place fertilizer near the base of plants. In the early 20th century, fertilizer application practices underwent significant transformation. This shift was driven by the adoption of technologies such as GPS, VRT, nanotechnology, satellite imagery, remote sensing (RS), sensors, data-driven models, mobile applications, software, and mechanized systems. Further advancements in mechanization and smart technologies, including mechanical spreaders, have enabled more uniform coverage across fields (Srikanthnaik, 2024). Later, fertigation, a technique of applying nutrients through irrigation systems to crops, was introduced as an efficient method (Pibars et al., 2022). GPS-enabled vehicles and aerial imagery have gained popularity for site-specific and depth-adjusting nutrient application, helping reduce environmental impact while improving crop yields. In addition, IoT-connected systems and decision support tools assist in tailoring fertilizer application based on soil type, moisture content, and other field-specific parameters (Kaviyarasan et al., 2024). However, GPS-based application methods have some major concerns, such as a high initial cost, which makes it difficult for small landholding farmers to adopt (Schimmelpfennig, 2016). These methods also require technical knowledge to operate, which poses a challenge for farmers who are uneducated or lack technical skills (Bhamini et al., 2025). Therefore, it is necessary to educate farmers on how to use emerging technologies (Mamabolo et al., 2025). To ensure that new innovations are accessible to all farmers, governments can provide subsidies for mechanization and training platforms that demonstrate new technologies as they arise (Getahun et al., 2024a).

2.7 Automated crop health monitoring and stress detection

Crop stress monitoring has evolved from traditional visual examination for wilting, discoloration, or abnormalities in plants to the use of advanced technologies. Modern approaches now incorporate ML and deep learning (DL) techniques to enable early detection of stress, minimizing damage to crop yield and enhancing precision in agricultural management. Among these technologies, RS techniques, including drone-based hyperspectral, multispectral, thermal imagery, and Normalized Difference Vegetation Index (NDVI), have been widely employed to detect stress in plants (Ojeilua et al., 2025; Stutsel et al., 2021). When integrated with ML and DL algorithms, these technologies significantly enhance the ability to analyze complex data and predict potential stress conditions before they escalate into serious yield losses. Sensors such as infrared thermometers and spectrometers are used to track canopy temperature for stress detection through real-time imagery and data analysis (Fan et al., 2024). Over the past few decades, molecular-based techniques have also been developed to identify stress at the biochemical level. Among the most commonly used methods are polymerase chain reaction (PCR and real-time PCR) and enzyme-linked immunosorbent assay; however, these techniques are often time-consuming and labor-intensive (Galieni et al., 2021).

Similar to plant/crop stress monitoring, monitoring the parameters associated with microclimate has also evolved from manual examination by experienced individuals to RS, aerial imagery, and IoT-based approaches that are being adopted in DA. Sensors such as soil moisture and temperature probes are used to track soil conditions, enabling stress detection through live data analysis. These IoT-based sensing technologies offer a real-time, noninvasive approach to detecting climate data (Eze et al., 2025). This enhances decision-making in precision agriculture by enabling timely interventions.

2.8 Harvesting

Harvesting has evolved significantly from the use of simple sickles and hand-powered tools to advanced machinery and combine harvesters, making the process faster and more efficient in the era of DA (Kaur et al., 2023). In earlier times, farmers relied on manual tools like sickles or cradles to cut down crop stalks, followed by threshing and winnowing to separate grains from the chaff. This method was highly labor-intensive and time-consuming, requiring substantial physical effort from farmers (Iron Solutions, 2020). The transition from manual harvesting to mechanized systems made progress in the mid-19th century with the invention of the mechanical reaper, which significantly accelerated the cutting process. However, threshing and winnowing still relied on manual labor, limiting overall efficiency (Morning Ag Clips, 2022) Subsequent innovations led to the development of combine harvesters, which were initially horse-drawn and later powered by steam and internal combustion engines. These machines revolutionized harvesting by integrating cutting, threshing, and winnowing into a single operation. This evolution significantly transformed the cultivation of lands, reducing labor demands and saving considerable time and effort for farmers (Lohan et al., 2018).

As harvesting technologies continued to evolve, the focus shifted toward precision, automation, and crop-specific engineering. Today's harvesters are equipped with GPS guidance, integrated sensors, and specialized attachments such as cotton strippers, pickers, and sugarcane harvesters, designed to minimize crop loss and reduce human intervention (Mishra et al., 2023). For delicate fruits like berries, robotic harvesters featuring computer vision and soft-touch grippers are emerging as a promising solution for gentle and selective picking (VPRC, 2023).

These technologies are gaining popularity as they mitigate labor shortages and minimize damage to fruits and vegetables during harvest. In addition, variable-rate harvesting technologies leverage GPS, yield monitors, and sensors to map spatial variability, enabling real-time adjustments and the creation of data-driven field maps. This precision-driven approach enhances decision-making and supports more sustainable and efficient harvesting practices.

3 Innovations and advances in precision and digital agricultural technologies

3.1 *Global Positioning System and GIS to address spatial variability*

GPS serves as foundational technology in precision agriculture, enabling location-specific management operations that enhance crop yields, optimize resource use, and support sustainable farming practices. Originally developed as a satellite-based navigation system, GPS was first adopted in agriculture for basic tasks such as positioning, field boundary mapping, and assessing soil health (Saleem et al., 2023). Over time, its role has expanded significantly with the rise of precision agriculture. Today, farmers rely on GPS to guide machinery with pinpoint accuracy, enabling site-specific operations and variable-rate application of inputs such as seeds, fertilizers, pesticides, and irrigation (Schumann, 2010). This precise spatial control not only improves operational efficiency but also minimizes resource waste and environmental impact.

Beyond navigation, GPS-enabled equipment also collects spatial data on key agronomic parameters such as soil properties, moisture levels, and weather conditions (Getahun et al., 2024b). This real-time information supports early decision-making in areas like irrigation scheduling, nutrient management, and pest control. Furthermore, GPS facilitates the use of auto-guided tractors and VRT, enabling precise input application and yield mapping while minimizing overlap and resource waste (Schumann, 2010). When integrated with agricultural software platforms, GPS data can be used to monitor crop performance and generate actionable insights, empowering farmers to make informed, data-driven decisions that enhance productivity and promote long-term sustainability (Saniga et al., 2024).

GIS serves as a powerful software tool that stores, analyzes, and visualizes spatial data to generate detailed maps (ESRI, 2025). By integrating GPS-derived information such as soil types, crop health, and yield data, GIS enables the creation of multilayered maps using inputs from RS technologies. These field variability maps allow farmers to precisely navigate their fields and apply inputs with high accuracy, adapted to localized conditions. Historically, it was difficult to align cultivation practices and yield outcomes with soil heterogeneity, limiting the effectiveness of site-specific management strategies. However, the integration of GPS and GIS has transformed this process. Farmers can now achieve highly accurate application of seeds, pesticides, herbicides, and fertilizers with improved control over chemical distribution (Mathenge et al., 2022). This advancement in precision technology has significantly enhanced agricultural productivity and sustainability (GPS, 2021).

3.2 Remote sensing and drones in agriculture

Remote sensing (RS) refers to the process of acquiring information about an object or area from a distance, without direct physical contact (NASA, 2021). It relies on the detection of electromagnetic radiation either reflected or emitted by objects on the Earth's surface (Read and Torrado, 2009). To capture and analyze this information, RS employs sensors mounted on satellites, aircraft, drones, and unmanned aerial vehicles (UAVs). These platforms generate high resolution multispectral or hyperspectral images that support enhanced crop monitoring, pest and disease detection, and the production of NDVI imagery (Mathenge et al., 2022; Praseartkul et al., 2023).

In precision agriculture, drones enhance RS capabilities by enabling variable-rate applications with improved spatial accuracy (Gade et al., 2025; Guebsi et al., 2024; Praseartkul et al., 2023). This targeted approach helps reduce input usage, minimizing water and fertilizer runoff, and thereby lowering environmental impact. In parallel, satellite imagery offers high-resolution coverage of extensive agricultural areas, supporting spatial differentiation in farming practices. Its long-term data records allow for tracking changes in land use, crop performance, and field variability across decades (Sekhar et al., 2024).

Examples include Sentinel-2, operated by the European Space Agency, which provides fine-scale multispectral imagery across thirteen spectral bands for land cover/use monitoring, climate change, and disaster monitoring (Phiri et al., 2020). Similarly, Landsat 8, a joint mission by national aeronautics and space administration (NASA) and the united states geological survey (USGS), offers a long-term dataset of terrestrial surfaces, including farmlands (Roy et al., 2014). Drones equipped with cameras or sensors can capture aerial photographs that reveal trends in pest activity, emerging crop diseases, and nutrient deficiencies, often before they are detectable by the human eye. These aerial images enable producers to respond promptly and precisely, minimizing delays that could affect crop yield. This capability supports site-specific actions, reducing excessive input use and contributing to more sustainable farming practices (Manifestly Checklists, 2024).

3.3 Internet of Things and wireless sensor networks for smart farming

Internet of things (IoT) is revolutionizing precision agriculture by enabling connectivity between physical devices such as sensors, drones, and machinery, and computer systems through internet-based communication (Mansoor et al., 2025; Kumar et al., 2024b). By automating farm operations, IoT systems significantly reduce the need for manual intervention. Particularly, these technologies facilitate the continuous collection of important field data, including soil moisture levels, climate conditions, surface and air temperatures, and relative humidity, which is used in real time for monitoring, control, and decision-making. This real-time data monitoring empowers farmers to make timely and informed crop management decisions (Rajak et al., 2023).

Moreover, the integration of IoT with farm equipment establishes a dynamic communication link between machines and computers, transforming conventional machinery into intelligent systems capable of autonomous operation (Kumar et al., 2024a). These smart machines can independently perform essential agricultural tasks such as tillage, seeding or transplanting, nutrient application, spraying, irrigation, and harvesting. Simultaneously, they record detailed information about soil and plant conditions, providing vital technical support for site-specific precision farming strategies (Neupane and Samadi, 2025).

These IoT systems function by collecting data from various sensors, including soil moisture sensors, temperature sensors, and weather stations, which monitor key environmental parameters (Mansoor et al., 2025). In addition to ground-based sensors, drones equipped with high-resolution cameras and advanced sensors capture data across larger areas of farmland. The information gathered from these devices is then analyzed to assess patterns in crop conditions and detect potential issues, enabling timely and informed decision-making (Ahmed and Shakoor, 2025).

One of the key applications of IoT in precision agriculture is automated irrigation, where water distribution is dynamically regulated based on real-time data. IoT systems utilize inputs from soil moisture sensors, temperature and humidity sensors, and evapotranspiration data to avoid excessive irrigation and minimize unnecessary water consumption. This targeted approach not only conserves water but also ensures optimal moisture levels for crop growth (Abdelmoneim et al., 2025b).

In addition to automated irrigation, IoT systems support site-specific nutrient application by examining nutrient content and accurately applying manure only when and where it is needed (Dhanaraju et al., 2022). This targeted approach minimizes both operational costs and environmental impact. Furthermore, pests and diseases can be detected in advance, enabling prompt and focused responses that prevent yield loss and reduce excessive use of fertilizers (Rajak et al., 2023).

Beyond crop management, IoT systems are also employed in monitoring livestock conditions, including activity, location, feeding schedules, and health status, using specialized sensors to support animal well-being (Tangorra et al., 2024). These technologies further extend to greenhouse environments, where they track and regulate climatic factors such as temperature, relative humidity, and light intensity to maintain optimal conditions required by crops, thereby enhancing yields (Manoharan et al., 2024).

Collectively, these applications of IoT contribute to improved operational performance and enable autonomous functions such as irrigation, fertilization, etc. By reducing labor costs and time, minimizing input waste, and ensuring precise resource application, IoT-driven agriculture promotes increased returns while maintaining consistency with sustainability goals.

3.4 Artificial intelligence and machine learning transforming farm decision-making

The integration of AI and ML into farming practices has transformed the agriculture sector, opening a new era of technological advancement. AI is a broad field that focuses on developing intelligent tools, machines, and computers that have the capability to simulate or mimic cognitive functions, similar to how humans observe, understand, and respond to verbal or written communication, assess data, and make decisions and suggestions (University of North Florida (UNF), 2025). ML, a subset of AI, enables machines to learn autonomously from past experiences rather than relying on manual programming (UNF, 2025). By utilizing algorithms to analyze vast amounts of data, ML extracts meaningful insights that inform precise and data-driven agricultural decisions.

In agriculture, AI and ML are applied by collecting data from various sources such as sensors, satellites, drones, and yield monitoring systems (Mishra and Mishra, 2023), as well as from disease symptoms in formats like text, video, images, or structured datasets. This data is then used to train ML models by feeding a portion of it into the system. The mathematical algorithms within these models learn patterns and trends from the input and generate logic, rules, feature extractions, or recognitions (Al-Zaiti et al., 2022). For example, variations in leaf color can indicate different plant conditions or diseases, which the model uses to predict outcomes and support decision-making (Islam et al., 2024).

These technologies enable precision farming through the site-specific application of inputs by analyzing various soil and crop characteristics. AI algorithms provide precise recommendations to farmers for tasks such as evaluating data from moisture sensors, climate conditions, and crop water requirements to generate site-specific irrigation schedules that optimize water use (OĞUZTÜRK, 2025). Fertilizer application rates and seed variety selection can also be determined based on soil types and climatic conditions (Abishek et al., 2023). These technologies also support predictive analytics and yield enhancement by evaluating both historical and real-time data related to crop and soil attributes.

Weed detection and control are facilitated through drone or satellite imagery, which distinguishes between crops and weeds (Shamshiri et al., 2024). AI algorithms analyze these images to identify weeds that compete with crops for nutrients and water and automatically apply herbicides to targeted areas, thereby promoting precision agriculture (Muchhadiya et al., 2024). Similarly, disease and pest management follow the same approach, allowing informed decisions before infestations affect crop yields (Garg, 2022).

AI also contributes to crop quality assessment by analyzing postharvest images based on parameters such as size, shape, color, and irregularities, enabling accurate sorting (Fadiji et al., 2023). Beyond crop production, AI and ML are applied in the animal and dairy industries for health monitoring, feeding schedule management, milking time optimization, behavior analysis, and disease prediction (Mishra and Mishra, 2023). Overall, AI and ML are revolutionizing agriculture by enabling precise, data-driven decisions that enhance productivity and sustainability. Their continued integration promises smarter farming systems and resilient food production for the future.

3.5 Robotics and automation in precision agriculture

Robotics and automation in agriculture go far beyond replacing manual labor; they are transforming the farming sector by enhancing productivity, precision, and sustainability. Agricultural robotics includes automated tractors for plowing, systems for seeding

Fig. 1.1 Precision apple harvesting using robotics. (Onishi et al., 2019).

and transplanting, and precision sprayers for pesticide application. Robotic harvesters use sensors to carefully pick fruits without causing damage, as shown in Fig. 1.1, while weeding robots detect and remove weeds with high accuracy (Cheng et al., 2023). UAVs and drones support these efforts by tracking crop conditions and spraying pesticides or disease treatments (Doggalli et al., 2024). Beyond open-field applications, automation in controlled environment agriculture enables operations such as irrigation scheduling, harvesting mature produce, and monitoring plant health using sensors and ML models (Farmonuat, 2025).

These robotic technologies have revolutionized agriculture by significantly increasing efficiency and precision, while drastically reducing the need for manual labor (Bazargani and Deemyad, 2024). They also simplify crop monitoring and optimization, contributing to more sustainable farm management. Robotic harvesting and automation minimize human error and operate consistently around the clock, ensuring reliable performance and improved productivity (Farmonuat, 2025). However, the adoption of these participant-developed technologies remains challenging in underdeveloped countries due to high initial investment costs, limited technological awareness among farmers, and the small average farm size (Gaurav et al., 2024). Continued innovation and accessibility efforts are essential to ensure these technologies benefit farming systems globally. Research committees use participatory technology development by highlighting low-cost solutions, lower operational complexity, and robust systems. These technologies are designed to be user-centric and require minimal training for farmers, who may have limited access to advanced tools but can effectively fulfill their major needs.

3.6 Blockchain for transparency and security in agriculture

Blockchain is a decentralized and immutable digital ledger system that records transactions across a network of interconnected computer systems (Ballamudi, 2016; Habib et al., 2022; Pakseresht et al., 2024). Each transaction is securely duplicated and shared throughout the blockchain network, ensuring transparency, traceability, and resistance to tampering (Pakseresht et al., 2024). In precision agriculture, blockchain plays a crucial role in enabling transparent farm-to-feed network traceability. The growing demand for smart, peer-to-peer systems capable of validating, encrypting, tracking, and evaluating farm-related data is driving interest in integrating blockchain with IoT technologies in smart agriculture. By replacing traditional methods of storing, organizing, and sharing farm data, blockchain offers a tamper-proof, decentralized, and reliable framework (Torky and Hassanein, 2020).

This secure system allows farm data to be recorded at every stage, from sowing to harvesting. Inputs such as seed variety, fertilizer type, and other relevant details are logged with timestamps, creating a secure and traceable supply chain (Demestichas et al., 2020). Such tracking not only benefits producers but also addresses increasing consumer interest in the origin of their food, driven by priorities such as food safety, environmental responsibility, regional market support, animal welfare, and labor rights. In the event of a product recall, blockchain enables rapid identification of the product's origin, making it easier to trace the source of the issue and ensure timely resolution (Eze et al., 2024). This technology offers a groundbreaking capability to track where and how food was cultivated, strengthening transparency and trust across the supply chain (Funk, 2023).

For instance, the technology company Project Provenance Ltd (using its PROVENANCE® platform) collaborated with the nonprofit organization Humanity United and the International Pole and Line Foundation and executed a six-month-long experimental project to trace tuna fish caught by fishermen in Indonesia. This project's aim was to use a smartphone app with blockchain technology for smart labeling to track and trace the source and legitimacy of the ethical sourcing label. The system applied data interoperability to monitor and identify products and certifications in a protected, continuous, user-friendly system. Point of sale (POS) was introduced for tracing the digital label across the distribution system. The fishermen were instructed to send a text message to log the fish; with this short message service (SMS) text message, a new digital asset was generated in the ledger with an immutable and distinct identification (ID) number. The fish would be attached to this unique ID physically using a bar code, radio frequency identification (RFID) tag, or other technologies. Then, this digital asset moved to the stock along with the catch, and a digital transaction was recorded on the blockchain. The fisherman's identity was also saved. Public software libraries were used to verify tracking and tracing by navigating the blockchain (Provenance, 2016; Motta et al., 2020).

As smart agriculture increasingly relies on data-driven technologies like IoT tools, sensor networks, and AI/ML models, the risk of cybercrime becomes a pressing concern. In this context, blockchain technologies help safeguard agricultural data from digital threats, offering a secure and resilient framework for protecting sensitive information and potentially being used in open agriculture data networks as well.

3.7 Decision support systems for data-driven farming

A decision support system (DSS) is a computer-based tool that synthesizes data to provide insights, helping growers make strategic decisions across various farming operations (Ramya, 2025). Alongside data analysis and interpretation, DSS plays a critical role in enhancing market responsiveness and operational efficiency. As innovative technologies continue to advance, smart farming and precision agriculture are becoming increasingly widespread, strengthening the importance of computer-assisted DSSs in modern agricultural practices (Rupnik et al., 2019).

DSS integrates data from diverse sources to provide site-specific recommendations for irrigation scheduling, fertilizer application, pest control, and other essential farm operations (Ramya, 2025). This proactive guidance helps reduce crop damage and mitigate environmental impact. The growing adoption of DSS is driven by emerging technologies such as AI and ML algorithms for predictive analytics, blockchain for ensuring data privacy and security, and IoT sensor networks that optimize data acquisition and real-time monitoring (Bhat et al., 2025). Incorporating DSS into precision agriculture empowers growers to make informed, timely decisions with greater confidence. As technology continues to evolve, these systems will remain central to sustainable and data-driven farming practices.

3.8 Integrating technologies and achieving data interoperability

To fully harness the potential of smart agriculture technologies, it is essential to integrate various components such as sensors, GPS modules, irrigation controllers, drones, and weather stations with farm management software or centralized computing systems. This integration enables seamless coordination among tools, allowing them to operate together smoothly.

A key aspect of this process is data interoperability, which refers to the ability of different systems and technologies to understand, exchange, and utilize information effectively (Dial Ventures, 2025). When integration and interoperability are achieved, they support precise farm operations, enhance decision-making, improve transparency, and boost overall efficiency and crop yields.

However, these benefits come with notable challenges. Limited availability of standardized data formats, communication barriers between various devices and protocols, and concerns around data security demand advanced technical expertise. These complexities can delay adoption, particularly among small and rural farming communities with limited access to digital infrastructure and training. Achieving smooth integration and data interoperability is key to unlocking the full potential of precision agriculture. Addressing technical challenges will be essential to ensure inclusive and scalable adoption across all farming communities.

3.9 Practical applications of precision agriculture technologies

Precision agriculture technologies are widely applied across key farming operations, with notable emphasis on irrigation management. Techniques such as sensor-based irrigation (e.g., soil moisture sensors), crop stress-based irrigation using the CWSI, precision irrigation systems, and evapotranspiration models enable more efficient water use and timely interventions (Achite et al., 2022; Boyaci et al., 2024; (Sui, 2016: Vilavan et al., 2024).

Beyond irrigation, these technologies play a vital role in nutrient management through VRT and soil test-based fertilizer recommendations (Cai et al., 2025). They also support pest and disease control through predictive alerts, thermal imaging, and AI-driven diagnostics (Gupta and Kumar Pal, 2025). In harvest and postharvest stages, precision tools facilitate yield mapping, moisture detection, and logistics optimization, contributing to improved productivity and resource efficiency across the agricultural value chain (Getahun et al., 2024c).

3.10 Socioeconomic and environmental impacts of precision agriculture

The transition from traditional to modern agriculture holds significant potential to meet the growing global food demand. Precision technologies are driving this shift by enhancing productivity while reducing resource consumption. Tools such as GPS and GIS-enabled variable rate applications, IoT sensors, and AI/ML integration have significantly improved resource efficiency. These technologies enable precise input management, optimizing the use of water, fertilizers, and pesticides, thereby reducing production costs and adverse environmental effects (Parra-López et al., 2025).

In terms of socioeconomic impact, precision agriculture has improved productivity and profitability. Site-specific input application allows farmers to achieve higher yields and better financial returns through efficient resource utilization (Getahun et al., 2024c). The integration of advanced technologies, including sensors, drones, and UAVs, facilitates data-driven decision-making, empowering farmers to manage operations with greater accuracy and confidence (Chen, 2025). As global populations grow and food scarcity intensifies, precision agriculture plays a critical role in ensuring consistent and resilient food production and distribution. Moreover, the adoption of these technologies has enhanced economic well-being by improving livelihoods and supporting rural development. Mechanization and automation have also reduced the reliance on manual labor, leading to increased time efficiency and lower labor costs (Bazargani and Deemyad, 2024).

Environmentally, precision agriculture offers substantial benefits. Traditional practices, such as excessive fertilizer use and intensive tillage, have contributed to soil degradation and erosion. In contrast, modern techniques like zero tillage and site-specific nutrient management help preserve soil structure, prevent erosion, and promote long-term soil health (Feliziani et al., 2025). Smart farming systems also reduce the use of inputs such as water, fertilizers, and pesticides, thereby lowering carbon footprints and mitigating the environmental impact of agriculture (Papadopoulos et al., 2024b). Advanced irrigation methods, including sensor-based scheduling, drip irrigation, and variable rate irrigation, enhance water use efficiency by minimizing losses from evaporation and runoff (Pourmohammad and Hajirad, 2025). These practices not only support sustainable water management but also contribute to climate change mitigation by reducing GHG emissions and promoting carbon sequestration in healthy soils.

4 Conclusion

Precision technologies have revolutionized modern agriculture by transitioning from uniform applications to site-specific and data-driven management. The integration of tools such as GPS, GIS, VRT, UAVs, RS, AI, ML, and DSS has optimized resource utilization, enhanced yield monitoring, and improved pest and disease detection. The adoption of IoT and Wireless Sensor Networks (WSNs) has enabled seamless connectivity between sensors and cloud-based platforms, automating operations and facilitating real-time decision-making. These advancements contribute to efficient, sustainable, and precise farming practices.

Despite these transformative benefits, several challenges hinder the widespread adoption of precision technologies. High initial investment costs for machinery, sensors, and software remain a major constraint, particularly for smallholder farmers and those in economically backward regions (Read and Torrado, 2009). Limited digital literacy, data management complexity, and interoperability issues between hardware and software further restrict effective use (Guebsi et al., 2024). In many rural and developing regions, poor internet connectivity and inadequate infrastructure exacerbate these limitations, restricting the performance of IoT-based systems. In addition, concerns surrounding data privacy, ownership, and potential misuse continue to discourage trust and adoption (Sekhar et al., 2024). Addressing these technical, economic, and infrastructural challenges is crucial to ensure equitable access and sustainable implementation.

Looking ahead, the future of precision agriculture lies in the integration of emerging innovations such as edge computing, 5G connectivity, and quantum computing, which promise faster data processing and improved connectivity. Climate-smart practices, blockchain-based traceability systems, and AI-driven analytics will play a crucial role in ensuring transparency, sustainability, and resource efficiency. Moreover, immersive technologies such as augmented reality (AR) and virtual reality (VR) can transform farmer education, equipment maintenance, and collaborative research through interactive digital environments. AR can further visualize real-time field data such as soil moisture, pest presence, and nutrient levels, empowering farmers to make site-specific input applications efficiently.

Overall, while certain barriers persist, precision agriculture continues to make a significant positive impact on the farming community. With appropriate regulations, capacity-building initiatives, and supportive policy frameworks, these challenges can be mitigated. Continued innovation and multidisciplinary collaboration among farmers, scientists, and technology developers will be essential to realizing the full potential of precision agriculture, ensuring resilient, efficient, and sustainable food production for the future.

References

Abdelmoneim, A.A., Kimaita, H.N., Al Kalaany, C.M., Derardja, B., Dragonetti, G., Khadra, R., 2025a. IoT sensing for advanced irrigation management: a systematic review of trends, challenges, and future prospects. Sensors (Basel, Switzerland) 25 (7), 2291.

Abdelmoneim, A.A., Kimaita, H.N., Al Kalaany, C.M., Derardja, B., Dragonetti, G., Khadra, R., 2025b. IoT sensing for advanced irrigation management: a systematic review of trends, challenges, and future prospects. Sensors (Basel, Switzerland) 25 (7), 2291.

Abishek, J., Kannan, P., Devi, M.N., Prabhaharan, J., Sampathkumar, T., Kalpana, M., 2023. Soil texture prediction using machine learning approach for sustainable soil health management. Int. J. Plant Soil Sci. 35 (19), 1416–1426.

Achite, M., Jehanzaib, M., Sattari, M.T., Toubal, A.K., Elshaboury, N., Wałęga, A., Kim, T.W., 2022. Modern techniques to modeling reference evapotranspiration in a semiarid area based on ANN and GEP models. Water 14 (8), 1210.

AGRIVI, 2022. Variable Rate Technology: everything You Need to Know. URL https://www.agrivi.com/blog/variable-rate-technology/ (accessed 7.17.25).

Ahmad, M.F., Ahmad, F.A., Alsayegh, A.A., Zeyaullah, M., AlShahrani, A.M., Muzammil, K., Hussain, S., 2024. Pesticides impacts on human health and the environment with their mechanisms of action and possible countermeasures. Heliyon 10 (7).

Ahmed, N., Shakoor, N., 2025. Advancing agriculture through IoT, Big Data, and AI: a review of smart technologies enabling sustainability. Smart Agric. Technol. 10, 100848.

Al-Zaiti, S.S., Alghwiri, A.A., Hu, X., Clermont, G., Peace, A., Macfarlane, P., Bond, R., 2022. A clinician's guide to understanding and critically appraising machine learning studies: a checklist for Ruling Out Bias Using Standard Tools in Machine Learning (ROBUST-ML). Eur. Heart J. Digit. Health 3 (2), 125–140.

Alexander, R., Nugent, C., Nugent, K., 2018. The dust bowl in the US: an analysis based on current environmental and clinical studies. Am. J. Med. Sci. 356 (2), 90–96.

American Society of Mechanical Engineers, 2012. Jethro Tull, URL https://www.asme.org/topics-resources/content/jethro-tull (accessed 10.29.25).

Angelakis, A.N., Valipour, M., Dietrich, J., Voudouris, K., Kumar, R., Salgot, M., Tsoutsos, T., 2022. Sustainable and regenerative development of water mills as an example of agricultural technologies for small farms. Water 14 (10), 1621.

Angelakis, A.N., Zaccaria, D., Krasilnikoff, J., Salgot, M., Bazza, M., Roccaro, P., Fereres, E., 2020. Irrigation of world agricultural lands: evolution through the millennia. Water 12 (5), 1285.

Balasundram, S.K., Shamshiri, R.R., Sridhara, S., Rizan, N., 2023. The role of digital agriculture in mitigating climate change and ensuring food security: an overview. Sustainability 15 (6), 5325.

Ballamudi, K.R., 2016. Blockchain as a type of distributed ledger technology. Asian J. Human. Art Literature 3 (2), 127–136.

Bazargani, K., Deemyad, T., 2024. Automation's impact on agriculture: opportunities, challenges, and economic effects. Robotics 13 (2), 33.

Benin, D., Snezhko, V., 2021. Water flow regulator for irrigation canals. In: E3S Web of Conferences, Vol. 264. EDP Sciences, pp. 03038.

Bhamini, K., Kumar, A., Magrey, A.H., Kulsoom, S., 2025. The role of GIS and GPS in precision fertilization: optimizing nutrient management for sustainable cropping. Int. J. Res. Agron. 8, 260–267.

Bhat, I.A., Ansarullah, S.I., Ahmad, F., Amir, S., Sidana, S., Sinha, A., Yazdani, G., 2025. Leveraging artificial intelligence in agribusiness: a structured review of strategic management practices and future prospects. Discov. Sustain. 6 (1), 565.

Boyaci, S., Kocięcka, J., Atilgan, A., Liberacki, D., Rolbiecki, R., Saltuk, B., Stachowski, P., 2024. Evaluation of crop water stress index (CWSI) for high tunnel greenhouse tomatoes under different irrigation levels. Atmosphere 15 (2), 205.

Cai, B., Shi, F., Geremew, B.A., Addis, A.K., Abate, M.C., Dessie, W., 2025. Precision agriculture techniques for optimizing chemical fertilizer use and environmental sustainability: a systematic review. Front. Agron. 7, 1665444.

Chataut, G., Bhatta, B., Joshi, D., Subedi, K., Kafle, K., 2023. Greenhouse gases emission from agricultural soil: a review. J. Agric. Food Res. 11, 100533.

Chen, X., 2025. The role of modern agricultural technologies in improving agricultural productivity and land use efficiency. Front. Plant Sci. 16, 1675657.

Cheng, C., Fu, J., Su, H., Ren, L., 2023. Recent advancements in agriculture robots: benefits and challenges. Machines 11 (1), 48.

Chinthalapudi, D.P., Pokhrel, S., Kingery, W.L., Shankle, M.W., Ganapathi Shanmugam, S., 2023. Exploring the synergistic impacts of cover crops and fertilization on soil microbial metabolic diversity in dryland soybean production systems using biolog EcoPlates. Appl. Biosci. 2 (3), 328–346.

Clapp, J., Ruder, S.L., 2020. Precision technologies for agriculture: digital farming, gene-edited crops, and the politics of sustainability. Glob. Environ. Polit. 20 (3), 49–69.

Dari, J., Presti, S.L., Brocca, L., 2025. Irrigation monitoring from satellite at hyper-high resolution: paving the way for remote-sensing-based agricultural water management support services. Agric. Water Manag. 317, 109627.

DeJong-Hughes, J., Daigh, A., 2021. University of Minnesota Extension, Upper Midwest Tillage Guide. University of Minnesota Extension Service and North Dakota State University. https://extension.umn.edu/soil-management-and-health/tillage-implements.

Demestichas, K., Peppes, N., Alexakis, T., Adamopoulou, E., 2020. Blockchain in agriculture traceability systems: a review. Appl. Sci. 10 (12), 4113.

Deming, D., 2019. The aqueducts and water supply of Ancient Rome. Ground Water 58 (1), 152.

Dhanaraju, M., Chenniappan, P., Ramalingam, K., Pazhanivelan, S., Kaliaperumal, R., 2022. Smart farming: internet of Things (IoT)-based sustainable agriculture. Agriculture 12 (10), 1745.

Dial Ventures, 2025. Data inter/intra operability in the agriculture and food industry. URL https://www.dialventures.com/library/data-inter-intra-operability-in-the-agriculture-and-food-industry (accessed 8.3.25).

DIG Corp, The History and Evolution of Irrigation Techniques. URL https://www.digcorp.com/the-history-and-evolution-of-irrigation-techniques/ (accessed 7.25.25).

Doggalli, G., Santhoshinii, E., Manojkumar, H.G., Srivastava, M., Ganesh, H.S., Barigal, A., Kundu, R., 2024. Drone technology for crop disease resistance: innovations and challenges. J. Sci. Res. Rep. 30 (8), 174–180.

ESRI, What is GIS?, Geographic Information System Mapping Technology. URL https://www.esri.com/en-us/what-is-gis/overview (accessed 10.29.25).

Evett, S.R., Colaizzi, P.D., Lamm, F.R., O'Shaughnessy, S.A., Heeren, D.M., Trout, T.J., Lin, X., 2020. Past, present, and future of irrigation on the US Great Plains. Trans. ASABE 63 (3), 703–729.

Eze, V.H.U., Eze, E.C., Alaneme, G.U., BUBU, P.E., Nnadi, E.O.E., Okon, M.B., 2025. Integrating IoT sensors and machine learning for sustainable precision agroecology: enhancing crop resilience and resource efficiency through data-driven strategies, challenges, and future prospects. Discov. Agric. 3 (1), 83.

Eze, V.H.U., Ugwu, C.N., Ogenyi, F.C., 2024. Blockchain-Enabled Supply Chain Traceability in Food Safety. Res. Output J. Biol. Appl. Sci. 3 (1), 46–51.

Fadiji, T., Bokaba, T., Fawole, O.A., Twinomurinzi, H., 2023. Artificial intelligence in postharvest agriculture: mapping a research agenda. Front. Sustain. Food Syst. 7, 1226583.

Fan, M., Stamford, J., Lawson, T., 2024. Using infrared thermography for high-throughput plant phenotyping. In: Photosynthesis: Methods and Protocols. Springer US, New York, NY, pp. 317–332.

Farmonuat, 2025. Modern Agriculture Technology: 7 Shocking New Advances. URL https://farmonaut.com/precision-farming/modern-agriculture-technology-7-shocking-new-advances (accessed 8.2.25).

Feliziani, G., Bordoni, L., Gabbianelli, R., 2025. Regenerative Organic Agriculture and Human Health: the Interconnection Between Soil, Food Quality, and Nutrition. Antioxidants 14 (5), 530.

Fox, J., Khalilian, A., Han, Y., Williams, P., Nafchi, A., Maja, J., Marshall, M., Barnes, E., 2018. Real-time, variable-depth tillage for managing soil compaction in cotton production. Open J. Soil Sci. 8, 147–161. doi:10.4236/ojss.2018.86012.

Funk, B., (2023). Blockchain Agriculture: revolutionizing Farming with Eden Green. URL https://www.edengreen.com/blog-collection/blockchain-technology-in-agriculture (accessed 8.3.25).

Gade, S.A., Madolli, M.J., García-Caparrós, P., Ullah, H., Cha-um, S., Datta, A., Himanshu, S.K., 2025. Advancements in UAV remote sensing for agricultural yield estimation: a systematic comprehensive review of platforms, sensors, and data analytics. Remote Sens. Appl. Soc. Environ. 37, 101418.

Galieni, A., D'Ascenzo, N., Stagnari, F., Pagnani, G., Xie, Q., Pisante, M., 2021. Past and future of plant stress detection: an overview from remote sensing to positron emission tomography. Front. Plant Sci. 11, 609155.

Garg, S., 2022. AI in Agriculture: Revolutionizing Farming Practices, ResearchGate.

Gaurav, M.A., Jyoti, Y.K., Mishra, N., Dutt, A., Gupta, P. (2024). Robotics and Automation in Modern Agricultural Practices.

Getahun, S., Kefale, H., Gelaye, Y., 2024a. Application of precision agriculture technologies for sustainable crop production and environmental sustainability: a systematic review. Sci. World J. 2024 (1), 2126734.

Getahun, S., Kefale, H., Gelaye, Y., 2024b. Application of precision agriculture technologies for sustainable crop production and environmental sustainability: a systematic review. Sci. World J. 2024 (1), 2126734.

Getahun, S., Kefale, H., Gelaye, Y., 2024c. Application of precision agriculture technologies for sustainable crop production and environmental sustainability: a systematic review. Sci. World J. 2024 (1), 2126734.

Ghazal, S., Munir, A., Qureshi, W.S., 2024. Computer vision in smart agriculture and precision farming: techniques and applications. Artif. Intell. Agric. 13, 64–83.

GPS, (2021). Agricultural Applications. URL https://www.gps.gov/applications/agriculture/ (accessed 8.1.25).

Guebsi, R., Mami, S., Chokmani, K., 2024. Drones in precision agriculture: a comprehensive review of applications, technologies, and challenges. Drones 8 (11), 686.

Gupta, G., Kumar Pal, S., 2025. Applications of AI in precision agriculture. Discov. Agric. 3 (1), 61.

Habib, G., Sharma, S., Ibrahim, S., Ahmad, I., Qureshi, S., Ishfaq, M., 2022. Blockchain technology: benefits, challenges, applications, and integration of blockchain technology with cloud computing. Future Internet 14 (11), 341.

Hassen, T.B., El Bilali, H., Allahyari, M.S., Chatti, C.B., 2025. Traditional irrigation knowledge for sustainable water resource management in arid Environments: insights from the MENA region. J. Arid Environ. 231, 105466.

Haug, B., 2017. Water and power: reintegrating the state into the study of Egyptian irrigation. Hist. Compass 15 (10), e12394.

Hossain, M.S., Ghosh, U.K., Islam, M.N., Khan, M.A.R., 2024. Precision agriculture practices for smart irrigation. In: Remote Sensing in Precision Agriculture: Transforming Scientific Advancement into Innovation. Earth Observation. pp. 175–188.

Iron Solutions, (2020). Harvest Equipment: a Brief History of the Combine. Iron Solutions. URL https://ironsolutions.com/a-brief-history-of-the-combine/ (accessed 7.26.25).

Ishizaki, S., Hirai, H., Sakagaito, T., Takeyama, T., Oido, N., Tamura, T., Umeda, M., 2024. Development of a transplanter-based transplanter for vegetable seedlings cultured in a cuttable nursery mat. J. Agric. Eng. 55 (2), 1571.

Islam, S., Samsuzzaman, Reza, M.N., Lee, K.H., Ahmed, S., Cho, Y.J., Chung, S.O., 2024. Image processing and support vector machine (SVM) for classifying environmental stress symptoms of pepper seedlings grown in a plant factory. Agronomy 14 (9), 2043.

Jahanbakht, M., Olsen, A., Marchant, R., Fillols, E., Azghadi, M.R. (2025). Advancements in Weed Mapping: a Systematic Review. *arXiv preprint arXiv:2507.01269*.

Kariyanna, B., Sowjanya, M., 2024. Unravelling the use of artificial intelligence in management of insect pests. Smart Agric. Technol. 8, 100517.

Kaur, B., Dimri, S., Singh, J., Mishra, S., Chauhan, N., Kukreti, T., Sharma, B., Singh, S.P., Arora, S., Uniyal, D., Agrawal, Y., Akhtar, S., Rather, M.A., Naik, B., Kumar, V., Gupta, A.K., Rustagi, S., Preet, M.S., 2023. Insights into the harvesting tools and equipment's for horticultural crops: from then to now. J. Agric. Food Res. 14, 100814.

Kaviyarasan, G., Rahale, S., Subramaniam, T., 2024. Revolutionizing agriculture: a deep dive into modern fertilizer application strategies. Recent Trends in Soil Science, SSSA, 119–137.

Kent, B., Hamer, G., 2004. Historical improvements in groundwater-pumping equipment and effects on farming in the United States. Henry PG Darcy Other Pioneers Hydraul. 207–219.

Kumar, H., Miller, S.A., Lyon, S.W., 2023b. Assessing nutrient concentrations and field-scale seepage load under an automated drainage water management system in Ohio. Smart Agric. Technol. 6, 100328.

Kumar, H., Srivastava, P., Lamba, J., Diamantopoulos, E., Ortiz, B., Morata, G., Bondesan, L., 2022. Site-specific irrigation scheduling using one-layer soil hydraulic properties and inverse modeling. Agric. Water Manag. 273, 107877.

Kumar, H., Srivastava, P., Lamba, J., Lena, B., Diamantopoulos, E., Ortiz, B., Bondesan, L., 2023a. A methodology to optimize site-specific field capacity and irrigation thresholds. Agric. Water Manag. 286, 108385.

Kumar, H., Srivastava, P., Lamba, J., Ortiz, B.V., Way, T.R., Sangha, L., Molinari, R., 2022. Within-field variability in nutrients for site-specific agricultural management in irrigated cornfield. J. ASABE 65 (4), 865–880.

Kumar, R., Kumar, A., Himanshu, S.K., 2024b. IoT and smart sensor applications in nutrient and irrigation water management. In: Agri-Tech Approaches for Nutrients and Irrigation Water Management. CRC Press. pp. 48–67.

Kumar, V., Sharma, K.V., Kedam, N., Patel, A., Kate, T.R., Rathnayake, U., 2024a. A comprehensive review on smart and sustainable agriculture using IoT technologies. Smart Agric. Technol. 8, 100487.

Lal, R., Reicosky, D.C., Hanson, J.D., 2007. Evolution of the plow over 10,000 years and the rationale for no-till farming. Soil Tillage Res. 93 (1), 1–12.

Liao, Q., Gu, S., Gao, S., Du, T., Kang, S., Tong, L., Ding, R., 2024. Crop water stress index characterizes maize productivity under water and salt stress by using growth stage-specific non-water stress baselines. Field Crops Res. 317, 109544.

Lohan, S.K., Jat, H.S., Yadav, A.K., Sidhu, H.S., Jat, M.L., Choudhary, M., Sharma, P.C., 2018. Burning issues of paddy residue management in north-west states of India. Renew. Sustain. Energy Rev. 81, 693–706.

Mamabolo, E., Mashala, M.J., Mugari, E., Mogale, T.E., Mathebula, N., Mabitsela, K., Ayisi, K.K., 2025. Application of precision agriculture technologies for crop protection and soil health. Smart Agric. Technol. 12, 101270.

Manifestly Checklists, The Role of Drones in Modern Farming, (2024). URL https://www.manifest.ly/blog/the-role-of-drones-in-modern-farming/ (accessed 8.1.25).

Manoharan, S., Lean, C.P., Li, C., Yuan, K.F., Kiat, N.P., Khan, M.R.B., 2024. IoT-enabled greenhouse systems: optimizing plant growth and efficiency. Malays. J. Sci. Adv. Technol. 4 (2), 169–179.

Mansoor, S., Iqbal, S., Popescu, S.M., Kim, S.L., Chung, Y.S., Baek, J.H., 2025. Integration of smart sensors and IOT in precision agriculture: trends, challenges and future prospectives. Front. Plant Sci. 16, 1587869.

Mathenge, M., Sonneveld, B.G., Broerse, J.E., 2022. Application of GIS in agriculture in promoting evidence-informed decision making for improving agriculture sustainability: a systematic review. Sustainability 14 (16), 9974.

McFadden, J., 2023. Variable rate technology adoption is on the rise | economic Research Service. URL https://www.ers.usda.gov/data-products/charts-of-note/chart-detail?chartId=107116 (accessed 7.17.25).

Mishra, H., Mishra, D., 2023. Artificial intelligence and machine learning in agriculture: transforming farming systems. Res. Trends Agric. Sci. 1, 1–16.

Mishra, P.K., Sharma, A., Prakash, A., 2023. Current research and development in cotton harvesters: a review with application to Indian cotton production systems. Heliyon 9 (2023), e16124.

Morning Ag Clips, (2022). The Mechanical Reaper: the impact of an important farming invention URL https://www.morningagclips.com/the-mechanical-reaper-the-impact-of-an-important-farming-invention/ (accessed 10.29.25).

Motta, G.A., Tekinerdogan, B., Athanasiadis, I.N., 2020. Blockchain applications in the agri-food domain: the first wave. Front. Blockchain 3, 6.

Muchhadiya, R.M., Kumawat, P.D., Gohil, B.S., Sakarvadia, H.L. (2024). Precision weed management using artificial intelligence tools and techniques for sustainable agriculture. *Agriculture & Food e-Newsletter*, *6*(1).

Mukherjee, A., Zade, S., Sidharth, S., Peerzade, W., Sundharan, M., 2024. Precision farming with automatic seed sowing machines. Agriculture Mag. 3 (12), 665–672.

NASA Earth data, Earth Observation Data Basics. Remote Sensing, URL https://www.earthdata.nasa.gov/learn/earth-observation-data-basics/remote-sensing (accessed 8.1.25).

National Oceanic and Atmospheric Administration, (2025). URL https://www.noaa.gov/news-release/gulf-of-america-dead-zone-below-average-scientists-find (accessed 8.28.25).

Neupane, A., Samadi, V., 2025. Empowering Precision Agriculture with the Internet of Things, Artificial Intelligence, and Robotics. Clemson Cooperative Extension, Land-Grant Press by Clemson Extension, Clemson (SC), pp. 1208. https://open.clemson.edu/forest_wildlife/53.

Nishio, T., (2019). Oregon State University, The invention of the rice transplanter by farmers. Ag-Small Farms/Commercial Ag. URL https://smallfarms.oregonstate.edu/smallfarms/invention-rice-transplanter-farmers.

OĞUZTÜRK, G.E., 2025. AI-driven irrigation systems for sustainable water management: a systematic review and meta-analytical insights. Smart Agric. Technol. 11, 100982.

Ojeilua, C.H., Eze, F.N., Ekechi, C.C., Atijosan, O.E., Ukasoanya, F.C., Chinwero, L.C., Ekechi, S.C., 2025. High-Resolution Stress Detection in Crops: integrating Satellite and Drone Remote Sensing for Resilient Agriculture. Afr. J. Agric. Sci. Food Res. 19 (1), 447–467.

Onishi, Y., Yoshida, T., Kurita, H., Fukao, T., Arihara, H., Iwai, A., 2019. An automated fruit harvesting robot by using deep learning. Robomech J. 6 (1), 1–8.

Pakseresht, A., Kaliji, S.A., Hakelius, K., 2024. Blockchain technology characteristics essential for the agri-food sector: a systematic review. Food Control 165, 110661.

Papadopoulos, G., Arduini, S., Uyar, H., Psiroukis, V., Kasimati, A., Fountas, S., 2024a. Economic and environmental benefits of digital agricultural technologies in crop production: a review. Smart Agric. Technol. 8, 100441.

Papadopoulos, G., Arduini, S., Uyar, H., Psiroukis, V., Kasimati, A., Fountas, S., 2024b. Economic and environmental benefits of digital agricultural technologies in crop production: a review. Smart Agric. Technol. 8, 100441.

Parra-López, C., Abdallah, S.B., Garcia-Garcia, G., Hassoun, A., Trollman, H., Jagtap, S., Carmona-Torres, C., 2025. Digital technologies for water use and management in agriculture: recent applications and future outlook. Agric. Water Manag. 309, 109347.

Pereira, G.P., Chaari, M.Z., Daroge, F., 2023. IoT-enabled smart drip irrigation system using ESP32. IoT 4 (3), 221–243.

Phiri, D., Simwanda, M., Salekin, S., Nyirenda, V.R., Murayama, Y., Ranagalage, M., 2020. Sentinel-2 data for land cover/use mapping: a review. Remote Sens. 12 (14), 2291.

Pibars, S.K., Gaballah, M.S., Mansour, H.A., Khalil, S.E., 2022. Management of fertigation and its effects on irrigation systems performance and crop yield. Int. J. Health Sci. 6 (S2), 1634–1659.

Pourmohammad, P., Hajirad, I., 2025. Innovative Irrigation Practices for Sustainable Agriculture: environmental Benefits and Implementation Challenges. J. Sustain. Dev. Nat. Resour. Manag. 1 (2), 1–11.

Praseartkul, P., Taota, K., Pipatsitee, P., Tisarum, R., Sakulleerungroj, K., Sotesaritkul, T., Cha-Um, S., 2023. Unmanned aerial vehicle-based vegetation monitoring of aboveground and belowground traits of the turmeric plant (Curcuma longa L.). Int. J. Environ. Sci. Technol. 20 (8), 8673–8686.

Provenance, (2016) From shore to plate: tracking tuna on the blockchain. URL https://legacy.provenance.org/tracking-tuna-on-the-blockchain (accessed 9.3.25).

Puntel, L.A., Bolfe, É.L., Melchiori, R.J.M., Ortega, R., Tiscornia, G., Roel, A., Balboa, G.R., 2022. How digital is agriculture in a subset of countries from South America? Adoption and limitations. Crop Pasture Sci. 74 (6), 555–572.

Rajak, P., Ganguly, A., Adhikary, S., Bhattacharya, S., 2023. Internet of Things and smart sensors in agriculture: scopes and challenges. J. Agric. Food Res. 14, 100776.

Ramya, B. (2025). Importance of Decision Support System in Agriculture. doi:10.13140/RG.2.2.24418.85446.

Ray, L.I., Swetha, K., Singh, A.K., Singh, N.J., 2023. Water productivity of major pulses—A review. Agric. Water Manag. 281, 108249.

Read, J.M., Torrado, M., 2009. Remote sensing. In: Kitchin, R., Thrift, N. (Eds.), International Encyclopedia of Human Geography. Elsevier, Oxford, pp. 335–346. doi:10.1016/B978-008044910-4.00508-3.

Roy, D.P., Wulder, M.A., Loveland, T.R., Ce, W., Allen, R.G., Anderson, M.C., Zhu, Z., 2014. Landsat-8: science and product vision for terrestrial global change research. Remote Sens. Environ. 145, 154–172.

Rupnik, R., Kukar, M., Vračar, P., Košir, D., Pevec, D., Bosnić, Z., 2019. AgroDSS: a decision support system for agriculture and farming. Comput. Electron. Agric. 161, 260–271.

Saleem, S.R., Levison, J., Haroon, Z., 2023. Environment: role of precision agriculture technologies. In: Precision Agriculture. Academic Press. pp. 211–229.

Saniga, N.S., Gangadhar, K., Sushma, H.A., Boraiah, B., (2024). Precision Agriculture: technologies And Application. pp. 44–58.

Šarauskis, E., Kazlauskas, M., Naujokienė, V., Bručienė, I., Steponavičius, D., Romaneckas, K., Jasinskas, A., 2022. Variable rate seeding in precision agriculture: recent advances and future perspectives. Agriculture 12 (2), 305.

Schimmelpfennig, D., 2016. Precision agriculture technologies and factors affecting their adoption. Amber Waves. U.S. Department of Agriculture, Economic Research Service, 11.

Schumann, A.W., 2010. Precise placement and variable rate fertilizer application technologies for horticultural crops. HortTechnology 20 (1), 34–40.

Sean, G.F.S., Seed Spreading Showdown: comparing the Pros and Cons of the Top Methods for Maximizing Your Pasture's Growth Potential – grayslake Feed Website. URL https://grayslakefeed.com/2023/04/10/seed-spreading-showdown-comparing-the-pros-and-cons-of-the-top-methods-for-maximizing-your-pastures-growth-potential/ (accessed 10.29.25).

Sekhar, M., Yadav, A., Yadav, D.K., 2024. Precision agriculture and remote sensing technologies. Mod. Concept Agron., 105–122.

Shamshiri, R.R., Rad, A.K., Behjati, M., Balasundram, S.K., 2024. Sensing and perception in robotic weeding: innovations and limitations for digital agriculture. Sensors 24 (20), 6743.

Sharma, A., Khar, S., 2024. Conceptualization and development of a semi-automatic vegetable transplanter prototype for small landholdings. Heliyon 10 (11), e31540.

Shepherd, M., Turner, J.A., Small, B., Wheeler, D., 2020. Priorities for science to overcome hurdles thwarting the full promise of the 'digital agriculture' revolution. J. Sci. Food Agric. 100 (14), 5083–5092.

Silva, J.A.O.S., Siqueira, V.S.D., Mesquita, M., Vale, L.S.R., Marques, T.D.N.B., Silva, J.L.B.D., Oliveira, H.F.E.D., 2024. Deep learning for weed detection and segmentation in agricultural crops using images captured by an unmanned aerial vehicle. Remote Sens. 16 (23), 4394.

Singh, P.K., Dey, P., Jain, S.K., Mujumdar, P., 2020. Hydrology and water resources management in ancient India. Hydrol. Earth Syst. Sci. Discuss. 2020, 1–20.

Srikanthnaik, J., 2024. Integrating precision technology in mechanized fertilizer sprayers: a step towards sustainable agriculture. Int J Agric Extension Soc. Dev. 7 (4), 648–651. doi:10.33545/26180723.2024.v7.i4h.1805.

Srivastava, R.K., Purohit, S., Alam, E., Islam, M.K., 2024. Advancements in soil management: optimizing crop production through interdisciplinary approaches. J. Agric. Food Res. 18, 101528.

Stutsel, B., Johansen, K., Malbéteau, Y.M., McCabe, M.F., 2021. Detecting plant stress using thermal and optical imagery from an unoccupied aerial vehicle. Front. Plant Sci. 12, 734944.

Sui, R., 2016. Use of soil moisture sensors for irrigation scheduling. In: Irrigation Show & Education Conference, Irrigation Association, pp. 1–6.

Surkes, S., 2021. Israel's drip irrigation pioneers aim to do away with flooded rice fields. The Times of Israel, 10 May 2021. Available at: The Times of Israel.

Taizy, (2020). The Development History of Transplanter Machine, 2020. URL https://nurseryseedingmachine.com/the-development-history-of-transplanter-machine/ (accessed 7.24.25).

Tangorra, F.M., Buoio, E., Calcante, A., Bassi, A., Costa, A., 2024. Internet of Things (IoT): sensors application in dairy cattle farming. Animals 14 (21), 3071.

Tarrant, A.R., Brainard, D.C., Tiemann, L.K., Hayden, Z.D., 2024. Weed control, soil health, and yield tradeoffs of between-bed management strategies in organic plasticulture vegetable production. Front. Sustain. Food Syst. 8, 1276415.

Thapa, B., Dura, R., 2024. A review on tillage system and no-till agriculture and its impact on soil health. Arch. Agric. Environ. Sci. 9 (3), 612–617.

Torky, M., Hassanein, A.E., 2020. Integrating blockchain and the internet of things in precision agriculture: analysis, opportunities, and challenges. Comput. Electron. Agric. 178, 105476.

UGA Extension, (2025). Mechanical Transplanters for Tree Fruit and Vegetable Seedlings. URL https://extension.uga.edu/publications/detail.html?number=B1576&title=mechanicaltransplanters-for-tree-fruit-and-vegetable-seedlings (accessed 7.24.25).

United Nations, 2017. Project Breakthrough: Digital Agriculture, feeding the future. Disrupive Technology Executive Briefs. United Nations Global Compact. Available at: https://breakthrough.unglobalcompact.org/site/assets/files/1332/hhw-16-0025-d_n_digital_agriculture.pdf.

University of Minnesota Extension, (2022) History of tillage and tillage research. URL https://extension.umn.edu/soil-management-and-health/history-tillage-and-tillage-research (accessed 10.28.25).

University of North Florida, 2025 Office of Faculty Excellence, Artificial Intelligence Definitions. URL https://www.unf.edu/ofe/ai/definitions.html (accessed 8.29.25).

Upadhyay, A., Singh, K.P., Jhala, K.B., Kumar, M., Salem, A., 2024. Non-chemical weed management: harnessing flame weeding for effective weed control. Heliyon 10 (12), e32776.

Vilavan, S., Das, D., Ullah, H., Gade, S.A., Ahmed, S.F., Cha-Um, S., Himanshu, S.K., 2024. Exploring the impacts of climate change and identifying potential adaptation strategies for sustainable rice production in Thailand's Lower Chao Phraya Basin through crop simulation modeling. Environ. Monit. Assess. 196 (12), 1192.

Visser, O., Sippel, S.R., Thiemann, L., 2021. Imprecision farming? Examining the (in) accuracy and risks of digital agriculture. J. Rural Stud. 86, 623–632.

VPRC, (2023). Automation in Agriculture: unleashing the Power of Robotic Harvesting – VPRC. URL https://vprc.sg/agriculture/automation-in-agriculture-unleashing-the-power-of-robotic-harvesting/ (accessed 10.29.25).

Waller, P., Yitayew, M., 2016. Center pivot irrigation systems. In: Irrigation and Drainage Engineering. Springer International Publishing, Cham, pp. 209–228.

Yannopoulos, S.I., Grismer, M.E., Bali, K.M., Angelakis, A.N., 2020. Evolution of the materials and methods used for subsurface drainage of agricultural lands from antiquity to the present. Water 12 (6), 1767.
Zhang, J., Yu, F., Zhang, Q., Wang, M., Yu, J., Tan, Y., 2024. Advancements of UAV and deep learning technologies for weed management in Farmland. Agronomy 14 (3), 494.
Zheng, M., Sun, Y., Mu, W., Bai, Y., Wang, Q., Lu, Z., Zhang, W., 2025. Effects of Subsurface Drip Irrigation Depth on Growth Characteristics and Yield Quality of Apples (Malus pumila Mill.) in Northwest China. Plants 14 (17), 2702.
Zhou, R., Yin, Y., 2023. Digital agriculture: mapping knowledge structure and trends. IEEE Access 11, 103863–103880.

Advancements in agricultural automation: Smart greenhouses and controlled environment agriculture

2

Jipanu Yamjabok[a], *Kanyanat Kasetsuntorn*[a], *Rujira Tisarum*[a], *Piyanan Pipatsitee*[a], *Sushil Kumar Himanshu*[b], *Avishek Datta*[b], *and Suriyan Cha-um*[a]

[a]National Center for Genetic Engineering and Biotechnology, National Science and Technology Development Agency, Pathum Thani, Thailand

[b]Agricultural Systems and Engineering, Faculty of Food, Agriculture and Natural Resources, Asian Institute of Technology, Pathum Thani, Thailand

1 Introduction

Global climate change poses a significant threat, adversely affecting biodiversity, environmental systems, and socio-economic components (Abbass et al., 2022; Ozdemir, 2022; Wahab et al., 2024). Extreme environmental fluctuations, such as severe temperature changes, flooding, drought, soil salinity, mineral toxicity, and ultraviolet irradiation, have been reported as major abiotic stresses derived from global warming disasters, leading to poor agricultural productivity (Zandalinas et al., 2021, 2022; Chaudhry and Sidhu, 2022; Yuan et al., 2023; Seth and Sebastian, 2024). In addition, the increasing global population and its corresponding demand for food production represent a critical challenge (Hossain et al., 2025; Liaqat et al., 2025). Traditional agricultural practices are inadequate for effectively addressing the complex challenges induced by climate change (Durham and Mizik, 2021; Misra and Ghosh, 2024). In recent years, the agricultural sector has undergone a substantial transformation driven by technological advancements and improved scientific knowledge to mitigate these issues (Anders et al., 2021; Mayer et al., 2022). Modern agricultural practices, including precision farming (Achour et al., 2021), automation (Kaya, 2025), and data-driven decision-making (Verma and Gawade, 2021), have been adopted to enhance resource efficiency, increase resilience to environmental stresses, and optimize crop production. Automation in agriculture enhances crop productivity and input use efficiency by reducing reliance on manual labor and minimizing human error. It enables precise management of resources such as water, fertilizers, and pesticides, leading to cost savings and sustainable practices. In addition, automation supports scalability and

Precision Technologies for Digital Agriculture. https://doi.org/10.1016/B978-0-443-40513-6.00021-4

continuous monitoring, allowing for timely interventions that improve crop quality and yield while addressing challenges posed by labor shortages and climate variability. Consequently, farmers can achieve increased crop yields, reduced production expenses, and enhanced sustainability in farming operations. Among various automation technologies in agriculture, smart greenhouses (Farooq et al., 2022a; Khan et al., 2020; Tripathy et al., 2021) are among the most widely adopted and popular solutions among farmers. A traditional greenhouse provides a controlled environment for crops, while a smart greenhouse integrates automation and sensor technologies for precise, real-time environmental management. Other promising modern agricultural practices incorporating innovative technologies include hydroponics (Verma and Gawade, 2021; Yang et al., 2024), vertical farming (Hosseini et al., 2021; Verma et al., 2025), and plant factories (Eaton et al., 2023; Graamans et al., 2018), all aimed at optimizing crop production. These soilless cultivation systems facilitate efficient utilization of space and resources through controlled environment agriculture (CEA) with precise application of nutrients and controlled climatic conditions. Hydroponics increases crop growth cycles and minimizes water consumption, while vertical farming enhances yield density by utilizing vertically stacked growing layers. Plant factories use advanced automation and environmental control systems to ensure consistent, year-round production of high-quality crops. In both smart greenhouses and plant factories, critical crop parameters, including water consumption, nutrient delivery, optimal temperature, light intensity, elevated CO_2 concentration, and relative humidity (RH), are continuously monitored and dynamically optimized throughout the entire growth cycle. These controlled environment systems, operating under closed or semiclosed conditions, aim to maximize crop productivity and enhance the net economic returns of plant production (Graamans et al., 2018; Achour et al., 2020; Song et al., 2024; Kaya, 2025). Nowadays, the concept of plant breeding programs tailored specifically for maximizing productivity under controlled environments is gaining significant attention. High-throughput phenotyping (HTP) platforms play a crucial role in this context by enabling the collection of large-scale, multidimensional data encompassing morphological, physiological, and biochemical traits through digital imaging technologies. These platforms offer nondestructive, accurate, and rapid protocols for screening plants, assessing their responses, and evaluating resilience to both biotic and abiotic stresses, thereby accelerating the development of optimized cultivars for CEA (Amitrano et al., 2022; Langstroff et al., 2022; Theerawitaya et al., 2023; Mansoor et al., 2025). Collectively, these methodologies represent a shift toward sustainable, resource-efficient agriculture that overcomes the inherent limitations of traditional farming practices.

2 Smart greenhouses

A greenhouse is a framed or inflatable structure covered with glass or plastic that provides a partially controlled environment for crop cultivation and is large enough to permit a person to enter and carry out cultural practices. Following the Green Revolution, emphasis shifted toward improving both crop quality and quantity to meet increasing

Fig. 2.1 Smart greenhouse.

food demands. Controlled environmental conditions are essential to achieve these goals, driving the development of protected agriculture. Greenhouses represent the most practical implementation of protected agriculture, utilizing a modified natural environment to optimize plant growth and yield (Thipe et al., 2017; Khan et al., 2021). A conventional greenhouse can be converted into a smart greenhouse (Fig. 2.1) by integrating advanced technologies such as sensors, automated climate control systems, and data analytics platforms (Terence and Purushothaman, 2020; Tripathy et al., 2021; Farooq et al., 2022b; Maraveas et al., 2022; Thilakarathne et al., 2025). This transformation involves installing devices to continuously monitor environmental parameters like temperature, humidity, light intensity, and CO_2 levels, coupled with automated systems for ventilation, irrigation, and nutrient application to the plants. In addition, connectivity through Internet of Things (IoT) enables real-time data collection and remote management, while AI-driven algorithms optimize growing conditions based on plant requirements. Compared to traditional farming practices, smart greenhouses offer significant benefits, including enhanced precision in resource use, higher crop yields, improved product quality, reduced labor dependency, and greater resilience to environmental fluctuations. As a result, they lead to more sustainable and efficient crop production by minimizing water and fertilizer waste and enabling year-round cultivation regardless of external weather conditions.

3 Hydroponic production in greenhouse environments

Hydroponics is a method of growing plants without soil, using liquid media to supply essential micro and macronutrients (Fig. 2.2). This system is broadly categorized into solution culture and medium culture, based on how nutrients are delivered to plant roots. The medium-culture hydroponics is more commonly used and relies on solid substrates to support seed germination and plant growth. These substrates include rockwool,

Fig. 2.2 Hydroponics in a greenhouse.

expanded clay, perlite, coconut chips, glass wool, and phenolic foam. However, each of the substrates has limitations. Rockwool produces fine dust that may pose health risks when inhaled and is nonbiodegradable. Perlite retains water poorly and requires frequent irrigation. Phenolic foam is also nonbiodegradable and may release toxic substances, contributing to plastic pollution. Importantly, these substrates only act as physical support and do not provide controlled nutrient delivery, which can lead to nutrient waste and environmental pollution in hydroponic systems. In solution culture, plant roots are placed directly in a nutrient solution. This can be either circulating or noncirculating. In circulating systems, the nutrient solution flows from a reservoir into the plant tank and back again. Examples include the nutrient film technique, where a shallow stream of nutrients flows over the roots, and the deep flow technique (DFT), where roots are fully submerged in the solution. In noncirculating systems, the solution stays in the tank without recirculation and is replaced when nutrient levels drop or when pH and electrical conductivity become unsuitable. Common noncirculating methods include the root dipping technique, where only the lower part of the roots touches the solution, and the Floating Technique, where roots are fully submerged. The capillary action technique delivers nutrients through capillary movement, either by placing the pot in a shallow container or using wicks to draw the solution from a reservoir. The ebb and flow method, similar to DFT, floods the root zone periodically and then drains the solution back, providing both nutrients and oxygen. These techniques offer flexible and efficient alternatives to traditional soil-based farming.

3.1 Advantages of the hydropic system

Hydroponics is an innovative farming method that offers many benefits over traditional agriculture. It improves resource use, supports year-round crop production, and increases yields. By using hydroponics, farmers can save water and nutrients, reduce the overuse of input resources, and grow crops more sustainably. The controlled environment of hydroponic systems allows precise management of plant growing conditions, leading to faster plant growth and higher productivity. Hydroponics also reduces soil erosion, limits the need for chemical inputs, and supports eco-friendly pest control. It enables urban farming in small spaces. As technology advances, combining hydroponics with smart systems like IoT, sensors, and automation makes the process more efficient. These tools allow real-time monitoring and control of inputs, reduce labor needs, and improve overall system performance. Smart hydroponic farming eliminates soil use, reduces water consumption, and applies nutrients directly to plant roots. It also helps maintain healthy crops in detail by tracking nutrient levels and plant health continuously. Integrated hydroponics with smart farming offers an impactful solution for sustainable, resilient, and localized food production.

3.2 Disadvantages of the hydropic system

The initial cost of a hydroponic setup is very high, making it an expensive option. Skilled and trained personnel are required to manage the system, as proper knowledge of plant growth and nutrition is essential. In closed systems, if soil-borne diseases or nematodes are introduced, they can spread rapidly to all growing beds through the shared nutrient tank. In addition, most plant varieties suitable for controlled environments still need further research and development to optimize their performance. Since plants respond very quickly to both good and poor nutrition, growers must observe and monitor them daily to ensure healthy growth.

4 Plant factories with artificial light

Since the early 2000s, plant factory technology (Fig. 2.3) has been used for the production of vegetables and fruits under controlled environmental conditions (Islam et al., 2021). In recent years, the development and implementation of plant factories have expanded, particularly across various regions in Asia, serving as a novel form of protected horticulture designed for large-scale food production (Kozai et al., 2021). These systems use closed-type growth chambers where environmental parameters are precisely regulated, enabling year-round cultivation of crops irrespective of external climatic conditions. CEA within plant factories offers several key advantages over conventional greenhouse and open-field cultivation. These include uniform and predictable plant growth, reduced production cycles, and increased yield. While greenhouses maintain semicontrolled environments, they remain partially influenced by external climatic parameters such as solar radiation, precipitation, and temperature (Islam et al., 2021; Kozai et al., 2021; Tian et al., 2022; Oh and Lu, 2023). Consequently, environmental variability within greenhouses can result in suboptimal growing conditions. In contrast, plant factories operate in fully enclosed systems that are completely isolated from the external environment, thereby ensuring higher stability and precision in environmental control. To achieve optimal plant growth and development within plant factories, it is essential to maintain several key ambient parameters within specific

Fig. 2.3 Plant factory with an artificial light setup.

thresholds. These include air temperature, RH, light intensity and spectrum, air velocity, and CO_2 concentration. Among these, light is particularly critical, as it serves as the primary energy source for photosynthesis and regulates numerous physiological processes. Air velocity plays a vital role in regulating the microclimate around plant surfaces, facilitating heat dissipation from leaf surfaces, and influencing the diffusion of CO_2 and water vapor through stomatal apertures. Proper airflow management ensures thermal regulation and supports effective gas exchange within the plant canopy. Temperature control is equally crucial, as it directly affects plant metabolism and CO_2 uptake. Elevated CO_2 concentrations, while potentially beneficial for photosynthesis, can also lead to increased transpiration rates and anatomical changes such as leaf thickening. RH significantly influences morphological traits, including leaf area expansion, root and shoot biomass, and bud formation. Moreover, RH indirectly affects stomatal conductance, thereby modulating gas exchange and water use efficiency.

5 Environmental, nutrient, and biological control in controlled environment agriculture

5.1 Plant nutrition supplementation

Unlike traditional farming, where soil plays a major role in nutrient delivery, CEA often relies on soilless growing methods such as hydroponics and plant factories. In these systems, nutrients are delivered directly to the plant roots through nutrient solutions, allowing for exact control over the type and concentration of each essential element (Hosseini et al., 2021; Richa et al., 2021; Yang et al., 2024). These nutrient formulations are carefully tailored to suit the specific needs of each plant species, growth stage, and environmental conditions, including light intensity and temperature. Sometimes, automated dosing systems and sensors are used to monitor and adjust pH, electrical conductivity, and nutrient levels in real time, ensuring that plants receive optimal nutrition without overuse or environmental runoff. This precision not only improves plant health and productivity but also conserves resources, making CEA a sustainable alternative to conventional agriculture (Verma and Gawade, 2021; Rajaseger et al., 2023; Yang et al., 2023). In a smart greenhouse, slow-release fertilizer with nano-coating, mixed together with substrates, is a good choice for some plant species, especially those with a long-life cycle (Chintakovid et al., 2022; Tisarum et al., 2025).

5.2 CO_2 enrichment

Carbohydrate is a primary product derived from photosynthesis. Under greenhouse and plant factory cultivation systems, CO_2 enrichment conditions (750–1500 μmol CO_2 mol^{-1}) are required for photosynthesis regulation (Cha-um et al., 2011; Tisarum et al., 2018), leading to the promotion of the CO_2 assimilation rate and its conversion into biomass as the final product of plant growth (Eaton et al., 2023; Osman et al., 2024; Villagran et al., 2025). Therefore, CO_2 assimilation rate is closely related to the light intensity and ambient RH, which regulate the opening and closing of stomatal pores (Chen et al., 2021, 2023a; Kaya, 2025).

5.3 Light quality, intensity, and duration

Light is a fundamental environmental factor essential for the photosynthesis of plants. Beyond photosynthesis, it also affects various physiological and developmental processes, including vegetative growth, flowering, stomatal conductance, biomass accumulation, and the production of bioactive compounds. When it comes to light, several key parameters should be considered, such as intensity, duration (photoperiod), wavelength (quality), direction, and shading. Each of these plays a specific role in regulating plant morphology and productivity (Table 2.1).

5.4 Temperature

Temperature is one of the most important environmental factors affecting a wide range of plant processes, from biomolecular to physiological levels, including photosynthesis, transpiration, respiration, flowering and fruiting, shoot and root growth, and the accumulation of pigments across plant organs. In terms of photosynthesis, increasing temperature generally promotes the rate of photosynthesis up to an optimal range; however, exceeding the optimal range can lead to a decline in

Table 2.1 Impact of light on the processes related to plant growth and productivity.

Light factor	Effect on plant	Examples
Light intensity	Increases photosynthesis up to the saturation point	• Strawberries require light intensity ≥100 μmol m^{-2} s^{-1} in ventilated indoor conditions for better fruit yield, size, and quantity (Verma et al., 2025) • Cannabis yield increases proportionally with higher indoor light intensity (Llewellyn et al., 2022) • Coriander shows the highest yield at 200 μmol m^{-2} s^{-1}; lower light (133 μmol m^{-2} s^{-1}) inhibits development (Wang et al., 2024)
Light duration (photoperiod)	Influences flowering and vegetative growth	• Strawberry showed improved growth, flowering, and yield under 6–8 h of artificial full-spectrum light (Verma et al., 2025) • Coriander experienced a significant yield decrease under an 8 h photoperiod compared to 16 h (Wang et al., 2024) • Tomato had enhanced growth, photosynthesis, and chlorophyll accumulation under a 20 h photoperiod in a plant factory (Wu et al., 2024)
Light quality (spectrum wavelength)	Different light wavelengths and their combinations affect key physiological processes in plants, including stomatal conductance, growth, photosynthesis, development, and flowering	• Lettuce, tomato, and cucumber show better growth under red+blue light; more blue light reduces plant height, leaf area, and shoot dry weight (Hogewoning et al., 2010; Naznin et al., 2019) • Stomatal conductance is lower under red light alone than under red+blue light, increasing transpiration and lowering leaf temperature (Matthews et al., 2020) • In lettuce, a red-to-far-red (R:FR) ratio of 1.6 significantly boosts fresh mass and growth (Yuanyuan et al., 2024) • In a plant factory, white+blue light (WB) improves leaf color and quality in red oakleaf and red butterhead lettuce, while white+far-red (WFR) enhances yield (Li et al., 2021a)
Shading	Shading delays flowering, reduces plant growth and yield, and decreases the accumulation of bioactive compounds in the harvested parts	• Tomato under shading showed delayed flowering and a 48% reduction in marketable yield. Greenhouse shading reduced total phenolics by 29% and flavonoids by 16% (Angmo et al., 2021) • Bell pepper developed taller stems and larger, thinner leaves under shade to improve light capture. Moderate shading (30%–47%) was optimal for growth, while high shade reduced photosynthesis despite lowering leaf temperature and transpiration (Díaz-Pérez, 2013)

photosynthetic activity due to thermal stress. Rising heat can cause partial denaturation or conformational changes in key photosynthetic enzymes such as RuBisCO (ribulose-1,5-bisphosphate carboxylase/oxygenase) or PEPC (phosphoenolpyruvate carboxylase or PEP carboxylase), leading to impaired photosynthetic efficiency and subsequent reductions in crop yield. In addition, under low-temperature conditions, photosynthetic activity slows down due to decreased enzyme kinetic rates, limiting carbon assimilation below the optimal thermal range. The optimum temperature range for photosynthesis depends on the photosynthetic pathway utilized by a given plant species. Generally, C4 plants exhibit higher photosynthetic rates at elevated temperatures (30°C–45°C) compared to C3 and Crassulacean Acid Metabolism (CAM) plants, which show potential optimal photosynthetic activity at 5°C–35°C and 10°C–25°C, respectively (Yamori et al., 2014). This difference is largely due to the temperature sensitivity of RuBisCO, the primary enzyme involved in carbon fixation. In C3 plants, RuBisCO becomes less efficient at higher temperatures due to increased photorespiration, which reduces its affinity for CO_2 assimilation. C4 plants overcome this limitation through both anatomical and biochemical adaptations that enhance photosynthetic efficiency under high-temperature conditions. Anatomically, C4 plants possess a specialized bundle sheath cell layer that functions as an additional site of photosynthesis (Sage and Sage, 2013). This enables a two-step carbon fixation process, with the first step carried out by PEPC in mesophyll cells, followed by carbon fixation by RuBisCO in the bundle sheath cells (Ludwig, 2016). This adaptation concentrates CO_2 around RuBisCO, significantly reducing photorespiration and enhancing photosynthetic efficiency, particularly under high light and temperature conditions (Table 2.2).

5.5 Relative humidity

RH is one of the key factors that directly influence plant transpiration and photosynthesis. Generally, the optimal RH for most plants lies between 50%–70%, although this range may vary depending on the climatic origin and species-specific adaptations. Under low levels of humidity or dry conditions, which are typically below 40%–50% RH, the atmospheric water potential becomes significantly lower than inside the leaf. This gradient drives the diffusion of water vapor out of the leaf through stomata. While transpiration helps generate a pull for water uptake from the roots to the leaves, excessive transpiration can cause rapid water loss. This reduces cell turgor pressure, leading to dehydration and wilting. To minimize water loss, plants respond by closing their stomata; however, this also limits CO_2 intake, thereby reducing photosynthetic efficiency. The combined effect of decreased transpiration, water uptake, and photosynthesis leads to a slower growth rate and a decline in crop yield (Taiz et al., 2015). Conversely, at high humidity (typically >70%–80% RH), the stomatal pores are closed, leading to a low transpiration rate and a low water potential gradient between the inside and outside of the leaf based on the transpiration stream. Consequently, the transpiration rate is

Table 2.2 Impact of temperature on the processes related to plant growth and productivity.

Impacts on the plant	Details	Examples
Photosynthesis	Increasing temperature generally enhances the rate of photosynthesis, but exceeding the optimal range causes a decline in photosynthetic activity (Yamori et al., 2014)	• C_4 plants: • Amaranth (*Amaranthus tricolor*) shows maximum capacity for photosynthesis at 35°C (Lin and Ehleringer, 1983) • Maize (*Zea mays*) grows best at daytime temperatures between 25°C and 33°C and nighttime temperatures between 17°C and 23°C, with the highest grain yield typically achieved around 25°C (Li and Zhang, 2022) • C_3 Plants: • Soybean (*Glycine max*) performs best at temperatures of 15°C–22°C during emergence, 20°C–25°C during flowering, and returns to an optimal range of 15°C–22°C at maturity (Liu et al., 2008; Zhang et al., 2016) • CAM plants: • Pineapple (*Ananas comosus*) thrives in optimal temperatures ranging from 20°C to 30°C, provided there is ample sunlight, high precipitation, and good air circulation (Mohd Ali et al., 2020)
Transpiration	Transpiration rates increase with temperature, potentially causing water stress if uptake is insufficient. At critical points, transpiration slows due to stomatal closure, which varies by plant species	• Drought-tolerant maize (*Zea mays* L.): Transpiration rate rises with increasing vapor pressure deficit (VPD) but increases only slightly beyond the breaking point due to stomatal closure. This breaking point shifts to higher VPD and transpiration levels when temperature increases (25°C–30°C) (Yang et al., 2012)
Respiration	Respiration rates increase with temperature up to a species-specific threshold, beyond which tissue damage and inactivation of respiratory enzymes occur, causing a decline or cessation of respiration (Crous et al., 2022; Scafaro et al., 2021)	• Warm conditions increase respiration in the leaves and the fruit of strawberry (*Fragaria* × *ananassa* Duch.) (Menzel, 2023) • Under a high temperature of 41°C, the leaf transpiration rate of rice increases significantly, but there is less damage to photosynthesis (Li et al., 2021b)

Flowering and fruiting	Temperature influences the timing of flowering and fruit development, with some plants requiring specific temperature conditions to initiate flowering	• In strawberry (*Fragaria* × *ananassa* Duch.), warm weather leads to the production of smaller fruits, which is linked to smaller flowers with fewer stamens, germinated pollen grains, carpels, and achenes per flower or fruit (Menzel, 2023) • In blackberry, a plant native to cold and temperate regions, winter chilling can promote flower bud initiation and enhance yield potential in partially induced plants (Sønsteby and Heide, 2023)
Shoot and root growth	Higher temperatures generally promote shoot and root growth, but growth is suppressed when temperatures exceed the optimal range. Low temperatures can inhibit or slow down plant growth	• In maize, the optimum temperature for shoot and root growth is around 30°C; growth declines gradually below this point and drops rapidly above it (Walne and Reddy, 2022)
Accumulation of pigments across plant organs	Temperature changes strongly influence pigment accumulation in plants, such as chlorophylls, carotenoids, anthocyanins, and betalains, with effects varying based on pigment type and the temperature sensitivity of related biosynthetic enzymes	• Winter wheat shows a decline in chlorophyll *a*, chlorophyll *b*, and carotenoids below 0°C, leading to leaf senescence (Zhang et al., 2023) • Apples accumulate more anthocyanins at cooler temperatures (12°C–17°C) compared to warmer ones (24°C–27°C) (Xie et al., 2012) • In red amaranth (Amaranthus tricolor L.), betacyanin levels peak during warmer months (28°C–29°C) and decline significantly in cooler months (18°C–19°C) (Khandaker et al., 2009)

Table 2.3 Relative humidity impacts on the processes related to plant growth and productivity.

Humidity level	Effect	Example/details
High relative humidity	• Stomata remain excessively open (some species)	Roses (*Rosa × hybrida*) grown at relative humidity levels above 85% exhibit increased water loss during leaf dehydration and impaired stomatal closure in the dark, leading to a reduced postharvest lifespan (Terfa et al., 2020)
	• Delay flowering and ripening (some species)	Tomato (*Solanum lycopersicum*), under increasing humidity, high temperatures (32°C–35°C), and dry climates, could effectively increase the rate of flowering and fruit setting (Wang et al., 2018b)
	• Increased risk of fungal or bacterial diseases	• Powdery mildew (*Podosphaera* spp.) in cucumber (Milod et al., 2021) • Downy mildew (*Peronospora* spp.) in sweet basil (*Ocimum basilicum*) (Cohen et al., 2017)

reduced, and plants can conserve water and maintain cell turgor, which enhances leaf expansion and biomass accumulation. However, the stomata remain excessively open in some plants, which leads to poor water use efficiency (WUE). For example, roses (*Rosa × hybrida*) grown at RH levels around 85% exhibit increased water loss during leaf dehydration and impaired stomatal closure in the dark, leading to a reduced postharvest lifespan (Terfa et al., 2020). Furthermore, the negative effects of high ambient humidity on flowering and ripening are evidently observed. For instance, in tomato (*Solanum lycopersicum*), increasing humidity under conditions of high temperature (32°C–35°C) and a dry climate could effectively increase the rate of flowering and fruit setting (Wang et al., 2018a). In addition, the high moisture in the air can promote the growth of plant pathogens, causing fungal or bacterial diseases in plants. For example, high RH (>80% RH) promotes the development of powdery mildew (*Podosphaera* spp.) in cucumber (Milod et al., 2021) and downy mildew (*Peronospora* spp.) in sweet basil (*Ocimum basilicum*) (Cohen et al., 2017). These infections further reduce plant health and productivity (Table 2.3).

5.6 Pollinator

In the greenhouse, the bumblebee is a good candidate pollinator, in addition to the wild type of bee fly species, which is also utilized (Zameer et al., 2022; Sheikh et al., 2024; Zhang et al., 2025). In recent years, the stem shaker, synthetic wind flow, and small drone applications have been reported as effective methods to facilitate pollination and increase fruit set, especially in tomatoes (Shukla et al., 2022; Hiraguri et al.,

2023). There is a limitation in the plant factory for applying pollination, even though parthenocarpic fruit induction using plant growth regulators is a common practice (Corti et al., 2024).

6 Breeding programs for plant production under simulated environmental conditions

To gain a better understanding of the genetic architecture underlying these traits, population screening and genetic studies (e.g., quantitative trait loci analyses, genome-wide association analyses) have been conducted in indoor-cultivated crops, resulting in the identification of extreme genotypes and candidate genes (Table 2.4). The extreme genotypes/varieties can be utilized as parental lines in a breeding program in order to develop new varieties with enhanced traits for an indoor cultivation system. However, this method is time-consuming, as it requires multiple crop generations (Van et al., 2022; Bhattarai et al., 2025). With the advancement of biotechnology and genomics technology, traits of interest can be improved by precisely modifying the reported candidate genes through precision breeding or gene editing, and new varieties can be generated within a few crop generations (Doudna and Charpentier, 2014; Hsu et al., 2014; Benke and Tomkins, 2017). To date, gene editing has been conducted in different crops to further increase crop yield and quality under indoor cultivation (Table 2.4).

7 High-throughput phenotyping platforms

Recently, crop requirements in the closed environmental system of plant production have been well established using HTP platforms (Fig. 2.4). It is a basic facility to collect big data for precision agriculture based on plant morphometric, physiological, and secondary metabolites throughout the whole life cycle of each crop species (Theerawitaya et al., 2023; Dhakal et al., 2024; Netrphan et al., 2024; Siangliw et al., 2024). Accurate environmental data in the precision agriculture system are also recorded in real-time and controlled environments (Bhujel et al., 2020; Hassan et al., 2021; Rosero-Montalvo et al., 2023). Subsequently, data related to water use efficiency, nutrient use efficiency, light and CO_2 response, optimal temperature, growth parameters, biomass accumulation, and final yield attributes are subjected to machine learning (ML) algorithms to develop predictive models for each crop (Arend et al., 2022; Jamil et al., 2022; Pipatsitee et al., 2023; Reza et al., 2023). Cloud computing linked with IoT is one of the most promising technologies to serve as cutting-edge innovation for precision agriculture (Zamora-Izquierdo et al., 2019; Kumar et al., 2024). Moreover, the warning system for bacterial and fungal disease infection and management in greenhouse cultivation, using big data from thermal infrared and hyperspectral imageries, is very useful for farmers in terms of

Table 2.4 Fundamental studies and gene-edited crop development for controlled-environment agriculture.

Category	Crop	Phenotype	Method	Findings or target gene	References
Population screening and genetic studies					
Yield and product quality	Tomato	Yield	Hybrid variety development	Increased yield up to 60%	Krieger et al. (2010)
	Spinach	Leaf traits	QTL analysis	3 candidate genes (Spo10792, Spo21018, and Spo21019)	Liu et al. (2021)
	Spinach	Shoot Biomass	GWAS study	5 SNP markers (chr4_28292655, chr6_1531056, chr6_37966006, chr2_18480277, chr4_47598760 on chromosomes 2 and 4)	Joshi et al. (2022)
	Spinach	Nutritional quality	Screening of 30 spinach varieties	Nutrient-rich accession 163 and 310	Rashid et al. (2022)
	Pepper	Fasciculate inflorescence	BSA and QTL analysis	4 candidate genes (Capana00g000667, Capana06g000476, Capana11g000671, Capana04g000175)	Lv et al. (2019)
	Pepper	Yield component traits	QTL analysis	24 QTLs	Alimi et al. (2013)
	Pepper	Multiple-flower-per-node	QTL and GWAS analysis	4 QTLs on chromosomes 1,2,7 and 11	Kim et al. (2022)
	Pepper	First flower node trait	BSA analysis	10 QTLs on chromosome 12	Zhang et al. (2018)
	Strawberry	Fruit uniformity	Multi-QTL analysis parental	1 QTL on chromosome 2B	Li et al. (2020)
	Pepper	Flowering time	QTL analysis	1 candidate gene Capana02g000700	Zhu et al. (2019)
Plant architecture	Pepper	Dwarfism	BSA analysis	1 single-base mutation in CaBRI1	Yang et al. (2020)
	Cucumber	Compact plant stature	QTL analysis	1 candidate gene CsERECTA	Chen et al. (2023b)
	Cucumber	Dwarfism	QTL analysis	1 candidate gene compact	Li et al. (2011)
	Strawberry	Runnerless trait	BSA analysis	candidate gene FveGA20ox4	Tenreira et al. (2017)

Nutrient/water utilization efficiency	Lettuce	WUE	Screening of 200 lettuce varieties	4 high WUE lettuce varieties (Slobolt, Grand Rapids, Western Green, and Australian)	Eriksen et al. (2016)
	Lettuce	NUE	QTL analysis of recombinant inbred lines	A major QTL on chromosome 7	Macias-González et al. (2021)
	Spinach	NUE	QTL analysis	39 QTLs across six linkage groups	Chan-Navarrete et al. (2016)
	Bell pepper	WUE	QTL analysis	2 QTLs on chromosome 10	Popovsky-Sarid et al. (2017)
Genetically modified crops displaying enhanced traits for controlled-environment agriculture					
Plant Architecture	Tomato	Smaller plant size than wild type	CRISPR	PROCERA	Tomlinson et al. (2019)
	Tomato	Decreased flowering time (leaves to first inflorescence)	CRISPR	SP5g	Soyk et al. (2017)
	Tomato	reduced shoot length	CRISPR	SP	Kwon et al. (2020)
	Tomato	jointless and elogated pedicel	CRISPR	Solyc12g038510	Roldan et al. (2017)
	Cabbage	Reduced plant height and dwarf stature	CRISPR	BolC.GA4.a	Lawrenson et al. (2015)
	Cabbage	Earlier flowering than wild type	TALEN	FRI	Sun et al. (2013)

Continued

Table 2.4 Fundamental studies and gene-edited crop development for controlled-environment agriculture—cont'd

Category	Crop	Phenotype	Method	Findings or target gene	References
Yield and product quality	Tomato	Enhanced CO_2 assimilation rate	Genetic transformation	SBPase	Ding et al. (2016)
	Tomato	Fruit softening	CRISPR	PL, PG2a, TBG4	Wang et al. (2019a)
	Tomato	Orange-ripened pericarp and altered ripening duration	CRISPR	AP2a, NOR, FUL1/TDR4, FUL2/MBP7	Wang et al. (2019b)
	Tomato	Significantly extended shelf life	CRISPR	ALC	Yu et al. (2017)
	Tomato	Pink tomato fruit	CRISPR	SlMYB12	Deng et al. (2018)
	Tomato	Increased lycopene content by 5.1-fold	CRISPR	SGR1, Blc, LCY-B1, LCY-B2, LCY-E	Li et al. (2018)
	Tomato	Increased ascorbate content	CRISPR	SlAPX4	Do et al. (2022)
	Tomato	Increased GABA content by 7–15-fold	CRISPR	SlGAD2, SlGAD3	Nonaka et al. (2017)
	Strawberry	Increased fruit size	CRISPR	FveRGA1	Zhou et al. (2021)
	Strawberry	Partially delayed anthocyanin accumulation	CRISPR	SlMYB10, SlCHS	Xing et al. (2018)
	Melon	Decreased ethylene evolution rate and increased shelf life	CRISPR	CmACO1	Nonaka et al. (2023)

QTL, quantitative trait loci; *GWAS*, genome-wide association study; *NUE*, nitrogen use efficiency; *WUE*, water use efficiency; *CRISPR*, clustered regularly interspaced short palindromic repeats; *TALEN*, transcription activator-like effector nucleases; *BSA*, bulk segregant analysis.

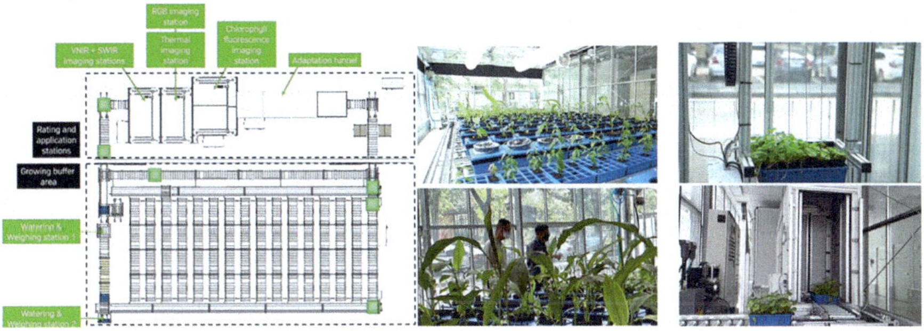

Fig. 2.4 Schematic layout of high-throughput phenotyping platforms.

prevention and elimination (Khan et al., 2020; Ouyang et al., 2022; Kounalakis et al., 2023; Mac et al., 2024).

8 Challenges and opportunities of smart greenhouses and CEA

Smart greenhouses, which utilize modern technologies such as IoT, automation, and artificial intelligence, present both significant opportunities and challenges. The smart greenhouse enables the precise application of irrigation and nutrients, controls micro-climatic conditions, leading to optimized resource use, higher crop yields, reduced labor needs, and helps farmers with year-round production. These systems also support sustainable practices by minimizing water and fertilizer overuse and by integrating with renewable energy sources, further reducing environmental impact. However, the adoption of smart greenhouses faces several challenges, including high initial investment costs, the need for technical expertise, dependence on stable internet and power infrastructure, and concerns about data security. In addition, maintaining sensor accuracy and ensuring interoperability between different devices can be a complex task. Other advanced agricultural practices, i.e., hydroponic, vertical farming, and plant factory systems with artificial light, utilize a continuous flow of water-soluble nutrient application arrangements, stacked layers, and LED lighting to grow crops without soil. Similar to the smart greenhouse, these practices also enable year-round crop production, higher crop yields, and reduced water overuse. They also avoid pests and diseases common in outdoor farming. However, initial setup costs are high due to sensors, actuators, system components, lighting, and automation systems. Energy consumption is also high, especially for artificial lighting and climate control. Technical skills are needed to manage nutrient levels, lighting schedules, and system maintenance. Limited crops can grow well in these systems. Despite these challenges, all these modern technologies offer strong potential for food production under indoor farming, harsh climates, and areas with limited land or water conditions. With better energy efficiency, automation, and support, these systems can help meet future food demands.

9 Conclusion

CEA, including smart greenhouses, hydroponic systems, and plant factories with artificial light, represents a transformative shift in modern crop production. These systems use advanced automation to manage agricultural practices more efficiently. By enabling precise control over environmental factors such as temperature, humidity, light, CO_2, and nutrient delivery, these systems offer significant advantages in resource use efficiency, crop quality, and year-round productivity. The integration of HTP platforms and advanced monitoring technologies further enhances the ability to assess plant performance in real time, facilitating data-driven decisions and accelerating breeding and cultivation strategies. Biological and environmental control methods help to maintain optimum plant growth conditions, reduce chemical inputs, and improve sustainability. The continued development and scaling of smart greenhouse technologies and plant factories hold significant potential for addressing food security, particularly in regions with limited arable land or unfavorable climates.

References

Abbass, K., Qasim, M.Z., Song, H., Murshed, M., Mahmood, H., Younis, I., 2022. A review of the global climate change impacts, adaptation, and sustainable mitigation measures. Environ. Sci. Pollut. Res. 29 (28), 4253942559.

Achour, Y., Ouammi, A., Zejli, D., Sayadi, S., 2020. Supervisory model predictive control for optimal operation of a greenhouse indoor environment coping with food-energy-water nexus. IEEE Access 8, 211562211575.

Achour, Y., Ouammi, A., Zejli, D., 2021. Technological progresses in modern sustainable greenhouses cultivation as the path towards precision agriculture. Renew. Sustain. Energy Rev. 147, 111251.

Alimi, N.A., Bink, M.C.A.M., Dieleman, J.A., Nicolaï, M., Wubs, M., Heuvelink, E., Magan, J., Voorrips, R.E., Jansen, J., Rodrigues, P.C., van der Heijden, G.W.A.M., Vercauteren, A., Vuylsteke, M., Song, Y., Glasbey, C., Barocsi, A., Lefebvre, V., Palloix, A., van Eeuwijk, F.A., 2013. Genetic and QTL analyses of yield and a set of physiological traits in pepper. Euphytica 190, 181–201.

Amitrano, C., Junker, A., D'Agostino, N., De Pascale, S., De Micco, V., 2022. Integration of high-throughput phenotyping with anatomical traits of leaves to help understanding lettuce acclimation to a changing environment. Planta 256 (4), 68.

Anders, S., Cowling, W., Pareek, A., Gupta, K.J., Singla-Pareek, S.L., Foyer, C.H., 2021. Gaining acceptance of novel plant breeding technologies. Trends Plant Sci. 26 (6), 575587.

Angmo, P., Phuntsog, N., Namgail, D., Chaurasia, O.P., Stobdan, T., 2021. Effect of shading and high temperature amplitude in greenhouse on growth, photosynthesis, yield and phenolic contents of tomato (*Lycopersicum esculentum* Mill.). Physiol. Mol. Biol. Plants 27, 1539–1546.

Arend, D., Psaroudakis, D., Memon, J.A., Rey-Mazón, E., Schüler, D., Szymanski, J.J., Scholz, U., Junker, A., Lange, M., 2022. From data to knowledge–big data needs stewardship, a plant phenomics perspective. Plant J. 111 (2), 335347.

Benke, K., Tomkins, B., 2017. Future food-production systems: vertical farming and controlled-environment agriculture. Sustain. Sci. Practice Policy 13 (1), 1326.

Bhattarai, K., Ogden, A.B., Pandey, S., Sandoya, G.V., Shi, A., Nankar, A.N., Jayakodi, M., Huo, H., Jiang, T., Tripodi, P., Dardick, C., 2025. Improvement of crop production in controlled environment agriculture through breeding. Front. Plant Sci. 15, 1524601.

Bhujel, A., Basak, J.K., Khan, F., Arulmozhi, E., Jaihuni, M., Sihalath, T., Lee, D., Park, J., Kim, H.T., 2020. Sensor systems for greenhouse microclimate monitoring and control: a review. J. Biosyst. Eng. 45 (4), 341361.

Chan-Navarrete, R., Dolstra, O., van Kaauwen, M., Lammerts van Bueren, E.T., van der Linden, C.G., 2016. Genetic map construction and QTL analysis of nitrogen use efficiency in spinach (*Spinacia oleracea* L.). Euphytica 208, 621–636.

Chaudhry, S., Sidhu, G.P.S., 2022. Climate change regulated abiotic stress mechanisms in plants: a comprehensive review. Plant Cell Rep. 41 (1), 131.

Cha-um, S., Chanseetis, C., Chintakovid, W., Pichakum, A., Supaibulwatana, K., 2011. Promoting root induction and growth of in vitro macadamia (*Macadamia tetraphylla* L. 'Keaau') plantlets using CO_2-enriched photoautotrophic conditions. Plant Cell Tissue Organ Culture 106 (3), 435444.

Chen, D., Mei, Y., Liu, Q., Wu, Y., Yang, Z., 2021. Carbon dioxide enrichment promoted the growth, yield, and light-use efficiency of lettuce in a plant factory with artificial lighting. Agron. J. 113 (6), 51965206.

Chen, D., Zhang, J., Sun, Z., Zhang, Z., Hu, J., 2023a. Multi-objective optimal regulation model and system based on whole plant photosynthesis and light use efficiency of lettuce. Comput. Electron. Agric. 206, 107617.

Chen, F., Yong, J., Zhang, G., Liu, M., Wang, Q., Zhong, H., Pan, Y., Chen, P., Weng, Y., Li, Y., 2023b. An LTR retrotransposon insertion inside CsERECTA for an LRR receptor-like serine/threonine-protein kinase results in compact (cp) plant architecture in cucumber. Theor. Appl. Genet. 136 (3), 31.

Chintakovid, N., Tisarum, R., Samphumphuang, T., Sotesaritkul, T., Cha-um, S., 2022. Evaluation of curcuminoids, physiological adaptation, and growth of *Curcuma longa* under water deficit and controlled temperature. Protoplasma 259 (2), 301315.

Cohen, Y., Ben Naim, Y., Falach, L., Rubin, A.E., 2017. Epidemiology of basil downy mildew. Phytopathology 107, 1149–1160.

Corti, C., Arcerito, F.R.M., de Landa, G.F., Mazzei, M.P., de Landa, M.F., Maggi, M., Galetto, L., 2024. Tomato production under greenhouse conditions: bumblebees or hormones? Sci. Hortic. 326, 112747.

Crous, K.Y., Uddling, J., De Kauwe, M.G., 2022. Temperature responses of photosynthesis and respiration in evergreen trees from boreal to tropical latitudes. New Phytol. 234, 353–374.

Deng, L., Wang, H., Sun, C., Li, Q., Jiang, H., Du, M., Li, C.B., Li, C., 2018. Efficient generation of pink-fruited tomatoes using CRISPR/Cas9 system. J. Genet. Genomics 45, 51–54.

Dhakal, A., Poland, J., Adhikari, L., Faryna, E., Fiedler, J., Rutkoski, J.E., Arbelaez, J.D., 2024. Implementing multi-trait genomic selection to improve grain milling quality in oats (*Avena sativa* L.). Plant Genome 17 (2), e20457.

Díaz-Pérez, J.C., 2013. Bell pepper (*Capsicum annum* L.) crop as affected by shade level: microenvironment, plant growth, leaf gas exchange, and leaf mineral nutrient concentration. HortScience 48 (2), 175182.

Ding, F., Wang, M., Zhang, S., Ai, X., 2016. Changes in SBPase activity influence photosynthetic capacity, growth, and tolerance to chilling stress in transgenic tomato plants. Sci. Rep. 6, 32741.

Do, J.H., Park, S.Y., Park, S.H., Kim, H.M., Ma, S.H., Mai, T.D., Shim, J.S.J., H, Y., 2022. Development of a genome-edited tomato with high ascorbate content during later stage of fruit ripening through mutation of *SlAPX4*. Front. Plant Sci. 13, 836916.

Doudna, J.A., Charpentier, E., 2014. The new frontier of genome engineering with CRISPR-Cas9. Science 346 (6213), 1258096.

Durham, T.C., Mizik, T., 2021. Comparative economics of conventional, organic, and alternative agricultural production systems. Economies 9 (2), 64.

Eaton, M., Shelford, T., Cole, M., Mattson, N., 2023. Modeling resource consumption and carbon emissions associated with lettuce production in plant factories. J. Clean. Prod. 384, 135569.

Eriksen, R.L., Knepper, C., Cahn, M.D., Mou, B., 2016. Screening of lettuce germplasm for agronomic traits under low water conditions. HortScience 51 (6), 669679.

Farooq, M.S., Javid, R., Riaz, S., Atal, Z., 2022a. IoT based smart greenhouse framework and control strategies for sustainable agriculture. IEEE Access 10, 9939499420.

Farooq, M.S., Riaz, S., Helou, M.A., Khan, F.S., Abid, A., Alvi, A., 2022b. Internet of things in greenhouse agriculture: a survey on enabling technologies, applications, and protocols. IEEE Access 10, 5337453397.

Graamans, L., Baeza, E., Van Den Dobbelsteen, A., Tsafaras, I., Stanghellini, C., 2018. Plant factories versus greenhouses: comparison of resource use efficiency. Agric. Syst. 160, 3143.

Hassan, S.I., Alam, M.M., Illahi, U., Al Ghamdi, M.A., Almotiri, S.H., Su'ud, M.M., 2021. A systematic review on monitoring and advanced control strategies in smart agriculture. IEEE Access 9, 3251732548.

Hiraguri, T., Shimizu, H., Kimura, T., Matsuda, T., Maruta, K., Takemura, Y., Takanashi, T., 2023. Autonomous drone-based pollination system using AI classifier to replace bees for greenhouse tomato cultivation. IEEE Access 11, 9935299364.

Hogewoning, S.W., Trouwborst, G., Maljaars, H., Poorter, H., van Ieperen, W., Harbinson, J., 2010. Blue light dose–responses of leaf photosynthesis, morphology, and chemical composition of *Cucumis sativus* grown under different combinations of red and blue light. J. Exp. Bot. 61, 3107–3117.

Hossain, A., Farhad, M., Aonti, A.J., Kabir, M.P., Hossain, M.M., Ahmed, B., Haq, M.I., Azim, J., 2025. Cereals production under changing climate. In: Challenges and Solutions of Climate Impact on Agriculture. Academic Press. pp. 6383.

Hosseini, H., Mozafari, V., Roosta, H.R., Shirani, H., van de Vlasakker, P.C., Farhangi, M., 2021. Nutrient use in vertical farming: optimal electrical conductivity of nutrient solution for growth of lettuce and basil in hydroponic cultivation. Horticulturae 7 (9), 283.

Hsu, P.D., Lander, E.S., Zhang, F., 2014. Development and applications of CRISPR-Cas9 for genome engineering. Cell 157 (6), 12621278.

Islam, S., Reza, M.N., Chowdhury, M., Chung, S.O., Choi, I.S., 2021. A review on effect of ambient environment factors and monitoring technology for plant factory. Precis. Agric. 3, 84.

Jamil, F., Ibrahim, M., Ullah, I., Kim, S., Kahng, H.K., Kim, D.H., 2022. Optimal smart contract for autonomous greenhouse environment based on IoT blockchain network in agriculture. Comput. Electron. Agric. 192, 106573.

Joshi, V., Shi, A., Mishra, A.K., Gill, H., DiPiazza, J., 2022. Genetic dissection of nitrogen induced changes in the shoot and root biomass of spinach. Sci. Rep. 12, 1–9.

Kaya, C., 2025. Intelligent environmental control in plant factories: integrating sensors, automation, and AI for optimal crop production. Food Energy Secur. 14 (1), e70026.

Khan, F.A., Ibrahim, A.A., Zeki, A.M., 2020. Environmental monitoring and disease detection of plants in smart greenhouse using internet of things. J. Phys. Commun. 4 (5), 055008.

Khan, N., Ray, R.L., Sargani, G.R., Ihtisham, M., Khayyam, M., Ismail, S., 2021. Current progress and future prospects of agriculture technology: gateway to sustainable agriculture. Sustainability 13 (9), 4883.

Khandaker, L., Akond, A.S.M.G.M., Oba, S., 2009. Air temperature and sunlight intensity of different growing period affects the biomass, leaf color and betacyanin pigment accumulations in red amaranth (*Amaranthus tricolor* L.). J. Cent. Eur. Agric. 10, 439–447.

Kim, Y., Kim, G.W., Han, K., Lee, H.Y., Jo, J., Kwon, J.K., Lemmon, Z., Lippman, Z., Kang, B.C., 2022. Identification of genetic factors controlling the formation of multiple flowers per node in pepper (*Capsicum* spp.). Front. Plant Sci. 13, 884338.

Kounalakis, N., Kalykakis, E., Kosmopoulos, D., Fasoulas, J., Sfakiotakis, M., 2023. A framework leveraging robotics and machine learning technologies for early disease and pest detection in greenhouse tomato crops. In: 2023 International Conference on Control, Artificial Intelligence, Robotics & Optimization (ICCAIRO). IEEE. pp. 74–81.

Kozai, T., Niu, G., Masabni, J.G. (Eds.), 2021. Plant Factory Basics, Applications and Advances. Academic Press, Elsevier, UK.

Krieger, U., Lippman, Z.B., Zamir, D., 2010. The flowering gene SINGLE FLOWER TRUSS drives heterosis for yield in tomato. Nat. Genet. 42 (5), 459463.

Kumar, V., Sharma, K.V., Kedam, N., Patel, A., Kate, T.R., Rathnayake, U., 2024. A comprehensive review on smart and sustainable agriculture using IoT technologies. Smart Agric. Technol. 8, 100487.

Kwon, C.T., Heo, J., Lemmon, Z.H., Capua, Y., Hutton, S.F., Van Eck, J., Park, S.J.L., B, Z., 2020. Rapid customization of Solanaceae fruit crops for urban agriculture. Nat. Biotechnol. 38 (2), 182188.

Langstroff, A., Heuermann, M.C., Stahl, A., Junker, A., 2022. Opportunities and limits of controlled-environment plant phenotyping for climate response traits. Theor. Appl. Genet. 135 (1), 116.

Lawrenson, T., Shorinola, O., Stacey, N., Li, C., Østergaard, L., Patron, N., Uauy, C., Harwood, W., 2015. Induction of targeted, heritable mutations in barley and *Brassica oleracea* using RNA-guided Cas9 nuclease. Genome Biol. 16, 1–13.

Li, B., Cockerton, H.M., Johnson, A.W., Karlström, A., Stavridou, E., Deakin, G., Harrison, R.J., 2020. Defining strawberry shape uniformity using 3D imaging and genetic mapping. Hortic. Res. 7, 1–13.

Li, X., Wang, Y., Chen, S., Tian, H., Fu, D., Zhu, B., Luo, Y., Zhu, H., 2018. Lycopene is enriched in tomato fruit by CRISPR/Cas9-mediated multiplex genome editing. Front. Plant Sci. 9, 559.

Li, Y., Yang, L., Pathak, M., Li, D., He, X., Weng, Y., 2011. Fine genetic mapping of cp: a recessive gene for compact (dwarf) plant architecture in cucumber, *Cucumis sativus* L. Theor. Appl. Genet. 123, 973–983.

Li, Y., Wu, L., Jiang, H., He, R., Song, S., Su, W., Liu, H., 2021a. Supplementary far-red and blue lights influence the biomass and phytochemical profiles of two lettuce cultivars in plant factory. Molecules 26, 7405.

Li, G., Chen, T., Feng, B., Peng, S., Tao, L., Fu, G., 2021b. Respiration, rather than photosynthesis, determines rice yield loss under moderate high-temperature conditions. Front. Plant Sci. 12, 678653.

Li, Z., Zhang, J., 2022. Effects of raised ambient temperature on the local and systemic adaptions of maize. Plants 11, 755.

Liaqat, W., Altaf, M.T., Jan, M.F., Ahmad, H., Khan, E.H., Barutçular, C., Cömertpay, G., Baloch, F.S., Mohamed, H.I., 2025. From global challenges to specific solutions: climate change impacts on agriculture with a focus on sorghum through bibliometric analysis. Cereal Res. Commun. 53 (4), 1917–1948.

Lin, Z.F., Ehleringer, J., 1983. Photosynthetic characteristics of *Amaranthus tricolor*, a C4 tropical leafy vegetable. Photosynth. Res. 4, 171–178.

Liu, X., Jin, J., Wang, G., Herbert, S.J., 2008. Soybean yield physiology and development of high-yielding practices in Northeast China. Field Crops Res. 105, 157–171.

Liu, Z., She, H., Xu, Z., Zhang, H., Li, G., Zhang, S., Qian, W., 2021. Quantitative trait loci (QTL) analysis of leaf related traits in spinach (*Spinacia oleracea* L.). BMC Plant Biol. 21 (1), 290.

Llewellyn, D., Golem, S., Foley, E., Dinka, S., Jones, A.M.P., Zheng, Y., 2022. Indoor grown cannabis yield increased proportionally with light intensity, but ultraviolet radiation did not affect yield or cannabinoid content. Front. Plant Sci. 13, 974018.

Ludwig, M., 2016. The roles of organic acids in C4 photosynthesis. Front. Plant Sci. 7, 647.

Lv, J., Liu, Y., Liu, Z., Wang, J., Ma, Y., Zhang, Z., Yang, S., Ou, L., Chen, W., Zou, X., 2019. Mapping and identifying candidate genes involved in the novel fasciculate inflorescence in pepper (*Capsicum annuum* L.). Mol. Breed. 39, 1–11.

Mac, T.T., Nguyen, T.D., Dang, H.K., Nguyen, D.T., Nguyen, X.T., 2024. Intelligent agricultural robotic detection system for greenhouse tomato leaf diseases using soft computing techniques and deep learning. Sci. Rep. 14 (1), 23887.

Macias-González, M., Truco, M.J., Smith, R., Cahn, M., Simko, I., Hayes, R., Michelmore, R.W., 2021. Genetics of robustness under nitrogen-and water-deficient conditions in field-grown lettuce. Crop Sci. 61 (3), 15821619.

Mansoor, S., Karunathilake, E.M., Tuan, T.T., Chung, Y.S., 2025. Genomics, phenomics, and machine learning in transforming plant research: advancements and challenges. Hortic. Plant J. 11 (2), 486503.

Maraveas, C., Piromalis, D., Arvanitis, K.G., Bartzanas, T., Loukatos, D., 2022. Applications of IoT for optimized greenhouse environment and resources management. Comput. Electron. Agric. 198, 106993.

Matthews, J.S.A., Vialet-Chabrand, S., Lawson, T., 2020. Role of blue and red light in stomatal dynamic behaviour. J. Exp. Bot. 71, 2253–2269.

Mayer, A., Kalt, G., Kaufmann, L., Röös, E., Muller, A., Weisshaidinger, R., Frehner, A., Roux, N., Smith, P., Theurl, M.C., Matej, S., Erb, K., 2022. Impacts of scaling up agroecology on the sustainability of European agriculture in 2050. EuroChoices 21 (3), 2736.

Menzel, C.M., 2023. A review of fruit development in strawberry: high temperatures accelerate flower development and decrease the size of the flowers and fruit. J. Hortic. Sci. Biotechnol. 98, 409–431.

Milod, N., Saad, G., Khalifa, Hanan, A., 2021. Effect of temperature and relative humidity on conidial germination of the causal agent of cucumber powdery mildew. J. Int. Med. Res. Health Sci. 1, 15–25.

Misra, S., Ghosh, A., 2024. Agriculture paradigm shift: a journey from traditional to modern agriculture. In: Biodiversity and Bioeconomy. Elsevier. pp. 113141.

Mohd Ali, M., Hashim, N., Abd Aziz, S., Lasekan, O., 2020. Pineapple (*Ananas comosus*): a comprehensive review of nutritional values, volatile compounds, health benefits, and potential food products. Food Res. Int. 137, 109675.

Naznin, M.T., Lefsrud, M., Gravel, V., Azad, M.O.K., 2019. Blue light added with red LEDs enhance growth characteristics, pigments content, and antioxidant capacity in lettuce, spinach, kale, basil, and sweet pepper in a controlled environment. Plants 8, 93.

Netrphan, S., Darwell, C.T., Wanichananan, P., Theerawitaya, C., Tongmark, K., Cha-um, S., Janta, S., Chakhonkaen, S., Sangarwut, N., Yamjabok, J., Sikaewtung, K., Kaewmungkun, K., Summat, P., Muangprom, A., 2024. Morphological diversity among *Andrographis paniculata* genotypes based on traits determined mainly by high-throughput phenotyping platform. Genet. Resour. Crop Evol. 72, 5717–5733.

Nonaka, S., Arai, C., Takayama, M., Matsukura, C., Ezura, H., 2017. Efficient increase of γ-aminobutyric acid (GABA) content in tomato fruits by targeted mutagenesis. Sci. Rep. 7 (1), 7057.

Nonaka, S., Ito, M., Ezura, H., 2023. Targeted modification of CmACO1 by CRISPR/Cas9 extends the shelf-life of *Cucumis melo* var. *reticulatus* melon. Front. Genome Ed. 5, 1176125.

Oh, S., Lu, C., 2023. Vertical farming-smart urban agriculture for enhancing resilience and sustainability in food security. J. Hortic. Sci. Biotechnol. 98 (2), 133140.

Osman, M., Qaryouti, M., Alharbi, S., Alghamdi, B., Al-Soqeer, A., Alharbi, A., Almutairi, K., Abdelaziz, M.E., 2024. Impact of CO_2 enrichment on growth, yield and fruit quality of F1 hybrid strawberry grown under controlled greenhouse condition. Horticulturae 10 (9), 941.

Ouyang, C., Hatsugai, E., Shimizu, I., 2022. Tomato disease monitoring system using modular extendable mobile robot for greenhouses: automatically reporting locations of diseased tomatoes. Agronomy 12 (12), 3160.

Ozdemir, D., 2022. The impact of climate change on agricultural productivity in Asian countries: a heterogeneous panel data approach. Environ. Sci. Pollut. Res. 29 (6), 82058217.

Pipatsitee, P., Praseartkul, P., Theerawitaya, C., Taota, K., Tisarum, R., Chungloo, D., Singh, H.P., Cha-um, S., 2023. Impact of temperature on centelloside content, growth characters, physio-morphological adaptations, and biochemical changes in Indian pennywort (*Centella asiatica*). J. Plant Growth Regul. 42 (10), 67766787.

Popovsky-Sarid, S., Borovsky, Y., Faigenboim, A., Parsons, E.P., Lohrey, G.T., Alkalai-Tuvia, S., Fallik, E., Jenks, M.A., Paran, I., 2017. Genetic and biochemical analysis reveals linked QTLs determining natural variation for fruit post-harvest water loss in pepper (*Capsicum*). Theor. Appl. Genet. 130, 445–459.

Rajaseger, G., Chan, K.L., Tan, K.Y., Ramasamy, S., Khin, M.C., Amaladoss, A., Haribhai, P.K., 2023. Hydroponics: current trends in sustainable crop production. Bioinformation 19 (9), 925.

Rashid, M., Yousaf, Z., Din, A., Munawar, M., Aftab, A., Riaz, N., Younas, A., Alaraidh, I.A., Okla, M.K., AbdElgawad, H., 2022. Assessment of mineral nutrient efficiency in genetically diverse spinach accessions by biochemical and functional marker strategies. Front. Plant Sci. 13, 889604.

Reza, M.N., Chowdhury, M., Islam, S., Kabir, M.S.N., Park, S.U., Lee, G.J., Cho, J., Chung, S.O., 2023. Leaf area prediction of pennywort plants grown in a plant factory using image processing and an artificial neural network. Horticulturae 9 (12), 1346.

Richa, A., Fizir, M., Touil, S., 2021. Advanced monitoring of hydroponic solutions using ion-selective electrodes and the internet of things: a review. Environ. Chem. Lett. 19 (4), 34453463.

Roldan, M.V.G., Périlleux, C., Morin, H., Huerga-Fernandez, S., Latrasse, D., Benhamed, M., Bendahmane, A., 2017. Natural and induced loss of function mutations in SlMBP21 MADS-box gene led to jointless-2 phenotype in tomato. Sci. Rep. 7 (1), 4402.

Rosero-Montalvo, P.D., Gordillo-Gordillo, C.A., Hernandez, W., 2023. Smart farming robot for detecting environmental conditions in a greenhouse. IEEE Access 11, 5784357853.

Sage, R.F., Sage, T.L., 2013. c4 plants. In: Levin, S.A. (Ed.), Encyclopedia of Biodiversity. second ed.. Academic Press, Waltham, pp. 361–381.

Scafaro, A.P., Fan, Y., Posch, B.C., Garcia, A., Coast, O., Atkin, O.K., 2021. Responses of leaf respiration to heatwaves. Plant Cell Environ. 44, 2090–2101.

Seth, P., Sebastian, J., 2024. Plants and global warming: challenges and strategies for a warming world. Plant Cell Rep. 43 (1), 27.

Sheikh, U.A.A., Ahmad, M., Aziz, M.A., Imran, M., Rahim, J., Roulston, T.A., Guo, S., Sun, C., 2024. Rearing of native bumblebee species *Bombus haemorrhoidalis* for greenhouse pollination in Pakistan. Agriculture 14 (4), 590.

Shukla, M.K., Singh, L., Vidya, S., Quasim, H., Bhandari, R., 2022. Pollination system for greenhouse flowering plants using nano drones. In: Advances in Mechanical Engineering and Technology: Proceedings of 6th International Conference on Advanced Production and Industrial Engineering (ICAPIE)-2021. Springer Singapore, Singapore, pp. 157162.

Siangliw, J.L., Ruangsiri, M., Theerawitaya, C., Cha-um, S., Poncheewin, W., Songtoasesakul, D., Thunnom, B., Ruanjaichon Ruanjaichon, V., Toojinda, T., 2024. Contrasting alleles of *OsNRT1. 1b* fostering potential in improving nitrogen use efficiency in rice. Plants 13 (20), 2932.

Song, S., Ong, E.J., Lee, A.M., Chew, F.T., 2024. How crop breeding programs can improve plant factories' business and environmental sustainability: insights from a farm level analysis. Sustain. Prod. Consum. 44, 298311.

Sønsteby, A., Heide, O.M., 2023. Temperature and daylength effects on growth and floral initiation in biennial-fruiting blackberry. Horticulturae 9, 1285.

Soyk, S., Müller, N.A., Park, S.J., Schmalenbach, I., Jiang, K., Hayama, R., Zhang, L., Van Eck, J., Jiménez-Gómez, J.M., Lippman, Z.B., 2017. Variation in the flowering gene SELF PRUNING 5G promotes day-neutrality and early yield in tomato. Nat. Genet. 49 (1), 162–168.

Sun, Z., Li, N., Huang, G., Xu, J., Pan, Y., Wang, Z., Tang, Q., Song, M., Wang, X., 2013. Site-specific gene targeting using transcription activator-like effector (TALE)-based nuclease in *Brassica oleracea*. J. Integr. Plant Biol. 55, 1092–1103.

Taiz, L., Zeiger, E., Møller, I.M., Murphy, A., 2015. Plant Physiology and Development, sixth ed.. Sinauer Associates, Sunderland, CT.

Tenreira, T., Lange, M.J.P., Lange, T., Bres, C., Labadie, M., Monfort, A., Hernould, M., Rothan, C., Denoyes, B., 2017. A specific gibberellin 20-oxidase dictates the flowering-runnering decision in diploid strawberry. Plant Cell 29 (9), 2168–2182.

Terence, S., Purushothaman, G., 2020. Systematic review of Internet of Things in smart farming. Trans. Emerg. Telecommun. Technol. 31 (6), e3958.

Terfa, M.T., Olsen, J.E., Torre, S., 2020. Blue light improves stomatal function and dark-induced closure of rose leaves (*Rosa hybrida*) developed at high air humidity. Front. Plant Sci. 11, 01036.

Theerawitaya, C., Praseartkul, P., Taota, K., Tisarum, R., Samphumphuang, T., Singh, H.P., Cha-um, S., 2023. Investigating high throughput phenotyping based morpho-physiological and biochemical adaptations of Indian pennywort (*Centella asiatica* L. Urban) in response to different irrigation regimes. Plant Physiol. Biochem. 202, 107927.

Thilakarathne, N.N., Bakar, M.S.A., Abas, P.E., Yassin, H., 2025. Internet of things enabled smart agriculture: current status, latest advancements, challenges and countermeasures. Heliyon 11 (3), e42136.

Thipe, E.L., Workneh, T., Odindo, A., Laing, M., 2017. Greenhouse technology for agriculture under arid conditions. In: Sustainable Agriculture Reviews. Springer International Publishing, Cham, pp. 3755.

Tian, Z., Ma, W., Yang, Q., Duan, F., 2022. Application status and challenges of machine vision in plant factory—A review. Inf. Process. Agric. 9 (2), 195211.

Tisarum, R., Samphumphung, T., Theerawitaya, C., Prommee, W., Cha-um, S., 2018. *In vitro* photoautotrophic acclimatization, direct transplantation and *ex vitro* adaptation of rubber tree (*Hevea brasiliensis*). Plant Cell, Tissue and Organ Culture 133 (2), 215223.

Tisarum, R., Bayaraa, U., Sotesaritkul, T., Samphumphuang, T., Singh, H.P., Cha-um, S., 2025. Impact of iso-osmotic acclimation in condition on *ex vitro* performance of white turmeric (*Curcuma zedoaria* [Cristm.] Roscoe). Vitro Cell. Dev. Biol. Plant 61 (2), 294305.

Tomlinson, L., Yang, Y., Emenecker, R., Smoker, M., Taylor, J., Perkins, S., Smith, J., MacLean, D., Olszewski, N.E., Jones, J.D.G., 2019. Using CRISPR/Cas9 genome editing in tomato to create a gibberellin-responsive dominant dwarf DELLA allele. Plant Biotechnol. J. 17, 132–140.

Tripathy, P.K., Tripathy, A.K., Agarwal, A., Mohanty, S.P., 2021. MyGreen: an IoT-enabled smart greenhouse for sustainable agriculture. IEEE Consum. Electron. Mag. 10 (4), 5762.

Van, V.T., Das, S., Hensel, G., Kim, J.Y., 2022. Genome editing and beyond: what does it mean for the future of plant breeding? Planta 255 (6), 130.

Verma, M.S., Gawade, S.D., 2021. A machine learning approach for prediction system and analysis of nutrients uptake for better crop growth in the hydroponics system. In: 2021 International Conference on Artificial Intelligence and Smart Systems (ICAIS). IEEE. pp. 150156.

Verma, P., Singh, G., Singh, S.K., Mirza, A.A., Bakshi, M., Anmol, L., Kumar, L., Rupesh, 2025. Modulating productivity of strawberries (*Fragaria ananassa* Duch.) through artificial full-spectrum light in indoor vertical farming. J. Soil Sci. Plant Nutr. 25, 1219–1233.

Villagran, E., Espitia, J.J., Amado, G., Rodriguez, J., Gomez, L., Velasquez, J.F., Gil, R., Baeza, E., Aguilar, C.E., Akrami, M., Arias, L.A., 2025. CO_2 Enrichment in protected agriculture: a systematic review of greenhouses, controlled environment systems, and vertical farms—Part 2. Sustainability 17 (7), 2809.

Wahab, F., Khan, M.J., Khan, M.Y., Mushtaq, R., 2024. The impact of climate change on agricultural productivity and agricultural loan recovery; evidence from a developing economy. Environ. Dev. Sustain. 26 (10), 2477724790.

Walne, C.H., Reddy, K.R., 2022. Temperature effects on the shoot and root growth, development, and biomass accumulation of corn (*Zea mays* L.). Agriculture 12, 443.

Wang, D., Samsulrizal, N.H., Yan, C., Allcock, N.S., Craigon, J., Blanco-Ulate, B., Ortega-Salazar, I., Marcus, S.E., Bagheri, H.M., Perez Fons, L., Fraser, P.D., Foster, T., Fray, R., Knox, J.P., Seymour, G.B., 2019a. Characterization of CRISPR mutants targeting genes modulating pectin degradation in ripening tomato. Plant Physiol. 179 (2), 544–557.

Wang, R., Tavano, E.C.D.R., Lammers, M., Martinelli, A.P., Angenent, G.C., de Maagd, R.A., 2019b. Re-evaluation of transcription factor function in tomato fruit development and ripening with CRISPR/Cas9-mutagenesis. Sci. Rep. 9, 1696.

Wang, F., Gao, Q., Ji, G., Wang, J., Ding, Y., Wang, S., 2024. Effects of light intensity and photoperiod on morphological development and photosynthetic characteristics of coriander. Horticulturae 10, 215.

Wang, S.S., Li, Y.L., Wen, X.Z., 2018a. Effect of increasing humidity on flowering, fruit-setting and pollen characteristics of tomato under heat stress. Acta Hortic. 1227, 305–312.

Wang, X., Xu, Y., Hu, Z., Xu, C., 2018b. Genomic selection methods for crop improvement: current status and prospects. Crop J. 6, 330–340.

Wu, S., Li, R., Bu, C., Zhu, C., Miao, C., Zhang, Y., Cui, J., Jiang, Y., Ding, X., 2024. Photoperiodic effect on growth, photosynthesis, mineral elements, and metabolome of tomato seedlings in a plant factory. Plants 13, 3119.

Xie, X.-B., Li, S., Zhang, R.-F., Zhao, J., Chen, Y.-C., Zhao, Q., Yao, Y.-X., You, C.-X., Zhang, X.-S., Hao, Y.-J., 2012. The bHLH transcription factor MdbHLH3 promotes anthocyanin accumulation and fruit colouration in response to low temperature in apples. Plant Cell Environ. 35, 1884–1897.

Xing, S., Jia, M., Wei, L., Mao, W., Abbasi, U.A., Zhao, Y., Chen, Y., Cao, M., Zhang, K., Dai, Z., Dou, Z., Jia, W., Li, B., 2018. CRISPR/Cas9-introduced single and multiple mutagenesis in strawberry. J. Genet. Genomics 45, 685–687.

Yamori, W., Hikosaka, K., Way, D.A., 2014. Temperature response of photosynthesis in C3, C4, and CAM plants: temperature acclimation and temperature adaptation. Photosynth. Res. 119, 101–117.

Yang, B., Zhou, S., Ou, L., Liu, F., Yang, L., Zheng, J., Chen, W., Zhang, Z., Yang, S., Ma, Y., Zou, X., 2020. A novel single-base mutation in CaBRI1 confers dwarf phenotype and brassinosteroid accumulation in pepper. Mol. Genet. Genomics 295, 343–356.

Yang, T., Altland, J.E., Samarakoon, U.C., 2023. Evaluation of substrates for cucumber production in the Dutch bucket hydroponic system. Sci. Hortic. 308, 111578.

Yang, T., Samarakoon, U., Altland, J., Ling, P., 2024. Influence of electrical conductivity on plant growth, nutritional quality, and phytochemical properties of kale (*Brassica napus*) and collard (*Brassica oleracea*) grown using hydroponics. Agronomy 14 (11), 2704.

Yang, Z., Sinclair, T.R., Zhu, M., Messina, C.D., Cooper, M., Hammer, G.L., 2012. Temperature effect on transpiration response of maize plants to vapour pressure deficit. Environ. Exp. Bot. 78, 157–162.

Yu, Q.H., Wang, B., Li, N., Tang, Y., Yang, S., Yang, T., Guo, X., Yan, C., Wang, P., Asmutola, Q., 2017. CRISPR/cas9-induced targeted mutagenesis and gene replacement to generate long-shelf life tomato lines. Sci. Rep. 7, 1–9.

Yuan, Y., Ton, B.L., Thomas, W.J., Batley, J., Edwards, D., 2023. Supporting crop plant resilience during climate change. Crop Sci. 63 (4), 18161828.

Yuanyuan, L.I., Cuinan, W.U., Xue, W.U., Shipu, X.U., Dingyi, Y.U.E., Lipan, H.U., Kai, C.A.O., 2024. Effects of red light to far-red light ratio on yield and quality of lettuce in a plant factory. J. Northwest A&F Univ. Nat. Sci. Ed. 52 (11), 73.

Zameer, S., Ali, M., Sajjad, A., Saeed, S., Matloob, A., Bashir, M.A., Reem, A.A., A., R., Hargis, B.M., Hashem, M., Alamri, S., Atta, S., 2022. Foraging behavior and visit optimization of bumblebees for the pollination of greenhouse tomatoes. J. King Saud Univ. Sci. 34 (1), 101744.

Zamora-Izquierdo, M.A., Santa, J., Martínez, J.A., Martínez, V., Skarmeta, A.F., 2019. Smart farming IoT platform based on edge and cloud computing. Biosyst. Eng. 177, 417.

Zandalinas, S.I., Fritschi, F.B., Mittler, R., 2021. Global warming, climate change, and environmental pollution: recipe for a multifactorial stress combination disaster. Trends Plant Sci. 26 (6), 588599.

Zandalinas, S.I., Balfagón, D., Gómez-Cadenas, A., Mittler, R., 2022. Plant responses to climate change: metabolic changes under combined abiotic stresses. J. Exp. Bot. 73 (11), 33393354.

Zhang, F., Jiang, N., Zhang, H., Huo, Z., Yang, Z., 2023. Effect of low temperature on photosynthetic characteristics, senescence characteristics, and endogenous hormones of winter wheat "Ji Mai 22" during the jointing stage. Agronomy 13, 2650.

Zhang, H., Shan, S., Zhou, Z., Gu, S., Li, Z., Khashaveh, A., Chen, H., An, J., Zhang, Y., 2025. Floral scent affects bumblebee pollinator preferences on the flowers of four tomato cultivars under greenhouse conditions. J. Econ. Entomol. 118, toaf102.

Zhang, L., Zhu, L., Yu, M., Zhong, M., 2016. Warming decreases photosynthates and yield of soybean [*Glycine max* (L.) Merrill] in the North China plain. Crop J. 4, 139–146.

Zhang, X., Wang, G., Chen, B., Du, H., Zhang, F., Zhang, H., Wang, Q., Geng, S., 2018. Candidate genes for first flower node identified in pepper using combined SLAF-seq and BSA. PLoS One 13, e0194071.

Zhou, J., Sittmann, J., Guo, L., Xiao, Y., Huang, X., Pulapaka, A., Liu, Z., 2021. Gibberellin and auxin signaling genes RGA1 and ARF8 repress accessory fruit initiation in diploid strawberry. Plant Physiol. 185 (3), 10591075.

Zhu, Z., Sun, B., Wei, J., Cai, W., Huang, Z., Chen, C., Cao, B., Chen, G., Lei, J., 2019. Construction of a high density genetic map of an interspecific cross of *Capsicum chinense* and *Capsicum annuum* and QTL analysis of floral traits. Sci. Rep. 9, 1054.

Data sharing, privacy management strategies, and security protocols for smart agricultural systems

3

Nerice Millet-Williams[a], *Keri Grant*[a], *and Hemendra Kumar*[a,b,c]
[a]College of Agriculture and Natural Resources, University of Maryland, College Park, MD, United States
[b]Digital and Precision Agriculture Lab, University of Maryland, College Park, MD, United States
[c]Central Maryland Research and Education Center, University of Maryland Extension, Upper Marlboro, MD, United States

1 Introduction

This chapter addresses the critical aspects of data sharing, privacy, and management strategies, alongside security protocols for smart agricultural systems. A systematic approach begins with establishing robust data governance frameworks that define clear policies for data ownership, access, and usage rights. This chapter illustrates successful strategies and protocols, ensuring both the effective use of data and the protection of field privacy and interests.

2 Digital privacy and security concerns in agriculture

Digital privacy is defined as the ability to decide what data is shared and what data is private. Privacy is a fundamental human right that fosters individual autonomy, dignity, and freedom. Privacy policies aim to define the sharing and ownership rights of data. Unfortunately, not all privacy policies promise complete privacy. The rapid advancements in technology have led to increased data collection, storage, and processing capabilities. Regulations have not kept up. Poor privacy practices and big data analytics can intrude upon personal rights through data mining and profiling. While privacy policies legally define data-sharing permissions, hackers may also seek access to your data, digital accounts, and more. Managing privacy and protecting personal information from hackers will keep your information safe, but in distinct ways.

Digital security is defined as the systems and protections in place to safeguard data and devices from unauthorized users. Digital security can be compromised by

Precision Technologies for Digital Agriculture. https://doi.org/10.1016/B978-0-443-40513-6.00013-5

exposing login information, shared internet connections, and other sensitive information. While new security technologies, such as smart sensors and mapping, are developing rapidly, persistent challenges remain in maintaining robust data security. Understanding and addressing these privacy and security concerns will enable farmers to protect their digital property while leveraging technological advancements for improved farm management and efficiency.

3 Basic components of digital privacy and security

Over the past five decades, the agriculture industry has undergone a significant transformation with advancements in machinery, technology, seeds, irrigation, and fertilizers that have enhanced productivity and efficiency. Agriculture is ever-evolving with digital technologies such as artificial intelligence (AI), analytics, and connected sensors. These innovations promise to further boost yields, optimize resource use, and improve sustainability. These tools utilize mass data collection and analysis to maximize productivity. Many high-tech farmers collect data on their soil, weather, fertilizer, pesticide, and crop yield on the cloud. This data allows users to minimize inputs and maximize outputs by managing fields on a smaller scale, increasing efficiency. However, now that important farm data is collected and stored electronically, there are new concerns to consider. Farmers must understand and manage data safety and privacy as they adopt these technologies. Proper data management practices will help ensure that personal and operational data are protected, supporting both the efficiency and security of their farming operations (McKinsey, 2024). Some of the key definitions shown in Table 3.1 are used in the agricultural and digital communities, which are important to understand for data privacy and sharing in smart agricultural systems.

Table 3.1 Important definitions of the keywords used in data sharing, privacy, and security.

Artificial intelligence (AI)	AI is artificial intelligence developed by using large amounts of data to "train" the computer algorithm. Agricultural technologies are developing that use AI to steer tractors, implement precision fertilizer application, identify diseases and weeds, and suggest management strategies.
Cloud	A data storage platform where the data can be accessed using the internet. The data is still stored physically and is usually located in large "warehouses" run by Google, Amazon, or other large companies.
Digital footprint	The trace you leave behind on the internet includes information from social media sites, location pings from devices, internet history, and more. In the agricultural field, data collected and saved on the cloud is part of your digital footprint, as it can be accessed through the internet.

Table 3.1 Important definitions of the keywords used in data sharing, privacy, and security—cont'd

Ransomware	Viral software that is maliciously installed into devices and networks. This software is designed to lock or delete files and functions of any digital tool. The attacker may ask for money in return for the removal of the software. If this software infiltrates a farm network, it may be able to lock devices, ground self-driving tractors, and stop scheduled irrigation.
Digital equity	Digital equity ensures fair access to technology, connectivity, and digital skills for all. In agriculture, it enables farmers to use data-driven tools for better productivity, regardless of age or background. Ensuring digital equity in agriculture must include strong cybersecurity and data protection measures.
IoT	The Internet of Things (IoT) encompasses a network of interconnected technologies that communicate via Bluetooth or share the same internet connection. This includes not only your smartphone and computer but also a variety of smart devices such as sensors, monitors, cameras, and smart lighting systems. Once a hacker gains access to any single device within this network, they may potentially infiltrate all other devices connected to the same internet network, posing a significant security risk.
Machine learning	A type of AI that uses training data to perform tasks and learn patterns.
AI	Artificial intelligence (AI) is a type of software that uses large amounts of relevant data and pattern recognition to perform traditionally human tasks. These tasks may include decision-making, problem-solving, and learning. New agricultural technologies may use AI to steer tractors, implement precision fertilizer application, identify diseases and weeds, and suggest management strategies.
Cloud	The cloud is a data storage platform where data is accessible using the internet. Data is usually stored physically at large "warehouses" run by Google, Amazon, or other large data management companies.
Digital footprint	A digital footprint is the information left on the internet as users interact with websites. This includes information from social media sites, location pings from devices, internet history, and more.
Ransomware	Ransomware is a category of viral software that is maliciously installed on devices and networks. This software is designed to lock or delete files and functions of any digital tool. The attacker may demand money in return for the removal of the software. If this software infiltrates a farm network, it may be able to lock devices, ground self-driving tractors, and stop scheduled irrigation.

Continued

Table 3.1 Important definitions of the keywords used in data sharing, privacy, and security—cont'd

Digital equity	Digital equity is the concept of fair access to technology, connectivity, and digital skills for all. In agriculture, it may look like all farmers using data-driven tools for better productivity, regardless of age or technological background. Ensuring digital equity in agriculture must include strong cybersecurity and data protection measures.
The Internet of Things (IoT)	The IoT encompasses a network of digital devices that communicate via Bluetooth or an internet connection. This includes smartphones, computers, and smart devices such as sensors, monitors, cameras, and smart lighting systems. Once a hacker gains access to any single device within this network, they can potentially infiltrate all other devices connected to the same internet network, posing a significant security risk.
Machine learning	Machine learning is a category of AI development that uses large amounts of data to learn patterns. Once software is successfully trained to recognize patterns, it will be able to perform tasks. For each task it performs and with additional data acquired, it will continue to learn patterns and improve its decision-making.
Smart sensors	Smart sensors are digital sensors with the ability to communicate through cables, Bluetooth, or Wi–Fi and interact with AI and analytic software. They may sense light, humidity, yield, location, etc.
Big data analytics	Big data analytics is an analytical process that uses huge amounts of data points to recognize patterns and predict outcomes. Data sets can include millions of data points, if not more. AI may also be used to help organize, clean, and analyze data.
Autonomous system	An autonomous system is a group of digital tools within the Internet of Things. Based on programming, commands are triggered by one device and sent to another. For example, a farmer may set his system to start irrigating an area when a moisture sensor detects dry soil.
Agricultural technology provider (ATP)	An agricultural technology provider is a company that interacts with farm sensors, collects, manages, and stores field data.

4 New technologies impacting agriculture

Smart agriculture is based on the concept of using data to improve production and manage fields at a granular level, turning vast amounts of data into actionable recommendations. These prescriptions are often developed with AI and big data analytics. However, a recent survey reveals significant concerns about AI among the public. Among those familiar with AI, 70% lack trust in companies to use AI responsibly, and roughly 80% worry that AI will be used in ways they are uncomfortable with or that were not originally intended (Pew Research Center, 2023a). Despite some optimism about AI potentially making life easier (62%), many respondents have substantial apprehensions regarding its applications (Pew Research Center, 2023b).

There are still many questions and concerns about AI, but it has become an important part of data analysis. AI models are trained by processing large amounts of data so they can learn to recognize patterns and make recommendations. In agriculture, tools like drones often use AI to perform tasks more efficiently based on the data they receive (Palli and Kumar, 2024). These models also learn from the results of their actions and continue to improve over time. This is why companies are so focused on collecting user data; the more data they have, the more accurate their systems become.

5 Methods and techniques

There are various methods for safeguarding privacy and security, ranging from basic password protection to detailed reviews of privacy policies. While privacy and security are closely related concepts, they require distinct management strategies to address their respective challenges effectively.

5.1 Methods and techniques—Privacy

In smart farming, both remote and on-site monitoring systems quickly and accurately collect data on various factors such as soil moisture, temperature, and crop health. This enables farmers to react promptly to environmental changes (Kumari and Saini, 2022). The opportunity for improved production also creates a risk of data mining by data management companies.

Protecting data privacy

Keeping farm data private is a top priority. There are several techniques available to ensure privacy and transparency:

1. **Controlled data sharing:** Farmers have control over who sees their data. Use strict policies to limit access to only those who need it, ensuring compliance with privacy laws (Al-Dhief et al., 2019). Read the privacy policies associated with equipment manufacturers and farm management programs used.
2. **Limiting data transfers:** Data does not need to be shared all the time. Send only essential data, reducing the risk of exposure to cyber threats (Sharma et al., 2020).
3. **Disabling data access during Cloud interactions**: Turning off data access while interacting with cloud services ensures that only authorized parties can view or manipulate sensitive information, thereby minimizing the risk of unauthorized access and enhancing privacy protections (Al-Dhief et al., 2019).

5.2 Methods and techniques—Security

Security measures

To protect farm data from cyber threats, we apply several security measures:

1. **Encryption:** Use secure methods to protect data as it moves between devices and storage systems, preventing unauthorized access (Bhatti and Humphreys, 2017).
2. **Verifying users and devices:** Ensure that only trusted users and devices can connect to the system, blocking unauthorized access attempts (Méndez et al., 2019). This can be achieved using passwords and access requests.
3. **Detecting unusual activity:** AI-powered monitoring helps identify strange patterns in data that may indicate a security issue, allowing quick responses to potential threats.

By combining these approaches, farmers can make smart farming more secure and efficient while keeping data safe. These methods help balance the benefits of advanced technology with the need for privacy and security in modern agriculture.

6 Common challenges and solutions

Digital privacy and security present distinct challenges, each requiring unique solutions. While both focus on protecting sensitive data, the nature of the threats they address differs. Privacy concerns primarily revolve around controlling access to personal information and preventing unauthorized data collection, often by corporations or third parties. In contrast, security focuses on safeguarding data from cyber threats such as hacking, malware, and unauthorized access. Understanding these differences is crucial for implementing effective protection strategies.

6.1 Common challenges and solutions—Privacy

Privacy

Precision agriculture depends on collecting and analyzing data from sources like remote sensing, satellite imagery, soil samples, and yield reports. This data helps farmers make informed decisions and improve farm management. However, while data collection offers many benefits, it also raises serious privacy concerns. Farmers risk losing control of their data if it is shared, sold, or accessed without permission. Service providers can use combined data from across the country to create valuable tools and predictions; yet, farmers may not be compensated for their contributions. Like personal data, farm data often lacks clear legal protections. The lack of transparency from Agricultural Technology Providers (ATPs) has raised concerns and may discourage some farmers from using precision agriculture technologies (Wiseman et al., 2022).

Privacy laws and regulations

Federal privacy laws and regulations in the United States play a crucial role in protecting personal information across various sectors. One of the most significant laws is the Health Insurance Portability and Accountability Act (HIPAA), which sets standards for the protection of health information and establishes rules for the use and disclosure of protected health information (PHI). Alongside HIPAA, the Federal Trade Commission (FTC) enforces rules, including the Health Breach Notification Rule (HBN Rule) and the Standards for Safeguarding Customer Information under the Gramm-Leach-Bliley Act (Safeguards Rule). These rules prevent unfair or deceptive practices in data privacy concerning financial data, health data, and digital health services. Recent years have seen an increased focus on the privacy of sensitive data, with regulations aimed at addressing emerging technologies and practices that could compromise consumer rights (WilmerHale, 2024) and (HIPAA Journal, 2024).

Moreover, as digital equity initiatives grow, there is a greater need for comprehensive federal privacy legislation to provide a unified framework for data protection. This includes considerations for biometric data, online tracking, and the collection of health-related information (WilmerHale, 2024; HIPAA Journal, 2024). Data collection across all sectors is booming without proper checks and balances to protect users.

In the United States, there is broad support for increased government regulation of personal data, with 72% of respondents advocating for more oversight, based on a survey conducted by the Pew Research Center. This sentiment crosses political party lines in the USA, as both Democrats (78%) and Republicans (68%) support greater regulation, consistent with findings from 2019 (Pew Research Center, 2024). However, public understanding of data privacy laws is low; 72% of Americans report little to no knowledge of these laws, and only 27% feel they understand them well (Shapero, 2024). It is important to understand privacy laws and regulations to know what protections are in place—and to recognize which types of data are not legally protected.

The Organization for Economic Co-operation and Development (OECD) Privacy Guidelines outline international principles for protecting privacy in the digital age. The OECD also recommends technological solutions to help safeguard consumer data. For example, encryption uses cryptographic techniques to secure communication and data storage. Privacy-enhancing technologies can protect privacy by minimizing or eliminating the collection of personally identifiable information. The OECD also emphasizes the use of anonymization techniques to de-identify data while still enabling meaningful analysis. In addition, it highlights blockchain technology as a potential solution for privacy concerns, offering decentralized and tamper-proof systems for certain applications. In Maryland, USA, the Maryland Online Data Privacy Act (MODPA) introduces rigorous restrictions on how companies collect and use consumer data, requiring businesses to minimize data collection and limit the sharing of sensitive information. MODPA also grants consumers several rights, such as accessing their data, correcting inaccuracies, requesting deletion, and opting out of targeted advertising and profiling. This bill was passed and approved by the Governor of Maryland on May 9, 2024 (Maryland General Assembly, 2024a, 2024b). This law is set to take effect on October 1, 2025, with full enforcement starting on April 1, 2026.

Privacy erosion and transparency

Privacy erosion refers to the gradual weakening or loss of personal privacy. It is caused by the increasing collection, use, and sharing of personal data by governments, corporations, and other entities. This process is often facilitated by data collection companies that hide their true intentions within long and complex privacy policies. These policies may mislead users or make it unclear whether the data is sold or analyzed for revenue. As data becomes more valuable and these technologies become more beneficial, companies may leverage looser privacy policies to increase their revenue. If users don't sign these policies, they may be excluded from accessing important tools or platforms. Farmers may have to decide between sharing their data or falling behind competitors. This dynamic exemplifies how privacy erosion can manifest. As location tracking and data mining become more prevalent, individuals may lose control over their personal information. This leads to concerns about privacy, autonomy, and trust.

Privacy solutions

The first step in ensuring privacy is to read the agreement with the ATP carefully. If there is confusion in understanding the agreement, reach out to the provider for clarification. Companies claim to offer free trials, such as planting prescriptions from seed companies, but they are not genuinely free. They charge customers for data. For example, seed companies might analyze data from yields, soil, rainfall, and other environmental factors to recommend specific seed varieties. Field data is often used to develop and sell recommendations or services, yet the farmers themselves typically receive no compensation for their contributions in providing this data.

In addition, if yield data and soil samples are integrated into mapping platforms like Climate FieldView or John Deere, these companies have detailed records of the most productive plots of land. This can have long-term implications. Understanding these elements can help protect field data and ensure users (farmers) retain control over their use (Australian Broadcasting Corporation News, 2022; Carbonell, 2016). Here are some key elements to check:

- **Data storage:** Know where your data is physically stored. ATPs may have data storage facilities or contract with another company, like Google. Understand the security measures of the data storage company regarding your information.
- **Data ownership**: Ensure legal rights to field data. Some agreements may stipulate that the ATP owns the data, limiting farmer control. For example, John Deere's data-sharing policies have raised concerns about who owns and can access data collected by their equipment (Australian Broadcasting Corporation News, 2022; Carbonell, 2016). If data ownership is not clearly defined, operators could lose the ability to control how field data is used or sold.
- **Access rights**: Retain access to data and understand how ATPs may access, share, or sell it. If privacy agreements allow ATPs to have unrestricted access, they might share or sell that data without notification. Unless stipulated in the privacy policies, data may be shared with seed companies, marketing firms, or foreign entities.

6.2 Common challenges and solutions—Security

Security

In today's digital age, security threats span various aspects of digital activity. Scammers relentlessly call, text, and email, and may also use compromised accounts to target connected organizations. Identity thieves phish for passwords and personal information to attack individuals and businesses alike. Protecting a business from digital attacks starts with protecting personal information and managing personal digital footprints. A digital footprint includes any trackable online activity. Separate actions can be connected through device location, shared internet connections, and more. This allows hackers to leverage access to unrelated accounts to gain digital access to your operational equipment or sensitive data.

Hackers digitally infiltrate businesses to steal information, install ransomware, or hack larger partnering companies. Ransomware is a virus that can disable or shut down digital operations. This can include breaking software, tractors, and sensors. Hackers can hold devices hostage until the victim pays the ransom (Check Point Software Technologies, 2023). It is uncommon for hackers to attack farms, but it can still have an impact. A successful ransomware attack can ground transportation, halt factories, and more, causing issues in other parts of a supply chain. Cyber Liability Insurance exists to protect against ransomware attacks. However, companies with this insurance might be considered targets.

Repercussions of ransomware

While the healthcare industry is often the primary target of digital attacks, the food and agriculture sector is increasingly at risk. In 2023, there were 75 reported ransomware complaints in this sector alone (Statista Research Department, 2023), with 17.6% of attacks proving successful (Search Logistics, 2024). These incidents show that no industry is immune—any business can suffer serious harm from ransomware. In 2024, the threat has grown even more severe. Ransomware attacks increased by 27% from the previous year, and malware continues to be the fastest-growing cyber threat, with 41% of enterprises reporting incidents. The average cost of recovering from a ransomware attack now approaches $2 million, and 8% of affected organizations choose to pay the ransom (Ransomware.org, 2024; Varonis, 2024). Beyond financial loss, such attacks can disrupt supply chains and cripple operations—especially in agriculture, where compromised sensors or systems can lead to overuse of water or pesticides, crop loss, and environmental damage (Gyamfi et al., 2024).

At the same time, protecting personal data remains critical. Identity theft is a growing issue, accounting for nearly 20% of complaints to the FTC in 2024 (Norton LifeLock, 2024; Comparitech, 2024). Victims lost a combined $8.8 billion to online fraud, with median losses around $650. Alarming trends also show that children are increasingly targeted, with approximately 915,000 cases of child identity theft reported (Comparitech, 2024). Together, these trends highlight the urgent need for stronger digital defenses, not only to protect individual privacy but also to safeguard critical industries like agriculture from the growing threat of cyberattacks.

Security solutions

1. **Understand your software**
    - Regularly check which apps are installed and remove unnecessary ones
    - Only allow trusted users to make changes to the software
    - Keep software up-to-date to patch security vulnerabilities
    - Review what data apps can access and limit unnecessary permissions
    - Know what the software does and why it needs certain access
    - Download software only from official or reputable sites
2. **Protecting your personal accounts**
    - Create strong passwords
    - Use different passwords for each account
    - Use secure password managers
    - Turn on two-factor authentication (2FA) whenever possible
3. **Don't fall for phishing**
    - Recognizing and avoiding phishing scams: It's crucial to identify phishing scams and steer clear of online fraud. Numerous resources are available to help you recognize these threats, including guides and tips on internet safety. For a comprehensive overview of how to avoid spam and phishing attacks, visit Maryland Online Internet Safety Resources (Marylanders Online, 2024). Utilizing such resources can enhance awareness and protection against potential scams.
    - After receiving strange messages, do not share any personal information or click on any provided links.
    - Check if the claims are true by contacting the organization directly. For example, after receiving an email from your bank about a large charge on a credit card, go directly to the bank's website or call the number provided by the bank—NOT the questionable email.
    - Install trusted antivirus software on your devices. This may help in case you do click on a fraudulent link.
4. **Manage your digital footprint**
    - Revisit privacy settings for online accounts
    - Keep track of cookies
    - Turn off location tracking
    - Limit information shared when possible, including email, phone number, name, etc.
    - Avoid questionable websites
    - Use a VPN to disrupt digital tracking
    - Reduce online accounts
5. **Secure your devices**
    - Keep devices in a safe place and password-protected.
    - Install trusted antivirus software on devices. Do research before deciding which software to choose.
    - Identify the Internet of Things (IoT) Devices: IoT encompasses a network of interconnected technologies that communicate via Bluetooth or share the same internet connection. This includes smartphones, computers, and a variety of smart devices such as sensors, monitors, cameras, and smart lighting systems. Once a hacker gains access to any single device within this network, they may potentially infiltrate all other devices connected to the same internet network, posing a significant security risk.
    - Research the security features of connected devices. They may come with firewalls, virus scanning, or encryption. Simpler devices are less likely to have these features.

- Enable 2FA: For software that supports it, enable 2FA to add an extra layer of security. This ensures that even if login credentials are compromised, unauthorized users can't access your accounts easily.

6. **Protect your data**
   - Data may be stored on local hard drives, cloud services, or a combination of both.
   - Local storage risks include hardware failure, physical damage, or physical theft. Back up data to prevent data loss.
   - Cloud services have physical locations in large data centers. These data centers may be run by the company you work with or contracted out to data companies like Google. Read the data security protocols of your cloud providers.

7 Perspectives, limitations, and future implications

7.1 Perspective

The integration of smart devices in agriculture offers numerous advantages, such as increased efficiency and improved decision-making. These devices generate substantial amounts of data, enabling precision farming. However, managing and securing this data presents various challenges. AI and machine learning have the potential to process data efficiently while strengthening security measures. These technologies can identify irregularities that might indicate cyber threats or system malfunctions. However, their application in agriculture is complex due to environmental variability and the limited computing capabilities of agricultural IoT devices (Sharma et al., 2020). Sharing security-related insights among farms can help mitigate threats by allowing collaborative defense strategies. Unfortunately, balancing information sharing with the need to protect data privacy remains an issue (Al-Dhief et al., 2019). Security strategies must address different layers of agricultural systems. There are cost considerations that may limit their use, and some security programs may disrupt autonomous systems and damage farm operations (Bhatti and Humphreys, 2017).

7.2 Limitations

Despite advancements in agricultural cybersecurity, multiple challenges persist. Smart farms rely on a variety of IoT devices with different security requirements, meaning that a single malware detection method may not be universally effective (Abdullah et al., 2021). While sharing threat intelligence among farms is beneficial, uncertainties remain regarding which data should be shared and how small-scale farmers can engage in such initiatives (Al-Dhief et al., 2019).

AI-based security models can be compromised if attackers manipulate sensor data or introduce misleading information during training, necessitating further research to strengthen AI resilience (Goodfellow et al., 2014). In addition, smart agricultural systems function in ever-changing environments, making it challenging to create consistent behavior models for anomaly detection (Zhang et al., 2020).

Existing cybersecurity intelligence systems need to be tailored to the unique threats associated with smart farming (Sharma et al., 2020). Traditional authentication methods, such as passwords, may not be practical for farms with extensive IoT networks. Future authentication methods must be user-friendly and scalable (Méndez et al., 2019).

Gateways connecting farm devices to the cloud must have effective firewalls and access controls. However, given their resource constraints, these security measures must be optimized for efficiency (Al-Dhief et al., 2019). Addressing these challenges is essential for ensuring the security and sustainability of smart farming systems.

7.3 Future implications

As technology continues to evolve at an accelerating pace, the significance of privacy and security will only increase. With the advent of new digital tools and technologies, safeguarding sensitive data becomes more complex and crucial. The growing reliance on cloud-based platforms, AI, and other digital innovations heightens the potential risks associated with breaches, necessitating the development of more robust security protocols and privacy regulations to keep pace with these advancements. Furthermore, as technologies become increasingly integrated into everyday operations, technological failures may have more severe consequences. The potential for widespread disruptions, whether due to data breaches, system malfunctions, or cyber-attacks, could jeopardize not only individual privacy but also the stability of critical sectors, including agriculture, healthcare, and finance. The complexity of these systems means that failure is not only more likely but could also carry much more significant and far-reaching impacts.

Looking toward the future, we may also reach a point where AI systems surpass human expertise in certain domains. In agriculture, for instance, AI-driven technologies may become more adept at optimizing farming techniques, crop management, and resource allocation than human farmers. This raises profound questions about the balance between human decision-making and machine-driven processes. The potential for AI to revolutionize farming practices presents exciting opportunities but also significant challenges regarding control, ethical decision-making, and the preservation of privacy and autonomy in sectors heavily reliant on technology. As these developments unfold, careful consideration will be needed to navigate the complex intersection of innovation, security, and societal impact.

8 Conclusion

While smart farming presents a transformative opportunity for agriculture, addressing security and data management concerns is critical. Farmers can protect their data privacy by becoming familiar with data security measures and staying updated on technological advances. Some organizations offer training to help farmers enhance these skills, enabling them to manage data more securely. Developing internal guidelines and training for farm workers on data protection practices is essential. In addition, understanding and scrutinizing legal agreements with ATPs is crucial, as many farmers are unaware of the terms regarding data usage. Reading and questioning these agreements can prevent data misuse and ensure better control over farm data (Wiseman et al., 2022).

As Maryland's farming community embraces digital tools to optimize operations, robust privacy and security measures are essential. Further research is needed to tailor cybersecurity solutions for agriculture, ensuring robust defense mechanisms against cyber threats. Understanding privacy agreements with ATPs, ensuring data ownership, and implementing strong security protocols are crucial steps. By fostering both digital access and data protection and privacy, Maryland farmers can confidently adopt new technologies while safeguarding their livelihoods in an increasingly interconnected world.

8.1 Key takeaways

1. **Read your data agreement carefully**
2. **Data sharing and data usage:**
   - Understand privacy policies. Be aware of where data is stored, with whom data is shared, and how data is being used
3. **Data storage**:
   - Understand the data storage platform
   - Ensure cloud providers have strong security measures and clear data policies
4. **Basic tracking**:
   - Track the activities in the data storage platform to know who is doing what
   - Be aware of cookies, ads, and privacy settings on devices
5. **AI considerations**:
   - Understand the limitations and potential biases of AI tools used in agriculture
6. **Farm-specific data**:
   - Know what happens to the data collected by yield monitors and other equipment

Nerice Millet-Williams: Writing—review & editing, Writing—original draft, Visualization, Software, Methodology, Conceptualization. **Keri Grant:** Writing—review & editing, Writing—original draft, Visualization, Software, Methodology, Conceptualization. **Hemendra Kumar:** Writing—review & editing, Writing—original draft, Visualization, Software, Methodology, Conceptualization, Validation, Supervision, Resources, Project administration, Investigation.

Disclose conflict of interest

The authors declare that they have no known competing financial interests or personal relationships that could have appeared to influence the work reported in this publication.

References

Abdullah, M., Aslam, R., Alam, M., 2021. Security in smart farming: a review of AI-based solutions. Comput. Electron. Agric. 181, 105971.

Al-Dhief, F.T., Latiff, L.A., Abdulkareem, K.H., Adamu, I., 2019. Cybersecurity threats in smart farming: a comprehensive review. IEEE Access 7, 122661–122673.

Australian Broadcasting Corporation News, 2022. Are big ag tech companies harvesting farmers' confidential data? ABC News. https://www.abc.net.au/news/2022-02-19/agriculture-data-protection/100840436.

Bhatti, J., Humphreys, T.E., 2017. Hostile control of ships via false GPS signals: demonstration and detection. Navigation 64 (1), 51–66.

Carbonell, I.M., 2016. The ethics of big data in big agriculture. Internet Policy Rev. 5 (1), 1–13. doi:10.14763/2016.1.405.

Census of Agriculture, 2017. Ag Data Transparent. www.agdatatransparent.com/ Accessed 24 Oct. 2024.

Check Point Software Technologies, 2023. Ransomware. Check Point. Retrieved October 21, 2024, from https://www.checkpoint.com/cyber-hub/threat-prevention/ransomware/

Comparitech, 2024. 30+ identity theft facts & statistics for 2024. Comparitech. Retrieved October 21, 2024, from https://www.comparitech.com/blog/vpn-privacy/identity-theft-statistics/

Goodfellow, I., Shlens, J., Szegedy, C., 2014. Explaining and harnessing adversarial examples. arXiv preprint arXiv:1412.6572.

Gyamfi, E.K., Kropczynski, J., Johnson, J.S., Yakubu, M.A., 2024. Internet of Things security and data privacy concerns in smart farming. In: 2024 IEEE World AI IoT Congress (AIIoT), Seattle, WA, USA, pp. 575–583. doi: 10.1109/AIIoT61789.2024.10578972.

HIPAA Journal, 2024. Federal data privacy legislation in 2024. HIPAA J. Retrieved October 21, 2024, from https://www.hipaajournal.com/federal-data-privacy-legislation-2024/.

Kumari, S., Saini, H., 2022. On-farm data security: practical recommendations for precision agriculture. Front. Sustain. Food Syst. 6, 884187.

Maryland General Assembly, 2024a. House Bill 1031: maryland Privacy Act. Retrieved from 2024 Regular Session—House Bill 1031 First Reader (maryland.gov). Legislation—HB0567.

Maryland General Assembly, 2024b. Senate Bill 541: Maryland Online Data Privacy Act. Retrieved from https://mgaleg.maryland.gov/mgawebsite/Legislation/Details/sb0541

Marylanders Online, 2024. Home. University of Maryland. Retrieved from https://marylandersonline.umd.edu/

McKinsey, 2024. Agriculture's Technology Future: How Connectivity Can Yield New Growth. Retrieved from. McKinsey & Company. https://www.mckinsey.com.

Méndez, D.A., Hernandez, P.J., Hernandez, C.A., 2019. Insider threats in smart farming: risks and mitigations. Comput. Secur. 87, 101563.

Norton LifeLock, 2024.Retrieved October 21, 2024, from. How Common Is Identity Theft in 2024? 24 Identity Theft Statistics. Norton LifeLock. https://us.norton.com/lifelock/identity-theft/2024-identity-theft-statistics.

Palli, A., Kumar, H., 2024, Integration of Smart Drones and AI into Agriculture—A Review. J. NACAA 17 (2). https://www.nacaa.com/journal/ebacfa77-5d94-44c8-ada3-8d891a3cef28.

Pew Research Center, 2023a.Retrieved from. Key Findings about Americans and Data Privacy. Pew Research Center. https://www.pewresearch.org/short-reads/2023/10/18/key-findings-about-americans-and-data-privacy/.

Pew Research Center, 2023b.Retrieved from. What the Data Says about Americans' Views of Artificial Intelligence. Pew Research Center. https://www.pewresearch.org/short-reads/2023/11/21/what-the-data-says-about-americans-views-of-artificial-intelligence/.

Pew Research Center, 2024.Retrieved from. Views of Data Privacy Risks, Personal Data, and Digital Privacy Laws in America. Pew Research Center. https://www.pewresearch.org.

Ransomware.org, 2024. State of Ransomware Report. Retrieved October 21, 2024, from https://ransomware.org/wp-content/uploads/2024/03/2024-State-of-Ransomware-Report_v1.pdf.

Search Logistics, 2024. Ransomware statistics. Search Logistics. Retrieved October 21, 2024, from https://www.searchlogistics.com/learn/statistics/ransomware-statistics/

Shapero, J., 2024. Bipartisan lawmakers renew push for data privacy bill amid concerns about TikTok, China, and U.S. internet users. USA Today. https://www.usatoday.com/story/news/politics/2024/04/08/bipartisan-data-privacy-internet-tiktok-congress/73243662007/

Sharma, P., Kumar, N., Park, J.H., 2020. Blockchain-based smart farming security: a survey. Sensors 20 (10), 2978.

Statista Research Department, 2023. Most targeted industries by ransomware attacks in the U.S. in 2022. Statista. Retrieved October 21, 2024, from https://www.statista.com/statistics/1323599/us-most-targeted-industries-by-ransomware-attacks/.

Varonis, 2024. Ransomware statistics. Varonis. Retrieved October 21, 2024, from https://www.varonis.com/blog/ransomware-statistics

WilmerHale, 2024. 2024 privacy law preview. WilmerHale. Retrieved October 21, 2024, from https://www.wilmerhale.com/en/insights/blogs/wilmerhale-privacy-and-cybersecurity-law/20240116-2024-privacy-law-preview

Wiseman, L., Linsner, T., Schuster, C., Monteiro, D., Caria, M., 2022. Protecting farmers' data privacy and confidentiality: recommendations and considerations. Front. Sustain. Food Syst. 6. Article 903230. doi:10.3389/fsufs.2022.903230.

Trustworthy artificial intelligence for digital agriculture

4

Prasad Deshpande, Wri Dutta, and Gaurav Jha
Department of Agronomy, Kansas State University, Manhattan, KS, United States

1 Introduction

Global investment in artificial intelligence (AI) is rapidly growing; spending reached $500 billion in 2024 (Manzoor, 2024) and is projected to exceed $632 billion by 2028 (Shirer, 2024), driven by advancements in AI applications, infrastructure, and services. Notably, agriculture has emerged as an essential sector within this expansion, reflecting increased recognition of the transformative potential of AI in addressing complex challenges for the global food-water-energy nexus. However, there is still skepticism regarding AI-driven decision-making. According to a survey (Megan, 2024), 90% of Americans have heard of AI, and 86% of respondents find AI text generators useful; only 48% of Americans support AI for household chores, and majorities disapprove of AI performing critical applications. Building confidence in AI tools will therefore require not only technical skill development but also user-centered design and participatory approaches that include farmers and stakeholders in the development and deployment of AI solutions.

Despite the increasing momentum behind AI-driven agricultural technologies, farmers and stakeholders often question the reliability of AI-driven predictions and automated decision-making, especially in high-stakes applications like crop management, yield forecasting, and resource optimization. This skepticism stems not only from concerns about prediction accuracy but also from a lack of transparency in how AI models arrive at their conclusions. Addressing these concerns about AI requires the development of interpretable, rigorous validation using real-world data and collaborative efforts between researchers, policymakers, and the farming community to build confidence in AI-powered agricultural solutions. There is also a concern about maintaining the privacy of users' data while using AI. This growing unease highlights the need not only for AI models with high-performing algorithms but also for *trustworthy AI*. Trustworthy AI in agriculture means building AI systems that farmers can understand and use with confidence and trust, even when things do not go as planned (like a sudden drought, equipment failure, an unexpected pest outbreak, etc.). *We define trustworthy AI in the agronomic context through three pillars: (i) lawful, as AI must follow data privacy laws and respect farmer data ownership; (ii) ethical, as AI tools should support fair access and avoid biased recommendations; and (iii) technically robust, as AI must hold up in the real world.*

Precision Technologies for Digital Agriculture. https://doi.org/10.1016/B978-0-443-40513-6.00020-2

This chapter introduces advancements in AI, focusing on trustworthiness, interpretability, and reliability in AI-driven applications by comparing them with conventional crop models and as an example of wheat yield prediciton. It also explores recent advancements in Geospatial AI (GeoAI). Lastly, the chapter synthesizes survey results to capture farmers' and stakeholders' perspectives on the adoption and limitations of trustworthy AI in agriculture. Rather than emphasizing mathematical formulations, the discussion centers on conceptual understanding and practical relevance. For a better understanding of these concepts, a popular and baffling problem in agronomy, i.e., crop yield estimation with a focus on the wheat crop, is used as a unifying example throughout different sections. The novelty of this chapter lies in consolidating advances in trustworthy AI– machine learning (ML) methods and linking them to practical agricultural challenges, with direct beneficiaries including farmers, policymakers, and researchers seeking reliable AI-based solutions.

2 Process-based models in agriculture

Process-based models in agriculture simulate the underlying biophysical, chemical, and environmental processes to understand and predict crop growth, soil dynamics, and resource interactions with high interpretability. Conventionally, wheat yield estimation is carried out using agricultural simulation systems known as biophysical process-based models. The details of a few process-based models are provided in Table 4.1.

The inputs for APSIM (Brown et al., 2014) can be categorized into genetic, environmental, management, and stochastic parameters. Genetic parameters include crop-specific traits such as wheat variety, phenology, biomass accumulation, leaf area index, and root depth, which determine the crop growth potential. Nongenetic parameters include environmental factors like climate data (daily temperature, rainfall, solar radiation, wind speed, and carbon dioxide levels) and soil properties (texture, bulk density, moisture retention, nutrient levels, and organic matter). Management parameters include farmer-driven practices such as sowing date, seeding rate, irrigation schedules, fertilization, tillage, and pest control. Finally, stochastic parameters account for uncertainties in the system, including variability in weather conditions, soil properties, and pest outbreaks, which affect yield outcomes. In addition, validation data from historical yields, field experiments, and remote sensing may help improve model accuracy and reliability. Other models also require similar input features.

Process-based models like APSIM are valued because of their interpretability. For example, APSIM generates detailed time-series data output on crop growth, soil moisture, nutrient levels, and yield, enabling users to trace how different variables evolve over time. The model also allows users to modify and inspect input parameters, such as genetic traits, soil properties, and management practices, making it possible to understand the *cause-and-effect* relationships within simulations. However, they are primarily deterministic (meaning they rely on known equations) and do not need to learn patterns from these processes based on AI training data. Thus, such models usually need less data to model the process, unlike AI models. The challenges in

Table 4.1 Summary of various process-based models.

Model	Biophysical inputs	Chemical inputs	Environmental inputs	Calibration needs	Output types
APSIM (Brown et al., 2014)	Plant biomass, canopy cover, root growth	N and C cycling, pesticide fate	Weather, soil hydrology, irrigation, tillage	High (modular, needs local calibration)	Yield, N leaching, soil C, water use, economic metrics
DSSAT (Jones et al., 2003)	Crop phenology, growth stages, LAI, root depth	Fertilizer type & rate, residue decomposition	Daily weather, soil profile, CO_2	Moderate to High (crop parameters, soil data)	Yield, biomass, soil moisture, N balance, phenology dates
AquaCrop (Steduto et al., 2009)	Canopy cover, harvest index, root depth	Limited (mostly fertigation)	Daily weather, irrigation regime, water table	Low to Moderate (simplified parameters)	Yield, water productivity, ET, soil water
EPIC (Williams et al., 1989)	Plant development, biomass, LAI	NPK application, soil fertility, manure	Long-term climate data, erosion factors	High (needs local agronomic & soil data)	Yield, erosion, runoff, nutrient loss, GHG
CERES (Ritchie et al., 1998)	Growth stages, leaf area, tillering	N fertilizer use, residue N	Daily weather, soil water, and temperature	Moderate (especially cultivar-specific)	Yield, phenology, soil water, & N dynamics
SWAT (Chen et al., 2021)	Land use, soil type, plant cover	Fertilizer, pesticide, manure	Climate, digital elevation model, stream flow	High (hydrology & management practices)	Streamflow, sediment yield, nutrient loading
CropSyst (Stöckle et al., 2003)	Canopy growth, phenology, tillage response	Fertilizer, pesticide, residue decomposition	Soil water, temperature, weather	Moderate to High (requires iterative tuning)	Yield, water use, N balance, soil erosion
WOFOST (van Ittersum et al., 2003)	Dry matter accumulation, leaf area index, root zone	None or very limited	Daily weather, soil physical properties	Moderate (simplified crop and soil modules)	Yield, biomass, water stress indicators
MONICA (Nendel et al., 2011)	Photosynthesis, biomass, crop-soil feedback	Full N cycle, GHG emissions	Weather, soil carbon, water balance	High (complex interactions, high-res inputs)	Yield, GHG emissions, nutrient fluxes, soil C/N

Note: The full forms of abbreviations are as follows, *N*, Nitrogen; *C*, Carbon; *LAI*, leaf area index; *ET*, Evapotranspiration; *GHG*, greenhouse gas; *NPK*, Nitrogen, Phosphorus, and Potassium; CO_2, carbon dioxide.

implementing process-based models include the unavailability of specific parameters and limitations in calibration processes. Moreover, the oversimplification of processes based on certain assumptions sometimes leads to limited accuracy of such methods. For example, many process-based models assume a uniform leaf distribution across the canopy, which can lead to errors in simulating light interception and photosynthesis, especially in row crops like maize or sunflowers with open canopies. In reality, canopy architecture varies with growth stage, row spacing, management strategies, soil fertility, and plant stress. Lastly, process-based models can only be developed if a physical and biological phenomenon exists and is already understood. Despite several limitations of the process-based models, the interpretability and mechanistic transparency of these models make them an important decision-support tool.

To guide the development of trustworthy AI for agricultural applications, we propose a core model characteristic that can promote interpretation, scientific validation, uncertainty awareness, data valuation, and privacy. Fig. 4.1 outlines a conceptual

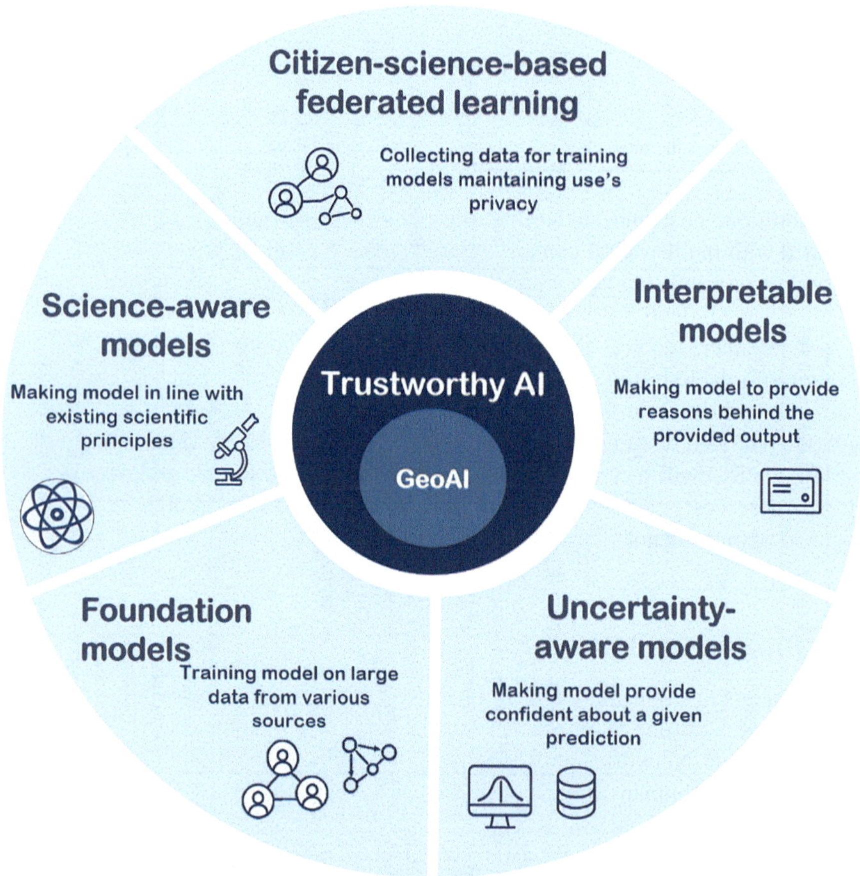

Fig. 4.1 Core components of Trustworthy AI framework for digital agriculture, unified through GeoAI.

framework integrating five major components of trustworthy AI, namely, science-aware models, interpretable models, uncertainty-aware models, FMs, and citizen science-based federated learning. These components form the foundation of a trustworthy AI system that has a geospatial reference to enable location-specific digital agriculture or GeoAI. The goal of GeoAI integration is to improve actionable insights for informed decision support.

3 Interpretable artificial intelligence

Interpretable AI focuses on making model predictions understandable to humans by highlighting how input features influence outcomes. This transparency is essential in agriculture, where trust and actionable insights are crucial for decision-making by farmers and stakeholders. Conventional AI models learn patterns and associations directly from historical data, making them highly adaptable to complex, nonlinear systems. Random forests, support vector machines, convolutional neural networks, and recurrent neural networks are examples of conventional AI models. For example, if an artificial neural network (ANN) is developed for forecasting wheat yield using weather and nutrient data as input features, the weights of the neural network neither have a direct physical meaning nor is there a one-to-one correspondence with input features. As a result, it becomes difficult to explain how specific weather or nutrient variables influenced the predicted wheat yield, which limits the model's transparency and trustworthiness for decision-making. Sometimes, conventional AI methods are prone to generating outputs that are physically not viable or illogical. For example, if a model is developed for forecasting crop yield and forecasts the yield as a negative value, domain experts can easily understand that it is physically impossible, as yield cannot be negative. In addition, the AI model may continue to predict increasing yield with rising nitrogen input, even beyond toxic levels, simply because it lacks knowledge of diminishing returns or nutrient toxicity. On the other hand, conventional AI models may not be aware of this fact. In reality, such models are unaware of the scientific phenomena being modeled since they only map input-output correlations based on the data they have observed as inputs.

3.1 Science-aware AI

Science-Aware Machine Learning refers to the integration of an AI model in line with the underlying scientific phenomena. Efforts are made to integrate scientific knowledge into AI models in different ways, from simple methods, such as changing the last layer of the neural network, to more complex approaches, such as enforcing the laws of conservation of energy, mass, or momentum. For example, the wheat yield (which can range from 1000 to 10,000 kg/ha; Asseng et al., 2020) cannot be negative; that condition can be enforced by modifying the last layer of a neural network.

Another way to make science-aware neural networks is by modifying the loss function (Ebert-Uphoff et al., 2021) to include physical or agronomic constraints. This leads to better model performance and sometimes interpretability as well. For example, if a neural network is used to predict wheat yield using weather and soil data, then with a regular loss function like mean squared error, it might sometimes give negative yield values, which are not possible in real life. To fix this, we can use a loss function that adds a penalty for negative outputs. One way is to use a custom version of mean squared error that increases the error when the predicted yield is less than zero. This teaches the model to avoid unrealistic results. Therefore, sometimes the modification of the loss function is needed for better performance. Relating this to a process-based simulation by the APSIM approach, which predicts wheat yield under drought conditions, the actual yield may be very low (e.g., 500 kg/ha). In this case, a small error (e.g., 100 kg/ha) could lead to a high percentage error (20%). This percentage error is crucial because, in low-yield conditions, even slight deviations significantly impact farmers' decision-making, food security, and economic outcomes. On the other hand, in high-yield conditions (e.g., 6000 kg/ha), the same error of 100 kg/ha would result in only a 1.67% error, which is relatively insignificant. To incorporate this issue, the loss function of the neural network can be modified as a custom weighted loss function. Another example is if a model is developed to predict soil pH. In the first case, the predicted pH is 8, whereas the actual pH is 9; in the other case, the predicted pH is 11, but the actual pH is 12. In both cases, the root mean squared error will be 1; however, due to the nonlinear nature of pH (neutral pH is 7, with minimum pH values as 1 and maximum as 14), the error in the second case is more severe than in the first case. In such a case, the loss function can be modified to accommodate this nonlinear nature by penalizing the extreme values more compared to values near pH 7. In that way, the conventional neural networks can be made science-aware just by modifying the loss functions. A comprehensive description of custom loss functions for neural networks in environmental science is provided by Cooperative Institute for Research in the Atmosphere, i.e., CIRA (Ebert-Uphoff et al., 2021).

The AI models can also assist in the calibration of process-based models. AI can be used prior to the process-based models (for example, clustering the inputs similar to input variables and using the process-based models for the clusters to avoid saving computational costs). AI can also be used later for postprocessing the output of process-based models; e.g., if the study aims to identify the effect of weather patterns on crop prices, the historical weather data can be fed to crop models (e.g., APSIM) to get the estimated crop yield. Furthermore, an AI model can be developed that estimates the crop prices using the estimated crop yield, as there is no physical relationship between crop yield and crop prices. This combined approach, where process-based models provide scientific understanding and AI models capture patterns, leads to more accurate and practical solutions for agricultural decisions based on farm data layers.

3.2 Probabilistic modeling in dynamic agricultural systems

Probabilistic AI models estimate not just a single outcome but a distribution of possible outcomes, reflecting real-world uncertainties. Unlike deterministic models, they account for variability in inputs and conditions, making them well-suited for dynamic agricultural systems. For example, soil moisture shows spatial variability depending on soil texture and topography. Moreover, there are temporal (and seasonal) variations due to irrigation and precipitation events. Typically, process-based models rely on averaging input conditions (e.g., temperature, rainfall, or soil properties) to generate outputs. This averaging can oversimplify the real-world variability and become a major source of deviation between the model outputs and actual field values. Moreover, the predictions from these models are deterministic, meaning that they always produce the same output for the same set of inputs. For instance, in the APSIM model, if we input a fixed set of parameters for wheat (such as sowing date, soil type, and weather conditions), the model will always predict the same yield, regardless of real-world uncertainties or variability that might affect yield in practice. This deterministic nature limits the ability of such models to capture the full range of outcomes observed in the field.

Significant research has focused on assessing the accuracy of yield prediction models, but a thorough investigation into the sources of uncertainty in these predictions is still lacking (Hao et al., 2021). Typically, uncertainties in model predictions arise from limitations in the model structure (both process-based and conventional AI), inaccuracies in input data, parameter estimation, and observational data used during calibration and validation processes (Vrugt et al., 2008). Instead of predicting a deterministic value, if the model estimates in a probabilistic way, the estimates become more informative. The probabilistic outputs may be just the probability of yield (e.g., the probability of wheat yield being 5000 kg/ha is 0.75). Sometimes, its associated confidence interval is calculated (e.g., a 95% confidence interval for wheat yield prediction is 4500–5500 kg/ha, i.e., there is a 95% chance the actual yield falls within this range). Such estimates provide an idea of the possible deviations of the wheat yield (model outputs) for situations where the input features are somewhat offset from their actual values.

3.3 Uncertainty-aware AI models

Another problem with conventional AI methods is their inability to deal with out-of -distribution samples. When an AI model trained on wheat yield data from the central parts of the United States of America (USA) is applied to a region like southern India, where climate, soil, and management practices differ significantly, it encounters out-of-distribution inputs. As a result, the model may produce inaccurate or biased predictions, despite appearing confident. This happens because the model has never seen such conditions during training. Thus, the model is not capable of understanding that the input data is erroneous or belongs to some unforeseen category. This is an example of out-of-distribution data. Addressing this issue is essential to ensure reliable predictions in diverse agricultural settings. Uncertainty-aware models aim to handle the limitations of input data (e.g., missing data and variability) and improve the model's generalization and risk-aware decision-making. These models are based on Bayesian methods, in which random values are generated based on some probability distribution that is learned during model training, e.g.,

Bayesian neural networks and Gaussian processes. Sometimes, ensemble methods are also used to estimate uncertainty (Bishop, 2016). However, due to these complex mathematical formulations, such models require larger computational resources compared to equivalent deterministic methods. Lastly, these models are capable of explaining when and why the model is not performing well (Deshpande et al., 2024).

3.4 Default-interpretable models in AI

As discussed, a primary concern while using conventional AI methods is the lack of interpretability. In other words, the model does not explain why, for the given set of inputs, it obtains a particular output. However, this concern is only valid for some AI models, such as neural networks. On the other hand, a few popular AI models are interpretable by default, e.g., decision trees and random forests. A decision tree mimics human decision-making with clear, rule-based logic. Each decision follows a sequence of if-then conditions, making it easy to trace how a prediction is made. For example, in a wheat yield prediction model, key factors such as rainfall, soil nutrients, and sowing date influence the outcome, representing climate, soil variability, and management practices, respectively. A simplified decision tree may first check if rainfall exceeds 500 mm; if yes, the next condition evaluates whether nitrogen application is greater than 100 kg/ha. If both conditions are met, the model predicts a high yield; otherwise, it predicts a medium yield. On the other hand, if rainfall is below 500 mm, the model directly classifies the yield as low (Fig. 4.2). A random forest builds many such decision trees using random subsets of data and features. Each tree makes its own prediction (e.g., high, medium, or low yield), and the final yield is based on the majority vote across trees. This reduces the chance of bias from any single tree and improves prediction reliability. This hierarchical structure allows users, such as farmers and agronomists, to understand how different factors contribute to yield predictions, making the model interpretable. Unfortunately, the interpretable models discussed above perform well for simple data without much nonlinearity and cannot always be used for modeling complex and dynamic phenomena (e.g., crop-climate interaction, weather forecasting).

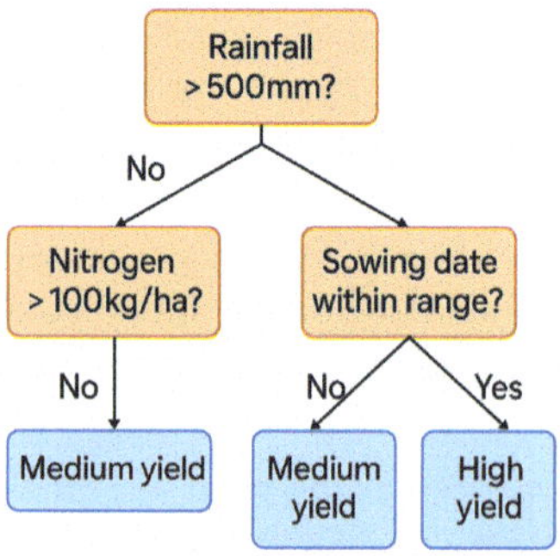

Fig. 4.2 Decision tree for wheat yield predictions.

3.5 Model-agnostic interpretability tools

For AI models that are not interpretable by default, some models are developed and built on top of the corresponding AI models to add a certain extent of interpretability. These models analyze the predictions of the original models for various simulated inputs to provide insights into how the predictions are made. Popular examples include Local Interpretable Model-agnostic Explanations (LIME) and SHapley Additive exPlanations (SHAP). LIME explains model predictions by approximating the complex model locally with an interpretable one. It perturbs the input data slightly and observes the changes in predictions to determine which features most influence the output for that specific instance. This helps build trust in individual predictions, especially in black-box models (Ribeiro et al., 2016).

The SHAP algorithm works by computing the contribution of each input feature to a prediction based on SHAP values from game theory (Lundberg and Lee, 2017). It fairly distributes the difference between the actual prediction and the average prediction among all features, showing how each one influences the model's output. One such example of wheat yield prediction (Srivastava et al., 2022) using SHAP presents the explainability analysis of the wheat yield prediction (Table 4.1). The model uses features like drained upper limit (DUL), windspeed, temperature, and wind speed to predict yield. If the model predicts a high yield for a specific case, SHAP can help answer *why* by showing how much each feature "pushed" the prediction up or down compared to the average prediction. The feature importance table (Table 4.1) highlights the most influential variables: DUL, wind speed, radiation, temperature, and other environmental factors. The plot shows that higher values of DUL correspond to higher yield values. This inference also aligns with the existing body of knowledge (Peichl et al., 2018). Moreover, the positive/negative effect of various input features also helps us achieve a better understanding of the process (Table 4.2).

4 Geospatial artificial intelligence

GeoAI is the integration of AI and machine learning techniques with geospatial data to analyze, model, and interpret spatial patterns and relationships (Song et al., 2023). Though the term GeoAI has recently evolved around 2017 (Li, 2024), AI has been used for geospatial datasets since the 1990s (Openshaw and Openshaw, 1997). GeoAI relies on various data sources, including georeferenced farm data, unoccupied aerial vehicles, and satellite imagery from Landsat, Sentinel, and Maxar; as well as weather and climate data from sources like National Oceanic and Atmospheric Administration (NOAA), European Centre for Medium-Range Weather Forecasts (ECMWF), and National Aeronautics and Space Administration (NASA). In addition, crowdsourced data, such as OpenStreetMap, Twitter, and Internet of Things (IoT) sensor networks, are becoming increasingly novel nowadays. The methods used in GeoAI vary from conventional geostatistical methods (e.g., kriging, inverse distance weighting) to advanced AI models (e.g., U-Net, GeoTransformer).

Table 4.2 Explainable AI for wheat yield prediction using the SHAP method (Modified from Srivastava, A.K., Safaei, N., Khaki, S., Lopez, G., Zeng, W., Ewert, F., Gaiser, T., Rahimi, J., 2022. Winter wheat yield prediction using convolutional neural networks from environmental and phenological data. Sci. Rep. 12, 3215. https://doi.org/10.1038/s41598-022-06249-w.). (A) Feature importance plot showing the impact of different variables on wheat yield predictions, where red indicates a positive impact and blue indicates a negative impact.

Rank	Variable	SHAP value	Impact type
1	DUL	0.155	Positive
2	Windspeed_10	0.085	Negative
3	Radiation_7	0.080	Negative
4	SAT	0.075	Negative
5	TempMax_8	0.070	Positive
6	Windspeed_17	0.068	Negative
7	HarvestDOY	0.065	Positive
8	Windspeed_11	0.062	Negative
9	BD	0.060	Positive
10	Precipitation_49	0.058	Negative
11	TempMin_10	0.055	Positive
12	Radiation_26	0.052	Positive
13	TempMax_9	0.050	Positive
14	TempMin_24	0.048	Negative
15	TempMin_8	0.045	Negative

Though the GeoAI concept is new in the agriculture sector, there are a few studies carried out. For instance, GeoAI is used in crop yield estimation and soil property prediction, and its effectiveness has been improved by fusing satellite imagery with AI models (Hosseini et al., 2023). In addition, it is employed to evaluate the environmental impact of agricultural land use (Gianquintieri et al., 2024). Moreover, large-scale image analysis and machine vision techniques are enhancing crop monitoring, disease detection, and sustainable farming practices (Li and Hsu, 2022). Lastly, the development of specialized GeoAI toolkits, such as Hyperfidelis (Sagan et al., 2024) supports precision agriculture by optimizing various farming operations.

4.1 Foundation models and their applications in agriculture

Over the past five years, GeoAI research has made significant progress. However, there are still notable gaps that require attention, including challenges in model explainability and difficulties in generalizing models to new geographic areas and data sources (Liu et al., 2023). The earlier-discussed trustworthy-AI methods provide a promising avenue for improving model explainability. Another emerging approach to enhance the generalizability of models is the use of FMs. FMs are large-scale AI models trained on vast amounts of diverse data, allowing them to serve as a general-purpose base for various downstream tasks. These models, such as GPT, BERT, and CLIP, leverage self-supervised learning and can be fine-tuned for specific applications, making them highly adaptable across different domains. Unlike conventional AI models, which require large task-specific labeled datasets and extensive expertise for development, FMs reduce the need for costly data labeling and can be applied across multiple agricultural AI tasks with minimal fine-tuning (Li et al., 2024).

The FMs for agriculture are still nascent (Guo et al., 2024; Li et al., 2024); however, the following are some examples. Large FMs like ChatGPT and GPT have been applied in agricultural robotics for robotic picking (Stella et al., 2023), providing AI-driven agricultural recommendations (Tzachor et al., 2023), and organizing unstructured agricultural metadata (Lu et al., 2023). Vision FMs, such as SAM, have been used for automatic chicken segmentation and bird tracking (Yang et al., 2024) as well as automatic leaf segmentation (Williams et al., 2024). Multimodal FMs, combining image-text data, have been utilized for cucumber disease recognition (Cao et al., 2023) and tomato picking in agricultural robotics (Stella et al., 2023). The earlier discussed models do not necessarily assume that the input features correspond to a geospatial dataset; i.e., there is a future scope for GeoAI in agriculture.

4.2 Federated learning and its applications in agriculture

FMs require vast amounts of datasets to achieve good performance. One of the challenges in developing FMs for agriculture is the complexity of its data. In agronomy, the 4 Vs-volume, variety, velocity, and veracity-highlight the challenges of managing large, diverse, rapidly growing, and sometimes uncertain data from sources like sensors, satellites, and field observations. Unlike text-based AI models such as ChatGPT, agricultural AI must process diverse geospatial multimodal data, including point data (e.g., plant height, temperature measured at a weather station), gridded datasets (e.g., yield maps, soil maps), and complex multidimensional data (e.g., hyperspectral imagery, soil health data in gridded format). This variability makes it difficult to develop highly accurate AI models that generalize well across different regions and farming conditions.

The datasets required for accurate AI models in agriculture can be either publicly available or privately sourced. Public datasets include satellite imagery, gridded weather data, and soil maps, which are generally available at a coarser scale (and thus may be less accurate compared to privately sourced data). On the other hand, private datasets-such as yield maps, soil test results, field-level management practices, and economic inputs-are often collected by farmers or agribusinesses. However, farmers may hesitate to share such private data due to concerns over privacy, data misuse, or potential economic disadvantages. In many cases, there is a lack of clarity around data ownership, rights of usage, and the economic value of agricultural data. This uncertainty can lead to distrust in data-sharing practices and hinder collaborative efforts to build robust, context-aware AI models for agriculture. Addressing data governance, privacy protections, and value-sharing mechanisms is essential to encourage participation and ensure the ethical use of farm-level data.

A promising solution to the issue of the challenge of maintaining AI is a citizen-science-based federated learning architecture, which could enhance AI adoption while preserving data privacy. In this approach, farmers contribute data locally, benefiting from AI-driven insights without compromising their sensitive information. A localized AI model processes farm-specific inputs while integrating knowledge from a global AI model, which is trained on aggregated data from multiple localized models. Crucially, the confidential farm data remains with the local models and is never directly shared with the global system, ensuring privacy while still improving AI performance.

This citizen-science-based federated learning architecture has broad applications in agriculture, from soil moisture estimation to yield prediction, enabling more reliable AI-driven decision-making. By addressing both data privacy and accuracy concerns, federated learning offers a practical pathway for increasing AI adoption in agriculture, ultimately improving productivity while maintaining farmers' trust.

5 Case study: How trustworthy is AI? Navigating challenges via regulations

As part of this exploratory work, we collected voluntary, anonymous input from a small group of Kansas-based agricultural professionals. Seventeen individuals from Kansas, primarily experienced crop producers and professionals involved in federally funded agriculture programs, provided voluntary, anonymous input through an informal reflection process. These respondents were mostly involved in crop farming for over 20 years, on farms spanning over 500 acres, and are a major part of the population who, in their near future, would implement and integrate AI as part of smart-farming techniques. Their viewpoints make up a useful corpus for comprehending the principles and methods common to AI development.

The responses can be broadly classified to serve three critical purposes. Firstly, the responders from different spectrums of agriculture, the government sector, and education reduce the bias in the AI assessment, offering an understanding of problems, beliefs, presumptions, and worries. Secondly, these responses offer viewpoints and insights on the quality, effect, unintended consequences, and ethical standards of this quickly evolving technology. The opinions of these participants may be used as markers of the kinds of justifications, factors, presumptions, procedures, difficulties, or "institutional logics" that shape this study, as well as the institutional cultures and values that are being literally encoded into AI models (Botha et al., 2017; Bronson, 2019). Thirdly, the responses to the survey also provide a starting point for upcoming comparative and follow-up research, which will make it easier to monitor how training and soft law governance mechanisms affect institutional culture, research methods, and values. These responses are only a first step toward a practice of ongoing "trustworthiness" in the research and development of AI technologies. To promote sincere and meaningful contemplation rather than biased answers, participation in the survey was completely optional and kept entirely private, and the form was designed to safeguard the identities of the participants. One limitation of this survey is the small sample size in comparison to the size of the population in Kansas. Along with that, the different domains between the authors and the survey responders may also provide some distance or gap in understanding. This is a pilot or a small proof-of-concept study that may be carried out more extensively, including partners and stakeholders, especially those who would eventually make use of new technologies (Silva et al., 2019). The pilot study helped, firstly, to enable the gathering of values, viewpoints, and ideas from a group of people coming from the domain of agriculture, at the forefront of AI technology and research. On the other hand, it would also help to provide the participants with a valuable chance to think openly and honestly about issues related to the goals, difficulties, and ethics of integrating AI into their area of expertise.

6 Issues related to data sharing and present regulations

AI plays a crucial role in the transition to Farming 4.0 (farming integrated with technological advances involving robotics, the IoT, and AI) (Mallinger and Baeza-Yates, 2024). The continuously evolving responsibilities, abilities, and requirements of farmers in this setting necessitate the development of frameworks and design criteria to comprehend the myriad implications of AI and to direct its appropriate application in order to preserve autonomy and minimize any potential negative impacts. There is a widespread belief that, although monitoring the advancement of AI technology is necessary, skepticism about the morality and safety of commercial technology prevails. Farmers are aware of the use of AI applications in today's agriculture, as they have actively depended on or used AI-based text generators (ChatGPT, Microsoft Copilot, etc.) to seek answers in their lives. They believe that AI can help reduce labor costs and improve crop yield prediction with better pest and disease control.

6.1 Incentivizing data sharing and ethical problems

Data is integral to AI progress, and the increasing demand for data has led to incentivization for secure acquisition (Goodhue et al., 1988; White et al., 2021). Ambiguity in the domains of private security, prediction accuracy, cost, and lack of understanding has deterred 60% of the surveyed population from integrating AI techniques into their farming operations. According to key findings, informal negotiations and trust between collaborators, partners, and participants enhance AI technology research more effectively than government regulation or control. Around 80% of the respondents agreed to contribute data for AI-model formulations only if they receive some benefits in return or if their data remains anonymous. The "power cube" concept of one-sided data dependency has led to bargaining tendencies (Cullen et al., 2014; Turner et al., 2020). There are several challenges in negotiating partnerships, with increased difficulty in securing data amidst growing concerns about ethical standards such as data privacy and transparency (Alexander et al., 2024). Secondly, commercialization has advanced more rapidly than academic research, becoming the primary driver of development in AI technologies (Birner et al., 2021).

New smart farm technologies, which often rely on data for testing and implementation, are becoming more and more popular due to a combination of regulatory, environmental, and economic concerns. This is how regulatory constraints encourage farmers and other agricultural participants to disclose their data in an indirect manner.

6.2 Regulations and their challenges

Survey results clearly indicate that the development of AI in smart precision agriculture suffers from the presence of minimal or skewed regulations. As the participants answered additional survey questions on difficulties and applications in daily life, it became evident that assuring the ethical study of new technologies is far more difficult to ensure if it involves no human presence, their success stories, or is not backed by government regulations. The absence of dedicated government intervention will only lead to more benefits for private organizations than public organizations, which is quite unrealistic.

As much as regulations and policy formulations are necessary for AI implementation, challenges in implementation have restricted research to a point of standstill. The Nagoya Protocol, a treaty for Access and Benefit Sharing endorsed in 2014 by United Nations and European Union (EU) members, based on prior informed consent between the involved communities, demonstrates a discrepancy between the organizations that create regulations and the people in charge of carrying them out and following them. It has comprehensive consequences for researchers in the United States (USA), even though it is an outside country (McCluskey et al., 2017). The Montréal Declaration for a Responsible Development of Artificial Intelligence (2018), the Digital Charter Impact Act, and the Artificial Intelligence and Data Act are some bills passed by the Canadian Government to regulate AI outside privacy legislation (Fjeld et al., 2020; Parliament of Canada, 2022). There is no international agreement governing AI systems and services outside of the EU. Nowadays, growers and producers in California are being encouraged by state legislation to use AI solutions in order to save expenses and keep their operations compliant.

6.3 Bridge of mutual trust rather than replacement

The negative past experiences of profit-seekers from improper data calibration and clarification emphasize the decision of 37% of surveyed individuals to seek more success stories before they are willing to implement and integrate AI. As Godfray et al. mentioned, sustainability, along with incentivizing extensive data access while incorporating competition and innovation in developing technologies, is the major governance challenge (Godfray et al., 2010). Small-scale initial profits, combined with perseverance, trust, and transparency from both ends, can lay the foundation for AI integration in smart farming (Greve and Ysa, 2023; Mintz and Du

Bois, 2002). However, the lack of knowledge and trust, which constituted 63% of the surveyed population preventing them from using AI, can only be addressed through rigorous and consistent testing, as economically waiting for long-term success restricts the innovation process (Akullo et al., 2018; Cullen et al., 2014). Detailed, interpretable, and implementable explanations of AI decisions provide greater understandability of AI algorithms (Dara et al., 2022; Doran et al., 2017; Slack et al., 2019).

7 Conclusion

Huge amounts of funding are spent globally on AI research. However, skepticism persists about using AI in critical applications, e.g., healthcare. Agriculture is no exception to this skepticism. The black-box nature of conventional AI models and privacy concerns are the primary reasons behind this hesitation. With examples from the agriculture domain, this chapter discusses advanced AI techniques that enhance the interpretability and trustworthiness of AI models. Moreover, GeoAI applications in the agriculture domain are outlined in this chapter. Lastly, future avenues for trustworthy AI are discussed.

Moving forward, addressing the challenges of trustworthiness requires multiple strategies: integrating probabilistic approaches to better represent uncertainty, leveraging interpretable AI models that align with domain knowledge, and adopting federated learning frameworks that safeguard farmers' privacy while enabling collective intelligence. FMs, though still nascent in agriculture, offer significant potential to reduce the cost of task-specific model development and expand the scope of AI applications, provided they are adapted carefully to agricultural contexts.

Ultimately, building trustworthy AI in agriculture requires a balance between technical robustness and social acceptance. By ensuring that AI systems are lawful, ethical, and transparent, and by involving farmers and stakeholders, we can foster greater confidence in AI solutions. Such approaches will not only enhance adoption but also unlock the transformative potential of AI to support sustainable food, water, and energy systems amid growing global challenges.

References

Akullo, D., Maat, H., Wals, A.E.J., 2018. An institutional diagnostics of agricultural innovation; public-private partnerships and smallholder production in Uganda. NJAS Wageningen J. Life Sci. 84, 6–12. doi:10.1016/j.njas.2017.10.006.

Alexander, C.S., Yarborough, M., Smith, A., 2024. Who is responsible for 'responsible AI'?: navigating challenges to build trust in AI agriculture and food system technology. Precis. Agric. 25, 146–185. doi:10.1007/s11119-023-10063-3.

Asseng, S., Guarin, J.R., Raman, M., Monje, O., Kiss, G., Despommier, D.D., Meggers, F.M., Gauthier, P.P.G., 2020. Wheat yield potential in controlled-environment vertical farms. Proc. Natl. Acad. Sci. 117, 19131–19135. doi:10.1073/pnas.2002655117

Birner, R., Daum, T., Pray, C., 2021. Who drives the digital revolution in agriculture? A review of supply-side trends, players and challenges. Appl. Econ. Perspect. Policy 43, 1260–1285. doi:10.1002/aepp.13145.

Bishop, C.M., 2016. Pattern Recognition and Machine Learning, Information Science and Statistics. Springer, New York, NY.

Botha, N., Turner, J.A., Fielke, S., Klerkx, L., 2017. Using a co-innovation approach to support innovation and learning: cross-cutting observations from different settings and emergent issues. Outlook Agric. 46, 87–91. doi:10.1177/0030727017707403.

Bronson, K., 2019. Looking through a responsible innovation lens at uneven engagements with digital farming. NJAS Wageningen J. Life Sci. 90–91, 1–6. doi:10.1016/j.njas.2019.03.001.

Brown, H.E., Huth, N.I., Holzworth, D.P., Teixeira, E.I., Zyskowski, R.F., Hargreaves, J.N.G., Moot, D.J., 2014. Plant modelling framework: software for building and running crop models on the APSIM platform. Environ. Model. Softw. 62, 385–398. doi:10.1016/j.envsoft.2014.09.005.

Cao, Y., Chen, L., Yuan, Y., Sun, G., 2023. Cucumber disease recognition with small samples using image-text-label-based multi-modal language model. Comput. Electron. Agric. 211, 107993. doi:10.1016/j.compag.2023.107993.

Chen, Y., Marek, G.W., Marek, T.H., Porter, D.O., Brauer, D.K., Srinivasan, R., 2021. Simulating the effects of agricultural production practices on water conservation and crop yields using an improved SWAT model in the Texas High Plains, USA. Agric. Water Manag. 244, 106574. doi:10.1016/j.agwat.2020.106574.

Cullen, B., Tucker, J., Snyder, K., Lema, Z., Duncan, A., 2014. An analysis of power dynamics within innovation platforms for natural resource management. Innov. Dev. 4, 259–275. doi:10.1080/2157930X.2014.921274.

Dara, R., Hazrati Fard, S.M., Kaur, J., 2022. Recommendations for ethical and responsible use of artificial intelligence in digital agriculture. Front. Artif. Intell. 5, 884192. doi:10.3389/frai.2022.884192.

Deshpande, P., Tripathi, S., Bhattacharya, A., 2024. Bayesian neural networks for satellite fog detection: quantifying epistemic and aleatoric uncertainties. Remote Sens. Earth Syst. Sci. doi:10.1007/s41976-024-00155-7.

Doran, D., Schulz, S., Besold, T.R., 2017. What Does Explainable AI Really Mean? A New Conceptualization of Perspectives. In: Proceedings of the First International Workshop on Comprehensibility and Explanation in AI and ML (CEX 2017). vol. 2071. https://ceur-ws.org/Vol-2071/CExAIIA_2017_paper_2.pdf.

Ebert-Uphoff, I., Lagerquist, R., Hilburn, K., Lee, Y., Haynes, K., Stock, J., Kumler, C., Stewart, J.Q., 2021. CIRA Guide to Custom Loss Functions for Neural Networks in Environmental Sciences—version 1. doi:10.48550/arXiv.2106.09757

Fjeld, J., Achten, N., Hilligoss, H., Nagy, A., Srikumar, M., 2020. Principled artificial intelligence: mapping consensus in ethical and rights-based approaches to principles for AI. SSRN Electron. J. doi:10.2139/ssrn.3518482.

Gianquintieri, L., Oxoli, D., Caiani, E.G., Brovelli, M.A., 2024. Implementation of a GEOAI model to assess the impact of agricultural land on the spatial distribution of PM2.5 concentration. Chemosphere 352, 141438. doi:10.1016/j.chemosphere.2024.141438.

Godfray, H.C.J., Beddington, J.R., Crute, I.R., Haddad, L., Lawrence, D., Muir, J.F., Pretty, J., Robinson, S., Thomas, S.M., Toulmin, C., 2010. Food security: the challenge of feeding 9 billion people. Science 327, 812–818. doi:10.1126/science.1185383.

Goodhue, D.L., Quillard, J.A., Rockart, J.F., 1988. Managing the data resource: a contingency perspective. MIS Q. 12, 373. doi:10.2307/249204.
Greve, C., Ysa, T., 2023. Introduction to the handbook on strategic public management. In: Greve, C., Ysa, T. (Eds.), Handbook on Strategic Public Management. Edward Elgar Publishing, pp. 2–9. doi:10.4337/9781789907193.00008.
Guo, W., Yang, Y., Wu, H., Zhu, H., Miao, Y., Gu, J., 2024. Big models in agriculture: key technologies, application and future directions. Smart Agric. 6, 1–13. doi:10.12133/j.smartag.SA202403015.
Hao, S., Ryu, D., Western, A., Perry, E., Bogena, H., Franssen, H.J.H., 2021. Performance of a wheat yield prediction model and factors influencing the performance: a review and meta-analysis. Agric. Syst. 194, 103278. doi:10.1016/j.agsy.2021.103278.
Hosseini, F.S., Seo, M.B., Razavi-Termeh, S.V., Sadeghi-Niaraki, A., Jamshidi, M., Choi, S.-M., 2023. Geospatial artificial intelligence (GeoAI) and satellite imagery fusion for soil physical property predicting. Sustainability 15, 14125. doi:10.3390/su151914125.
Jones, J.W., Hoogenboom, G., Porter, C.H., Boote, K.J., Batchelor, W.D., Hunt, L.A., Wilkens, P.W., Singh, U., Gijsman, A.J., Ritchie, J.T., 2003. The DSSAT cropping system model. Eur. J. Agron. Modell. Crop. Syst. Sci. Software Appl. 18, 235–265. doi:10.1016/S1161-0301(02)00107-7.
Li, J., Xu, M., Xiang, L., Chen, D., Zhuang, W., Yin, X., Li, Z., 2024. Large Language Models and Foundation Models in Smart Agriculture: Basics, Opportunities, and Challenges. doi:10.48550/arXiv.2308.06668
Li, W., 2024. GeoAI in Social Science. doi:10.48550/ARXIV.2401.05398
Li, W., Hsu, C.-Y., 2022. GeoAI for large-scale image analysis and machine vision: recent progress of artificial intelligence in geography. ISPRS Int. J. Geo-Inf. 11, 385. doi:10.3390/ijgi11070385.
Liu, Y., Zhang, Q., Dong, W., Li, Z., Liu, T., Wei, W., Zuo, M., 2023. Autoformer-based model for predicting and assessing wheat quality changes of pesticide residues during storage. Foods 12, 1833. doi:10.3390/foods12091833.
Lu, G., Li, S., Mai, G., Sun, J., Zhu, D., Chai, L., Sun, H., Wang, X., Dai, H., Liu, N., Xu, R., Petti, D., Li, C., Liu, T., Li, C., 2023. AGI for Agriculture. doi:10.48550/arXiv.2304.06136.
Lundberg, S.M., Lee, S.-I., 2017. A unified approach to interpreting model predictions. In: Advances in Neural Information Processing Systems. Curran Associates, Inc.
Mallinger, K., Baeza-Yates, R., 2024. Responsible AI in farming: a multi-criteria framework for sustainable technology design. Appl. Sci. 14, 437. doi:10.3390/app14010437.
Manzoor, E., 2024. Comparing Major Companies' AI Spending in 2024 and the Challenge of Productionizing AI Solutions. AIM Councils. accessed 3.3.25 https://council.aimresearch.co/comparing-major-companies-ai-spending-in-2024-and-the-challenge-of-productionizing-ai-solutions/.
McCluskey, K., Barker, K.B., Barton, H.A., Boundy-Mills, K., Brown, D.R., Coddington, J.A., Cook, K., Desmeth, P., Geiser, D., Glaeser, J.A., Greene, S., Kang, S., Lomas, M.W., Melcher, U., Miller, S.E., Nobles, D.R., Owens, K.J., Reichman, J.H., Da Silva, M., Wertz, J., Whitworth, C., Smith, D., 2017. The U.S. culture collection network responding to the requirements of the Nagoya protocol on access and benefit sharing. mBio 8, e00982-17. doi:10.1128/mBio.00982-17.
Megan, F., 2024. Survey reveals concerns and adoption trends around AI's rising influence [WWW Document. In: School of Communication and Information |. Rutgers University.

URL. accessed 3.3.25. https://comminfo.rutgers.edu/news/survey-reveals-concerns-and-adoption-trends-around-ais-rising-influence.

Mintz, S.W., Du Bois, C.M., 2002. The anthropology of food and eating. Annu. Rev. Anthropol. 31, 99–119. doi:10.1146/annurev.anthro.32.032702.131011.

Nendel, C., Berg, M., Kersebaum, K.C., Mirschel, W., Specka, X., Wegehenkel, M., Wenkel, K.O., Wieland, R., 2011. The MONICA model: testing predictability for crop growth, soil moisture and nitrogen dynamics. Ecol. Model. 222, 1614–1625. doi:10.1016/j.ecolmodel.2011.02.018.

Openshaw, S., Openshaw, C., 1997. Artificial Intelligence in Geography. Wiley, Chichester, West Sussex.

Parliament of Canada, 2022. Government Bill (House of Commons) C-27 (44-1)—First Reading—Digital Charter Implementation Act, 2022. Parliament of Canada.

Peichl, M., Thober, S., Meyer, V., Samaniego, L., 2018. The effect of soil moisture anomalies on maize yield in Germany. Nat. Hazards Earth Syst. Sci 18, 889–906. doi:10.5194/nhess-18-889-2018.

Ribeiro, M.T., Singh, S., Guestrin, C., 2016. "Why Should I Trust You?": explaining the predictions of any classifier. In: Proceedings of the 22nd ACM SIGKDD International Conference on Knowledge Discovery and Data Mining, KDD '16. Association for Computing Machinery New York, NY, USA,, pp. 1135–114410.1145/2939672.2939778

Ritchie, J.T., Singh, U., Godwin, D.C., Bowen, W.T., 1998. Cereal growth, development and yield. In: Tsuji, G.Y., Hoogenboom, G., Thornton, P.K. (Eds.), Understanding Options for Agricultural Production. Springer, Netherlands, Dordrecht, pp. 79–98. doi:10.1007/978-94-017-3624-4_5.

Sagan, V., Coral, R., Bhadra, S., Alifu, H., Al Akkad, O., Giri, A., Esposito, F., 2024. Hyperfidelis: a software toolkit to empower precision agriculture with GeoAI. Remote Sens. 16, 1584. doi:10.3390/rs16091584.

Shirer, M., 2024. Worldwide Spending on Artificial Intelligence Forecast to Reach $632 Billion in 2028, according to a New IDC Spending Guide [WWW Document]. accessed 3.3.25. IDC: The premier global market intelligence company. https://www.idc.com/getdoc.jsp?containerId=prUS52530724.

Silva, L.M.D., Bitencourt, C.C., Faccin, K., Iakovleva, T., 2019. The role of stakeholders in the context of responsible innovation: a meta-synthesis. Sustainability 11, 1766. doi:10.3390/su11061766.

Slack, D., Friedler, S.A., Scheidegger, C., Roy, C.D., 2019. Assessing the Local Interpretability of Machine Learning Models. doi:10.48550/ARXIV.1902.03501. (arXiv Preprint).

Song, Y., Kalacska, M., Gašparović, M., Yao, J., Najibi, N., 2023. Advances in geocomputation and geospatial artificial intelligence (GeoAI) for mapping. Int. J. Appl. Earth Obs. Geoinf. 120, 103300. doi:10.1016/j.jag.2023.103300.

Srivastava, A.K., Safaei, N., Khaki, S., Lopez, G., Zeng, W., Ewert, F., Gaiser, T., Rahimi, J., 2022. Winter wheat yield prediction using convolutional neural networks from environmental and phenological data. Sci Rep 12, 3215. doi:10.1038/s41598-022-06249-w.

Steduto, P., Hsiao, T.C., Raes, D., Fereres, E., 2009. AquaCrop—the FAO crop model to simulate yield response to water: I. Concepts and underlying principles. Agron. J. 101, 426–437. doi:10.2134/agronj2008.0139s.

Stella, F., Della Santina, C., Hughes, J., 2023. How can LLMs transform the robotic design process? Nat Mach Intell 5, 561–564. doi:10.1038/s42256-023-00669-7.

Stöckle, C.O., Donatelli, M., Nelson, R., 2003. CropSyst, a cropping systems simulation model. Eur. J. Agron. Modell. Crop. Syst. Sci. Software Appl. 18, 289–307. doi:10.1016/S1161-0301(02)00109-0.

Turner, J.A., Horita, A., Fielke, S., Klerkx, L., Blackett, P., Bewsell, D., Small, B., Boyce, W.M., 2020. Revealing power dynamics and staging conflicts in agricultural system transitions: case studies of innovation platforms in New Zealand. J. Rural Stud. 76, 152–162. doi:10.1016/j.jrurstud.2020.04.022.

Tzachor, A., Devare, M., Richards, C., Pypers, P., Ghosh, A., Koo, J., Johal, S., King, B., 2023. Large language models and agricultural extension services. Nat Food 4, 941–948. doi:10.1038/s43016-023-00867-x.

van Ittersum, M.K., Leffelaar, P.A., van Keulen, H., Kropff, M.J., Bastiaans, L., Goudriaan, J., 2003. On approaches and applications of the Wageningen crop models. Eur. J. Agron. Modell. Crop. Syst. Sci. Software Appl. 18, 201–234. doi:10.1016/S1161-0301(02)00106-5.

Vrugt, J.A., ter Braak, C.J.F., Clark, M.P., Hyman, J.M., Robinson, B.A., 2008. Treatment of input uncertainty in hydrologic modeling: doing hydrology backward with Markov chain Monte Carlo simulation. Water Resour. Res. 44. doi:10.1029/2007WR006720

White, E.L., Thomasson, J.A., Auvermann, B., Kitchen, N.R., Pierson, L.S., Porter, D., Baillie, C., Hamann, H., Hoogenboom, G., Janzen, T., Khosla, R., Lowenberg-DeBoer, J., McIntosh, M., Murray, S., Osborn, D., Shetty, A., Stevenson, C., Tevis, J., Werner, F., 2021. Report from the conference, 'identifying obstacles to applying big data in agriculture. Precis. Agric. 22, 306–315. doi:10.1007/s11119-020-09738-y.

Williams, D., Macfarlane, F., Britten, A., 2024. Leaf only SAM: A segment anything pipeline for zero-shot automated leaf segmentation. Smart Agric. Technol. 8, 100515. doi:10.1016/j.atech.2024.100515.

Williams, J.R., Jones, C.A., Kiniry, J.R., Spanel, D.A., 1989. The EPIC crop growth model. Trans. ASAE 32, 0497–0511. doi:10.13031/2013.31032.

Yang, X., Dai, H., Wu, Z., Bist, R., Subedi, S., Sun, J., Lu, G., Li, C., Liu, T., Chai, L., 2024. An innovative segment anything model for precision poultry monitoring. Comput. Electron. Agric. 222, 109045. doi:10.1016/j.compag.2024.109045.

Precision farming with the Internet of Everything

5

Ahmet B. Kilic[a], *Fatih E. Bilgen*[a], *and Ozgur B. Akan*[a,b]
[a]Center for neXt-generation Communications (CXC), Department of Electrical and Electronics Engineering, Koç University, Istanbul, Turkey
[b]Internet of Everything (IoE) Group, Electrical Engineering Division, Department of Engineering, University of Cambridge, Cambridge, United Kingdom

1 Introduction

Feeding a projected 9.7 billion people by 2050, while mitigating agriculture's contributions to deforestation, biodiversity loss, and greenhouse-gas emissions, demands a radical rethink of how we grow food (United Nations Department of Economic and Social Affairs, Population Division, 2022). Conventional Green Revolution inputs, such as fertilizers, pesticides, and heavy machinery, have delivered yield gains at the expense of soil health and ecological resilience. At the same time, an explosion of low-power sensors, autonomous robots, nanoscale devices, and microsatellites has given rise to Agriculture 4.0 and the broader Internet of Things (IoT) movement in farming (Babar and Akan, 2025). Yet plants, fungi, and bacteria have been exchanging biochemical messages for hundreds of millions of years, forming a de facto communication network that silently orchestrates nutrient flows and stress responses. The key insight of this chapter is that digital networks need not replace these biological systems; they can extend, interpret, and reinforce them. We therefore frame precision farming through the lens of the Internet of Everything (IoE). In the agricultural context, the IoE is an integrative paradigm that connects digital devices, living organisms, and natural ecosystems into a unified communication and decision-making network. It brings together biological networks, such as plants, fungi, and microbes that naturally exchange signals; cyber-physical networks, such as sensors, drones, satellites, and vehicles that capture, process, and act on data; and human and institutional actors, such as farmers, cooperatives, and policymakers, who interpret and implement knowledge. In this way, the IoE in agriculture is not simply an overlay of digital devices but a holistic framework where living and nonliving entities jointly contribute to resilient, adaptive, and sustainable food systems. Building on this definition, the remainder of the chapter is structured as follows.

- Section 2 unpacks the Internet of Living Things (IoLT) by detailing the Internet of Plants (IoP), Internet of Fungi (IoF), Internet of Bacteria (IoB), and the integrative Internet of Bio-Nano Things (IoBNT).

Precision Technologies for Digital Agriculture. https://doi.org/10.1016/B978-0-443-40513-6.00005-6

- Section 3 reviews the cyber-physical infrastructure under the hood of the Internet of Agricultural Things (IoAT), such as the Internet of Drones (IoD), Internet of Vehicles (IoV), and Internet of Space Things (IoST), which couple living signals to cloud analytics and autonomous actuators.
- Section 4 presents the Internet of Agent Plants (IoAP), a hierarchical architecture in which embedded agent plants and a central agent plant (CAP) propose concrete services such as disease forewarning, calorie-based supply balancing, and proactive fire mitigation.
- Section 5 discusses interdisciplinary challenges, including biocompatible sensing and communication, protocol standardization, and field-scale robustness, that must be addressed before IoE agriculture can reach its full potential.

By integrating insights from plant physiology, microbial ecology, nanotechnology, and communication engineering, the chapter aims to equip researchers and practitioners with a coherent framework for designing next-generation bio-digital agro-ecosystems (see Fig. 5.1).

2 Internet of Living Things in agriculture

Precision farming increasingly recognizes that below-ground ecosystems function as an interconnected communications web, called the IoF, IoB, and IoP. The IoF refers to the vast mycelial networks formed by soil fungi (especially mycorrhizal fungi) that interlink plant roots, while the IoB denotes the complex signaling and networking among soil bacterial communities. Together with plant signaling, these biological networks form an IoLT in the soil, enabling information exchange and coordination across species.

This section examines how plants, fungi, and bacteria communicate within and between their networks, and how these interactions with plants create a harmonious, integrated system. We focus on the communication aspects: how signals, nutrients, and information propagate through living communities, and how this facilitates symbiotic cooperation among all three domains.

2.1 Internet of Plants

Plants, once thought passive and isolated, are now recognized as responsive and interconnected organisms that sense environmental changes such as light, temperature, and touch, and communicate stress to neighbors (Trewavas, 2017; Midzi et al., 2022). This collective signaling forms the IoP, a natural communication network where plants exchange information through chemical, electrical, mycorrhizal, and acoustic modalities (Volkov and Shtessel, 2018). The mechanisms underlying the IoP are numerous and complex. However, the discovery of this network challenges the traditional perception of plants as single entities and instead portrays them as active participants in a wider ecological network.

Plants are exposed to various stressors, which are mainly classified as biotic and abiotic. Abiotic stressors include drought, high temperatures, and salinity, whereas

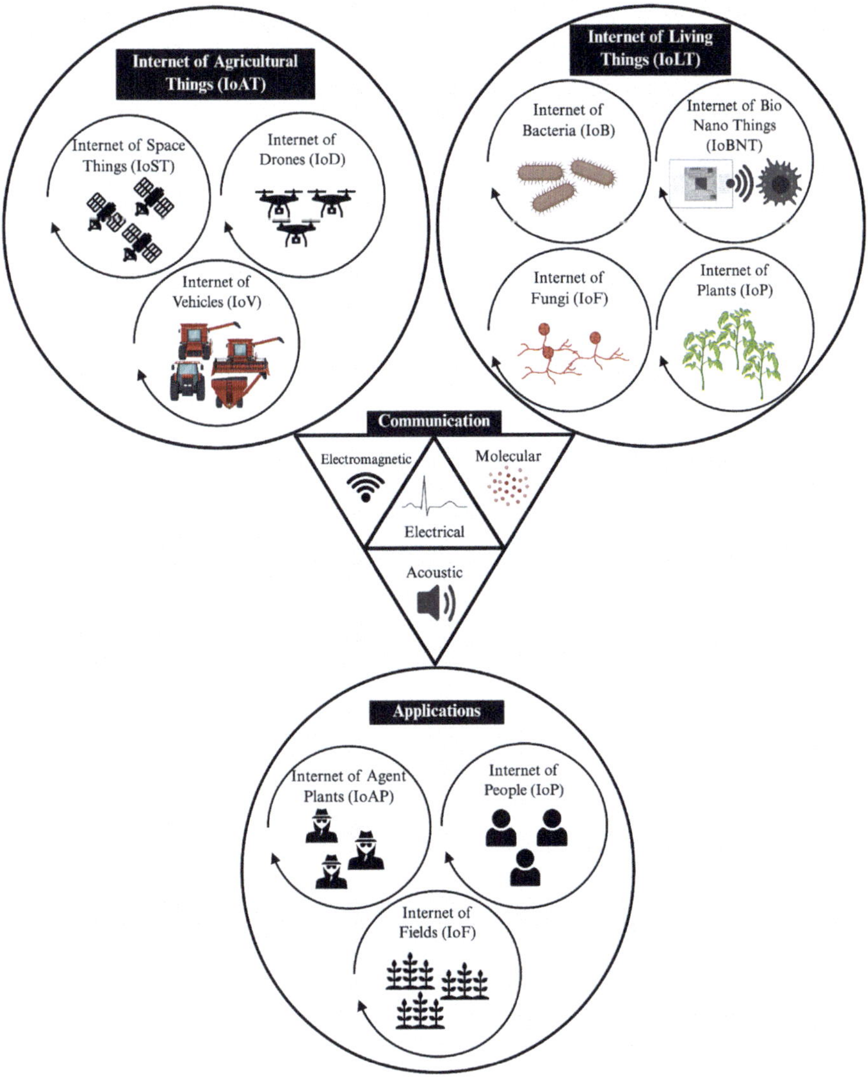

Fig. 5.1 Overview of the paper's structure. Section 2 and Section 3 depict two complementary domains: the Internet of Living Things (IoLT) and the Internet of Agricultural Things (IoAT). These domains are linked to each other by diverse communication modalities. Section 4 outlines potential real-life applications.
Created with https://BioRender.com.

biotic stressors include insects, bacteria, viruses, fungi, nematodes, arachnids, and weeds. These stressors result in plants communicating with each other and warning neighboring plants about approaching danger (Midzi et al., 2022). Therefore, understanding communication between plants is essential for understanding the stresses they experience in the context of the IoP.

Various modalities, including chemical, electrical, mycorrhizal, and auditory signaling, are used by plants to communicate with each other (Volkov and Shtessel, 2018). Each modality has a unique mechanism, benefit, and disadvantage. Gaining a solid understanding of each of these modalities is essential for comprehending the IoP.

The use of volatile organic compounds (VOCs) is referred to as chemical communication between plants. Chemical signaling via VOCs conveys the type and severity of stress while also deterring herbivores and attracting beneficial organisms (Delory et al., 2016; Kilic and Akan, 2024). This communication modality occurs between plants both aboveground and belowground (Bais et al., 2004; Midzi et al., 2022). An illustration of chemical communication can be found in Fig. 5.2.

Chemical communication is a relatively well-studied mode of interaction between plants. However, to fully decipher the IoP, a comprehensive understanding of chemical signaling, particularly belowground, is needed, as research in this area is limited, and the mode of communication is more complex (Bais et al., 2004; Delory et al., 2016).

By connecting different plants below the soil, fungi enable a communication modality between plants. Fungal species colonize various host plants and interact with each other, thus creating complex networks between them (Boyno and Demir, 2022). Plants communicate with other plants of the same or different species and developmental stages by using these fungal networks to exchange nutrients and

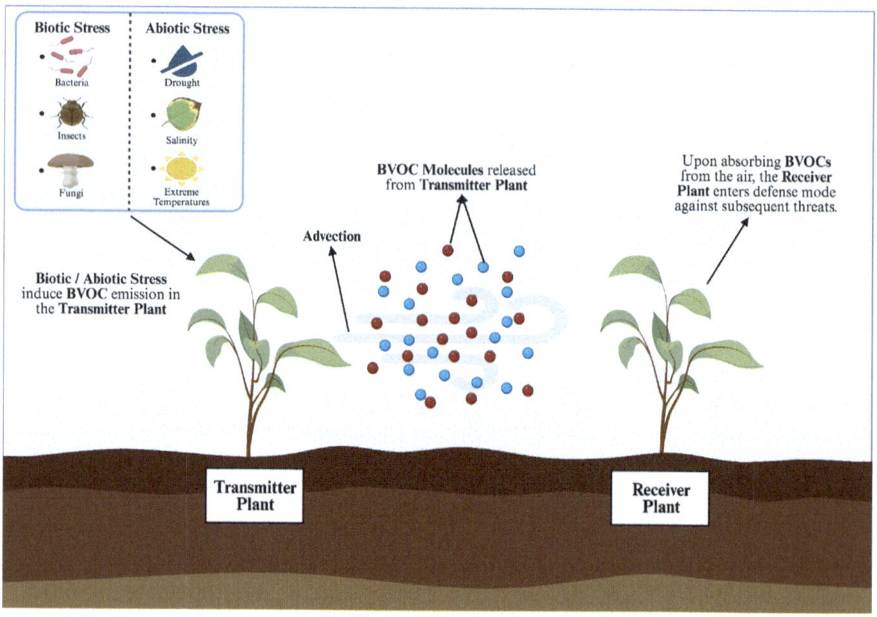

Fig. 5.2 Aboveground chemical signaling between plants via volatile organic compounds (VOCs) enables stress-induced communication and defense priming in neighboring individuals. Created with https://BioRender.com.

information (Simard et al., 2012). Although mycorrhizal networks have been experimentally shown to be one of the modalities used by plants to communicate with each other, the exact mechanism is not yet understood (Boyno and Demir, 2022; Gilbert and Johnson, 2017).

Plants also transmit electrical signals above and belowground: action potentials triggered by mild stimuli, variation potentials by physical damage, and system potentials by chemical stress, with amplitude and type encoding stimulus strength (Vodeneev et al., 2015; Zimmermann et al., 2009). For all electrical signals, research suggests that the conveyed information increases the resistance of the receiving plant (Vodeneev et al., 2015).

Finally, a recently identified communication modality in plants is the use of acoustic signals. It has been shown that plants respond to acoustic stimuli and emit acoustic signals that carry information about the type of stress and the species of the plant (Gagliano et al., 2012; Khait et al., 2023). However, a clear example of acoustic communication between plants has not yet been demonstrated. In addition, the mechanisms of acoustic emission and perception by plants are not yet understood (Demey et al., 2023).

Together, these modalities form a dynamic IoP in which plants coordinate responses and interact with other organisms. In agriculture, such networks could enable early detection of stress for sustainable interventions, while in natural ecosystems, they may inform climate resilience strategies. Integrated within the broader IoE, the IoP offers new opportunities for resource management, environmental monitoring, and ecologically responsible development.

With this basis, the IoP becomes an important part of the larger IoE, which connects digital networks to biological systems. By incorporating IoP principles into this framework, new prospects for improving resource management, environmental observation, and ecologically responsible development at all scales are presented.

2.2 Internet of Fungi

Mycorrhizal fungi weave colossal hyphal networks that knot the roots of multiple plant species, forming the functional equivalent of an underground internet. A single mycelium can connect dozens of neighboring hosts, trading mineral nutrients and water for photosynthate while simultaneously transporting information through soil (Frew et al., 2025). Because the hyphae of one fungal individual infiltrate several root systems at once, the resulting common mycelial network (CMN) becomes shared infrastructure for whole plant communities.

This CMN infrastructure is indicative of several plant actions. Work on mycorrhizal fungi and plant symbioses revealed that herbivore attack cues propagate through CMNs. When aphids fed on one plant, volatile-mediated defenses were induced in adjacent, uninfected plants only if the neighbors were linked by intact hyphae (Babikova et al., 2013). Thus, the fungi transmit an early-warning message that primes distant hosts before insects arrive. Other studies have shown that CMNs can relay other chemical status reports, such as pathogen presence, nutrient scarcity, and even drought stress as well (Dervash et al., 2024).

Fungal networks transmit information through several well-studied modalities. Pressure-driven cytoplasmic streaming in thick conduit hyphae redistributes carbon, nitrogen, and phosphorus to where they are most needed (Schmieder et al., 2019). Diffusible chemical messengers, such as hormones, reactive oxygen species, and volatile compounds, spread through the cytoplasm, coordinating growth inside the mycelium and leaking into mycorrhizal links to warn neighboring plants of stress (Ullah et al., 2024). Localized calcium waves triggered by damage travel only a few millimeters before damping, so stress is managed rapidly and locally without suspending colony-wide development (Itani et al., 2023). Fungi also propagate action-potential-like voltage spikes along hyphae at centimeters per minute; variations in amplitude, duration, and frequency hint at an electrical code that modulates fungal behavior (Thomas and Cooper, 2022; Adamatzky, 2022; Fukasawa et al., 2024). Together, these dynamic channels give the mycelium a flexible toolkit for redistributing resources, coordinating growth, and sharing stress signals within the colony and with surrounding plants.

IoF is a living, multimodal communications grid. Hyphal networks connect different plants, relay biochemical alerts, conduct electrical impulses, and coordinate their own vast bodies through calcium oscillations and nutrient shuttling. This integrative signaling confers ecological advantages such as shared resources, early-warning defense, and adaptive growth while offering conceptual templates for resilient, decentralized communication technologies. By decoding the fungal lexicon and mapping its infrastructural logic, we not only illuminate a hidden layer of ecosystem connectivity but also unlock new paradigms for bio-inspired networking.

2.3 Internet of Bacteria

Bacteria in soil exist not as isolated cells but as communities that constantly communicate and coordinate (Koca and Akan, 2025). One of the most well-studied modes of bacterial communication is quorum sensing (QS), a chemical signaling system that bacteria use to assess their population density and synchronize behavior. Through QS, cells secrete small molecules; when these auto-inducers accumulate to a threshold, they bind cognate receptors and reprogram community gene expression (Mukherjee and Bassler, 2019). QS operates like a biological medium-access protocol: only when the shared channel has a high signal concentration do cells launch collective tasks such as biofilm formation, virulence, or bioluminescence (Duddy and Bassler, 2021). A universal molecule called AI-2 even lets unrelated species, plants, animals, and some viruses join the conversation (González and Venturi, 2013; Hughes et al., 2010; Wang et al., 2020; Vendeville et al., 2005).

At short range, cells move information physically: a flagellated bacterium boosts a nutrient gradient and descends when it loses the trail, effectively sampling the field one step at a time (Van Gestel et al., 2021). Inside biofilms, conductive pili and cytochrome nanowires pass electrons over micrometer distances (Ueki, 2021), and slow potassium waves can coordinate colonies centimeters apart (Liu et al., 2015).

DNA itself serves as a routing protocol. In conjugation, an F-pilus connects two neighbors and hands over a plasmid like a direct cable (Virolle et al., 2020). Bacteriophages can mis-package host DNA and carry it to the next cell, a biological form of packet switching (Chiang et al., 2019). Tiny membrane vesicles broadcast nucleic acids and proteins to many receivers, offering multicast delivery (Marinacci et al., 2023). Taken together, all of these form a living communication stack.

2.4 Internet of Bio-Nano Things

The IoBNT is a revolutionary communication framework that consists of heterogeneous networks made up of artificial micro/nano-scale devices, also known as Bio-Nano Things, and biological entities (Civas et al., 2023). These devices interface with traditional communication networks to enable novel environmental applications. The integration of IoBNT with the IoP, IoF, and IoB is important for precision farming. Unlike these domain-specific networks, IoBNT facilitates integration across biological systems by embedding smart, nano-scale components capable of monitoring, interpreting, and acting upon molecular-level changes. IoBNT enables a paradigm change in precision farming by monitoring and adjusting to real-time variations through the incorporation of nano-biosensors in the biological space. Nano-biosensors are the key elements in IoBNT, giving rise to highly sensitive, compact, and cost-effective devices for sensing agricultural parameters (Johnson et al., 2021). These sensors utilize various nanomaterials, such as carbon nanotubes, metal nanoparticles, and nanowires, to achieve enhanced sensitivity through advanced transduction methods, including electrochemical and optical signaling (Srivastava et al., 2018).

The capacity of these nano-biosensors to detect environmental stressors such as pollution, heavy metal ions, pathogens, and microbial load is one of their primary characteristics. In addition, they support continuous monitoring of conditions such as humidity, temperature, and traceability within agricultural systems (Johnson et al., 2021). When included in IoBNT systems, these capabilities provide real-time diagnostics and information sharing among networked devices, which enhances crop management and decision-making. For example, sensors on plant leaves can detect hydrogen peroxide signal waves (Johnson et al., 2021). Plants use these waves to communicate internally and trigger their defense mechanisms. Monitoring these signals provides crucial information, enabling early identification of pest attacks or environmental stress before serious damage occurs (Johnson et al., 2021). Moreover, IoBNT technologies borrow from successful applications in the health sector and are now adapted to agricultural uses for food safety monitoring (Srivastava et al., 2018). This includes detecting preservatives, antibiotics, toxins, and even analyzing aroma and gas composition in food products (Srivastava et al., 2018). Overall, IoBNT enables multiplex and real-time sensing, promotes resource-saving farming practices, and enhances sustainability. All of these contribute to precision farming. IoBNT enables cross-disciplinary understanding by integrating plant physiology, nanotechnology, and environmental monitoring. For instance, detection of extracellular DNA by nanosensors offers information on virus use of host plants and molecular information on biotic stress defense systems (Zhu et al., 2024). Alternatively, site-specific delivery of fungicides and nutrients that improve plant health without associated concurrent excess environmental toxicity is offered using nanotechnology (Nizamani et al., 2024). These operations are combined in IoBNT designs, where arrays of nanosystems repetitively monitor an array of ecological and biochemical data. Through a conjoining of disease diagnosis from the biological sciences, nutritional management from the chemical sciences, and pathogen identification, IoBNT formulates a shared vision for field health.

Despite its potential, IoBNT faces important challenges that must be addressed before large-scale deployment in agriculture. The fabrication and deployment of nanosensors remain costly, and ensuring their durability in diverse soil and climate conditions is still an open issue. Many nanomaterials raise concerns about toxicity, environmental persistence, and unintended effects on nontarget organisms, which complicate field approval and regulatory acceptance. Reliable power supply at the nano-scale is also a limitation, as most nanosensors cannot harvest sufficient energy for long-term operation. Finally, the standardization of data formats and secure integration of IoBNT outputs with existing agricultural IoT systems is far from solved. These limitations highlight the need for cross-disciplinary research not only in device design but also in safety assessment, ecological compatibility, and regulatory frameworks.

3 Agricultural Internet of Things

Agriculture is undergoing a revolutionary digital transformation, generically referred to as Agriculture 4.0, which aims to address some of the world's largest challenges, including deforestation, depletion of natural resources, and climate change (Mitra et al., 2022). Central to this transformation is the integration of IoT technologies into agriculture. This convergence has led to the development of the IoAT, a system that connects devices, sensors, and equipment to support real-time data exchange, remote monitoring, and automated control of agricultural processes (Babar and Akan, 2025).

As per Babar and Akan (2025), growers are turning more toward IoT-based systems in an effort to meet the increasing demands for productive and sustainable food production. IoAT enables this shift through real-time monitoring, enhanced data collection, and advanced automation. Through reduced inputs of time and human power, the systems offer an optimized mechanism that supports foresight-driven decision-making and prompt responses to crop and environmental factors, enhancing productivity and reducing wastage (Lakshmanan et al., 2020).

IoAT-enabled smart agriculture also has advantages over conventional practices. Among these are improved water conservation, more precise use of fertilizers and pesticides, healthier crop yields, and lower operational costs. The framework also enables farming in unconventional locations, such as cities or deserts, contributing to food security regardless of geography. IoAT technologies also offer important benefits for the environment. They help lower greenhouse gas emissions, prevent soil erosion, and reduce the overall harm farming causes to nature. In addition, real-time data helps farmers better understand what is happening in their fields. This allows them to respond more accurately to the needs of their crops (Babar and Akan, 2025).

IoAT utilizes a variety of networked systems and equipment. The IoAT ecosystem functions in or as part of the primary movers of precision farming, which include satellite-based systems, autonomous tractors, drones, and bio-integrated energy-harvesting technologies. Altogether, these technologies demonstrate how digital agriculture is transforming traditional practices and propelling the agendas of Agriculture 4.0 (Babar and Akan, 2025). Fig. 5.3 illustrates how the IoAT is utilized to support precision farming.

Nevertheless, there are various challenges in implementing IoAT in agriculture. Many farms in remote areas face connectivity issues. Data security concerns still exist, and incompatibility among devices manufactured by different companies makes it hard to obtain a single, coherent system. Addressing these issues can enable IoAT to reach its full potential and operate at increased scales to create a positive contribution to agriculture (Babar and Akan, 2025).

Beyond these technical hurdles, cost and adoption remain significant barriers. Smallholder farmers, who make up a large proportion of global food producers, often lack the capital to invest in connected machinery or subscription-based digital

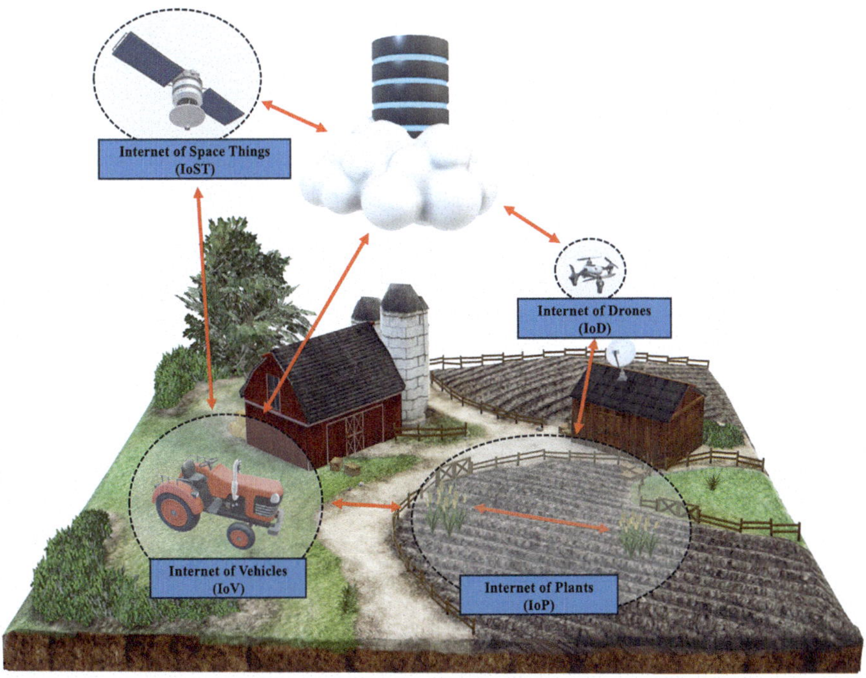

Fig. 5.3 Illustration of IoAT applications aimed at achieving precision farming.

services. Training and technical support are equally critical, as sophisticated systems may fail if end-users do not have the skills or resources to operate and maintain them. Without inclusive strategies, IoAT may widen the gap between technologically advanced farms and those left behind.

3.1 Internet of Drones

The IoD is a significant development in farm automation, using intelligent, real-time management and communication between autonomous drones. These artificial intelligence (AI)-empowered drones transcend traditional limitations such as processing capacity and line-of-sight flight. As a result, they can gather information on their own, carry out farming tasks, and help make timely decisions. The IoD is an appropriate positioning of the bigger picture of IoT in precision farming for improving productivity, resource efficiency, and environmental sustainability. By combining drones with smart sensors and wireless networks, the IoD enables real-time monitoring of crops, animals, and environmental variables, advancing global food security and sustainable practices (Babar and Akan, 2025).

A wide range of applications for drones has been realized in precision farming. Unmanned aerial vehicles (UAVs) are used to monitor plant growth, detect crop diseases, and evaluate soil characteristics with high spatial and temporal resolution (Boursianis et al., 2022). For instance, UAVs equipped with multispectral cameras can capture vegetation indices such as the normalized difference vegetation index (NDVI), which aids in crop biomass estimation as well as stress levels (Boursianis et al., 2022). Plant growth monitoring and phenotyping at the individual level are made possible through 3D reconstruction modeling (Boursianis et al., 2022), while thermal imaging is used to monitor soil moisture and irrigation (Boursianis et al., 2022). In addition, machine learning algorithms and drone systems facilitate coordinated operations such as selective spraying and weed detection (Boursianis et al., 2022; Babar and Akan, 2025), yield estimation, and pollination support.

IoD communication relies on drone-to-drone, drone-to-ground station, and cloud-based communication, through which drones can transmit data to remote servers for storage and analysis alongside other information. Drones can exchange information through low-power wide-area networks such as Long Range (LoRa) and Narrowband Internet of Things (NB-IoT), both known for their long range and low energy consumption (Boursianis et al., 2022). These communication protocols enable drones to function as part of a distributed intelligent system that operates alongside fixed IoT devices and cloud analytics, promoting scalable and resilient farm monitoring (Boursianis et al., 2022).

Despite the potential, IoD has its drawbacks, such as battery life constraints, regulatory problems, and vulnerability to cyberattacks. The dynamic topology and mobility of drone networks contribute to the difficulty in providing secure and reliable communication (Babar and Akan, 2025). Blockchain-based security architectures and the standardization of communication protocols have been proposed to alleviate these problems (Babar and Akan, 2025).

In addition, cost and scalability present major barriers. Agricultural drones equipped with multispectral cameras, thermal sensors, and AI processors are expensive to purchase and maintain, which limits their accessibility for smallholder farmers. Maintenance of fleets in harsh outdoor environments, including dust, wind, and variable weather, requires specialized expertise and adds to operational expenses. Without affordable business models, IoD adoption risks being concentrated among large-scale industrial farms.

With continued research and integration with AI, edge computing, and secure data sharing, the IoD will be a cornerstone of tomorrow's precision farming systems, but only if these cost, ecological, and regulatory challenges are addressed alongside technical advancements.

3.2 Internet of Vehicles

The IoV refers to a dynamic network of connected and autonomous vehicles that can communicate with one another and with surrounding infrastructure. By integrating Vehicle Ad-hoc Networks with the IoT, IoV enhances communication between vehicles, sensors, humans, and cloud data centers (Babar and Akan, 2025). IoV allows intelligent farming machinery, remote sensors, and cloud platforms to work together by sharing data in real-time. Hence, it relies on the technology of IoT and advances precision farming another step further.

Autonomous vehicles play a central role in precision farming because they allow for immediate, automated agriculture on a large farm. Automated tractors, for instance, can be set up to travel particular routes (Stentz et al., 2002). After that, these routes are split up and assigned to a fleet of tractors, each of which has sensors mounted on board that can identify obstructions like humans, animals, or other machinery. When something is in the way, the tractors stop and wait for further instructions sent wirelessly (Stentz et al., 2002). These vehicles also demonstrate a high level of precision in path tracking and have the ability to monitor their own performance. Similarly, autonomous harvesting systems have been developed to detect and cut the peduncles of crops using robotic arms guided by computer vision models such as Mask R-CNN (Zhang et al., 2019). These robots employ specialized grippers that harvest without damaging the fruit, illustrating the suitability of IoV-enabled machinery for crop handling in unstructured environments (Zhang et al., 2019). Together, these systems exemplify the transition toward data-driven decision-making through constant, mobile sensing and task execution.

Crucial to these capabilities is the communication infrastructure that underlies the IoV. IoV-enabled machines utilize vehicle-to-vehicle (V2V) and vehicle-to-everything (V2X) communication protocols to establish uninterrupted interactions among vehicles, control systems, and surrounding devices. While V2V allows tractors and harvesters to communicate directly with each other, V2X extends communication to infrastructure, sensors, and cloud systems. This facilitates distributed decision-making across several units and guarantees compatibility with larger IoT-based agricultural systems. Agricultural systems are therefore better able to respond to field conditions in real time and are more coordinated. There are several opportunities for IoV application in agriculture. IoV can improve farming operations by using real-time data, particularly in the areas of resource and route management, which will increase fuel economy. Predictive maintenance, a crucial component of IoV, will be essential for reducing equipment downtime and increasing overall productivity. In addition, IoV systems play a key role in ensuring food safety by making it easier to trace products from the field to the consumer. They also improve on-farm safety by sending real-time alerts about machinery conditions and worker activities. IoV allows for more sustainable agriculture thanks to its efficiency in reducing waste and making optimal use of resources (Babar and Akan, 2025).

Despite the benefits it brings, there are also disadvantages to the integration of IoV in farming. The high initial capital investment and technical expertise requirements are significant deterrents to the utilization of these technologies for the majority of farmers. Due to connectivity issues, real-time communication can be limited in remote or rural areas. Furthermore, IoV networks pose security and standardization challenges because of their distributed and dynamic nature. Robust cybersecurity and stable frameworks are needed to ensure the secure operation of autonomous farm vehicles (Babar and Akan, 2025). Nevertheless, the continued evolution of IoV promises to further scale and secure the future of precision farming.

3.3 Internet of Space Things

The IoST transplants the IoT vision into low-Earth orbit by transforming satellites into both data relays and remote-sensing nodes for agriculture (Babar and Akan, 2025). Small, inexpensive CubeSats constitute the orbital segment, acting simultaneously as "backhaul in the sky" and "eyes in the sky": they relay data from on-farm sensors while carrying multispectral, thermal, or radar payloads for Earth observation (Kak and Akyildiz, 2022). Terrestrial layers include IoST hubs, which are gateway ground stations that bridge satellites and legacy networks, and devices such as soil-moisture probes, micro-weather stations, livestock tags, and autonomous machinery. The multitier design enables accurate local measurements to be fused with wide-area crop surveillance, realizing an IoE vision for precision farming (Kak and Akyildiz, 2022).

Research on space-ground connectivity centers on crafting ultra-efficient, delay-tolerant links that allow coin-cell-powered sensors to communicate directly with orbit. LoRaWAN experiments have already proven the principle that frames remain intelligible after an 832 km hop to a CubeSat, which demonstrates that even the lightest nodes can bypass terrestrial towers (Parada et al., 2023). Building on such successes, standard-setting bodies are closing the gap between terrestrial and nonterrestrial IoT. The 3GPP has elevated NB-IoT into space, enabling a single device to transmit through the nearest cell tower when coverage exists or fall back to an overhead satellite with only a firmware tweak (Kua et al., 2021). The forthcoming 5G and 6G releases go further by formalizing nonterrestrial networks, in which swarms of nano-satellites act as aerial base-stations for massive machine-type communication (Chen et al., 2023). Some architectures insert a middle tier of high-altitude platform stations, such as solar drones and stratospheric balloons, whose LoRa gateways have already relayed tens of kilobits per second of farm data from connectivity deserts. Yet, despite this versatility, current economics and deployment speed still tip the balance toward rapidly launched CubeSat constellations as the workhorse of planet-wide agricultural IoT (Parada et al., 2023).

To reduce bandwidth demand and latency, IoST platforms increasingly embed edge-computing logic in orbit. Satellites preprocess raw imagery or sensor telemetry, filtering noise, aggregating packets, or running lightweight convolutional neural networks to flag anomalies before forwarding concise alerts or feature vectors to the cloud (Wei et al., 2019). This space-ground continuum off-loads central servers, cuts transmission costs, and enables near-real-time agronomic decision support even where terrestrial backhaul is absent.

The resulting communications infrastructure underpins several high-value agricultural services. First, remote field monitoring: soil moisture probes, micro-climate stations, and camera traps in connectivity black spots can stream data continuously, letting farmers trigger irrigation or frost protection regardless of geography (Parada et al., 2023). Second, integrated sensing: fusing CubeSat imagery with in-situ measurements and numerical weather forecasts yields a multiscale picture of crop health, enabling variable-rate fertilization, early pest or disease alerts, and improved yield forecasting (De La Parte et al., 2023). Third, livestock and asset tracking: IoT collars or machinery tags outside cellular range upload location, physiology, or utilization metrics via satellite, extending supply-chain visibility to vast rangelands (Akhigbe et al., 2021). Finally, climate resilience: distributed flood gauges, fire detectors, and soil-moisture networks feed near-real-time data through IoST even when terrestrial infrastructure fails, supporting early-warning systems and informing climate-smart practices worldwide (Chen et al., 2023). Collectively, these advances position IoST as a cornerstone of globally equitable, data-driven precision farming, translating orbital connectivity and sensing breakthroughs into tangible on-farm value.

Despite these advantages, IoST faces several limitations that could hinder widespread adoption. Launching and maintaining satellite constellations is expensive, and even low-cost CubeSats require constant replenishment due to their short orbital lifetimes. Bandwidth is limited, and although edge-computing helps reduce congestion, many applications may still struggle with latency or restricted data rates compared to terrestrial systems. Regulatory and geopolitical issues also complicate deployment, since spectrum allocation and orbital rights are managed internationally and often contested. Furthermore, farmers in developing regions may lack affordable access to IoST services, raising equity concerns. Finally, the accumulation of space debris poses a long-term sustainability risk for satellite-based agriculture. Overcoming these challenges will be essential for IoST to transition from experimental pilots to a reliable global infrastructure for precision farming.

4 Applications

Having shown in Section 2 that soil biota already form a self-organizing IoLT, and in Section 3 that cyber-physical platforms, such as the IoAT, IoD, IoV, and IoST, can tap into and augment that living web, we are now ready to move to the application layer. Put differently: once biological and electronic networks are both in place, what concrete agro-ecosystem services can they deliver?

This section answers this question by introducing the IoAP. An agent plant is an artificially created, plant-like structure embedded in crop fields. Its core functions are: (1) communicating with surrounding plants, (2) relaying the gathered information to a CAP, (3) influencing neighboring plants to achieve specific goals based on commands from the CAP, and (4) harvesting energy by mimicking mycorrhizal fungi–plant symbiosis (Bilgen and Akan, 2024). A CAP supervises a field-scale cluster of agent plants, collects their data, and coordinates responses. It does not poll every crop directly; instead, it communicates only with its agent plants, which monitor their local micro-populations via molecular, electrical, and optical cues. Establishing this hierarchical structure unlocks several applications (see Fig. 5.4).

4.1 Early disease detection and precision treatment

Plant pathogens spread quickly, while pesticide application spreads environmental harm even faster. Because each agent plant maintains a one-to-one molecular dialog with its assigned crops, it can spot early anomalies such as volatile-profile shifts,

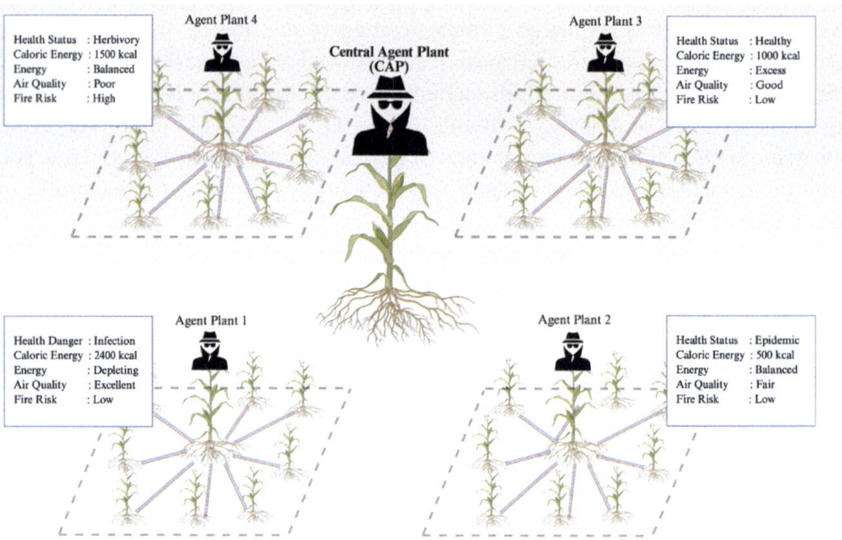

Fig. 5.4 Schematic of the IoAP, where a central agent plant (CAP) coordinates clusters to monitor crops and enable disease detection, wildfire mitigation, and calorie-aware balancing. Created with https://BioRender.com.

reactive oxygen species bursts, or electrical-pattern changes well before visual symptoms appear. Therefore, CAP provides disease detection, diagnosis, and precisely targeted treatment while reducing environmental harm.

4.2 Proactive wildfire mitigation

Heat and drought-stress signatures that precede ignition are readily sensed through plant bio-electrical and VOC channels, continuously harvested by agent plants, and streamed to the CAP.

These signals feed a field-scale fire-risk index. When the index exceeds a threshold, the CAP can trigger targeted irrigation, dispatch drones to apply water-retaining hydrogels at hotspots, or alert civil-protection services via IoST backhaul. Such minutes-ahead intervention can stop small ignitions from escalating into landscape-scale wildfires, which is an increasingly urgent need in hotter, drier climates.

4.3 Calorie-aware supply-demand balancing

Global food chains rarely optimize for nutritional energy; they optimize for market prices. With IoAP connectivity, each agent plant can estimate the caloric value currently ripening within its zone and report that figure to the CAP. When the CAPs of several fields are linked through an Internet of Fields backbone, the aggregate stock of available kilocalories becomes visible at district, regional, or even national scale. Simultaneously, an Internet of People architecture (Miranda et al., 2015) can estimate population-level calorie demand. Matching the two streams lets policymakers allocate planting, harvesting, and distribution tasks more surgically than today, reducing overproduction in one area and shortages in another, and thereby easing pressure on land and fertilizer inputs.

4.4 Field examples and case studies

While the IoAP is presented as a forward-looking architecture, several real-world examples already show how elements of the IoE can operate in agriculture. In biological networks, experiments with bean plants connected by mycorrhizal fungi demonstrate that stress cues from aphid infestation can travel through fungal networks and prime neighboring plants for defense, offering a natural template for early-warning systems in farming (Babikova et al., 2013). Greenhouse trials with tomato and tobacco plants have revealed that stressed individuals emit airborne sounds that can be detected and classified, suggesting acoustic biosensing as a future tool for crop stress monitoring (Khait et al., 2023).

Digital components of IoE are also emerging in practice. In vineyards, drones equipped with multispectral cameras have been used to calculate vegetation indices such as NDVI, enabling early detection of nutrient deficiencies and water stress. Drone-based selective spraying has been piloted to reduce pesticide input while maintaining yield quality, showing how IoD platforms can close the loop between sensing and action (Boursianis et al., 2022). Robotic strawberry harvesters in

California have been tested with computer vision models such as Mask R-CNN, allowing automated detection and cutting of peduncles without damaging fruit, and demonstrating the potential of IoV platforms for delicate crop-handling tasks in real time (Zhang et al., 2019).

At larger scales, the IoST is advancing from concept to field trials. CubeSat-enabled LoRa communication has been demonstrated at an 832 km orbit, proving that even low-power farm sensors can transmit directly to satellites without terrestrial towers (Parada et al., 2023). Together, these examples show that IoE in agriculture is no longer confined to theory. From fungal warning signals and plant bioacoustics to drones, autonomous tractors, and nanosatellite relays, the building blocks of IoE are already in practice, pointing toward the feasibility of fully integrated bio-digital agro-ecosystems.

5 Challenges and future directions

The convergence of biology, nanotechnology, and communication theory for agriculture presents both opportunities and challenges. IoLT and its branches, such as IoP, IoF, IoB, and IoBNT, are in the early stages of interdisciplinary synergy. Their success depends on robust models that connect plant physiology, microbial ecology, molecular signaling, sensor networks, and digital infrastructure.

A central challenge is interdisciplinary integration. Biological systems operate on different timescales and principles than digital ones. Plant and fungal signaling unfolds over minutes or hours and depends heavily on biochemical and environmental context. Translating these stochastic, analog processes into digital, quantifiable signals for machine learning or control systems is difficult. In addition, each biological subsystem, including plants, fungi, and bacteria, relies on distinct signaling molecules and methods, which require customized sensing and interpretation solutions.

Scalability and ruggedness further complicate deployment. While nano-biosensors and micro-embedded devices succeed in controlled settings, scaling to large farms is difficult. Ensuring long-term sustainability by powering devices without causing ecological disruption and maintaining stability in diverse soils, pH levels, and microbial communities remains unresolved. Ethical and environmental risks of engineered microbes or invasive nanodevices in soil must also be carefully assessed.

Interoperability and standardization are equally vital. Precision farming within the IoE vision depends on seamless data exchange between living and digital systems, yet current protocols are not designed for biological contexts. Standardized models are needed to interpret biological information, modulate signals, and securely transfer data. Harmonized ontologies must preserve biological complexity while enabling communication efficiency.

Despite these hurdles, IoE-enabled precision farming holds great promise. In such systems, drones coordinate with sensor-tagged plants, fungi, and bacteria to track plant health, tailor fertilization, and monitor conditions in real time. Satellites provide macro-level data, while soil-implanted biosensors capture fine-grained information on nutrients, microbes, and stress signals. Autonomous vehicles deliver context-specific treatments, such as precision irrigation and microbial inoculation, based on integrated bio-cyber data streams.

Achieving this vision requires investment in translational research, biocompatible materials, and multiscale modeling. An integrated research ecosystem that spans synthetic biology, plant sciences, communication engineering, and environmental monitoring will be key to making IoE-based precision farming both sustainable and scalable.

6 Conclusion

This chapter has argued that true precision farming will emerge only when living and digital internets are treated as complementary, interlinked strata of a single information ecology. The IoLT, spanning plants, fungi, and bacteria, already performs decentralized sensing, computation, and actuation through chemical and electrical signaling, while the IoT, encompassing drones, vehicles, and satellites, provides the bandwidth, compute power, and mobility to observe and influence those biological processes at scale. By bringing these domains together under the vision of the IoE, agriculture can be reimagined as a coherent system where biological networks and cyber-physical infrastructures coevolve.

The central finding of this chapter is that precision farming should not be understood merely as an accumulation of sensors and devices, but as an architecture that fuses living communication networks with digital platforms. The proposed IoAP illustrates how this integration can enable practical services such as precise, molecularly guided treatments, calorie-aware yield balancing, and wildfire suppression triggered by stress signatures at the plant level. These examples demonstrate that the convergence of biology and digital networks can deliver both efficiency gains and ecological resilience.

Looking forward, the IoE in agriculture opens a pathway toward food systems that are simultaneously productive, adaptive, and sustainable. Achieving this vision requires progress in nanoscale biosensing, biocompatible energy harvesting, secure edge–cloud coordination, and ethical frameworks for deploying engineered organisms and devices in open fields. If these challenges are met, IoE-enabled precision farming could evolve into a planetary infrastructure that links soil biochemistry to orbital constellations, supporting resilient food production under climate volatility and even extending agriculture into new habitats beyond Earth.

Acknowledgements

Funding: This work was supported in part by the AXA Research Fund (AXA Chair for Internet of Everything at Koç University).

References

Adamatzky, A., 2022. Language of fungi derived from their electrical spiking activity. R. Soc. Open Sci. 9 (4). doi:10.1098/rsos.211926.

Akhigbe, B.I., et al., 2021. IoT technologies for livestock management: a review of present status, opportunities, and future trends. Big Data Cogn. Comput. 5 (1), 10. doi:10.3390/bdcc5010010.

Babar, A.Z., Akan, O.B., 2025. Sustainable and Precision Agriculture with the Internet of Everything (IoE). https://arxiv.org/abs/2404.06341.

Babikova, Z., et al., 2013. Underground signals carried through common mycelial networks warn neighbouring plants of aphid attack. Ecol. Lett. 16 (7), 835–843. doi:10.1111/ele.12115.

Bais, H.P., et al., 2004. How plants communicate using the underground information superhighway. Trends Plant Sci. 9 (1), 26–32. doi:10.1016/j.tplants.2003.11.008.

Bilgen, F.E., Akan, O.B. (2024) 'Mycorrhizal fungi and plant symbiosis for energy harvesting in the internet of plants. In: *Proceedings of the 11th Annual ACM International Conference on Nanoscale Computing and Communication*, pp. 35–40. doi:10.1145/3686015.3689365.

Boursianis, A.D., et al., 2022. Internet of things (IoT) and agricultural unmanned aerial vehicles (UAVs) in smart farming: a comprehensive review. Internet Things 18, 100187. doi:10.1016/j.iot.2020.100187.

Boyno, G., Demir, S., 2022. Plant-mycorrhiza communication and mycorrhizae in inter-plant communication. Symbiosis 86 (2), 155–168. doi:10.1007/s13199-022-00837-0.

Chen, X., Xu, Z., Shang, L., 2023. Satellite Internet of Things: challenges, solutions, and development trends. Front. Inf. Technol. Electron. Eng. 24 (7), 935–944. doi:10.1631/fitee.2200648.

Chiang, Y.N., Penadés, J.R., Chen, J., 2019. Genetic transduction by phages and chromosomal islands: the new and noncanonical. PLoS Pathog. 15 (8), e1007878. doi:10.1371/journal.ppat.1007878.

Civas, M., et al., 2023. graphene and related materials for the internet of bio-nano things. APL Mater. 11 (8). doi:10.1063/5.0153423.

De La Parte, M.S.E., et al., 2023. Spatio-temporal semantic data management systems for IoT in agriculture 5.0: challenges and future directions. Internet Things 25, 101030. doi:10.1016/j.iot.2023.101030.

Delory, B.M., et al., 2016. Root-emitted volatile organic compounds: can they mediate belowground plant-plant interactions? Plant Soil 402 (1–2), 1–26. doi:10.1007/s11104-016-2823-3.

Demey, M.L., Mishra, R.C., Van Der Straeten, D., 2023. Sound perception in plants: from ecological significance to molecular understanding. Trends Plant Sci. 28 (7), 825–840. doi:10.1016/j.tplants.2023.03.003.

Dervash, M.A., et al., 2024. Fungal Internet: the natural networking systems. SpringerBriefs Microbiol., 77–83. doi: 10.1007/978-3-031-66293-5_7.

Duddy, O.P., Bassler, B.L., 2021. Quorum sensing across bacterial and viral domains. PLoS Pathog. 17 (1), e1009074. doi:10.1371/journal.ppat.1009074.

Frew, A., Varga, S., Klein, T., 2025. Mycorrhizal networks: understanding hidden complexity. Funct. Ecol. 39 (6), 1322–1327. doi:10.1111/1365-2435.70063.

Fukasawa, Y., et al., 2024. Electrical integrity and week-long oscillation in fungal mycelia. Sci. Rep. 14 (1). doi:10.1038/s41598-024-66223-6.

Gagliano, M., et al., 2012. Out of sight but not out of mind: alternative means of communication in plants. PLoS ONE 7 (5), e37382. doi:10.1371/journal.pone.0037382.

Gilbert, L., Johnson, D., 2017. Plant–plant communication through common mycorrhizal networks. Adv. Bot. Res. 83–97. doi:10.1016/bs.abr.2016.09.001.

González, J.F., Venturi, V., 2013. A novel widespread interkingdom signaling circuit. Trends Plant Sci. 18 (3), 167–174.

Hughes, D.T., et al., 2010. Chemical sensing in mammalian host–bacterial commensal associations. Proc. Natl. Acad. Sci. USA 107 (21), 9831–9836. doi:10.1073/pnas.1002551107.

Itani, A., et al., 2023. Local calcium signal transmission in mycelial network exhibits decentralized stress responses. PNAS Nexus 2 (3). doi:10.1093/pnasnexus/pgad012.

Johnson, M.S., Sajeev, S., Nair, R.S. (2021) 'Role of nanosensors in agriculture. In: *International Conference on Computational Intelligence and Knowledge Economy (ICCIKE*, pp. 58–63. doi:10.1109/iccike51210.2021.9410709.

Kak, A., Akyildiz, I.F., 2022. Towards automatic network slicing for the internet of space things. IEEE Trans. Netw. Serv. Manag. 19 (1), 392–412. doi:10.1109/tnsm.2021.3117692.

Khait, I., et al., 2023. Sounds emitted by plants under stress are airborne and informative. Cell 186 (7), 1328–1336.e10. doi:10.1016/j.cell.2023.03.009.

Kilic, A.B., Akan, O.B., 2024. End-to-End mathematical modeling of stress communication between plants, https://arxiv.org/abs/2410.11790.

Koca, B.Y., Akan, O.B., 2025. Bacterial communications and computing in internet of everything (IoE). IEEE Commun. Surv. Tutor. 27 (3), 1839–1866. doi:10.1109/COMST.2024.3451302.

Kua, J., et al., 2021. Internet of things in space: a review of opportunities and challenges from satellite-aided computing to digitally-enhanced space living. Sensors 21 (23), 8117. doi:10.3390/s21238117.

Lakshmanan, R., et al., 2020. Automated smart hydroponics system using internet of things. Int. J. Power Electron. Drive Syst./Int. J. Electr. Comput. Eng. 10 (6), 6389. doi:10.11591/ijece.v10i6.pp6389-6398.

Liu, J., et al., 2015. Metabolic co-dependence gives rise to collective oscillations within biofilms. Nature 523 (7562), 550–554. doi:10.1038/nature14660.

Marinacci, B., et al., 2023. Latest update on outer membrane vesicles and their role in horizontal gene transfer: a Mini-Review. Membranes 13 (11), 860. doi:10.3390/membranes13110860.

Midzi, J., et al., 2022. Stress-induced volatile emissions and signalling in Inter-Plant communication. Plants 11 (19), 2566. doi:10.3390/plants11192566.

Miranda, J., et al., 2015. From the internet of things to the internet of people. IEEE Internet Comput. 19 (2), 40–47. doi:10.1109/mic.2015.24.

Mitra, A., et al., 2022. Everything you wanted to know about smart agriculture, https://arxiv.org/abs/2201.04754.

Mukherjee, S., Bassler, B.L., 2019. Bacterial quorum sensing in complex and dynamically changing environments. Nat. Rev. Microbiol. 17 (6), 371–382. doi:10.1038/s41579-019-0186-5.

Nizamani, M.M., et al., 2024. Revolutionizing agriculture with nanotechnology: innovative approaches in fungal disease management and plant health monitoring. Sci. Total Environ. 928, 172473. doi:10.1016/j.scitotenv.2024.172473.

Parada, R., et al, 2023. LORA-based low-cost nanosatellite for emerging communication networks in complex scenarios. Aerospace 10 (9), 754. doi:10.3390/aerospace10090754.

Schmieder, S.S., et al., 2019. Bidirectional propagation of signals and nutrients in fungal networks via specialized hyphae. Curr. Biol. 29 (2), 217–228.e4. doi:10.1016/j.cub.2018.11.058.

Simard, S.W., et al., 2012. Mycorrhizal networks: mechanisms, ecology and modelling. Fungal Biol. Rev. 26 (1), 39–60. doi:10.1016/j.fbr.2012.01.001.
Srivastava, A.K., Dev, A., Karmakar, S., 2018. Nanosensors and nanobiosensors in food and agriculture. Environ. Chem. Lett. 16 (1), 161–182. doi:10.1007/s10311-017-0674-7.
Stentz, A., et al., 2002. A system for semi- autonomous tractor operations. Auton. Robots 13 (1), 87–104. doi:10.1023/a:1015634322857.
Thomas, M.A., Cooper, R.L., 2022. Building bridges: mycelium–mediated plant–plant electrophysiological communication. Plant Signal. Behav. 17 (1). doi:10.1080/15592324.2022.2129291.
Trewavas, A., 2017. The foundations of plant intelligence. Interface Focus 7 (3), 20160098. doi:10.1098/rsfs.2016.0098.
Ueki, T., 2021. Cytochromes in extracellular electron transfer in *Geobacter*. Appl. Environ. Microbiol. 87 (10), e03109-20. doi:10.1128/AEM.03109-20.
Ullah, A., Gao, D., Wu, F., 2024. Common mycorrhizal network: the predominant socialist and capitalist responses of possible plant–plant and plant–microbe interactions for sustainable agriculture. Front. Microbiol. 15. doi:10.3389/fmicb.2024.1183024
United Nations Department of Economic and Social Affairs, Population Division, 2022. World Population Prospects 2022: Summary of Results. United Nations, New York.
Van Gestel, J., et al., 2021. Short-range quorum sensing controls horizontal gene transfer at micron scale in bacterial communities. Nat. Commun. 12 (1). doi:10.1038/s41467-021-22649-4.
Vendeville, A., Winzer, K., Heurlier, K., Hardie, K.R., 2005. Making "sense" of metabolism: autoinducer-2, LUXS and pathogenic bacteria. Nat. Rev. Microbiol. 3 (5), 383–396. doi:10.1038/nrmicro1146.
Virolle, C., Goldlust, K., Djermoun, S., Bigot, S., Lesterlin, C., 2020. Plasmid transfer by conjugation in Gram-negative bacteria: from the cellular to the community level. Genes 11 (11), 1239. doi:10.3390/genes11111239.
Vodeneev, V., Akinchits, E., Sukhov, V., 2015. Variation potential in higher plants: mechanisms of generation and propagation. Plant Signal. Behav. 10 (9), 1057365. doi:10.1080/15592324.2015.1057365.
Volkov, A.G., Shtessel, Y.B., 2018. Electrical signal propagation within and between tomato plants. Bioelectrochemistry 124, 195–205. doi:10.1016/j.bioelechem.2018.08.001.
Wang, S., Payne, G.F., Bentley, W.E., 2020. Quorum sensing communication: molecularly connecting cells, their neighbors, and even devices. Annu. Rev. Chem. Biomol. Eng. 11 (1), 447–468. doi:10.1146/annurev-chembioeng-101519-124728.
Wei, J., Han, J., Cao, S., 2019. Satellite IoT edge intelligent computing: a research on architecture. Electronics 8 (11), 1247. doi:10.3390/electronics8111247.
Zhang, T., et al., 2019. An autonomous fruit and vegetable harvester with a low-cost gripper using a 3D sensor. Sensors 20 (1), 93. doi:10.3390/s20010093.
Zhu, C., et al., 2024. A self-reporting nanosensor for detecting bacterial exploitation in plants through extracellular DNA. Nano Lett. 24 (48), 15464–15472. doi:10.1021/acs.nanolett.4c05168.
Zimmermann, M.R., et al., 2009. System potentials, a novel electrical long-distance apoplastic signal in plants, induced by wounding. Plant Physiol. 149 (3), 1593–1600. doi:10.1104/pp.108.133884.

Section II

Computer vision, digital tools, and sensing technologies for modern agriculture

Harnessing remote sensing and geographic information systems in precision and digital agriculture

6

Subhankar Debnath[a], *Mallappa Jadiyappa Madolli*[a,b], *Shubham Gade*[a], *Aliza Adhikari*[a], and *Sushil Kumar Himanshu*[a]
[a]Agricultural Systems and Engineering, Faculty of Food, Agriculture and Natural Resources, Asian Institute of Technology, Pathum Thani, Thailand
[b]Department of Agricultural Engineering, University of Agricultural Sciences, Bengaluru, Karnataka, India

Abbreviations

CWSI	crop water stress index
ET	evapotranspiration
EVI	enhanced vegetation index
GIS	geographic information systems
GNDVI	green normalized difference vegetation index
LAI	leaf area index
LiDAR	light detection and ranging
METRIC	mapping evapotranspiration at high resolution with internalized calibration
MODIS	moderate resolution imaging spectroradiometer
NDVI	normalized difference vegetation index
NIR	near-infrared
NWI	normalized water index
PA	precision agriculture
RDVI	renormalized difference vegetation index
RS	remote sensing
SAR	synthetic aperture radar
SAVI	soil-adjusted vegetation index
SEBAL	surface energy balance algorithm for land
SPAD	soil plant analysis development
TIR	thermal infrared
UAVs	unmanned aerial vehicles
VIS	visible
VRI	variable rate irrigation
WI	water index

Precision Technologies for Digital Agriculture. https://doi.org/10.1016/B978-0-443-40513-6.00011-1

1 Introduction

Agriculture plays a crucial role in sustaining human life by providing essential nutrients for food. However, the agricultural sector faces several challenges that threaten both its productivity and sustainability. A key challenge is the rapidly increasing global population, which is expected to exceed 9 billion by 2050 (Debnath et al., 2018). This ever-increasing population growth significantly increases food demand, requiring significant advancements in agricultural practices to ensure global food security. In addition, climate change, through temperature shifts, changing precipitation patterns, and more frequent extreme weather events, negatively impacts crop growth and yields (Debnath et al., 2021a; Santosh et al., 2023). Climate-related uncertainties make agricultural production highly unpredictable, emphasizing the need for resilient and adaptable farming systems. The traditional farming methods, which depend on manual labor and conventional techniques, are often unable to meet the growing agricultural demands and challenges. These methods are typically inefficient in water usage, soil management, and pest control. Consequently, there is a pressing need for modern technology to enhance the efficiency and sustainability of agricultural practices.

Recent technological advancements have led to the development of smart agriculture. This approach uses modern tools and data to manage variability in fields. Precision agriculture (PA) is a key method that optimizes inputs like water, fertilizers, and pesticides (Kumar et al., 2022a, 2022b; Gade et al., 2025). It helps increase crop yields while reducing environmental impact. PA is part of digital agriculture, which includes technologies such as big data, artificial intelligence, Internet of Things, blockchain, remote sensing (RS), and geographic information systems (GIS). These tools improve decision-making, boost productivity, and support sustainability in the agricultural value chain.

In agriculture, RS techniques provide real-time data on crop health, soil conditions, and weather, enabling timely interventions. GIS plays a vital role in modern agriculture by supporting land use planning and decision-making. It helps planners design effective crop rotation and zoning plans. GIS collects and analyzes data on soil conditions, crop health, and weather patterns. This data helps predict crop yields accurately. It also helps in identifying suitable sites for agricultural projects. In addition, GIS supports environmental impact assessments, ensuring sustainable agricultural development. In the past, numerous studies have demonstrated that GIS and RS are essential tools for mapping natural resources across various spatial scales, ranging from large to small (Abbasi et al., 2023; Croft et al., 2020; Dash, 2023; Imtiaz et al., 2025; Lambert et al., 2018; Ma et al., 2019; Mondal et al., 2023; Rossini et al., 2013; Su et al., 2022; Zhao et al., 2023). For example,(Croft et al., 2020) mapped the spatial variability of leaf chlorophyll content in fields with different nitrogen fertilizer levels using multispectral Landsat-8 Operational Land Imager (OLI) data at 30 m resolution. Imtiaz et al. (2025) evaluated the use of high-resolution multispectral imagery (Sentinel-2A and PlanetScope imagery) along with machine learning (ML) techniques for potato yield prediction. Key vegetation indices (normalized difference vegetation index (NDVI), green NDVI (GNDVI), soil-adjusted vegetation index (SAVI), enhanced vegetation index (EVI), etc.) and spectral bands relevant to crop monitoring were extracted and used in ML models. The gradient tree boosting technique with Sentinel-2A and harvester data showed better performance in the study. The approach enables real-time, field-scale yield prediction maps, supporting site-specific management and sustainable farming. Lambert et al. (2018) assessed the use of unmanned aerial vehicles (UAVs) for mapping weeds in wheat fields. Aerial imagery was collected from 26 fields using both Red, Green, and Blue (RGB) and R_{mod} (670–750 nm) spectral bands. The study indicated that weed density showed a strong correlation with R_{mod} image intensity. However, prediction performance across fields was inconsistent, indicating limited generalizability of the models.

This chapter aims to present a comprehensive overview of the application of GIS and RS technologies in PA, while also addressing the potential challenges and limitations associated with their use.

2 Conceptual framework of remote sensing

RS operates by detecting and analyzing electromagnetic radiation that is either emitted or reflected from terrestrial objects or surfaces from a distance. The process begins with an energy source, which may be natural (i.e., solar radiation) or artificial, as in the case of active sensors like radar systems (Fig. 6.1). When electromagnetic energy reaches the Earth's surface, it is partially absorbed, reflected, or transmitted. These interactions depend on the target's properties, such as composition, surface roughness, texture, and

Fig. 6.1 Key elements of the remote sensing process: (A) energy source, (B) radiation and atmosphere, (C) interaction with target, (D) sensor detection, (E) data transmission and processing, (F) interpretation and analysis, and (G) application.

moisture content. The reflected or emitted radiation is then captured by remote sensors, which are typically deployed on platforms such as satellites, aircraft, or UAVs. These sensors are capable of detecting electromagnetic energy across multiple spectral bands, including visible (VIS), infrared, and microwave regions. The captured data are recorded as digital imagery, where each pixel represents the intensity of radiation measured at specific wavelengths. Once acquired, the data are either transmitted to ground receiving stations in real-time or stored onboard for subsequent retrieval. Preprocessing steps are essential before analysis and typically include radiometric calibration, geometric correction, and atmospheric compensation to ensure data accuracy and consistency. After preprocessing, advanced analytical techniques and algorithms are applied to extract quantitative and qualitative information about the Earth's surface or specific target features. These outputs support a wide range of applications, including land use classification, environmental monitoring, PA, and resource management. This illustrates the overall workflow of RS applications in agriculture. In this context, solar radiation serves as the primary source of electromagnetic energy, which is directed toward the Earth's surface. The portion of this energy that is reflected from agricultural fields is captured by RS platforms and subsequently analyzed to support data-driven decision-making in agricultural management. Table 6.1 presents the specifications of the thermal camera and the multispectral camera sensors used in PA, while Table 6.2 provides an overview of satellite-based products frequently utilized in agricultural RS for monitoring and analysis.

Table 6.1 Specification of the thermal camera and the multispectral camera used in precision agriculture.

Camera	Resolution	Weight (g)	Spectral band	Type of camera
FLIR A325as	320 × 240	700	7.5–13.0 µm	Thermal
FLIR T450sc	320 × 240	880	7.5–13.0 µm	
Gobi-384	384 × 288	500	8.0–14.0 µm	
ICI 320x	320 × 240	150	7.0–14.0 µm	
ICI 7640 P-Series	640 × 480	127.6	7.0–14.0 µm	
InfraTec mobileIR M4	160 × 120	265	8.0–14.0 µm	
Optris PI400	382 × 288	320	7.5–13.0 µm	
Pearleye LWIR	640 × 480	790	8.0–14.0 µm	
Photon 320	324 × 256	97	7.5–13.5 µm	
Tamarisk 640	640 × 480	121	8.0–14.0 µm	
Tau 640	640 × 512	110	7.5–13.5 µm	
Thermal-Eye 4500AS	640 × 480	108	7.0–14.0 µm	
Thermoteknix MIRICLE 370K	640 × 480	166	8.0–12.0 µm	
Thermovision A40 M	320 × 240	1400	7.5–13.5 µm	

Table 6.1 Specification of the thermal camera and the multispectral camera used in precision agriculture—cont'd

Camera	Resolution	Weight (g)	Spectral band	Type of camera
ADC Micro	2048 × 1536	200	520–600 nm, 630–690 nm, 760–900 nm	Multispectral
Buzzard	1280 × 1024	250	500, 550, 675, 700, 750, and 780 nm	
MiniMCA6	1280 × 1024	700	490, 550, 680, 720, 800, and 900 nm	
RedEdge	280 × 960	150	475, 560, 668, 717, and 840 nm	
Sentera Quad	1248 × 950	170	655, 725 and 800 nm	
Sequoia (MS)	1280 × 960	72	550, 660, 735, and 790 nm	
XNiteCanon SX230	4000 × 3000	223	385–470 nm, 500–570 nm, 670–770 nm	

Table 6.2 Most commonly used satellite products for agricultural remote sensing.

Satellite product	Wavelength	Wavebands present	Spatial resolution	References
Landsat-8	0.43–12.5 μm	Blue, Green, Red, NIR, SWIR, PAN	15 m (Pan), 30 m (Multi), 100 m (Thermal)	(Mundada and Jain, 2025)
PlanetScope imagery	0.44–0.86 μm	Blue, Blue, Green, Yellow, Red, Red Edge, NIR	~3–5 m	(Imtiaz et al., 2025; Singh et al., 2025)
MODIS (Terra and Aqua)	0.4–14.4 μm	NIR, SWIR, TIR	250 m, 500 m, 1000 m	(Wang, 2025)
Sentinel-2	0.443–2.190 μm	Blue, Blue, Green, Red, Vegetation Red Edge, NIR, SWIR	10 m, 20 m, 60 m (depending on the band)	(da Costa et al., 2025; Imtiaz et al., 2025)
WorldView-3 imagery	0.40–12.5 μm	Blue, Green, Red, NIR, SWIR, CAVIS, PAN	0.31 m (Pan), 1.24 m (MS), 3.7 m (SWIR)	(Fevgas et al., 2025)

3 Types of RS used in precision agriculture

RS technologies play a key role in modern agricultural practices by facilitating the acquisition and analysis of spatial and temporal data related to crop conditions, land characteristics, and environmental parameters. A variety of RS techniques are employed in agriculture, each offering unique capabilities for monitoring and management. These techniques are outlined below:

3.1 Optical sensing

Optical RS utilizes sensors that detect reflected electromagnetic radiation in the VIS, near-infrared (NIR), and thermal infrared spectral regions. These data are instrumental in assessing vegetation health, canopy structure, and land cover classification. Optical imagery is primarily acquired through satellite platforms or aerial systems and is widely used for phenological monitoring and land surface analysis.

3.2 Multispectral sensing

Multispectral sensors capture data across discrete spectral bands of the electromagnetic spectrum. These sensors are particularly effective in vegetation monitoring, as they allow for the derivation of vegetation indices such as the NDVI, GNDVI, and EVI. Such indices are critical for quantifying plant growth, biomass estimation, and monitoring crop development stages.

3.3 Hyperspectral sensing

Hyperspectral RS involves the acquisition of data across hundreds of narrow, continuous spectral bands. This high spectral resolution enables precise identification of the biochemical and biophysical properties of crops and soils. Applications in agriculture include early disease detection, nutrient deficiency assessment, and detailed crop type classification. Hyperspectral data facilitate a granular level of spectral discrimination that surpasses multispectral capabilities.

3.4 Thermal sensing

Thermal sensing detects emitted radiation from surfaces to measure their temperature. In agricultural contexts, thermal sensors are employed to monitor canopy temperature, enabling indirect assessments of crop water stress, irrigation scheduling, and evapotranspiration (ET) estimation. This technique is particularly valuable in precision irrigation management and drought monitoring.

3.5 Radar sensing

Radar systems operate in the microwave domain and are capable of acquiring data under all weather and lighting conditions. Synthetic aperture radar (SAR) sensors measure the backscatter intensity of microwave signals, which is influenced by surface roughness, vegetation structure, and moisture content. In agriculture, SAR data support applications such as soil moisture mapping, biomass estimation, crop classification, and phenological stage assessment.

3.6 Light detection and ranging

Light detection and ranging (LiDAR) technology uses laser pulses to determine distances by measuring the time delay between emission and return. This results in high-resolution, three-dimensional point cloud data representing surface elevation and object structure. In agriculture, LiDAR is applied to estimate crop height, assess canopy structure, generate high-precision digital elevation models, and support site-specific management practices in PA.

3.7 Unmanned aerial vehicles

UAVs, or drones, equipped with various sensors (e.g., RGB, multispectral, thermal, and LiDAR) offer high spatial and temporal resolution data collection over localized areas. Fig. 6.2 shows a high spatial resolution image captured by the RCB camera of the drone. UAV-based RS supports real-time crop monitoring, disease and pest detection, stress mapping, and targeted intervention. The flexibility and cost-effectiveness of UAV platforms make them particularly valuable for PA at field and sub-field scales.

Fig. 6.2 High-resolution images captured by a drone.

4 Application of RS and GIS in PA

In PA, RS data is acquired through satellite or UAV-based imaging systems. The captured imagery can be either stored on a cloud server for remote access or processed locally. Utilizing the spectral signatures of target features, the imagery enables tasks such as crop monitoring, soil analysis, and stress detection (Fig. 6.3). More details about each application are discussed below.

4.1 Nutrient management

Timely and accurate fertilizer application is critical to maximizing crop growth and yields while minimizing environmental harm caused by nutrient losses to groundwater and surface water (Kumar et al., 2022a). Traditionally, fertilizers are applied uniformly at planting and during key growth stages. However, fertilizer requirements vary both spatially and temporally due to differences in soil properties, topography, management practices, weather, and hydrology (Kumar et al., 2022a; Hendricks et al., 2019). Capturing this variability for PA applications is challenging with conventional tools like chlorophyll meters. RS technologies, using vegetation indices such as NDVI and SAVI, have shown strong correlations with plant chlorophyll content, photosynthetic activity, and overall productivity (Argento et al., 2021; Croft et al., 2020). These indices help to map spatial variability in crop nutrient status, aiding site-specific nutrient management. Tables 6.3 and 6.4 summarize various vegetation indices used in PA for nutrient management. Fig. 6.4 shows the NDVI index variation in a rice field. Recently, tractor-mounted remote sensors such as the GreenSeeker, Yara N-Sensor, and Crop Circle have enabled real-time, variable-rate fertilizer applications based on crop reflectance data (Ali et al., 2017). These sensors, typically mounted in front of the spray boom, calculate vegetation indices like NDVI to assess nitrogen

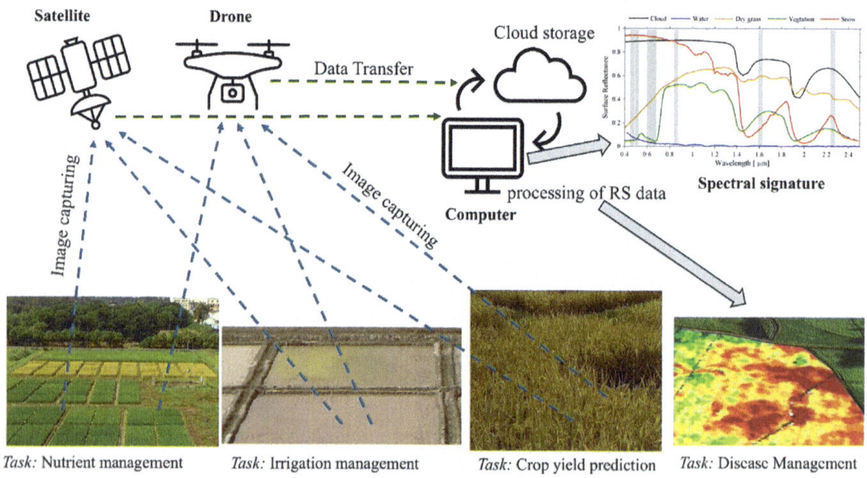

Fig. 6.3 Simple workflow showing the applications of RS/GIS in precision agriculture.

Table 6.3 Multispectral band vegetation indices used in precision agriculture.

Vegetation index	Definition	References
Normalised green	$\frac{G}{(NIR+R+G)}$	(Shafi et al., 2019)
Normalised red	$\frac{R}{(NIR+R+G)}$	(Shafi et al., 2019)
Ratio vegetation index	$\frac{NIR}{R}$	(Sishodia et al., 2020)
Green red vegetation index	$\frac{NIR}{G}$	(Demir and Başyiğit, 2023)
Difference vegetation index	$NIR-R$	(Sishodia et al., 2020)
Green difference vegetation index	$NIR-G$	(Avola et al., 2019)
Normalised difference vegetation index	$\frac{(NIR-R)}{(NIR+R)}$	(Haboudane et al., 2004; Sishodia et al., 2020)
Green normalised difference vegetation index	$\frac{(NIR-G)}{(NIR+G)}$	(Sishodia et al., 2020)
Soil-adjusted vegetation index	$\frac{1.5(NIR+R)}{(NIR+R+0.5)}$	(Sishodia et al., 2020)
Optimized soil-adjusted vegetation index	$\frac{(NIR-R)}{(NIR+R+0.16)}$	(Sishodia et al., 2020)
Green optimized soil-adjusted vegetation index	$\frac{(NIR-G)}{(NIR+G+0.16)}$	(Mao et al., 2020)
Modified soil-adjusted vegetation index 2	$\frac{(2.NIR+1)-\sqrt{(2.NIR+1)^2-8(NIR-RED)}}{2}$	(Haboudane et al., 2004)

Note: G represents reflectance in the green wavelength, NIR denotes near-infrared reflectance, and R indicates reflectance in the red wavelength.

needs and relay the information to nutrient applicators. Algorithms convert these indices into recommended nitrogen application rates, often by comparing field data to a reference index from a well-fertilized (nitrogen-rich) strip. Tools such as the Nitrogen Fertilizer Optimization Algorithm have been successfully implemented to guide in-season nitrogen applications for various crops (Zhang et al., 2020). Despite the availability of these technologies, adoption remains low among farmers, especially in large-scale commercial settings. One major barrier is the limited evidence of consistent economic benefits in terms of yield or profitability. To enhance the

Table 6.4 Hyperspectral narrow-band vegetation indices used in precision agriculture.

Vegetation index	Definition	References
Greenness index	$\frac{R_{554}}{R_{677}}$	(Croft et al., 2020)
Simple ratio	$\frac{NIR}{RED}$	(Croft et al., 2020)
Normalized difference vegetation index (NDVI)	$\frac{(R_{800}-R_{680})}{(R_{800}+R_{680})}$	(Croft et al., 2020)
Green NDVI	$\frac{(R_{801}-R_{550})}{(R_{801}+R_{550})}$	(Croft et al., 2020)
Plant senescence reflectance index (Adjusted)	$\frac{R_{800}}{R_{680}}$	(Pismennaya and Deryugin, 2025)
Normalized difference index	$\frac{(R_{780}-R_{710})}{(R_{780}-R_{680})}$	(Cao and Yang, 2023)
Modified chlorophyll absorption in reflectance index	$(R_{700}-R_{670})-0.2(R_{700}-R_{550})\frac{R_{700}}{R_{670}}$	(Haboudane et al., 2004)
Transformed chlorophyll absorption in reflectance index	$3\left[(R_{700}-R_{670})-0.2(R_{700}-R_{550})\frac{R_{700}}{R_{670}}\right]$	(Rubio-Delgado et al., 2021)
Optimized soil-adjusted vegetation index	$\frac{1.16(R_{800}-R_{670})}{(R_{800}+R_{670}+0.16)}$	(Croft et al., 2020)

Triangular vegetation index	$60(R_{750}-R_{550})-100(R_{670}-R_{550})$	(Haboudane et al., 2004)
Renormalized difference vegetation index	$\frac{(R_{800}-R_{670})}{\sqrt{(R_{800}+R_{670})}}$	(Haboudane et al., 2004)
Modified simple ratio	$\frac{\left(\frac{R_{800}}{R_{670}}-1\right)}{\sqrt{\left(\frac{R_{800}}{R_{670}}+1\right)}}$	(Haboudane et al., 2004)
Modified soil-adjusted vegetation index	$0.5\left[2R_{800}+1-\sqrt{(2R_{800}+1)^2-8(R_{800}-R_{670})}\right]$	(Haboudane et al., 2004)
Modified triangular vegetation index	$1.44(R_{800}-R_{550})-3(R_{670}-R_{550})$	(Haboudane et al., 2004)
Modified chlorophyll absorption in reflectance index 2	$\frac{[3.75(R_{800}-R_{670})-1.95(R_{800}-R_{670})]}{\sqrt{(2R_{800}+1)^2+(6R_{800}-5\sqrt{R_{670}})-0.5}}$	(Haboudane et al., 2004)

Note: R denotes the reflectance value at the specified wavelength (in nanometers) indicated by the subscript. NIR specifically refers to reflectance in the NIR spectral region.

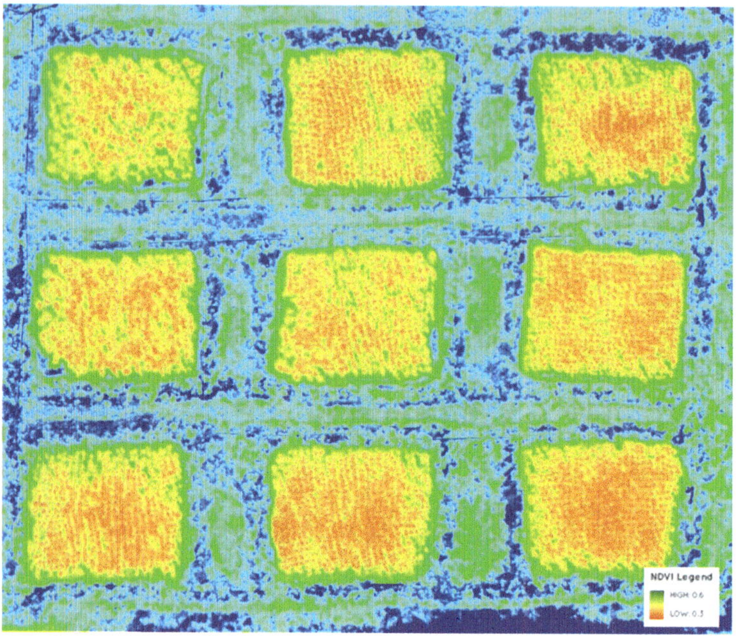

Fig. 6.4 NDVI index variation of rice yield.

efficiency and adoption of these systems, ongoing research is exploring the use of UAVs and advanced remote sensors across diverse crops and climatic zones. For instance,(Zhu et al., 2022) conducted UAV flight missions to capture crop images during the jointing, heading, and grain-filling stages of maize growth from 2018 to 2020. The MicaSense RedEdge-M multispectral camera recorded data in five spectral bands (blue, green, red, red edge, and NIR). Flights had 85% front and 75% side overlap, and reflectance panels were photographed before each mission for radiometric calibration. All images were taken at 60 m height with nadir (0°) direction between 11:00 and 13:00 under stable sunlight and low wind conditions (<5 m/s). The study concluded that the selected UAV-derived indicators, when processed using a random forest model, proved effective for reliably estimating the interactions between fertilizer and irrigation on maize growth. Shou et al. (2007) evaluated the feasibility of using high-resolution satellite imagery to assess nitrogen status in winter wheat across the North China Plain. Correlation analysis showed that both broadband vegetation indices and individual band reflectance values were strongly associated with sap nitrate concentration, soil plant analysis development readings, total nitrogen concentration, and aboveground biomass. It is important to note that vegetation indices reflect a range of biotic and abiotic influences such as soil, moisture, disease, and other environmental stressors. Thus, when interpreting vegetation indices for nutrient management, one must account for potential confounding factors like drought or pest pressure to avoid misjudging plant nutrient needs.

4.2 Irrigation management

The timing and rate of irrigation are critical for reducing crop water stress and optimizing growth and yield at the same time (Fan et al., 2022; Kumar et al., 2022a, 2023; Himanshu et al., 2023; Ale et al., 2020). Farmers use various irrigation practices depending on factors such as water availability, farm infrastructure, local/regional regulations, farm size, and knowledge. While many farmers rely on uniform irrigation schedules based on their experience, larger farms use soil moisture sensors for more precise irrigation (Lena et al., 2022). Various advisory services also guide irrigation based on local climate data. However, most of these conventional methods overlook variability within fields, leading to inefficient water use. RS enables variable rates of irrigation by collecting images at different stages of the growing season to assess key indicators of crop water demand, such as ET, soil moisture, and crop water stress. These indicators help to apply water more precisely, reduce stress, enhance yields uniformly, and minimize water and nutrient losses.

Conventional methods of ET measurement (e.g., weighing lysimeter) are generally expensive and do not provide spatially variable ET estimates resulting from differences in land use, soils, topography, and other hydrologic processes (Debnath et al., 2015). RS data is widely used to estimate ET, which is needed to determine crop water requirements and to schedule irrigation (Alves Varella et al., 2015; Bonfante et al., 2019; Sapkota et al., 2025; Stone et al., 2016). Popularly used methods, such as the surface energy balance algorithm for land and mapping ET at high resolution with internalized calibration, fully solve the energy balance equation by determining the sensible heat flux (H) based on the identification of dry and wet pixels in the image (Allen et al., 2011). Empirical constants derived from dry and wet pixels and measured radiometric surface temperature are then used to determine pixel-wise surface-air temperature differences and sensible heat flux. Recently, several studies have evaluated energy balance approaches with satellite data (e.g., Landsat 7 and 8) to estimate ET for irrigation management in PA (Mokhtari et al., 2019; Nouri et al., 2017; Pareeth and Karimi, 2023). However, the coarse spatio-temporal resolution of thermal satellite data limits the application of surface energy balance approaches in PA.

Crop coefficient-based approaches for estimating ET involve establishing a relationship between vegetation indices (e.g., NDVI, SAVI) and the crop coefficient (K_c) (Abbasi et al., 2023; Munusamy et al., 2025). A more detailed method uses remotely sensed crop biophysical parameters such as leaf area index (LAI), canopy height, and albedo to estimate unknown variables in the Penman-Monteith equation for ET calculation. These ET estimates are typically integrated into soil water balance models to determine irrigation requirements. Bhatti et al. (2020) used Landsat and UAV data to estimate K_c-based ET and variable rate irrigation (VRI) needs for corn and soybean in Nebraska. Stone et al. (2016) showed that NDVI-derived K_c and ET can guide VRI scheduling effectively, achieving yields comparable to in-situ soil moisture-based techniques. Bonfante et al. (2019) used Sentinel-2 data to derive LAI and albedo, applied the Penman–Monteith equation, and calculated dynamic irrigation needs through a decision support system for corn production in Italy.

Various electromagnetic bands, such as optical, thermal, and microwave bands, are used to calculate multiple indices for precision water management (Tables 6.3 and 6.4). Dangwal et al. (2015) monitored the water stress of wheat using multitemporal Landsat data, the vegetation water stress index, and the land surface wetness index water stress factor. The study found that the land surface wetness index water stress factor performed better than vegetation water stress index for water stress detection. Elvanidi et al. (2017) demonstrated that remotely acquired crop reflectance and temperature measurements can effectively assess crop water status. The study highlighted key spectral regions as 490–510, 530–560, 660–670, and 730–760 nm for useful water stress monitoring. In addition, indices such as the photochemical reflectance index, modified red simple ratio index, and modified ratio NDVI showed strong correlations with substrate water content and the crop water stress index (CWSI). The study concluded that combining reflectance and temperature data could serve as an early warning system for water stress, supporting optimal crop growth. Rossini et al. (2013) presented a method for mapping water stress in a maize field using hyperspectral RS imagery. Among several vegetation indices, the Photochemical Reflectance Index using 570 nm (PRI_{570}) showed the strongest correlation with active chlorophyll fluorescence and relative leaf water content, making it effective for detecting early water stress before structural damage occurred. Ihuoma and Madramootoo (2019) explored the potential of using vegetation indices (NDVI, renormalized difference vegetation index, Optimized Soil-Adjusted Vegetation Index (OSAVI), Water Index (WI), and normalized water index) to detect water stress and optimize irrigation water use efficiency in greenhouse-grown tomato crops. The study showed that WI, OSAVI, and WI/NDVI were the most sensitive indices for detecting water stress. However, no single index proved sufficient for accurately estimating water stress. As a result, there is a need for innovative data integration methods that combine multiple indices. Future research should focus on integrating thermal and narrow-band hyperspectral data and developing advanced, cost-effective analysis tools to support precise irrigation scheduling and decision-making. The CWSI, derived from thermal RS, is a widely used tool for estimating irrigation needs. Sapkota et al. (2025) found that CWSI is as effective as ET in identifying irrigation management zones, with both indicators effectively capturing vegetation responses to irrigation during the summer season.

4.3 Crop monitoring and yield

Monitoring crop growth and yield is essential to understanding how crops respond to environmental conditions and agronomic practices, as well as to developing effective management strategies for field operations or interventions (Debnath et al., 2021b). Among the key indicators of crop growth monitoring are LAI and biomass. Traditional in-situ methods for estimating LAI (physical and optical) and biomass (often through destructive sampling) are labor-intensive, time-consuming, and provide limited spatial coverage (Fei et al., 2022). These limitations can be addressed using RS, which offers valuable spatial information on crop growth metrics like LAI and biomass. Crop yield estimation using RS typically follows two approaches: the model-based approach and the empirical approach. In the model-based approach, remotely sensed biophysical parameters (e.g., LAI) are used as inputs in crop growth models to simulate yield and biomass, whereas statistical relationships (e.g., regression models) are established in the empirical approach between remotely sensed indices (e.g., NDVI, LAI) and measured yield/biomass from representative fields. These models can then be applied to estimate yield across larger areas. Numerous studies have applied RS imagery to estimate crop yield (Amankulova et al., 2023; Fei et al., 2022; Ji et al., 2022; Bawa et al., 2023). Such insights enable site-specific management, including variable rate applications of water, nutrients, and pesticides, thereby improving productivity and resource-use efficiency. RS has been widely used to estimate LAI and biomass for a range of crops, including row crops, orchards, and vines (Amankulova et al., 2023; Kang et al., 2022; Zhao et al., 2023). These estimations typically involve developing regression or ML models using reference data, such as ground-measured LAI and corresponding vegetation indices. However, accurate LAI estimation from reflectance data can be challenging, particularly during early growth stages when soil background effects dominate. To address this, researchers have developed modified vegetation indices that minimize soil interference (Zhen et al., 2020). Recently, red-edge-based indices have shown strong potential for improving LAI estimation across various crops (Dong et al., 2019). In the context of PA, understanding spatial variability in crop yield is critical for evaluating responses to management and environmental stress.

4.4 Disease management

Plant diseases can lead to substantial crop losses and reduced farm profits. Early detection and accurate mapping of disease spread are critical for minimizing damage and preventing further transmission (Hashim et al., 2020). Traditional field scouting, while commonly used, is labor-intensive, time-consuming, and prone to human error. Moreover, it often struggles to identify diseases at early stages when symptoms are subtle or not yet visible. Some diseases may not produce visible signs at all, or symptoms may appear too late for effective intervention. Conventional scouting methods also fall short in capturing the spatial extent and severity of disease outbreaks. RS offers a more efficient and scalable solution for early-stage disease monitoring. Technologies such as RGB, multispectral, hyperspectral, and thermal imaging have been successfully used to detect diseases across various crops (Hashim et al., 2020; Ju et al., 2023; Kaya and Gürsoy, 2023; Thomas et al., 2018). For instance,(Ju et al., 2023) acquired multispectral imagery of the wheat canopy using an UAV, selected optimal vegetation indices using both the K-means algorithm and genetic algorithm, and developed a wheat leaf rust monitoring model using a backpropagation neural network. The developed model achieved R^2 of 0.902 and an Root Mean Square Error (RMSE) of 5.45% on the training set, and an R^2 of 0.784 with an RMSE of 4.76% on the validation set. Kaya and Gürsoy (2023) present a novel deep learning approach for plant disease detection that leverages the fusion of RGB and segmented images. A multiheaded DenseNet-based architecture was designed to process both image types simultaneously. Using fivefold cross-validation, the model achieved an average accuracy of 98.17%, with corresponding recall, precision, and F1-score values of 98.17%, 98.16%, and 98.12%, respectively. The fusion-based method effectively distinguishes plant diseases with varying visual characteristics. The high accuracy and low standard deviation demonstrate the model's robustness, making it a strong candidate for integration into automated plant disease detection and early warning systems. However, interpreting spectral data can be challenging because both biotic (e.g., disease) and abiotic (e.g., water or nutrient stress) factors can influence plant reflectance. This overlap makes it difficult to isolate the effects of individual stressors. While general vegetation indices like NDVI are commonly used, they may not offer sufficient specificity for accurate disease detection under field conditions. The development of disease-specific spectral indices can improve accuracy, especially during critical crop stages such as flowering, where timely intervention is essential to prevent significant yield or quality losses (Mahlein et al., 2013).

4.5 Weed management

Traditional weed management methods, which involve the uniform application of herbicides across entire fields, are often inefficient and contribute to increased risks of off-site pesticide loss (Davis and Pradolin, 2016). In contrast, applying herbicides at variable rates based on actual need can significantly improve treatment efficiency, lower input costs, and reduce environmental pollution (Gundy and Dille, 2022). RS technologies have been widely utilized to map weed patches in agricultural fields, enabling site-specific weed management strategies (Huang et al., 2025). Weeds can be distinguished from crops based on their unique spectral signatures, which are linked to their distinct phenological or morphological traits. In recent years, ML has become a powerful tool for image classification in weed mapping, offering high levels of accuracy and efficiency (Sa et al., 2018; Su et al., 2022). UAVs have emerged as a preferred RS platform for weed detection due to their ability to capture high-resolution imagery (as fine as 5 cm), which is necessary for accurate mapping (Huang et al., 2018). For example,(Sa et al., 2018) proposed an approach for dense semantic weed classification using multispectral imagery acquired by a micro aerial vehicle. The study used the recently developed SegNet architecture, which enables dense semantic labeling while supporting multiple input image channels and incorporating class balancing, tailored to sugar beet and weed datasets. Lambert et al. (2018) evaluated the effectiveness of unmanned aerial systems for mapping grass weeds in wheat fields. The study investigated the accuracy of aerial imagery to estimate in-field weed densities using statistical models.

5 Current progress, knowledge gaps, and implementation challenges

RS offers significant potential across nearly all stages of PA, from land preparation to harvesting. The integration of high spatial-resolution, multitemporal satellite imagery with low-cost UAVs and commercially available ground-based proximity sensors has transformed the operational landscape of PA. Numerous analytical approaches, including empirical models, regression techniques, and a broad spectrum of ML algorithms, have been employed to harness RS data for PA applications. Various vegetation indices derived from multispectral and hyperspectral imagery have been developed and validated for tasks such as variable-rate fertilizer application, irrigation scheduling, disease detection, weed mapping, and yield forecasting. Despite these advancements, several critical challenges hinder the large-scale adoption of RS technologies in both commercial and smallholder agricultural systems. While much satellite data is freely available, effective utilization often demands advanced technical skills and specialized software for image preprocessing, calibration, and analysis. For example, radiometric, geometric, and atmospheric corrections are essential preprocessing steps that require domain expertise. Moreover, many PA applications, particularly those related to disease and weed detection, require imagery with high spatial, spectral, and temporal resolutions. Publicly available satellite datasets often lack this level of detail. In addition, factors such as cloud cover and inconsistent solar irradiance can compromise image usability, especially in optical RS. In such contexts, UAV-based imagery serves as a viable low-cost alternative, especially for small-scale farms. However, deploying UAVs also involves regulatory compliance (e.g., drone pilot licensing), operational planning, and data processing using specialized software. Hyperspectral sensors mounted on UAVs or satellites can capture detailed crop biophysical information, yet they remain costly and computationally intensive to process. To address these limitations, there is a growing need to develop advanced information and communication technologies, as well as spectral decomposition and chemometric techniques, to convert complex sensor data into actionable agronomic insights. ML methods show strong potential to derive continuous spatial and temporal datasets from instantaneous satellite observations. When combined with knowledge-driven, physically based models, these hybrid approaches can enhance decision support systems for PA. Despite the extensive research on RS applications in agriculture, there remains a lack of standardized, reproducible frameworks that are robust across diverse climatic, soil, crop, and management conditions. The accuracy of RS-based analyses is influenced by multiple factors, including the spatial, spectral, and temporal resolution of imagery; atmospheric conditions; crop type and phenological stage; and the analytical methodologies employed (e.g., regression, ML, or physically based models). For instance, surface energy balance models used for ET estimation often exhibit significant spatial and temporal variability, introducing uncertainty into irrigation scheduling and other PA decisions. Further research is needed to quantify and understand the spatio-temporal uncertainties associated with RS-based estimates of ET, soil moisture, disease stress, and other crop health indicators. Importantly, crop spectral signatures are influenced by a combination of site-specific variables (e.g., soil type, topography), management practices, and concurrent biotic and abiotic stressors (e.g., nutrient deficiency, water stress, pest and disease pressure). As a result, detection models that perform well under controlled conditions may not generalize effectively to real-world field environments where multiple interacting stressors exist. Given the complexity of RS data processing and the expertise required, there is a pressing need for the development of simplified, reliable workflows for real-time image processing, analysis, and decision-making. Major gaps persist in the creation of user-friendly tools and frameworks that can translate RS data into actionable insights for end-users. The development of accurate, scalable, and accessible RS solutions will be key to enabling broader adoption in both commercial and noncommercial PA systems.

6 Conclusions

The growing pressures from population expansion and the increasing global demand for agricultural production underscore the critical need for enhanced agricultural resource management. Achieving sustainable agricultural practices requires accurate, real-time data regarding the types, quality, quantity, and distribution of key input resources. RS technologies present an invaluable tool for obtaining these data, thereby facilitating improvements in agricultural production systems. Globally, agricultural surveys are increasingly deployed to collect empirical data on crops, rangelands, livestock, and other agricultural resources. This data is essential for the effective management of limited and dwindling natural resources. Surveys on the principles of PA help in optimizing resource allocation and distribution across diverse economic sectors. RS technologies, leveraging the biophysical properties of crops and soils, hold significant potential to transform the detection, characterization, and monitoring of agricultural productivity. The integration of RS-derived insights with ground-truth data enhances the robustness and accuracy of decision-making processes in agricultural management. The application of RS within the domain of PA is set to revolutionize agricultural practices. By providing spatially explicit, real-time information on crop health, yield forecasts, irrigation optimization, and nutrient management, RS enables informed decision-making. This, in turn, helps farmers optimize resource utilization, increase production efficiency, and promote environmental sustainability. Despite the challenges inherent in the adoption and integration of RS technologies, ongoing advancements in research and technology continue to enhance their applicability. As these innovations advance, RS will increasingly enable more accessible, efficient, and sustainable agricultural practices, ultimately benefiting both farmers and the global agricultural industry.

References

Abbasi, N., Nouri, H., Didan, K., Barreto-Muñoz, A., Borujeni, S.C., Opp, C., Nagler, P., Thenkabail, P.S., Siebert, S., 2023. Mapping vegetation index-derived actual evapotranspiration across croplands using the Google Earth engine platform. Remote Sens. 15 (4), 1017. doi:10.3390/rs15041017.

Ale, S., Omani, N., Himanshu, S.K., Bordovsky, J.P., Thorp, K.R., Barnes, E.M., 2020. Determining optimum irrigation termination periods for cotton production in the Texas High Plains. Trans. ASABE 63 (1), 105–115.

Ali, M.M., Al-Ani, A., Eamus, D., Tan, D.K.Y., 2017. Leaf nitrogen determination using non-destructive techniques—A review. J. Plant Nutr. 40 (7), 928–953. doi:10.1080/01904167.2016.1143954.

Allen, R., Irmak, A., Trezza, R., Hendrickx, J.M.H., Bastiaanssen, W., Kjaersgaard, J., 2011. Satellite-based ET estimation in agriculture using SEBAL and METRIC. Hydrol. Process. 25 (26), 4011–4027. doi:10.1002/hyp.8408.

Alves Varella, C.A., Gleriani, J.M., Dos Santos, R.M., 2015. Precision Agriculture and Remote Sensing Sugarcane: Agricultural Production, Bioenergy and Ethanol. Elsevier Inc., Brazil, pp. 185–203. doi:10.1016/B978-0-12-802239-9.00009-8.

Amankulova, K., Farmonov, N., Mukhtorov, U., Mucsi, L., 2023. Sunflower crop yield prediction by advanced statistical modeling using satellite-derived vegetation indices and crop phenology. Geocarto Int. 38 (1) 2197509. doi:10.1080/10106049.2023.2197509.

Argento, F., Anken, T., Abt, F., Vogelsanger, E., Walter, A., Liebisch, F., 2021. Site-specific nitrogen management in winter wheat supported by low-altitude remote sensing and soil data. Precis. Agric. 22 (2), 364–386. doi:10.1007/s11119-020-09733-3.

Avola, G., Di Gennaro, S.F., Cantini, C., Riggi, E., Muratore, F., Tornambè, C., Matese, A., 2019. Remotely sensed vegetation indices to discriminate field-grown olive cultivars. Remote Sens. 11 (10), 1242. doi:10.3390/rs11101242.

Bawa, A., Samanta, S., Himanshu, S.K., Singh, J., Kim, J., Zhang, T., Chang, A., Jung, J., DeLaune, P., Bordovsky, J., Barnes, E., Ale, S., 2023. A support vector machine and image processing based approach for counting open cotton bolls and estimating lint yield from UAV imagery. Smart Agric. Technol. 3, 100140. doi:10.1016/j.atech.2022.100140.

Bhatti, S., Heeren, D.M., Barker, J.B., Neale, C.M.U., Woldt, W.E., Maguire, M.S., Rudnick, D.R., 2020. Site-specific irrigation management in a sub-humid climate using a spatial evapotranspiration model with satellite and airborne imagery. Agric. Water Manag. 230, 105950. doi:10.1016/j.agwat.2019.105950.

Bonfante, A., Monaco, E., Manna, P., De Mascellis, R., Basile, A., Buonanno, M., Cantilena, G., Esposito, A., Tedeschi, A., De Michele, C., Belfiore, O., Catapano, I., Ludeno, G., Salinas, K., Brook, A., 2019. LCIS DSS—an irrigation supporting system for water use efficiency improvement in precision agriculture: a maize case study. Agric. Syst. 176, 102646. doi:10.1016/j.agsy.2019.102646.

Cao, J., Yang, H., 2023. A dynamic normalized difference index for estimating soil organic matter concentration using visible and near-infrared spectroscopy. Ecol. Indic. 147, 110037. doi:10.1016/j.ecolind.2023.110037.

Croft, H., Arabian, J., Chen, J.M., Shang, J., Liu, J., 2020. Mapping within-field leaf chlorophyll content in agricultural crops for nitrogen management using Landsat-8 imagery. Precis. Agric. 21 (4), 856–880. doi:10.1007/s11119-019-09698-y.

da Costa, O.P., Inácio, F.D., da Silvada Silva, J.E., Barboza, T.O.C., da Silvada Silva, W.H.B., Lacerda, L.N., Dos Santos, A.F., 2025. Impact of atmospheric corrections on satellite imagery for corn yield prediction using machine learning. Smart Agric. Technol. 12, 101216. doi:10.1016/j.atech.2025.101216.

Dangwal, N., Patel, N.R., Kumari, M., Saha, S.K., 2015. Monitoring of water stress in wheat using multispectral indices derived from Landsat-TM. Geocarto Int. 31 (6), 682–693. doi:10.1007/s11119-016-9492-3.

Dash, P.K., 2023. Remote sensing as a potential tool for advancing digital soil mapping. In: Remote Sensing of Soils: Mapping, Monitoring, and Measurement. Elsevier, India, pp. 357–370. doi:10.1016/B978-0-443-18773-5.00021-1.

Davis, A.M., Pradolin, J., 2016. Precision herbicide application technologies to decrease herbicide losses in furrow irrigation outflows in a Northeastern Australian cropping system. J. Agric. Food Chem. 64 (20), 4021–4028. doi:10.1021/acs.jafc.5b04987.

Debnath, S., Adamala, S., Raghuwanshi, N.S., 2015. Sensitivity analysis of FAO-56 Penman-Monteith method for different agro-ecological regions of India. Environ. Process. 2 (4), 689–704. doi:10.1007/s40710-015-0107-1.

Debnath, S., Mishra, A., Mailapalli, D.R., Raghuwanshi, N.S., 2018. Quantifying yield gap for rice cropping systems in lower gangetic plains. Paddy Water Environ. 16 (3), 601–615. doi:10.1007/s10333-018-0653-z.

Debnath, S., Mishra, A., Mailapalli, D.R., Raghuwanshi, N.S., 2021a. Identifying most promising agronomic adaptation strategies to close rainfed rice yield gap in future: a model-based assessment. J. Water Clim. Change 12 (6), 2854–2874. doi:10.2166/WCC.2021.094.

Debnath, S., Mishra, A., Mailapalli, D.R., Raghuwanshi, N.S., Sridhar, V., 2021b. Assessment of rice yield gap under a changing climate in India. J. Water Clim. Change 12 (4), 1245–-1267. doi:10.2166/wcc.2020.086.

Demir, S., Başyiğit, L., 2023. Evaluating bare soil properties and vegetation indices for digital farming applications from UAV-based multispectral images. Data Sci. Appl. 6 (1), 5–10.

Dong, T., Liu, J., Shang, J., Qian, B., Ma, B., Kovacs, J.M., Walters, D., Jiao, X., Geng, X., Shi, Y., 2019. Assessment of red-edge vegetation indices for crop leaf area index estimation. Remote Sens. Environ. 222, 133–143. doi:10.1016/j.rse.2018.12.032.

Elvanidi, A., Katsoulas, N., Bartzanas, T., Ferentinos, K.P., Kittas, C., 2017. Crop water status assessment in controlled environment using crop reflectance and temperature measurements. Precis. Agric. 18 (3), 332–349. doi:10.1007/s11119-016-9492-3.

Fan, Y., Himanshu, S.K., Ale, S., DeLaune, P.B., Zhang, T., Park, S.C., Colaizzi, P.D., Evett, S.R., Baumhardt, R.L., 2022. The synergy between water conservation and economic profitability of adopting alternative irrigation systems for cotton production in the Texas High Plains. Agric. Water Manag. 262, 107386. doi:10.1016/j.agwat.2021.107386.

Fei, S., Li, L., Han, Z., Chen, Z., Xiao, Y., 2022. Combining novel feature selection strategy and hyperspectral vegetation indices to predict crop yield. Plant Methods 18, 119. doi:10.1186/s13007-022-00949-0.

Fevgas, G., Lagkas, T., Argyriou, V., Sarigiannidis, P., 2025. New vegetation stress assessment approach via WorldView-3 imagery, validated with UAV thermal imaging. Int. J. Remote Sens. 46 (12), 4764–4780. doi:10.1080/01431161.2025.2513562.

Gade, S.A., Madolli, M.J., García-Caparrós, P., Ullah, H., Cha-um, S., Datta, A., Himanshu, S.K., 2025. Advancements in UAV remote sensing for agricultural yield estimation: a systematic comprehensive review of platforms, sensors, and data analytics. Remote Sens. Appl. Soc. Environ. 37, 101418 . doi:10.1016/j.rsase.2024.101418.

Gundy, G.J., Dille, J.A., 2022. Implementing variable-rate herbicide applications based on soil physical properties in grain sorghum. Precis. Agric. 23 (3), 768–790. doi:10.1007/s11119-021-09860-5.

Haboudane, D., Miller, J.R., Pattey, E., Zarco-Tejada, P.J., Strachan, I.B., 2004. Hyperspectral vegetation indices and novel algorithms for predicting green LAI of crop canopies: modeling and validation in the context of precision agriculture. Remote Sens. Environ. 90 (3), 337–352. doi:10.1016/j.rse.2003.12.013.

Hashim, I.C., Shariff, A.R.M., Bejo, S.K., Muharam, F.M., Ahmad, K., Hashim, H., 2020. Application of thermal imaging for plant disease detection. IOP Conf. Ser. Earth Environ. Sci. 540 (1), 012052. doi:10.1088/1755-1315/540/1/012052.

Hendricks, G.S., Shukla, S., Roka, F.M., Sishodia, R.P., Obreza, T.A., Hochmuth, G.J., Colee, J., 2019. Economic and environmental consequences of overfertilization under extreme weather conditions. J. Soil Water Conserv. 74 (2), 160–171. doi:10.2489/jswc.74.2.160.

Himanshu, S.K., Ale, S., Bell, J., Fan, Y., Samanta, S., Bordovsky, J.P., Gitz, D.C., Lascano, R.J., Brauer, D.K., 2023. Evaluation of growth-stage-based variable deficit irrigation strategies for cotton production in the Texas High Plains. Agric. Water Manag. 280, 108222. doi:10.1016/j.agwat.2023.108222.

Huang, Y., Reddy, K.N., Fletcher, R.S., Pennington, D., 2018. UAV low-altitude remote sensing for precision weed management. Weed Technol. 32 (1), 2–6. doi:10.1017/wet.2017.89.

Huang, Y., Sun, X., Rai, N., Yao, H., Reddy, K.N., Jenkins, J., 2025. Remote sensing for precision weed management. Pest Manag. Sci. 81 (8), 4190–4199. doi:10.1002/ps.8833.

Ihuoma, S.O., Madramootoo, C.A., 2019. Sensitivity of spectral vegetation indices for monitoring water stress in tomato plants. Comput. Electron. Agric. 163, 104860. doi:10.1016/j.compag.2019.104860.

Imtiaz, F., Farooque, A.A., Randhawa, G.S., Wang, X., Esau, T.J., Garmdareh, S.E.H., Acharya, B., 2025. Optimizing potato yield mapping and prediction: integrating satellite-based remote sensing and machine learning for sustainable agriculture. Comput. Electron. Agric. 237, 110636. doi:10.1016/j.compag.2025.110636.

Ji, Z., Pan, Y., Zhu, X., Zhang, D., Wang, J., 2022. A generalized model to predict large-scale crop yields integrating satellite-based vegetation index time series and phenology metrics. Ecol. Indic. 137, 108759. doi:10.1016/j.ecolind.2022.108759.

Ju, C., Chen, C., Li, R., Zhao, Y., Zhong, X., Sun, R., Liu, T., Sun, C., 2023. Remote sensing monitoring of wheat leaf rust based on UAV multispectral imagery and the BPNN method. Food Energy Secur. 12 (4), e477. doi:10.1002/fes3.477.

Kang, Y., Gao, F., Anderson, M., Kustas, W., Nieto, H., Knipper, K., Yang, Y., White, W., Alfieri, J., Torres-Rua, A., Alsina, M.M., Karnieli, A., 2022. Evaluation of satellite Leaf Area Index in California vineyards for improving water use estimation. Irrig. Sci. 40 (4–5), 531–551. doi:10.1007/s00271-022-00798-8.

Kaya, Y., Gürsoy, E., 2023. A novel multi-head CNN design to identify plant diseases using the fusion of RGB images. Ecol. Inform. 75, 101998. doi:10.1016/j.ecoinf.2023.101998.

Kumar, H., Srivastava, P., Lamba, J., Diamantopoulos, E., Ortiz, B., Morata, G., Bondesan, L., 2022b. Site-specific irrigation scheduling using one-layer soil hydraulic properties and inverse modeling. Agric. Water Manag. 273, 107877.

Kumar, H., Srivastava, P., Lamba, J., Lena, B., Diamantopoulos, E., Ortiz, B., Bondesan, L., 2023. A methodology to optimize site-specific field capacity and irrigation thresholds. Agric. Water Manag. 286, 108385.

Kumar, H., Srivastava, P., Lamba, J., Ortiz, B.V., Way, T.R., Sangha, L., Takhellambam, B.S., Morata, G., Molinari, R., 2022a. Within-field variability in nutrients for site-specific agricultural management in irrigated cornfield. J. ASABE. 65 (4), 865–880. doi:10.13031/ja.15042.

Lambert, J.P.T., Hicks, H.L., Childs, D.Z., Freckleton, R.P., Gonzalez-Andujar, J., 2018. Evaluating the potential of Unmanned Aerial Systems for mapping weeds at field scales: a case study with Alopecurus myosuroides. Weed Res. 58 (1), 35–45. doi:10.1111/wre.12275.

Lena, B.P., Bondesan, L., Pinheiro, E.A.R., Ortiz, B.V., Morata, G.T., Kumar, H., 2022. Determination of irrigation scheduling thresholds based on HYDRUS-1D simulations of field capacity for multilayered agronomic soils in Alabama, USA. Agric. Water Manag. 259, 107234. doi:10.1016/j.agwat.2021.107234.

Ma, Y., Minasny, B., Malone, B.P., Mcbratney, A.B., 2019. Pedology and digital soil mapping (DSM). Eur. J. Soil Sci. 70 (2), 216–235. doi:10.1111/ejss.12790.

Mahlein, A.-K., Rumpf, T., Welke, P., Dehne, H.-W., Plümer, L., Steiner, U., Oerke, E.-C., 2013. Development of spectral indices for detecting and identifying plant diseases. Remote Sens. Environ. 128, 21–30. doi:10.1016/j.rse.2012.09.019.

Mao, Z.H., Deng, L., Duan, F.Z., Li, X.J., Qiao, D.Y., 2020. Angle effects of vegetation indices and the influence on prediction of SPAD values in soybean and maize. Int. J. Appl. Earth Obs. Geoinf. 93, 102198. doi:10.1016/j.jag.2020.102198.

Mokhtari, A., Noory, H., Pourshakouri, F., Haghighatmehr, P., Afrasiabian, Y., Razavi, M., Fereydooni, F., Naeni, A.S., 2019. Calculating potential evapotranspiration and single crop coefficient based on energy balance equation using Landsat 8 and Sentinel-2. ISPRS J. Photogramm. Remote Sens. 154, 231–245. doi:10.1016/j.isprsjprs.2019.06.011.

Mondal, B.P., Sahoo, R.N., Das, B., Chakraborty, D., Kumar, N., Raj, G.B., 2023. Digital soil mapping: concepts, methods, and applications—remote sensing and GIS perspectives. In: Remote Sensing of Soils: Mapping, Monitoring, and Measurement. Elsevier, India, pp. 195–212. doi:10.1016/B978-0-443-18773-5.00018-1.

Mundada, S., Jain, P., 2025. Predicting soil organic carbon with ensemble learning techniques by using satellite images for precision farming. Sci. Rep. 15, 28760. doi:10.1038/s41598-025-09804-3.

Munusamy, S., Rajasekaran, E., Saraswat, D., Kustas, W.P., Bambach, N., McElrone, A.J., Castro, S.J., Prueger, J.H., Alfieri, J.G., Alsina, M.M., 2025. The utility and applicability of vegetation index-based models for the spatial disaggregation of evapotranspiration. Irrig. Sci. 43 (4), 803–818. doi:10.1007/s00271-024-00963-1.

Nouri, H., Faramarzi, M., Sobhani, B., Sadeghi, S.H., 2017. Estimation of evapotranspiration based on surface energy balance algorithm for land (SEBAL) using Landsat 8 and modis images. Appl. Ecol. Environ. Res. 15 (4), 1971–1982. doi:10.15666/aeer/1504_19711982.

Pareeth, S., Karimi, P., 2023. Evapotranspiration estimation using Surface Energy Balance Model and medium resolution satellite data: an operational approach for continuous monitoring. Sci. Rep. 13 (1), doi:10.1038/s41598-023-38563-2.

Pismennaya, E., Deryugin, A., 2025. Vegetation Indices for Analyzing Sugar Beet Hybrid Variations on Leached Chernozem in the Stavropol Upland, Vol. 1534. Springer Science and Business Media LLC, pp. 120–129. doi:10.1007/978-3-031-98127-2_12.

Rossini, M., Fava, F., Cogliati, S., Meroni, M., Marchesi, A., Panigada, C., Giardino, C., Busetto, L., Migliavacca, M., Amaducci, S., Colombo, R., 2013. Assessing canopy PRI from airborne imagery to map water stress in maize. ISPRS J. Photogramm. Remote Sens. 86, 168–177. doi:10.1016/j.isprsjprs.2013.10.002.

Rubio-Delgado, J., Pérez, C.J., Vega-Rodríguez, M.A., 2021. Predicting leaf nitrogen content in olive trees using hyperspectral data for precision agriculture. Precis. Agric. 22 (1), 1–21. doi:10.1007/s11119-020-09727-1.

Sa, I., Chen, Z., Popovic, M., Khanna, R., Liebisch, F., Nieto, J., Siegwart, R., 2018. weedNet: dense semantic weed classification using multispectral images and MAV for smart farming. IEEE Robot. Autom. Lett. 3 (1), 588–595. doi:10.1109/lra.2017.2774979.

Santosh, D.T., Debnath, S., Maitra, S., Sairam, M., Sagar, L.L., Hossain, A., Moulick, D., 2023. Alleviation of climate catastrophe in agriculture through adoption of climate-smart technologies. In: Chatterjee, U., Shaw, R., Kumar, S., Raj, A.D., Das, S. (Eds.), Climate Crisis: Adaptive Approaches and Sustainability. Springer Nature, Switzerland, pp. 307–332. doi:10.1007/978-3-031-44397-8_17.

Sapkota, A., Roby, M., Peddinti, S.R., Fulton, A., Kisekka, I., 2025. Comparative analysis of evapotranspiration (ET), crop water stress index (CWSI), and normalized difference vegetation index (NDVI) to delineate site-specific irrigation management zones in almond orchards. Sci. Hortic. 339, 113860. doi:10.1016/j.scienta.2024.113860.

Shafi, U., Mumtaz, R., García-Nieto, J., Hassan, S.A., Zaidi, S.A.R., Iqbal, N., 2019. Precision agriculture techniques and practices: from considerations to applications. Sensors (Switzerland). 19 (17), 3796. doi:10.3390/s19173796.

Shou, L., Jia, L., Cui, Z., Chen, X., Zhang, F., 2007. Using high-resolution satellite imaging to evaluate nitrogen status of winter wheat. J. Plant Nutr. 30 (10), 1669–1680. doi:10.1080/01904160701615533.

Singh, R., Kaur, S., Joshi, D.R., Iboyi, J., Dar, E., Sharma, L.K., Singh, H., 2025. Estimating cotton biomass and nitrogen content using high-resolution satellite and UAV data fusion with machine learning. Smart Agric. Technol. 12, 101191. doi:10.1016/j.atech.2025.101191.

Sishodia, R.P., Ray, R.L., Singh, S.K., 2020. Applications of remote sensing in precision agriculture: a review. Remote Sens. 12 (19), 1–31. doi:10.3390/rs12193136.

Stone, K.C., Bauer, P.J., Sigua, G.C., 2016. Irrigation management using an expert system, soil water potentials, and vegetative indices for spatial applications. Trans. ASABE. 59 (3), 941–948. doi:10.13031/trans.59.11550.

Su, J., Yi, D., Coombes, M., Liu, C., Zhai, X., McDonald-Maier, K., Chen, W.-H., 2022. Spectral analysis and mapping of blackgrass weed by leveraging machine learning and UAV multispectral imagery. Comput. Electron. Agric. 192, 106621. doi:10.1016/j.compag.2021.106621.

Thomas, S., Kuska, M.T., Bohnenkamp, D., Brugger, A., Alisaac, E., Wahabzada, M., Behmann, J., Mahlein, A.K., 2018. Benefits of hyperspectral imaging for plant disease detection and plant protection: a technical perspective. J. Plant Dis. Prot. 125 (1), 5–20. doi:10.1007/s41348-017-0124-6.

Wang, J., 2025. Identification of soil texture and color using machine learning algorithms and satellite imagery. Sci. Rep. 15, 30982. doi:10.1038/s41598-025-17166-z.

Zhang, K., Yuan, Z., Yang, T., Zhenzhou, L., Cao, Q., Tian, Y., Zhu, Y., Cao, W., Liu, X., 2020. Chlorophyll meter-based nitrogen fertilizer optimization algorithm and nitrogen nutrition index for in-season fertilization of paddy rice. Agron. J. 112 (1), 288–300. doi:10.1002/agj2.20036.

Zhao, D., Zhen, J., Zhang, Y., Miao, J., Shen, Z., Jiang, X., Wang, J., Jiang, J., Tang, Y., Guofeng, W., Disney, M., Yinghai, K., 2023. Mapping mangrove leaf area index (LAI) by combining remote sensing images with PROSAIL-D and XGBoost methods. Remote Sens. Ecol. Conserv. 9 (3), 370–389. doi:10.1002/rse2.315.

Zhen, Z., Chen, S., Qin, W., Yan, G., Gastellu-Etchegorry, J.P., Cao, L., Murefu, M., Li, J., Han, B., 2020. Potentials and limits of vegetation indices with BRDF signatures for soil-noise resistance and estimation of leaf area index. IEEE Trans. Geosci. Remote Sens. 58 (7), 5092–5108. doi:10.1109/TGRS.2020.2972297.

Zhu, W., Rezaei, E.E., Nouri, H., Sun, Z., Jing, L., Danyang, Y., Siebert, S., 2022. UAV-based indicators of crop growth are robust for distinct water and nutrient management but vary between crop development phases. Field Crops Res. 284, 108582. doi:10.1016/j.fcr.2022.108582.

Integration of computer vision for smart agriculture: Fundamentals and applications

7

Pooja Chouhan[a], *Narendra Singh Chandel*[a], *Subir Kumar Chakraborty*[a], *Dilip Jat*[a], *Abhishek Upadhyay*[b], *and Dilwar Singh Parihar*[c]
[a]Central Institute of Agricultural Engineering, Bhopal, India
[b]Dr. Rajendra Prasad Central Agricultural University, Samastipur, Bihar, India
[c]International Rice Research Institute, Varanasi, India

1 Introduction

Agriculture has been the cornerstone of human civilization, serving as the earliest and most enduring industry essential to the survival and development of societies. Over millennia, it has evolved into a multifaceted sector, interwoven with the cultural, economic, and social fabric of nations. However, as the global population continues to grow and climate change exacerbates environmental uncertainties, agriculture faces unprecedented challenges. Addressing these challenges requires not only traditional farming practices but also leveraging the transformative potential of modern digital tools and sensing technologies.

India exemplifies the centrality of agriculture in a nation's economy and society. Home to over 55% of its population who depend on farming for their livelihood, the agricultural sector forms the backbone of rural India. Key crops like rice, wheat, sugarcane, and oilseeds dominate the country's agricultural landscape, alongside nonfood crops such as rubber, cotton, and jute. Agriculture not only provides employment to over 60% of the population but also contributes approximately 17%–18% to the nation's Gross Domestic Product (GDP), highlighting its significance in India's socio-economic framework. Globally, India ranks second in farm output, emphasizing the strategic role of agriculture in ensuring food security and economic stability (Singh and Sahni, 2019).

The agricultural sector, despite its prominence in India and other developing nations, has always lagged behind in embracing digital innovations. However, the winds are changing radically. From now on, as the agriculture sector advances with artificial intelligence (AI), machine learning (ML), and computer vision technologies, it is entering the epoch of the digital age (Upadhyay et al., 2025b). These technologies assume critical importance in rethinking farming practices to address the pressing issues of modern farming namely, food security and the impacts of climate change on agriculture.

Precision Technologies for Digital Agriculture. https://doi.org/10.1016/B978-0-443-40513-6.00004-4

Climate change constitutes a two-fold challenge yet-faced by the agricultural sector worldwide food insecurity for the growing population and the impacts of climate change on agriculture (Bali et al., 2023). The increased temperature, unpredictable precipitation patterns, and more frequent extreme weather events have started to destabilize traditional farming practices. This has already led to food shortages and famines in regions such as Sub-Saharan Africa and South Asia (Ryan et al., 2023). In India—a country often too dependent on monsoon-fed agriculture—this has further aggravated uncertainty, overshadowing crop yields and placing a heavy burden on millions of farming-related livelihoods.

Technological advancement is an important factor in this regard. The use of digital tools and sensing technologies, powered by AI and computer vision, provides actionable insights that empower farmers (Chandel et al., 2025). These tools enable precise resource management in irrigation and reduce dependence on chemical input fertilizers and pesticides. Their integration with AI systems in agriculture to reduce pesticide use by 60% and cut water use by 50% was considered important among the many expectations and visions of the World Economic Forum. These advances increase productivity and greatly contribute to environmental sustainability, conserving critical resources for the benefit of future generations (Bechar and Vigneault, 2016).

Historically, agriculture has been slow in technological adoption compared to other industries. However, the current digital transformation brings new tools and techniques to the forefront. Among these, computer vision and digital sensing technologies have emerged as game-changers, providing solutions to age-old problems in crop management, disease detection, and resource optimization.

Computer vision, which pertains to AI, links visual information reaching the machine and creates a computational model of the perception of those pictures. Once integrated with digital sensing tools, it transforms traditional agricultural practices, including automatic monitoring of crop health, disease detection, and optimization of fertilizer and pesticide applications. In this way, for instance, advanced algorithms can analyze leaf images and recognize disease in its early stages, thus offering timely interventions to prevent extensive crop damage.

Computer vision, a subdiscipline of AI, is working miracles in the healthcare, security, and manufacturing industries. Its application in agriculture is comparatively revolutionary since it could be used to resolve some of the biggest challenges facing farmers. Employing visual data interpretation and analysis by machines, it supports large-scale applications ranging from disease detection to yield estimation. Powered by smart algorithms, it has the capability to detect plant diseases, identify pests, and ascertain crop status with incredible precision.

Traditional plant disease diagnosis methods are mainly based on visual examination and are hence time-consuming, prone to errors, and heavily reliant on an expert's knowledge. On the other hand, the integration of computer vision with deep learning (DL) models provides solutions for these tasks, enabling accurate and scalable outcomes by automating the process. Moreover, real-time field monitoring through advanced image processing techniques, such as segmentation and object detection (Chouhan et al., 2020), can ensure timely interventions so that yield losses do not occur. The use of computer vision in agriculture is about more than just efficiency; it is about sustainability as well. With tools for precision agriculture, farmers can apply inputs, such as adjustments to substances like water, fertilizers, and pesticides, which directly support the principle of minimizing waste and reducing environmental impact. Additionally, the automation of labor-intensive processes like harvesting and sorting results in less human effort and significantly higher productivity.

This chapter discusses a few applications of computer vision in agriculture involving digital tools and sensing technologies. The chapter covers an exhaustive review of the state-of-the-art in computer vision plant disease detection, noting methodologies and their case studies. The chapter will also consider AI-driven systems as they address larger-scale issues in agriculture, such as resource optimization and environmental sustainability. Pulling together recent research and practical applications, this chapter will hopefully make the strongest case for articulating the prospects of computer vision in agriculture.

2 Computer vision-based sensing technology

The AI system, computer vision, has functioned as an extra set of eyes for machines with the help of modern cameras and computational algorithms, which are developed to mimic human vision (Upadhyay et al., 2025a). With the combination of capturing hands-on flow by cameras and computational systems, it has enhanced the automation ability of AI, especially for several agricultural tasks. These technologies facilitate the transmission of crop, livestock, and environmental farm data to virtual platforms, enabling object detection, tracking, and the analysis of complex visual information for automated agricultural operations. In the last few decades, expert systems relying on these technologies have been incorporated into various agricultural operations and have reported great success, improved efficiency, and high accuracy, as noted by researchers (Rico-Fernández et al., 2019). The hierarchy of key processes in computer vision is illustrated in Fig. 7.1.

The computerized systems are primarily based on Red, Green, and Blue (RGB) color models, where each pixel is represented using three wavelengths of light—red, green, and blue. These systems are capable of automatic grading and inspection of

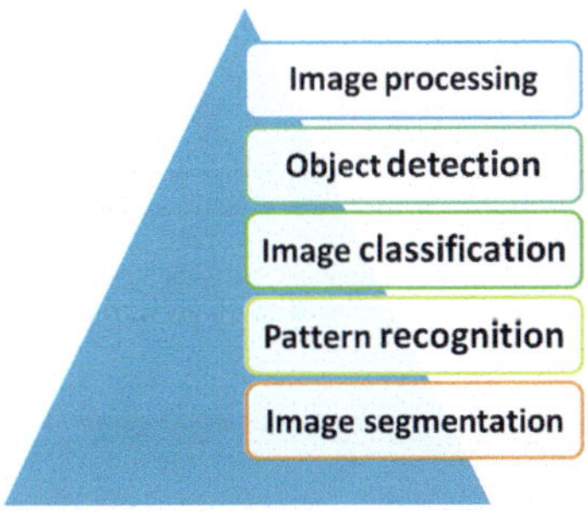

Fig. 7.1 Hierarchy of key processes in computer vision.

features such as flatness, shape, color, size, and defects. Such systems have been applied in agricultural activities, such as grading and sorting of fruits. However, when it comes to detecting certain defects, traditional systems fall short of performing their tasks in some areas. Traditional systems are yet to perform tasks efficiently due to the resemblance of some defects to the natural color and texture of crop surfaces.

2.1 Image acquisition techniques in computer vision

RGB image

RGB imaging uses conventional digital cameras to collect visible light reflected off objects, which it translates into digital images (Fig. 7.2). Visible light is usually characterized by the 400–700 nm range in the electromagnetic spectrum, which includes the red, green, and blue bands. It is suggested that this imaging process has been extensively utilized by researchers for precision agriculture. Zhang et al. (2022) utilized a digital SLR camera to detect diseases in grapevines. Similarly, Khan et al. (2021) employed an RGB camera mounted on an unmanned aerial vehicle (UAV) to capture images of coriander fields, facilitating the identification of spraying areas for effective weed control.

Multispectral computer vision

Multispectral systems, on the other hand, render monochromatic images for two or more wavelength bands. This offers a straightforward yet effective means of analyzing agricultural conditions. These systems are, therefore, particularly useful for monitoring large areas, such as fields or orchards, where general insight is required into either the health or hydration of the plants. With its application within the broader agricultural landscape, computer vision has transformed tasks in agriculture from manual labor and human expertise to automated processes. Modern computer vision systems involve complex AI applications coupled with cameras and sensors that surpass human precision and insight. All of this necessitates high-quality training datasets to be curated, a fact that limits the development of accurate and effective computer-vision models (Fig. 7.3).

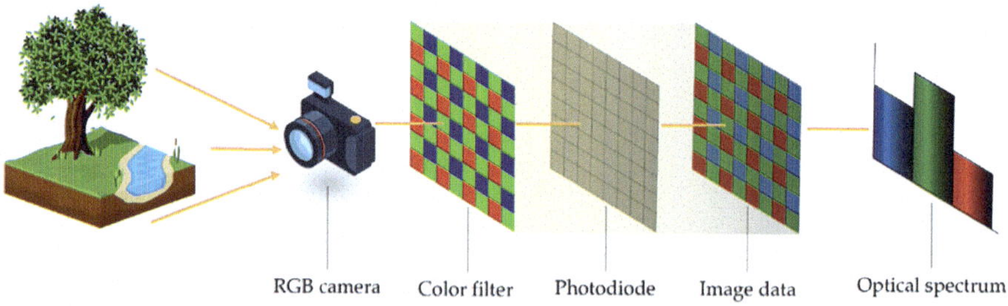

Fig. 7.2 Overview of the RGB camera.

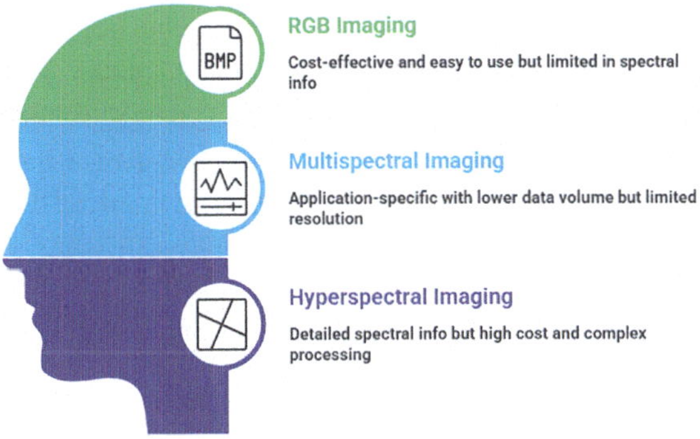

Fig. 7.3 Advancements in computer vision.

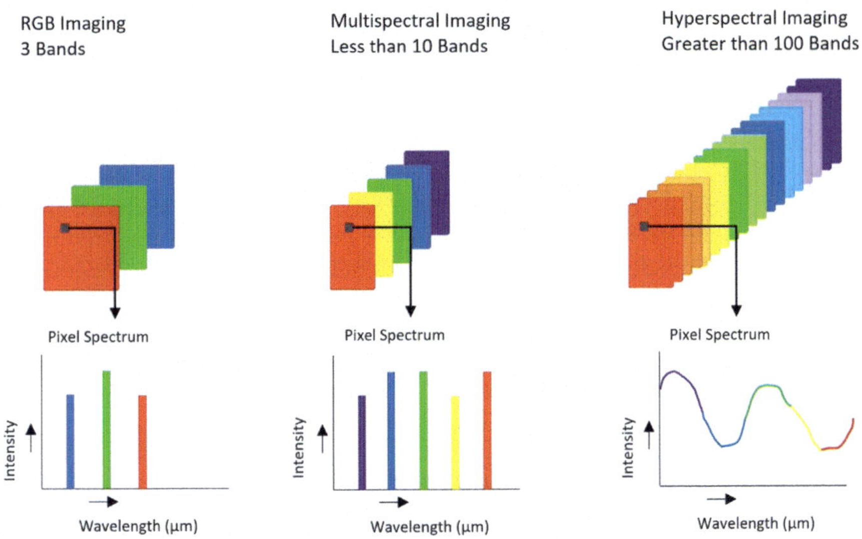

Fig. 7.4 Image acquisition techniques with bands.

Hyperspectral computer vision

Hyperspectral imaging is an imaging technique merged with spectroscopic techniques aimed at extracting detailed spectral information for each and every pixel in an image (Lorente et al., 2012). A hypercube, or data cube, is formed; in other words, it packages a considerable amount of different species included in the two-dimensional image. These systems employ techniques such as point scanning, line scanning, and area scanning for the generation of hyperspectral image cubes (Fig. 7.4). Such data provide unique views of plant status and help by capturing slight changes in plant health, soil composition, and environmental stresses.

Aerial photography

Drones are capable of capturing aerial images through structured flight paths and stitching the images together to generate maps of the entire study area, with sufficient image overlap to achieve higher precision (Alsalam et al., 2017). Moreover, the advent of drones integrated with computer vision technology has created an immense revolution in monitoring vital crop growth stages and has delivered high efficiency and productivity compared to conventional systems. Recent research outlined by Egi et al. (2022) considers the implications of such integration highly relevant. In AI-powered drones targeting this purpose, tomato flowers and fruiting stages were detected and quantified as prime indicators for evaluating crop yield. Such information is undoubtedly useful for farmers, being utilized as time-sensitive data to help make informed decisions to amplify yield.

2.2 Image preprocessing and augmentation

The raw image information captured may contain noise, distortions, or inconsistencies, which can affect the performance of computer vision models. The first and foremost step is to enhance image quality and standardize the input data to make the analysis straightforward and accurate. This involves operations such as resizing, cropping, noise reduction, and contrast enhancement to ensure uniform dimensions and clarity across the dataset. It is followed by normalizing or scaling the pixel values, which greatly facilitates convergence for the models being trained. Augmentation is then effectively employed to artificially enlarge the size of the training datasets, thereby increasing the generalization of the models built. This includes operations such as rotation, flips, translations, brightness adjustments, or adding Gaussian noise, thus replicating real-world variations. All these preprocessed and augmented images form a robust dataset that a model can use for effective interpretation and processing of varied agricultural scenarios.

2.3 Types of data labeling in agriculture

The next step after this step was data labeling, which is very important in the field of applying computer vision to agriculture-related tasks. The process of labeling coins will enable AI models to train for different applications. Some of the next few types of labels include:

1. **Detection of crop in bounding boxes**
   Bounding box annotation creates veridical perception models capable of distinguishing crops from weeds by identifying and eliminating unwanted weed types beforehand. In this case, crop detection streamlines the process of precision weeding and the automation of harvest.
2. **Analysis of size and shape through key points**
   Key point annotation is done, as its name suggests, to assign a landmark value to specified points on objects such as fruits and vegetables; it is useful for assessing size and form, maturity, and the control of growth stages.
3. **Semantic segmentation for crop inspection**
   Semantic segmentation assigns a class label to every pixel in an image; hence, there is potential for applications where there is significant interest in crop detection, monitoring

cow respiration, and classification of agricultural datasets. By segmenting and classifying visual data, crop health monitoring, pest detection, and yield estimation all become possible with the utmost precision.

4. **Polylines for classifying crop lanes**
   Polyline annotation is used to delineate crop lanes, irrigation patterns, and other linear features in large farms. This method supports the development of robust models for disease detection, irrigation analysis, and harvest prediction, contributing to more efficient farm management.
5. **Geographic information system and geospatial data annotation**
   Geographic information system and geospatial data annotation enhance AI-driven farming drones and automated systems. By labeling and interpreting spatial data, these models can monitor farm conditions, track crop health, and assess the functionality of agricultural equipment. This results in improved farm productivity and operational efficiency.

2.4 Model development and training

Model development and training are thus highly germane steps in creating the effective functioning of a computer vision system. This is an iterative process that starts with the selection of an appropriate model architecture based on the specific application, such as convolutional neural networks (CNN) for image classification or Mask R-CNN for detection and segmentation. The dataset is then divided into training, validation, and testing subsets to facilitate an unbiased evaluation. During the training process, for each iteration, the model learns a pattern or feature from the input data by optimizing necessary parameters using techniques such as backpropagation and a variant of gradient descent. Hyper parameter optimization seeks to further tune model performance for parameters like the learning rate, batch size, or number of layers. During training, regular validation of model performance is often needed to periodically check for overfitting and to make the model more generalizable. Finally, testing is conducted using appropriate measures of accuracy, precision, recall, and general robustness. The developed model is then fine-tuned and deployed for real-world agricultural applications (Fig. 7.5).

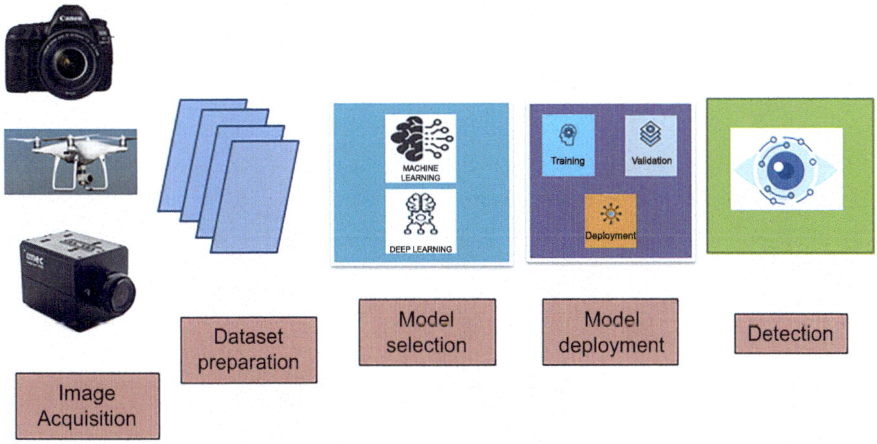

Fig. 7.5 Process of model development and deployment.

3 Application of computer vision in agriculture

Key applications of computer vision in agriculture highlight its contribution to improved precision and operational efficiency (Fig. 7.6).

3.1 Grading of fruits and vegetables

Fruits and vegetables are subject to deterioration in their quality as a result of diseases, physical damage, or over ripening. For the maintenance of food quality, efficient sorting and grading are crucial in reducing food waste. Diseases such as scab, rot, and blotch, which affect fruits during harvesting and subsequent handling phases, cause severe losses both in yield and quality. These setbacks, therefore, highlight the reliability and objectivity required in assessing the quality of fruits and vegetables, especially given the rising demand for better-quality food products by consumers. In view of these challenges, computer vision systems have established themselves as nondestructive and automated methods for cost-effective quality assessment. A widely used image processing-based inspection modality evaluates the grading, sorting, and disease detection processes for fruits and vegetables. Grading systems play a pivotal role in postharvest quality assessment. Arakeria (2016) developed an efficient tomato grading system utilizing computer vision techniques. This system consists of two main phases: hardware and software. The hardware automates the process of capturing images and sorting fruits into appropriate containers, reducing manual handling. Meanwhile, the software employs image processing methods to assess ripeness and detect defects, achieving a 96.47% efficiency rate in evaluating tomato quality. Similarly, Ali and Thai (2017) proposed a system that combines mechanical and electrical components to evaluate fruit quality based on external factors like decay and surface defects, offering a more time-efficient and accurate

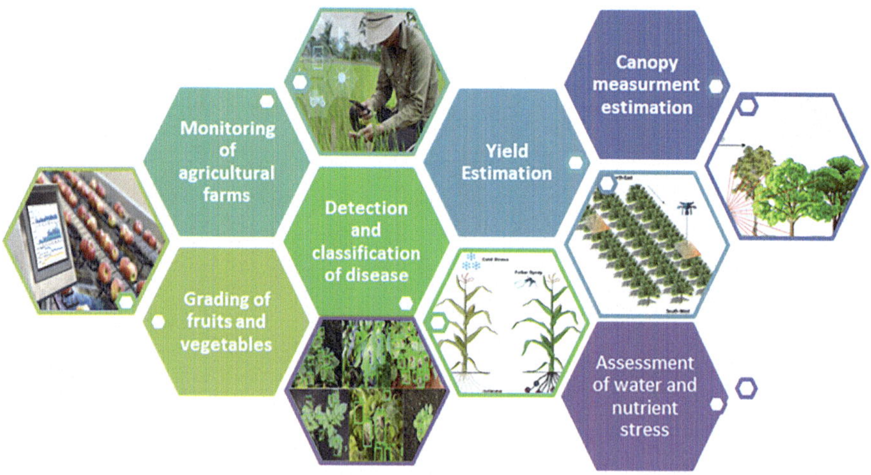

Fig. 7.6 Application of computer vision in agriculture.

grading process. Computer vision has emerged as a disruptive tool for agriculture, enabling efficient image interpretation and analysis for knowledge extraction. Image classification, segmentation, and object detection have become important areas to enhance agricultural imaging applications, often coupled with neural networks and DL. DL automatically extracts features, thus eliminating the need for manual feature extraction, and is highly useful in RGB and thermal imagery analysis for crop-stress detection. Chakraborty et al. (2022) elucidated the application of deep-learning models for preharvest and postharvest assessments, utilizing RGB images.

Recent advancements in computer vision have significantly improved fruit grading accuracy. Ismail and Malik (2022) developed a real-time system using a Raspberry Pi with a camera and touchscreen, capable of segmenting and grading multiple fruits in a single image. Trained on apple and banana datasets with the Efficient Net model, the system achieved average accuracies of 99.2% for apples and 98.6% for bananas. A stacking ensemble further improved recognition rates, while real-time testing showed accuracies of 96.7% for apples and 93.8% for bananas. These AI-based grading systems reduce manual effort and provide scalable, efficient, and automated solutions for postharvest quality assessment.

3.2 Detection and classification of disease

ML and computer vision techniques provide great promise for the early detection (Fig. 7.7) of plant diseases—this is central to halting the spread of infections and minimizing the need for harmful pesticides. Crop diseases are a leading threat to global food security, causing enormous economic losses and yield reductions. However, innovation in AI technologies is providing farmers with novel tools to combat these emerging challenges (Upadhyay et al., 2024, 2025c). By implementing ML and computer vision in the field, a farmer can detect and diagnose aspects of disease so that precise interventions can be undertaken to promote sustainable agriculture and reduce the impact of diseases.

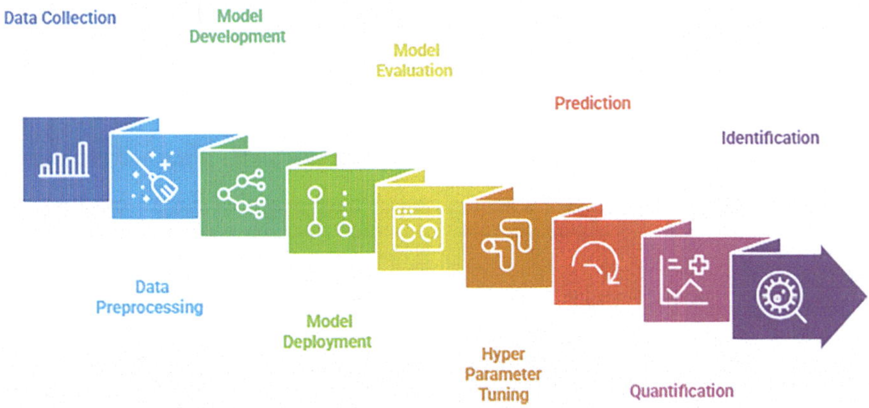

Fig. 7.7 Deep learning-based disease detection process.

Computer vision, with an added twist from ML, has emerged as a transformative technology that allows for crop monitoring on a large scale. For instance, Ahmed and Reddy (2021) proposed a mobile system for plant leaf disease diagnosis that uses ML. In this system, CNNs are used as the DL engine to classify the diseases of plants into 38 distinct classes. The researchers created and compiled an extensive dataset of healthy and diseased plant leaf images, comprising (Ahmed and Reddy, 2021) 96,206 images for training, validation, and testing purposes.

The Android-based application provides farmers with an interface that allows the capture of infected plants directly in the field. The results of the classification are available to users within seconds of identifying an infection and also include an interpretation confidence score. This helps farmers make urgent decisions about the next possible course of action. This tool not only aids in crop health management but also reduces the likelihood of pesticide or fertilizer misapplication. The performance evaluation of the system demonstrated a promising 94% accuracy in disease identification across 14 crop species, underscoring its credibility and potential for real-world application. The convergence of AI with accessible tools, such as smartphones, is effectively democratizing advanced agricultural diagnostics, providing farmers worldwide with an efficient way to combat crop diseases. This development represents a significant leap in the sustainable agricultural landscape: a farmer, equipped with actionable insights, can ensure food security and optimize resource utilization (Fig. 7.8).

Fig. 7.8 Heat map of grape, potato, and coffee leaf (Zhou et al., 2024).

3.3 Assessment of water and nutrient stress

Water and nitrogen stresses are abiotic stresses resulting from an excess or absence of water and nitrogen in plants. This nitrogen level can considerably affect the yield of the crop. Within AI and computer vision, some of the major applications in agriculture include the estimation of nitrogen in plants and the creation of prescription maps for accurate site-specific fertilizer application. The ideal level of nitrogen varies with regard to crop species. Likewise, water stress in plants can be analyzed using automated techniques, promising an intelligent irrigation system that applies water based on real plant needs, thereby minimizing water waste.

Nitrogen deficiency is closely associated with leaf color and texture, thus rendering imaging techniques effective in predicting nitrogen levels. Numerous authors who sought the determination of nitrogen levels were cited in the literature. Haider et al. (2021) analyzed some color characteristics using RGB, and this was followed by a comparison against reference values to predict the nitrogen content of each leaf. Similarly, Zhao et al. (2021) employed plant growth indices and canopy cover analysis to assess nitrogen levels in winter wheat. Beyond RGB, noninvasive techniques, including visible-NIR (Near Infrared), hyperspectral, and thermal imaging, provide high-throughput means of assessing water and nutrient stress, offering practical alternatives to labor-intensive invasive methods. Thermal imaging, in particular, has proven effective for detecting soil water stress, although sensor cost and environmental factors such as wind and humidity can limit its wider application.

DL approaches have further enhanced stress detection capabilities. Chandel et al. (2021) applied models such as AlexNet, GoogLeNet, ResNet-50, and Inception V3 to classify soybean, maize, wheat, and okra crops as water-stressed or nonstressed (Fig. 7.9), demonstrating the strong potential of these techniques. Timely detection is crucial for efficient irrigation management and minimizing water loss.

In a two-year study on winter wheat, Chandel et al. (2021) employed an in-house thermal-RGB imaging system to capture images under flood and sprinkler irrigation at 100%, 75%, 50%, and 25% of crop evapotranspiration across critical growth stages (20, 35, 70, 95, and 108 days after sowing). ResNet-50 achieved an accuracy of 96.9% with RGB imagery and 98.4% with thermal imagery. In addition, a Deep Learning–

Fig. 7.9 Sample training dataset images of water-stressed and nonstressed wheat and maize crops (Chandel et al., 2024).

Long Short Term Memory (DL-LSTM) model demonstrated 96.7% predictive accuracy, highlighting the effective combination of computer vision and thermal imaging for stress assessment.

A subsequent study by Chandel et al. (2024) focused on okra, analyzing 3200 thermal-RGB images under similar irrigation treatments using ResNet-50 and MobileNetV2. Thermal images consistently outperformed RGB, with ResNet-50 achieving 87.9% and MobileNetV2 84.3% accuracy, compared to 78.6% and 74.1% for RGB images, respectively. Flood irrigation produced the highest yield (10,666 kg/ha) and water use efficiency (1.16 kg/m^3). These studies underscore the growing significance of thermal and RGB imaging combined with DL in the precise assessment of water and nutrient stress, supporting precision irrigation strategies and sustainable crop production.

These researchers highlight the growing importance of thermal and RGB imagery through refined DL techniques in underwriting assessments for water and nutrient stresses. Such technologies propel agrobotics to implement operational insights that drive precision agriculture in irrigation scheduling and resource allocation (Fig. 7.9) thereby addressing some setbacks to sustainable farming.

To provide a clearer comparison of methodologies, key findings, limitations, and suggested improvements, Table 7.1 summarizes the critical analysis of disease detection and stress assessment studies further.

3.4 Canopy measurement

Accurate canopy height measurement is critical for assessing crop health and plant vigor, as it directly correlates with photosynthetic capacity, biomass accumulation, and yield potential. Canopy height serves as a vital biophysical indicator, providing

Table 7.1 Comparative analysis and condensation of disease detection and stress assessment studies.

S. No.	Area of study	Methods	Key findings	Limitations	Suggested improvements
1.	Disease detection	CNN, ResNet50, YOLOv5, Mask R-CNN	High accuracy (85%–95%) in controlled datasets	Small dataset sizes, lack of real-world validation, high computational demand	Benchmark datasets needed; lightweight models for mobile/edge deployment
2.	Stress assessment	NDVI, RGB + Deep Learning, UAV-based imaging	Effective for early stress detection	Overlapping indicators with disease detection; high cost of UAVs	Combine stress + disease detection in integrated models; explore low-cost multispectral sensors

insights into plant growth stages, stress responses, and overall agricultural productivity. Traditional measurement methods, such as manual ruler-based sampling or handheld sensors, are labor-intensive, time-consuming, and impractical for large-scale operations. To address these limitations, recent advancements in remote sensing and semiautomated computational models have revolutionized canopy height estimation, enabling scalable, efficient, and precise monitoring. Valluvan et al. (2023) developed a semiautomated framework for estimating crop canopy height using drone-captured aerial imagery. Their model integrates Structure-from-Motion photogrammetry, dense point cloud reconstruction, and polygon triangulation to derive 3D canopy structures. Deployed in maize fields in semiarid regions of central South India, the system demonstrated strong agreement with ground truth measurements, achieving a correlation coefficient (r) of 0.85 and a root mean square error of 14.17 cm. These results validate the model's reliability for large-scale applications under heterogeneous field conditions. Remote sensing technologies, such as Light Detection and Renging (LiDAR) and multispectral optical imagery, have further enhanced canopy measurement by enabling nondestructive, high-resolution data collection across vast agricultural landscapes. LiDAR generates precise 3D point clouds to quantify structural heterogeneity, while optical sensors track canopy dynamics through spectral indices like Normalized Difference Vegetation Index (NDVI). These methods minimize manual labor while improving accuracy and temporal resolution key requirements for precision agriculture. For instance, Ara et al. (2021) highlight how granular, sensor-derived data supports scientific decision-making in irrigation, fertilization, and pest control, directly optimizing yields and resource efficiency. The shift from conventional field measurements to remote sensing reflects broader trends toward data-driven agriculture. However, challenges remain in standardizing methodologies across diverse crops and environments. Future research should prioritize algorithmic generalizability, computational efficiency, and integration with multisource agronomic variables (e.g., soil moisture, nutrient levels) to refine predictive models. Innovations in UAV-mounted sensors and edge computing could further democratize access to canopy measurement tools, empowering farmers in resource-limited regions (Rathore et al., 2026).

3.5 Yield estimation

Yield estimation is a critical component of modern precision agriculture, enabling farmers to optimize resource allocation (e.g., water, fertilizers), plan harvesting schedules, and forecast market prices. In India, however, the agricultural sector faces persistent challenges in improving crop productivity. Nearly 60% of crops depend on monsoon rainfall for irrigation, leading to inefficiencies in water management and unpredictable yield outcomes. These factors contribute to poor growth in the sector. To address this, advancements in information technology (IT), such as remote sensing, ML, and IoT-based soil sensors, are revolutionizing yield estimation by interfacing with smart agriculture. These tools provide data-driven insights into crop health, soil conditions, and weather patterns, empowering farmers to make informed decisions and bridge the productivity gap. Crop yield prediction has garnered substantial attention in various agricultural research initiatives and organizations. Such predictions analyze historical crop yield data, helping farmers in planning and decision-making and, in essence, providing the government with critical information that aids in developing effective farm policies regarding crop insurance, supply chain management, and resource allocation. Understanding historical trends in crop cultivation, area cultivated, and possible market prices can lead to the development of income growth profiles for farmers, thereby laying a foundation for agricultural economic stability and sustainability. Different kinds of studies have been undertaken in Table 7.2, as demonstrated below.

4 Computer vision in agricultural robotics

Combining computer vision and robots has led to some smart automated solutions for farming, such as controlling weed infestation, planting, and harvesting, with minimal human intervention. Traditional agriculture is heavily mechanized, with most harvesting being done manually. This dependence usually translates to exorbitant labor costs and low efficiency. The incorporation of computer vision technology into agricultural robots has paved a new path for advanced intelligent harvesting systems (Fig. 7.11). These technological advancements include automated machinery for harvesting and picking robots equipped with computer vision to identify crops for accurate harvesting. Hence, agriculture is tilting toward significant automation in terms of labor displacement, efficiency, and productivity.

Many studies have been conducted on the use of computer vision for automated crop harvesting, with several successful real-world applications reported. An interesting example is the automated cherry-picking system developed by Zhang et al. (2017). The system utilized robotic vision techniques, in which images were preprocessed using a median filter, threshold segmentation based on Otsu's algorithm, area-threshold noise reduction, and the Hough transform method. The developed approach achieved an efficiency of over 96%, reducing harvesting costs while improving operational productivity.

In contrast, crops such as sweet peppers present unique challenges for robotic harvesting systems. Detection of sweet peppers is difficult due to occlusion from leaves and branches, variations in plant height, and similarities between crop and background colors. To address these challenges, McCool et al. (2016) proposed a vision-based crop detection and segmentation system using local binary pattern (LBP) features. Their approach achieved an average detection accuracy of 66.8%, with LBP identifying approximately 65.2% of sweet peppers under field conditions. Although the results demonstrated promising progress, further research is still required to improve detection accuracy and operational efficiency under variable environmental conditions.

Huang et al. (2021) reported through their research that the conventional manual harvesting method is juxtaposed with slightly mechanical techniques that include

Table 7.2 Challenges and solutions in crop yield estimation.

S. No.	Problem	Explanation	Proposed solution	References
1.	Dependence on monsoon rainfall and poor irrigation management	Nearly 60% of Indian crops rely on monsoon rainfall, leading to water mismanagement, unpredictable yield outcomes, and poor productivity growth.	Use IoT-based soil moisture sensors, remote sensing data, and smart irrigation scheduling to optimize water use. The integration of IT tools enables precision resource allocation.	Osman et al. (2021), García et al. (2020), Benzaouia et al. (2023)
2.	Unreliable traditional visual inspections for yield estimation	Farmers often rely on manual visual inspections to estimate yield, which are prone to human error, inefficiency, and poor accuracy.	Computer vision with deep learning models (YOLO, SegNet, UAV-based frameworks) provides automated, accurate, and scalable yield estimation.	Daiaeddine et al. (2025), Osman et al. (2021), Subeesh et al. (2024)
3.	Lack of data-driven decision-making and poor policy support	Farmers and policymakers lack access to reliable historical yield data, leading to weak farm planning, ineffective crop insurance, and poor supply chain management.	Machine learning and big data analytics for yield prediction using historical yield, spectral indices (NDVI, GNDVI, OSAVI), and climate data help develop farm policies, insurance schemes, and income growth models.	Zhang et al. (2020)
4.	Postharvest logistical inefficiencies	High labor costs, container placement issues, and crop wastage affect overall farm profitability.	Smart harvesting frameworks with object detection, tracking (YOLO+Deep SORT), and GPS-based geospatial yield mapping optimize agricultural processes. Optimization algorithms reduce logistical overhead.	Osman et al. (2021)
5.	Scalability challenges in diverse cropping systems	Traditional yield estimation methods fail to scale for orchards, greenhouses, and large-scale farms.	UAVs+deep learning models (e.g., YOLOv7 and fruit size-to-weight mapping) enable scalable orchard management and demonstrate a near-perfect correlation with actual harvest results (Fig. 7.10)	Subeesh et al. (2024)

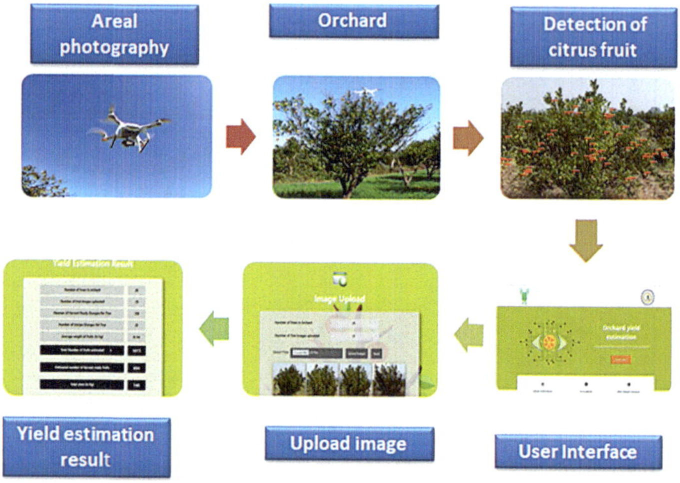

Fig. 7.10 Yield estimation with DeepYield.

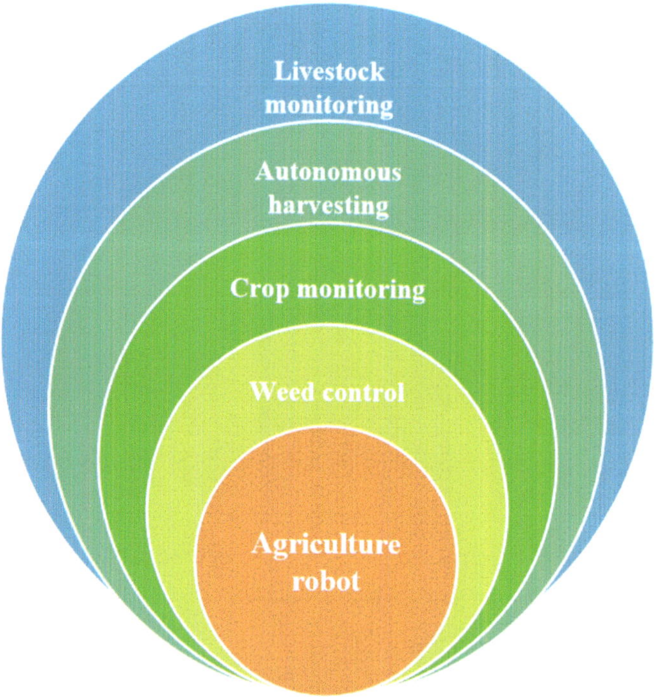

Fig. 7.11 Application of robotics in precision agriculture.

bending, twisting, and lifting. They found that, with the aid of sensors to measure the forces and motion applied during the experimental procedure, bending stands out as the most efficient technique when applying minimal force upon attaining

Fig. 7.12 Computer vision-based robotics in a polyhouse.
Source: Google.

a perfect understanding of the harvest (100% success score). In light of this, there is a probable avenue for robotic systems to adapt their techniques to optimize picking and accomplish the proposed feats with respect to efficiency, as well as reducing the picking strain imposed on an operative during harvesting actions (Fig. 7.12).

4.1 Swarm robotics and coordinated field operations

Similar to collaborative robotics, swarm robotics represents a futuristic and transformative approach to agricultural automation. By leveraging the capabilities of multiple robots, swarm robotics enables farmers to achieve incredible efficiency and precision suited to large fields.

4.2 Swarm drones for pollination and pest control

Where natural pollinators are lacking, swarm drones—with computer vision systems—can conduct targeted pollination. The drones identify flowers and effectively deliver pollen, leading to increased yields. Swarm drones also detect pest hotspots using image-based analysis and collaboratively implement pest control measures with pinpoint accuracy, while also minimizing pesticide use and environmental impact.

4.3 Collaborative field operations

Under swarm robotics techniques, multiple robots apply AI and computer vision to coordinate field operations like planting, watering, and harvesting. Autonomous tractors, robotic planters, and advanced irrigation bots work together simultaneously, increasing operational efficiencies and allowing for minimal human involvement. This coordinated approach is particularly beneficial for large-scale farms where labor is reduced, thus enhancing productivity and sustainability.

5 Merits and challenges

5.1 Merits of using computer vision

i. **Enhanced efficiency:** Computer vision trained with DL can automate many tasks in agriculture, including plant monitoring, disease detection, and resource management. The result is increased speed, lower labor costs, and greater efficiency in the use of water, fertilizers, and pesticides.
ii. **Improved precision:** High-accuracy image analysis helps farmers with tasks such as targeted spraying in cases of specific infestations, minimizing waste, reducing environmental effects, and increasing yield quality and quantity.
iii. **Scalable solutions:** Today's computer vision systems, powered by cloud and edge computing, enable them to be virtually limitless. These systems can scale to accommodate growing data volumes and diverse agricultural contexts, ranging from small farms to the largest industrial-scale operations.
iv. **Real-time decision-making:** The inclusion of edge computing enables the instantaneous analyzing of video feeds or other data, allowing for immediate decision-making. Because farmers can act when conditions dictate, such as when applications of nutrients or water are needed, productivity and efficient resource allocation improve.
v. **Support for sustainability goals:** Providing insights into soil health, carbon sequestration, and biodiversity, computer vision enables a range of sustainable agricultural practices. This, in turn, allows for the effective monitoring and reduction of various environmental behaviors and supports the implementation of progressive ideas in sustainable agriculture.
vi. **Improved crop and livestock management:** Computer vision assists in monitoring livestock health and behavior and enables the early detection of abnormalities. This technology also allows for monitoring growth patterns and phenotyping in crops to improve breeding strategies and yields.

5.2 Challenges in the implementation of computer vision with deep learning

i. **High processing power requirements**
DL demands very powerful Graphics Processing Unit (GPU)s or specialized hardware in order to carry out intense computations at scale. Looking forward, scaling the systems is becoming quite costly for small companies because regular upgrades of infrastructure need to be incorporated to keep them performing at this level (Fig. 7.13).
ii. **The need for excessive amounts of data**
To learn effectively, large amounts of data must be collected, labeled, and managed. This data collection-process is resource-intensive and calls for the requisite infrastructure in terms of storage and data pipelines for smooth processing. The low quality and biased

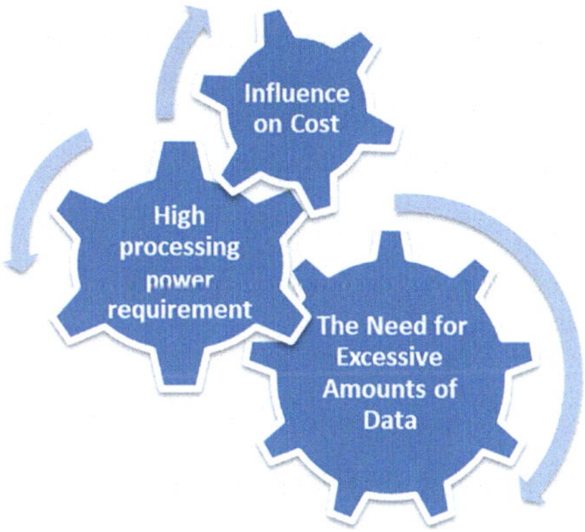

Fig. 7.13 Major challenges in the integration of computer vision.

nature of indices of assessments for targeted ML tasks remain difficult and should continuously be challenged by the body.

- **Cloud solutions:** Advantages: Scaling on demand, load control, and ease of updating work flows from maintenance to execution. Bigger delay times affect on-line processing; operational costs escalate because of increased dimensions of the model and data volume; therefore, meeting privacy and security standards requires extensive encryption, following laws like the General Data Protection Regulation (GDPR).

iii. Edge device solutions

Edge computing cuts latency because data processing is done locally on devices equipped with AI accelerators. Nevertheless:

- Resource restrictions: Limited memory and power place limitations on the complexity of models that can be deployed.
- The need for maintenance: Frequent requests for updates make revisions quite challenging for individuals in remote locations.
- The challenge of real-world performance: Annular factors that are attributed to greater lighting as well as weather variations can greatly interfere with the accuracy of models, thus necessitating extensive field testing and optimization.

iv. Deployment essentials

- **Data quality:** The performance will be hampered without a reliable, well-labeled, and diverse dataset for success in generalization. Processing aids and augmentation improve its robustness.
- **Code review and versioning:** Regular development review and documentation will ensure the transparency of the development and deployment process, aiding in debugging as well as team collaboration.
- **Hyper parameter optimization:** By fine-tuning parameters like learning rates and batch sizes, overfitting is avoided, and accuracy is improved.
- Infrastructure is set up in such a way as to ensure its reliability in terms of operating hardware, software, and network capacity.
- Continuous model management through regular re-training of the model with fresh data maintains the model's efficiency, especially in dynamic environments like agriculture.

v. **Influence on cost**
Whether using cloud services or edge device processing and storage, the cost of running these modes for processing and/or storage could become very high; hence, this poses a challenge for farmers whose livelihoods are shrouded in secrecy because some processes might not be economically viable in the long run.
vi. **Privacy and security**
Robust measures must be in place to mitigate the breach of sensitive agricultural data, including but not limited to yield predictions, drugs, or proprietary techniques, as dictated by regulatory compliance.

6 Conclusion and future pathway

Computer vision is becoming a powerful tool in agriculture, with the ability to tackle big challenges like food security, climate change, and sustainable resource use. It can be applied at every stage of farming, from soil preparation and crop monitoring to harvesting and storage, by providing farmers with useful information through image analysis, object detection, and plant health monitoring. For computer vision to work well in real farming, it requires good-quality data that is well-labeled, reliable, and secure. Issues such as data privacy, sharing, and system compatibility are still major concerns, especially when combined with other technologies like IoT, cloud computing, or robotics. If these issues are not addressed, large-scale adoption will remain difficult.

In the future, DL models are expected to become lighter, faster, and easier to explain, making them suitable for use on farms with limited resources. The use of autonomous machines, drones, and vision-based robots will help farmers perform tasks like spraying, irrigation, and harvesting with less labor, lower costs, and greater efficiency.

Computer vision will also bring progress in crop breeding and plant phenotyping, helping researchers quickly identify traits such as drought tolerance or disease resistance. New tools like augmented reality may allow farmers to use smart glasses or devices to see real-time crop health, repair alerts, or harvesting information directly in the field.

From an environmental point of view, computer vision can support sustainable and regenerative farming by monitoring soil health, biodiversity, and carbon storage. These applications will not only help conserve natural resources but also support climate-smart agriculture. To achieve these goals in agriculture, future work should focus on several key areas. First, there is a need to build open, diverse, and standardized datasets for different crops and environments to ensure reliable and generalized outcomes. Second, researchers should design efficient models capable of running in real time without relying on expensive hardware, making the technology more accessible. In addition, it is essential to address adoption challenges such as high costs, lack of farmer training, limited digital infrastructure, and issues of trust in technology. Finally, integrating systems such as AI, Internet of Things (IoT), robotics, and edge computing will be crucial in developing complete and practical smart farming solutions. Computer vision is not just a support tool; it is a key driver of digital and sustainable agriculture. With improvements in performance, scalability, and adoption, it has the potential to transform global farming into systems that are more efficient, resilient, and environmentally friendly.

References

Ahmed, A.A., Reddy, G.H., 2021. A mobile-based system for detecting plant leaf diseases using deep learning. AgriEngineering, 3 (3), 478–493.

Ali, M.A., Thai, K.W., 2017. Automated fruit grading system. In: 2017 IEEE 3rd International Symposium in Robotics and Manufacturing Automation (ROMA). IEEE, pp. 1–6.

Alsalam, B.H.Y., Morton, K., Campbell, D., Gonzalez, F., 2017. Autonomous UAV with vision based on-board decision making for remote sensing and precision agriculture. In 2017 IEEE Aerospace Conference (pp. 1–12). IEEE.

Ara, I., Turner, L., Harrison, M.T., Monjardino, M., DeVoil, P., Rodriguez, D., 2021. Application, adoption and opportunities for improving decision support systems in irrigated agriculture: A review. Agric. Water Manag. 257, 107161.

Arakeri, M.P., 2016. Computer vision based fruit grading system for quality evaluation of tomato in agriculture industry. Procedia Comput. Sci. 79, 426–433.

Bali, A., Chouhan, S.S., Kumar, G., et al., 2023. Adaptive fault-tolerant control for pure-feedback stochastic nonlinear systems with sensor and actuator faults. Circuits Syst. Signal Process. 42, 5838–5867. doi:10.1007/s00034-023-02366-7.

Bechar, A., Vigneault, C., 2016. Agricultural robots for field operations: concepts and components. Biosyst. Eng. 149, 94–111. doi:10.1016/j.biosystemseng.2016.06.014.

Benzaouia, M., Hajji, B., Mellit, A., Rabhi, A., 2023. Fuzzy-IoT smart irrigation system for precision scheduling and monitoring. Comput. Electron. Agric. 215, 108407.

Chakraborty, S.K., Chandel, N.S., Jat, D., Tiwari, M.K., Rajwade, Y.A., Subeesh, A., 2022. Deep learning approaches and interventions for futuristic engineering in agriculture. Neural Comput. Appl. 34, 20539–20573.

Chandel, N.S., Chakraborty, S.K., Jat, D., Chouhan, P., 2024. Smart farming management system: pre and post-production interventions. In: Chouhan, S.S., Saxena, A., Singh, U.P., Jain, S. (Eds.), Artificial Intelligence Techniques in Smart Agriculture. Springer. doi: 10.1007/978-981-97-5878-4_5.

Chandel, N.S., Chakraborty, S.K., Rajwade, Y.A., Dubey, K., Tiwari, M.K., Jat, D., 2021. Identifying crop water stress using deep learning models. Neural Comput. Appl. 33, 5353–5367.

Chandel, N.S., Jat, D., Chakraborty, S.K., Upadhyay, A., Subeesh, A., Chouhan, P., et al., 2025. Deep learning assisted real-time nitrogen stress detection for variable rate fertilizer applicator in wheat crop. Comput. Electron. Agric. 237, 110545.

Chouhan, S.S., Singh, U.P., Jain, S., 2020. Applications of computer vision in plant pathology: a survey. Arch. Comput. Methods Eng. 27, 611–632. doi:10.1007/s11831-019-09324-0.

Daiaeddine, M.J., Badrouss, S., El Harti, A., Bachaoui, E.M., Biniz, M., Mouncif, H., 2025. UAV imagery, advanced deep learning, and YOLOv7 object detection model in enhancing citrus yield estimation. Foods Raw Mater. 13 (2), 242–253.

Egi, Y., Hajyzadeh, M., Eyceyurt, E., 2022. Drone-computer communication based tomato generative organ counting model using YOLO V5 and deep-sort. Agriculture, 12 (9), 1290.

García, L., Parra, L., Jimenez, J.M., Lloret, J., Lorenz, P., 2020. IoT-based smart irrigation systems: An overview on the recent trends on sensors and IoT systems for irrigation in precision agriculture. Sensors, 20 (4), 1042.

Haider, T., Farid, M.S., Mahmood, R., Ilyas, A., Khan, M.H., Haider, S.T.A., et al., 2021. A computer-vision-based approach for nitrogen content estimation in plant leaves. Agriculture 11 (8), 766.

Huang, M., He, L., Choi, D., Pecchia, J., Li, Y., 2021. Picking dynamic analysis for robotic harvesting of Agaricus bisporus mushrooms. Comput. Electron. Agric. 185, 106145.

Ismail, N., Malik, O.A., 2022. Real-time visual inspection system for grading fruits using computer vision and deep learning techniques. Inf. Process. Agric. 9 (1), 24–37.

Khan, S., Tufail, M., Khan, M.T., Khan, Z.A., Iqbal, J., Wasim, A., 2021. Real-time recognition of spraying area for UAV sprayers using a deep learning approach. PLOS ONE 16 (4), e0249436.

Lorente, D., Aleixos, N., Gómez-Sanchis, J., Cubero, S., García-Navarrete, O.L., Blasco, J., 2012. Recent advances and applications of hyperspectral imaging for fruit and vegetable quality assessment. Food Bioprocess Technol. 5, 1121–1142.

McCool, C., Dayoub, F., Lehnert, C., Perez, T., Upcroft, B., 2016. Visual detection of occluded crop for automated harvesting. In: 2016 IEEE International Conference on Robotics and Automation (ICRA). IEEE, pp. 2506–2512.

Osman, Y., Dennis, R., Elgazzar, K., 2021. Yield estimation and visualization solution for precision agriculture. Sensors 21 (19), 6657.

Robotics & machine vision in farming, n.d. Retrieved from https://www.imveurope.com/feature/how-robots-machine-vision-are-revolutionising-farming.

Rathore, J., Walsh, O.S., Gardezi, M., Mehla, M.K., Vardhan, K., 2026. UAVs in Agriculture: A Review of Challenges, Limitations, Stakeholder Perspectives, and Data Privacy.

Rico-Fernández, M.P., Rios-Cabrera, R., Castelán, M., Guerrero-Reyes, H.I., Juarez-Maldonado, A., 2019. A contextualized approach for segmentation of foliage in different crop species. Comput. Electron. Agric. 156, 378–386.

Ryan, M., Isakhanyan, G., Tekinerdogan, B., 2023. An interdisciplinary approach to artificial intelligence in agriculture. NJAS Impact Agric. Life Sci. 95, 1.

Singh, R.S., Sahni, R.K., 2019. Transformation of Indian agriculture through mechanization. Economic Affairs 64 (2), 297–303. doi:10.30954/0424-2513.2.2019.4.

Subeesh, A., Kumar, S.P., Chakraborty, S.K., Upendar, K., Chandel, N.S., Jat, D., et al., 2024. UAV imagery coupled deep learning approach for the development of an adaptive in-house web-based application for yield estimation in citrus orchard. Measurement 234, 114786.

Upadhyay, A., Chandel, N., Chakraborty, S., 2024. Disease control measures using vision-enabled agricultural robotics. In: Artificial Intelligence for Agriculture. Springer. doi:10.1007/978-981-99-8684-2_10.

Upadhyay, A., Chandel, N.S., Singh, K.P., Chakraborty, S.K., Nandede, B.M., Kumar, M., et al., 2025c. Deep learning and computer vision in plant disease detection: a comprehensive review of techniques, models, and trends in precision agriculture. Artif. Intell. Rev. 58 (3), 92.

Upadhyay, A., Gautam, P.V., Shukla, P., Sahu, R., Kumar, A., Umashanker, U., Ghritalahre, B., Prasad, A., Mansuri, S., 2025a. Harnessing Artificial Intelligence and Machine Learning for Sustainable Agriculture Transformation. ACEPSW. doi:10.58532/nbennurACEPSW6.

Upadhyay, A., Patel, A., Chandel, N., Chakraborty, S., Bhalekar, D., 2025b. Leveraging AI and ML in precision farming for pest and disease management: benefits, challenges, and future prospects. In: AI in Smart Farming. Springer. doi:10.1007/978-981-96-2413-3_23.

Valluvan, A.B., Raj, R., Pingale, R., Jagarlapudi, A., 2023. Canopy height estimation using drone-based RGB images. Smart Agric. Technol. 4, 100145.

Zhang, P.P., Zhou, X.X., Wang, Z.X., Mao, W., Li, W.X., Yun, F., ... Tan, C.W., 2020. Using HJ-CCD image and PLS algorithm to estimate the yield of field-grown winter wheat. Sci. Rep. 10 (1), 5173.

Zhang, Q., Chen, S., Yu, T., et al., 2017. Cherry recognition in natural environment based on the vision of picking robot. IOP Conf. Ser. Earth Environ. Sci. 61, 012021.

Zhang, Z., Qiao, Y., Guo, Y., He, D., 2022. Deep learning based automatic grape downy mildew detection. Front. Plant Sci. 13, 872107.

Zhao, B., Zhang, Y., Duan, A., Liu, Z., Xiao, J., Liu, Z., et al., 2021. Estimating the growth indices and nitrogen status. Front. Plant Sci. 12, 619522.

Zhou, Y., Yan, H., Ding, K., Cai, T., Zhang, Y., 2024. Few-shot image classification of crop diseases based on vision–language models. Sensors 24 (18), 6109.

Emergence and application of the Internet of Living Things in precision farming

8

Meenakshi Sharma[a,b] *and Prashant Kaushik*[c]
[a]Department of Chemistry, Ranchi University, Ranchi, Jharkhand, India
[b]Joita Bioseed AI Pvt. Ltd., Kurukshetra, Haryana, India
[c]Chaudhary Charan Singh Haryana Agricultural University, Hisar, Haryana, India

Abbreviations

AI	artificial intelligence
BLE	bluetooth low energy
EIS	electrical impedance spectroscopy
GIS	geographic information systems
GPS	global positioning systems
IoLT	internet of living things
IoT	internet of things
LPWAN	low-power wide-area network
MEMS	micro-electromechanical systems
ML	machine learning
NB-IoT	narrowband IoT
NFC	near field communication
PA	precision agriculture
ROI	return on investment
SMEs	small and medium-sized enterprises
UAVs	unmanned aerial vehicles
VOCs	volatile organic compounds
VRA	variable-rate application
WSN	wireless sensor networks

1 Introduction

The 21st century presents humanity with a complex paradox: an expanding global population set against the backdrop of finite arable land and increasingly scarce natural resources (Godfray et al., 2010). Projections indicate that food production must increase by an estimated 70% by 2050 to meet the needs of a projected 9.7 billion people (Alexandratos and Bruinsma, 2012). This challenge is amplified by the impacts of climate change, which include erratic weather, a higher prevalence of pests and diseases, and intensified water stress, all of which threaten agricultural productivity and food security (Wheeler and von Braun, 2013). Traditional farming methods, often

Precision Technologies for Digital Agriculture. https://doi.org/10.1016/B978-0-443-40513-6.00019-6

characterized by generalized resource application, are proving inadequate. These practices frequently lead to the overapplication of fertilizers and pesticides, inefficient water use, and a significant environmental footprint (Tilman et al., 2002).

In response, agriculture has shifted from intuition-based practices toward a data-driven paradigm known as "precision farming" or "smart agriculture" (Liang et al., 2019). This approach uses advanced technologies to observe, measure, and respond to variability in crops and livestock (Gebbers and Adamchuk, 2010). The core premise is to apply the right treatment at the right place, at the right time, and in the right amount to optimize resource use, minimize waste, and enhance productivity. Early precision farming relied on global positioning systems (GPS), geographic information systems (GIS), and remote sensing to map field variability and guide machinery (Zhang et al., 2002).

The advent of the Internet of Things (IoT) has amplified these capabilities. IoT devices, ranging from environmental sensors to automated machinery, gather vast amounts of data on soil conditions, weather, and crop growth (Gade et al., 2025; Kumar et al., 2024; Tzounis et al., 2017). This data stream provides actionable insights that help farmers make informed decisions about irrigation, fertilization, and pest control. However, IoT systems have a fundamental limitation: they primarily monitor the environment *around* biological entities, inferring their health from external conditions rather than directly assessing their internal physiological states (Mishra et al., 2021). This gap between environmental data and direct biological insight is where the Internet of Living Things (IoLT) emerges as the next frontier.

This chapter aims to provide a comprehensive analysis of the IoLT within the context of precision agriculture. The primary objectives are to: (1) clearly define IoLT and differentiate its core principles and data focus from the conventional IoT; (2) detail the essential technological components, from biosensors to data analytics platforms, that constitute a functional IoLT system; (3) explore the diverse and transformative applications of IoLT in plant, livestock, and soil management; and (4) critically evaluate the technical, economic, and ethical challenges that must be addressed to realize the full potential of IoLT in creating more sustainable and productive agricultural systems.

2 Understanding the Internet of Living Things

To grasp the potential of IoLT in agriculture, it is essential to define it, distinguish it from the more established IoT, and identify the core components and unique biological data it generates (Yue et al., 2020).

2.1 Defining IoLT: Beyond IoT

The IoT refers to a network of interconnected physical objects embedded with sensors and software that exchange data over the internet (Gubbi et al., 2013). In agriculture, this primarily involves inanimate objects like soil moisture sensors, weather stations, and drones that collect environmental and operational data (Tzounis et al., 2017).

The IoLT extends this concept to living organisms themselves (Yue et al., 2020). IoLT is a network of biologically integrated sensors, actuators, and communication technologies that enable real-time monitoring, analysis, and interaction with the physiological states and biological processes of living entities (Mishra et al., 2021). Unlike IoT, which centers on the physical environment, IoLT delves into the biological world, extracting insights directly from plants, animals, and microorganisms (as illustrated in Fig. 8.1).

The key distinction lies in the nature of the "things" being monitored. While IoT gathers physical parameters like temperature and location, IoLT focuses on biological parameters (Fig. 8.1). These include physiological signals (e.g., plant sap pH, animal heart rate), biochemical markers (e.g., specific proteins, volatile organic compounds), and microbial activity (e.g., metabolic rates, gene expression) (Nadeem et al., 2021; Saeed et al., 2022). This direct interaction with biological systems enables a more nuanced and proactive approach to agricultural management, allowing for interventions long before visible symptoms of stress or disease appear (Mishra et al., 2021).

2.2 *Core components of IoLT systems*

A robust IoLT system in agriculture is composed of several interconnected layers, each playing a distinct role in data acquisition, transmission, and processing (summarized below).

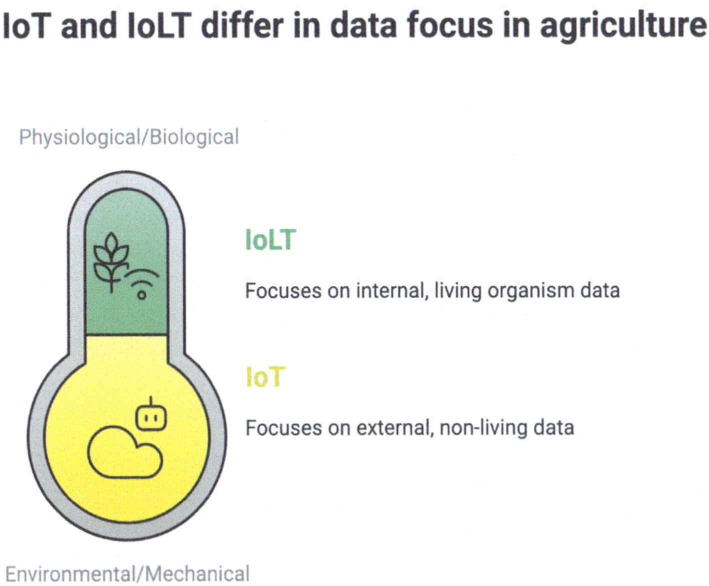

Fig. 8.1 Conceptual diagram illustrating the distinct data focus of the Internet of Things (IoT) versus the Internet of Living Things (IoLT) in agriculture. IoT systems primarily collect external data (e.g., environmental, mechanical), while IoLT systems focus on internal data sourced directly from the physiological and biological processes of living organisms. Source: Authors' own creation.

- **Biosensors:** These devices detect specific biological or chemical signals directly from living organisms and convert them into measurable electrical outputs. Examples include plant-based sap pH sensors, animal-based wearable heart rate monitors, and soil-based microbial respiration sensors (Kwong et al., 2021; Nadeem et al., 2021; Patel et al., 2022).
- **Bio-actuators:** These are systems that interact with or induce a change in biological systems based on received data. Examples include targeted nutrient delivery systems for plants or automated drug delivery systems for livestock (Yue et al., 2020).
- **Communication networks:** This infrastructure transmits data from biosensors to processing units and relays commands to bio-actuators. Technologies include Low-Power Wide-Area Networks (LPWANs) like LoRaWAN and NB-IoT, 5G, and Bluetooth Low Energy (Citoni et al, 2019).
- **Edge & cloud computing:** Edge computing provides localized processing near the data source to reduce latency, while cloud platforms are used for storing and analyzing large volumes of data using machine learning and artificial intelligence (AI) models (Ahmad et al., 2024).
- **User interface & decision support systems:** These are applications and dashboards that present processed information to farmers in an understandable format, facilitating informed decision-making through mobile apps, alerts, and prescriptive recommendations (Liang et al., 2019).
- **Power management:** Continuous energy for sensors and modules is crucial, often relying on energy-harvesting techniques like solar panels or highly efficient batteries (Zou et al., 2024).

3 The evolution of precision farming

The development of precision agriculture marked a pivotal shift from uniform field management to a data-driven approach that accounts for in-field variability. This evolution was enabled by key technologies like GPS for precise location mapping and GIS for analyzing spatial data (Zhang et al., 2002). The subsequent integration of IoT sensors represented the next major step, creating an interconnected farm ecosystem. However, these technologies still operate at a distance from the biological processes themselves, a limitation that IoLT is poised to overcome (Johnson and Lee, 2023).

4 Applications of IoLT in precision farming

4.1 Real-time plant health monitoring

IoLT offers significant advantages for the continuous, noninvasive monitoring of plant health. It enables proactive management by detecting stress factors long before they become visible.

- **Biosensors for plant physiology:** Micro-sensors embedded in stems can measure sap flow, providing a direct indicator of water stress and allowing for precise irrigation (Nadeem et al., 2021). Electrical impedance spectroscopy sensors can detect subtle changes in the electrical properties of plant tissues that signal disease or nutrient deficiencies (He et al., 2025).

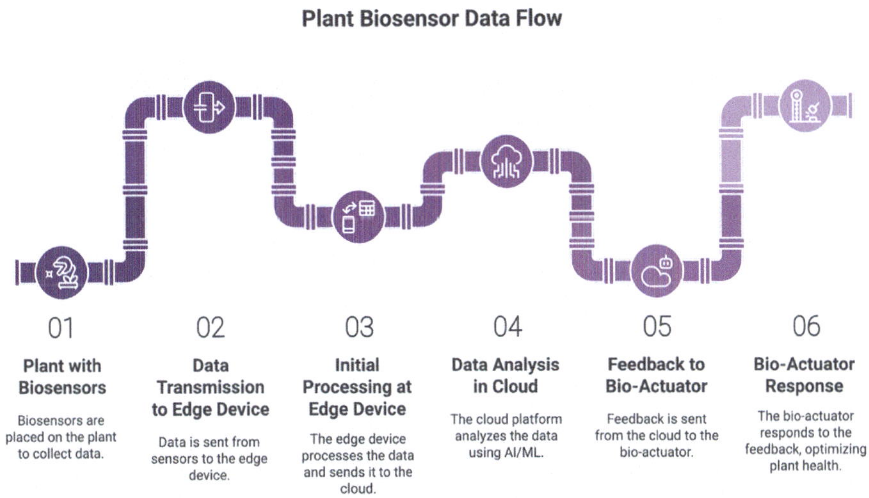

Fig. 8.2 A six-step workflow of an IoLT-enabled closed-loop system for plant health monitoring. The process begins with (1) data collection by biosensors on the plant, followed by (2) transmission to an edge device for (3) initial processing. (4) The data is then analyzed in the cloud using AI/ML algorithms, generating (5) feedback that is sent to a bio-actuator, which (6) elicits a direct response to optimize plant health.
Source: Authors' own creation.

- **Predictive analytics for crop management:** Data from IoLT biosensors, combined with environmental data, can be fed into AI models to predict the onset of stress, disease, or pest infestations (Saeed et al., 2024). This facilitates highly targeted treatments, reduces the reliance on broad-spectrum pesticides, and enables personalized crop nutrition at the individual plant level (Fig. 8.2).

5 Technological enablers and architecture of IoLT systems

The success of IoLT hinges on the convergence of several cutting-edge technologies.

- **Advanced biosensors and bio-actuators:** A key trend is the miniaturization of sensors, moving toward nano-sensors and micro-electromechanical systems that allow for noninvasive integration with plants and animals. The development of flexible and even biodegradable sensors is a critical area of research (Santos et al., 2024).
- **Communication protocols:** Efficient and robust communication networks are essential. LPWANs are well-suited for agricultural deployments due to their long-range and low power consumption, while 5G promises the ultra-low latency needed for real-time control of bio-actuators (Citoni et al., 2019; Tzounis et al., 2017).
- **Edge computing and AI:** The large volume of data generated by IoLT makes edge computing indispensable for initial processing, which reduces bandwidth requirements and latency (Chen and Jiang, 2019). AI and ML algorithms are crucial for pattern recognition, predictive modeling, and optimizing bio-feedback loops.

6 Challenges and future outlook of IoLT in precision farming

Despite its promise, the widespread adoption of IoLT faces several hurdles. Technical challenges include sensor longevity and biocompatibility in harsh agricultural environments, power management for distributed sensor networks, and the need for data standardization and interoperability (Mishra et al., 2021). Economically, the high upfront investment can be a significant barrier for small and medium-sized farms, necessitating clear evidence of a tangible return on investment (Rutten et al., 2020). Furthermore, robust regulatory and ethical frameworks are needed to address data ownership, privacy, animal welfare, and the environmental impact of sensor disposal (Rana and Sharma, 2025).

The future of agriculture with IoLT points toward a symbiotic system that is both biologically informed and highly responsive. This will lead to hyperpersonalized agriculture, where inputs are tailored to individual plants and animals. Farming will become more predictive and prescriptive, shifting from reactive problem-solving to proactive intervention. The creation of "digital twins" for biological systems—digital replicas of individual plants or entire farm ecosystems, continuously updated with real-time IoLT data—will enable advanced simulation and optimization (Pisharody et al., 2023). Addressing the current challenges through sustained interdisciplinary research and supportive policies will be crucial to harnessing the full power of IoLT.

7 Conclusion

The IoLT represents a paradigm shift in precision agriculture, moving beyond monitoring the environment to engaging directly with the biology of organisms. By leveraging advanced biosensors and bio-actuators, IoLT provides a real-time stream of physiological, biochemical, and behavioral data from plants, animals, and soil microbes. This direct biological interface transforms farming from a practice of reacting to external conditions and visible symptoms to one of intimately understanding and proactively managing internal biological processes. The applications are transformative, enabling hyperpersonalized crop nutrition, early disease detection in livestock, and the optimization of the soil microbiome, thereby enhancing efficiency, reducing waste, and minimizing environmental impact.

While the potential of IoLT is immense, its widespread adoption requires overcoming significant technical, economic, and ethical hurdles. Issues of sensor durability, data interoperability, high upfront costs, and the need for clear regulatory frameworks for data privacy and animal welfare must be systematically addressed. Moreover, the environmental footprint of these new technologies requires careful consideration to ensure a truly sustainable outcome. Nevertheless, the trajectory of agricultural innovation points unequivocally toward a future where IoLT is an indispensable tool. Continued research, interdisciplinary collaboration, and supportive policies will be essential to harness the full, transformative power of IoLT and cultivate a healthier, more productive, and fundamentally sustainable agricultural future.

References

Ahmad, Z., Khan, A., Singh, R., 2024. Edge-AI Integration in IoLT for Real-Time Agricultural Analytics. Springer.

Alexandratos, N., Bruinsma, J., 2012. World Agriculture Towards 2030/2050: The 2012 Revision. FAO, Rome.

Chen, M., Jiang, X., 2019. Edge computing in smart agriculture. IEEE Internet Things J. 6 (5), 8711–8727.

Citoni, B., Fioranelli, F., Imran, M.A., Abbasi, Q., 2019. Internet of things and LoRaWAN-enabled future smart farming. IEEE Internet Things Mag. 2 (4), 14–19. doi:10.1109/IOTM.0001.1900043.

Gade, S.A., Madolli, M.J., García-Caparrós, P., Ullah, H., Cha-um, S., Datta, A., Himanshu, S.K., 2025. Advancements in UAV remote sensing for agricultural yield estimation: a systematic comprehensive review of platforms, sensors, and data analytics. Remote Sens. Appl. Soc. Environ. 23529385. 37. Available from: https://www.sciencedirect.com/science/journal/23529385. doi:10.1016/j.rsase.2024.101418.

Gebbers, R., Adamchuk, V.I., 2010. Precision agriculture and food security. Science. 10959203. 327 (5967), 828–831. doi:10.1126/science.1183899.

Godfray, H.C.J., Beddington, J.R., Crute, I.R., Haddad, L., Lawrence, D., Muir, J.F., Pretty, J., Robinson, S., Thomas, S.M., Toulmin, C., 2010. Food security: the challenge of feeding 9 billion people. Science. 10959203. 327 (5967), 812–818. doi:10.1126/science.1185383.

Gubbi, J., Buyya, R., Marusic, S., Palaniswami, M., 2013. Internet of Things (IoT): a vision, architectural elements, and future directions. Future Gener. Comput. Syst. 0167739X. 29 (7), 1645–1660. doi:10.1016/j.future.2013.01.010.

He, S., Zhang, S., Liu, C., Wang, J., Liu, H., Yu, D., 2025. Advanced detection of early plant stress based on electrical impedance spectroscopy. Biosens. Bioelectron. 255, 118934.

Johnson, M., Lee, H., 2023. From precision to prediction: the evolving role of data in smart farming. J. Agric. Inform. 14 (2), 1–15.

Kumar, R., Kumar, A., Himanshu, S.K., 2024. IoT and smart sensor applications in nutrient and irrigation water management. In: Agri-Tech Approaches for Nutrients and Irrigation Water Management. CRC Press, Thailand, pp. 48–67. Available from: https://www.routledge.com/Agri-Tech-Approaches-for-Nutrients-and-Irrigation-Water-Management/Gupta-Himanshu-Gupta/p/book/9781032450230. doi:10.1201/9781003441175-3.

Kwong, W.Y., Sun, X., Huang, J., Kwong, K.H., 2021. Internet of Living Things (IoLT) in animal health and disease control. IEEE Internet Things J. 8 (15), 12053–12061.

Liang, C., Liang, B., Yu, K., Zhao, J., 2019. Advances in smart agriculture based on Internet of Things and artificial intelligence. J. Clean. Prod. 239, 117941.

Mishra, S., Sharma, M., Tiwari, A., Choudhury, S., 2021. Internet of living things (IoLT) in agriculture: a review on current trends and future directions. J. Phys. Conf. Ser. 1957 (1), 012015.

Nadeem, M., Li, J., Yang, Z., Chen, Y., Wu, H., Khan, M.A., Zhang, W., 2021. Advanced biosensors for plant water status monitoring in precision agriculture: a review. Sensors 21 (9), 3122.

Patel, V.K., Pandey, A.C., Kumar, V., Kumar, R., Kumar, A., Singh, R., Kumar, S., 2022. Recent advances in biosensors for soil health monitoring: a review. Sens. Int. 3, 100155.

Pisharody, S., Ma, Y., Hu, B., 2023. Digital twin for agricultural systems: a review. Comput. Electron. Agric. 209, 107936.

Rana, M., Sharma, S., 2025. Emerging security and privacy challenges in IoLT-based smart agriculture. In: Next-Generation IoT-Based Smart Agriculture. Springer, Cham, pp. 235–250.

Rutten, C.J., Hogeveen, H., Steeneveld, W., 2020. Monitoring infectious diseases in dairy cows using sensor data: a review. Vet. Res. 51 (1), 1–13.

Saeed, A., Al-Khayyat, A., Al-Hamami, S., Al-Khayyat, F., 2022. Volatile organic compounds (VOCs) as biomarkers for plant stress detection in agriculture: a review. Sensors 22 (13), 4872.

Saeed, A., Al-Khayyat, A., Al-Hamami, S., Al-Khayyat, F., 2024. A review of volatile organic compounds (VOCs) as biomarkers for plant stress detection in agriculture. Sensors 24 (5), 1482.

Santos, F., Costa, J., Martins, R., 2024. Biodegradable and flexible biosensors for sustainable IoLT applications. Adv. Sustain. Syst. 8 (3), 2300451.

Tilman, D., Cassman, K.G., Matson, P.A., Naylor, R., Polasky, S., 2002. Agricultural sustainability and intensive production practices. Nature. 00280836. 418 (6898), 671–677. doi:10.1038/nature01014.

Tzounis, A., Katsaprakakis, N.M., Saridakis, E., Kampianakis, M., 2017. Internet of Things in agriculture, recent advances and future challenges. Comput. Electron. Agric. 153, 341–354.

Wheeler, T., von Braun, J., 2013. Climate change impacts on global food security. Science. 0036-8075. 341 (6145), 508–513. doi:10.1126/science.1239402.

Yue, X., Yang, S., Cheng, X., Zhang, Y., Li, J., 2020. Internet of living things: an overview of the biological IoT. IEEE Access 8, 174780–174797.

Zhang, N., Wang, M., Wang, N., 2002. Precision agriculture—a worldwide overview. Comput. Electron. Agric. 01681699. 36 (2-3), 113–132. doi:10.1016/s0168-1699(02)00096-0.

Zou, H., Shi, J., Zhang, S., Li, Y., Sun, Y., 2024. A review of energy harvesting technologies for smart agriculture. J. Clean. Prod. 434, 139086.

Enhancing crop management with digital twin technology

9

Rohit Anand[a], *Dattatray G. Bhalekar*[b], *Karishma Kumari*[c], *and Roaf Ahmad Parray*[d]
[a]Department of Food, Agricultural, and Biological Engineering, The Ohio State University, Columbus, OH, United States
[b]Department of Biological Systems Engineering, Center for Precision and Automated Agricultural Systems, Washington State University, Prosser, WA, United States
[c]Department of Agronomy, Horticulture and Plant Science, South Dakota State University, Brookings, SD, United States
[d]Division of Agricultural Engineering, ICAR-Indian Agricultural Research Institute, New Delhi, India

1 Introduction

The agricultural sector is undergoing a profound transformation driven by advancements in digital technologies (DTs). Among these innovations, digital twin technology (DTT) stands out as a revolutionary approach with the potential to redefine crop management practices. Dr. Michael Grieves introduced the concept of digital twins in the early 2000s during his work on product lifecycle management (Grieves and Vickers, 2016). Originally conceptualized in the manufacturing and aerospace industries, digital twins are virtual representations of physical systems or objects that enable real-time monitoring, simulation, and analysis (Wang, 2020). Over time, advancements in data analytics, Internet of Things (IoT) devices, and artificial intelligence (AI) have expanded the capabilities of digital twins, making them indispensable for industries. These virtual replicas are continuously updated with data from their physical counterparts, creating a dynamic feedback loop that enhances decision-making processes (Fig. 9.1).

The adoption of DTT in agriculture is rapidly increasing, as it proves to be a powerful tool for addressing the complex challenges of crop management. This is especially important because the agricultural sector today faces critical issues, including shrinking farmland, rising costs for inputs like fertilizers and water, and the urgent need to reduce greenhouse gas emissions. However, traditional crop management methods often lack the precision and flexibility needed to effectively tackle these problems. In contrast, this technology offers a smarter and more innovative solution by enabling better decision-making, efficient resource utilization, and sustainable farming practices (Verdouw et al., 2021; Zhang et al., 2025; Manivasagam, 2025). By integrating data

Precision Technologies for Digital Agriculture. https://doi.org/10.1016/B978-0-443-40513-6.00006-8

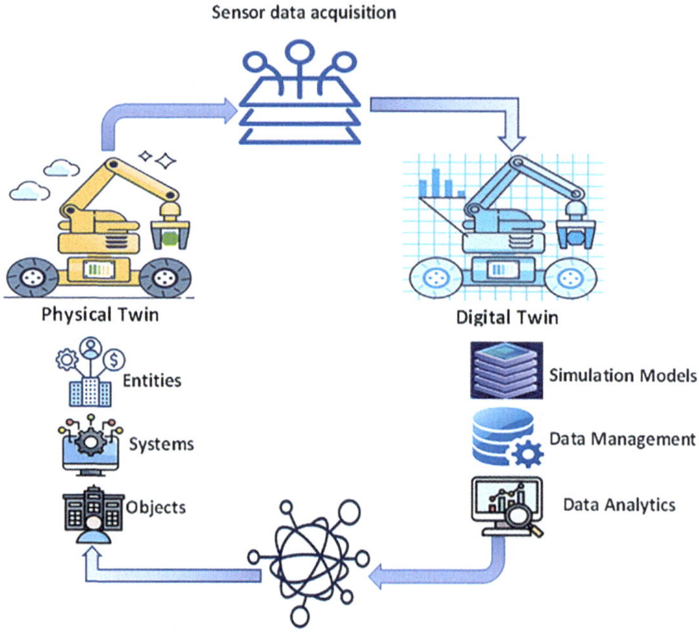

Fig. 9.1 The generic representation of a DT scheme's dynamic integration of physical and digital domains in agriculture.

from diverse sources, such as sensors, drones, and satellites, digital twins create a comprehensive and dynamic view of crop health, growth, and environmental conditions. Furthermore, this technology empowers stakeholders to simulate various scenarios, predict potential outcomes, and optimize farming strategies, ultimately leading to enhanced productivity and long-term sustainability (Botín-Sanabria et al., 2022).

The incorporation of this technology into agriculture signifies a critical leap forward in crop management to meet the challenges of the 21st century. It offers the promise of enhanced productivity, reduced environmental impact, and greater resilience in the face of an uncertain future.

This chapter will explore the intricacies of DTT, including its physical-digital integration, key components, and cross-industry advantages. Practical applications in crop management, such as precision farming, real-time monitoring, disease detection, and yield prediction, will be discussed. Additional sections will address benefits for farmers, adoption challenges, and case studies highlighting real-world impacts. Finally, emerging trends and prospects will provide a comprehensive understanding of how DTT is set to transform agriculture.

2 Digital twins: Physical-digital integration

The DTT is based on the unbroken connection of the physical and digital worlds, whereby the physical counterpart, system, or process is represented in a virtual space. Such integration enables supervision, evaluation, and improves performance at a

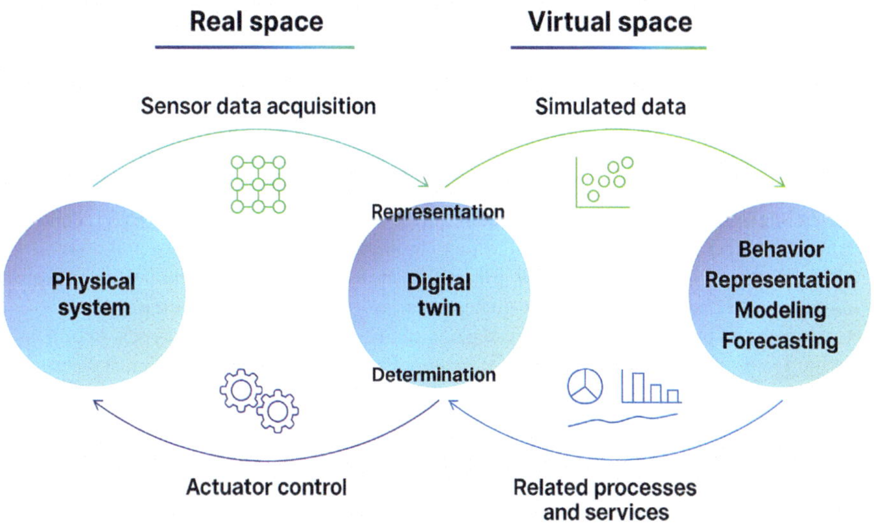

Fig. 9.2 Data interaction between a physical object and its virtual counterpart in a digital twin (Piddubna, 2024).

real-time rate, narrowing the gap that exists between activities performed in the real world and those carried out in a virtual space in a bid to enhance decision-making and operations (Peladarinos et al., 2023). The physical-digital integration in digital twins relies on key components that enable accurate data acquisition, processing, and simulation, forming the foundation of this transformative technology (Fig. 9.2).

3 Key components of digital twins

1. **Sensors:** They play a critical role in collecting real-time data from the physical entity or system. These devices measure various parameters, such as temperature, humidity, pressure, motion, and more, depending on the specific application. In agriculture, for example, sensors monitor soil moisture, nutrient levels, crop health, and weather conditions, providing essential input for the digital twin.
2. **Data collection:** The data collected by sensors is transmitted to a central repository or cloud platform. Advanced communication technologies, such as the IoT, wireless networks, and edge computing, enable seamless data transfer. In this phase, raw data from multiple sources is aggregated and stored for further analysis.
3. **Data processing:** It involves analyzing and refining raw data to extract meaningful insights. This is achieved using advanced analytics tools, machine learning (ML) algorithms, and AI. The processed data forms the foundation for creating accurate virtual models that replicate the physical system's behavior and characteristics.
4. **Simulation:** It is a key aspect of digital twins, allowing stakeholders to test and predict various scenarios without impacting the physical system. By replicating real-world conditions and interactions, simulations help optimize performance, predict outcomes, and mitigate potential risks. In agriculture, for instance, simulations can model crop growth under different irrigation schedules or assess the impact of pest infestations.

4 Advantages of digital twins across agriculture

Digital twins offer significant benefits across various industries, revolutionizing operations, enhancing productivity, and driving innovation. Key advantages are given below:

1. **Real-time monitoring and control:** It provides real-time visibility into the performance and condition of physical systems. This enables timely intervention, such as predictive maintenance in manufacturing or an immediate response to environmental changes in agriculture.
2. **Enhanced decision-making:** By offering a comprehensive and accurate representation of physical systems, digital twins enable data-driven decision-making. Simulations and predictive models help stakeholders evaluate scenarios and choose optimal strategies.
3. **Cost savings:** It reduces costs by minimizing the need for physical prototypes, optimizing resource use, and preventing downtime. In agriculture, this translates to efficient water and fertilizer application, thereby lowering input costs.
4. **Improved sustainability:** It supports sustainable practices by optimizing resource consumption and reducing waste. For example, in urban planning, digital twins aid in designing energy-efficient infrastructure, while in agriculture, they enable precision farming to minimize environmental impact.
5. **Innovation and customization:** By providing a virtual testing environment, digital twins accelerate innovation and facilitate the development of customized solutions. In the automotive industry, for instance, they aid in designing vehicles with improved performance and safety features.
6. **Risk management:** It allows stakeholders to anticipate and address potential risks through simulations and predictive analytics. This capability is particularly valuable in sectors like healthcare and disaster management, where proactive measures can save lives.

The ability of digital twins to integrate physical systems with digital models has positioned them as a cornerstone of modern technology. From enhancing operational efficiency to driving innovation, the adoption of this technology is reshaping industries, including agriculture, where they play a pivotal role in optimizing crop management and promoting sustainable practices (Fig. 9.3).

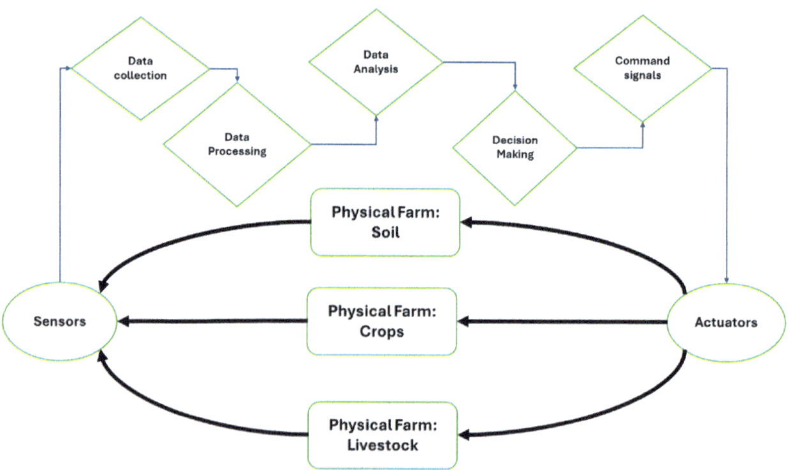

Fig. 9.3 A schematic representation of a digital twin (DT) integration in agriculture.

5 Application of digital twins in crop management

This technology offers advanced real-time monitoring, simulation, and predictive analysis tools that transform traditional crop management. By creating virtual replicas of farms, fields, and crops, farmers can make data-driven decisions, optimize resource usage, and mitigate risks (Fig. 9.4). Below are some key applications of digital twins in crop management:

5.1 Precision farming and real-time monitoring

DTT is being adopted in precision farming and real-time crop monitoring applications. By creating virtual models of agricultural systems, this technology allows farmers to simulate and fine-tune essential tasks like irrigation, fertilization, and pest control (Escribà-Gelonch et al., 2024). These digital replicas combine data from IoT sensors, satellite imagery, and precision farming tools to provide a detailed, real-time picture of crop conditions (Benny et al., 2024; Bloch et al., 2022; Chowdhury et al., 2025). With the support of cloud, fog, and edge computing, farmers can access and analyze data instantly, helping them make smarter decisions and test different scenarios to see potential outcomes (Kalyani et al., 2024). In addition, when paired with crop simulation models, digital twins can track crop health, forecast yields, and adjust predictions based on real-world field data (Bloch et al., 2022). While the benefits for farmers and agricultural businesses are significant, challenges like handling complex data and managing multiple operations simultaneously still

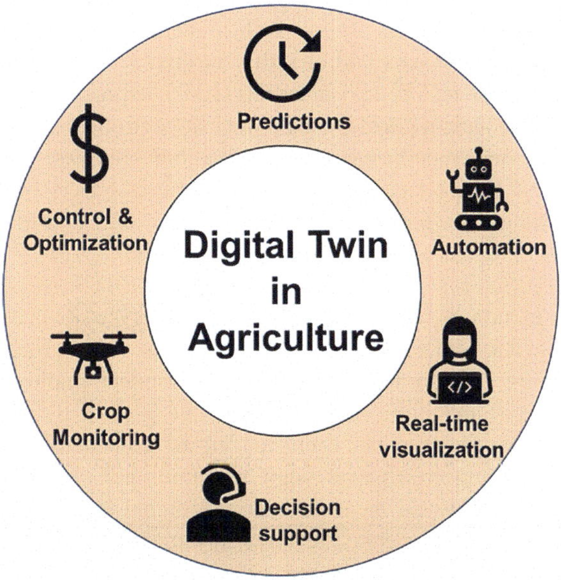

Fig. 9.4 Schematic representation of digital twin applications in agriculture.

need to be addressed (Escribà-Gelonch et al., 2024). Despite these hurdles, digital twins hold immense promise for transforming how we manage agriculture and boosting the efficiency of crop production.

A prominent application of digital twin farming is to enhance precision agriculture by providing in-depth insights into crop health. It allows real-time monitoring of critical parameters, including weather, soil moisture, crop conditions, and nutrient levels. This continuous monitoring empowers farmers to make data-driven decisions regarding irrigation, fertilization, and harvesting, with reduced input loss and increased yields (Skobelev et al., 2020). The combination of remote sensing, IoT devices, and data analytics enables a holistic view of soil, water, and crop health. In addition, advanced ML techniques, including convolutional neural networks, facilitate early detection of crop diseases and nutrient deficiencies, providing valuable information for crop protection. Moreover, these technologies enable farmers to simulate various scenarios, such as irrigation scheduling or pest control, to evaluate their efficacy before implementation. These simulations help optimize resource use, minimize input loss, and enhance operational efficiency (Zhang et al., 2020).

5.2 Modeling crop growth and predicting yields

One of the biggest challenges in agriculture is anticipating how crops will grow under varying conditions. Digital twins offer a solution by modeling crop growth through simulations. These models incorporate historical data, environmental factors, and weather forecasts, allowing farmers to predict yields and make informed decisions to optimize growth, minimize risks, and improve productivity.

The integration of DTs in agriculture, particularly through crop digital twin systems (CDTS), is revolutionizing crop growth modeling and yield prediction. DTs automate operational workflows and quantify uncertainties, enabling accurate predictions of crop parameters. The IoT underpins CDTS, acting as a real-time platform for data collection, processing, and transmission. In controlled-environment agriculture, IoT supports yield monitoring, disease detection, resource optimization, and quality enhancement by precisely controlling microclimatic factors like temperature, humidity, light, and gas concentrations (Rehman et al., 2022; Navarro et al., 2020).

Crop growth simulation models, such as World Food Studies (WOFOST) (Hijmans et al., 1994), Crop Environment Resource Synthesis (CERES) (Ritchie, 1985), and Agricultural Production Systems sIMulator (APSIM) (McCown et al., 1996) for multiple crops, and RiceGrow for single crops (Tang et al., 2009), replicate physiological processes using environmental inputs. These models employ principles like growing degree days and light energy to simulate growth dynamics, providing decision-making tools for crop management. Virtual crop models further enhance predictions by simulating crop morphology, structure, and physiology, creating an accurate link between physical environments and digital counterparts for effective yield prediction and management. The three key methodologies, which are rule-based, instrument-based, and image-based, offer unique contributions. For instance, rule-based models like L-System (Prusinkiewicz et al., 1996) rely on empirical growth patterns, while instrument-based methods (Xhimitiku et al., 2021) use 3D scanning technologies to provide accurate reconstructions. Image-based methods (Lin et al., 2022) utilize noninvasive sensors like Red, Green, Blue (RGB) or Red, Green, Blue and Depth (RGBD) cameras, light detection and ranging (LiDAR) techniques, and spectral imaging to monitor key crop parameters. These parameters are critical for evaluating crop growth trajectories in a digital framework. The adoption of techniques like image-based sensing (Lin et al., 2022) enables precise, scalable, and cost-effective monitoring of crop health and performance.

5.3 Disease detection and pest control

Crop diseases and pests can cause substantial crop loss, often spreading rapidly before detection. In such scenarios, digital twins may act as an early warning system by continuously analyzing real-time data and providing alerts to farmers. The digital twins technique is capable of detecting anomalies, such as changes in leaf color or temperature, that indicate the onset of disease or pest infestations. These insights help farmers deploy targeted treatments, reducing the spread of damage while preserving environmental health (Anand et al., 2024a, Anand et al. (2024b); Bhalekar et al., 2024a, 2024b).

Presently, various sensors such as ultrasonic, RGB, and LiDAR are being used for variable-rate chemical applications based on target detection (Bhalekar et al., 2023). However, these semiautonomous spray technologies still require human intervention, creating threats to operator safety. The recent advancements in AI, sensing, and robotics technology can offer autonomous disease management solutions (Wang et al., 2022, Dhar et al., 2025). The Digital Twin Modeling approach for pest management incorporates advanced state-of-the-art technologies to deliver a holistic solution. The typical digital twin-based pest management framework is organized into multiple layers. The perceptual layer utilizes various sensing platforms such as RGB cameras, spectral sensors, LiDAR, and weather sensors, integrated strategically with IoT architecture. Data collected using these sensing platforms is wirelessly transmitted using communication technologies such as long range (LoRa), long range wide area network (LoRaWAN), and 5G, passing through the transport layer and stored in the database of cloud platforms. In the application layer, the collected data is processed to develop a virtual model for pest prediction. AI and ML algorithms are employed to clean, analyze, and visualize this data, supporting Integrated Pest Management decision-making (Kumari et al., 2025, Dai et al., 2024).

5.4 Water management and optimization strategies

Water scarcity is a significant global issue affecting agriculture. Digital twins address this challenge by providing insights into water usage and irrigation needs. By simulating different irrigation strategies and rainfall scenarios, digital twins enable farmers to optimize water usage. This ensures precise and sustainable water use, ultimately leading to cost savings and improved yields (García-Tejero et al., 2011).

For example, the use of evapotranspiration models in digital twins helps estimate crop water needs more accurately (Syed et al., 2024). This helps farmers apply the right amount of water at the right time, avoiding overirrigation or underirrigation. Digital twins also enable the use of precision irrigation techniques, where water is delivered in precise amounts to specific parts of the field. By incorporating remote sensing technologies, such as drones and satellite imagery, into the digital twin system, farmers can monitor variability in soil moisture and adjust irrigation accordingly. This approach has been shown to significantly reduce water wastage and increase crop productivity (Ghazal et al., 2024; Parewai et al., 2025). Pesantez et al. (2022) stated that integrating digital twins with water resource management systems allows for better decision-making in conserving water and improving overall farm resilience to water scarcity.

Thus, digital twins offer a transformative approach to water management in agriculture by providing data-driven insights and optimization strategies. Through improved monitoring, simulation, and precision irrigation, farmers can conserve water, reduce costs, and enhance crop productivity, contributing to more sustainable agricultural practices.

5.5 Soil health and nutrient management

Soil health is critical in agriculture, and maintaining it requires careful management of nutrients and organic matter. Digital twins create a detailed virtual model of soil properties, such as pH levels, nutrient availability, and organic content. By simulating the effects of fertilizer application and crop rotation, farmers can develop strategies that enhance soil health, prevent erosion, and sustain long-term agricultural productivity. Several studies have demonstrated how digital twins can integrate soil nutrient levels with weather forecasts and crop-specific needs to predict optimal fertilizer application times (Silva et al., 2023). This approach helps farmers avoid nutrient deficiencies or excesses, which can harm both crops and the environment. By simulating soil-nutrient interactions, this technology helps farmers determine the most effective fertilizer application rates and timings. For example, using nutrient cycling models, digital twins can predict how different fertilizers will affect soil nutrient levels and crop growth (Mummaneni et al., 2024). The ability to model these dynamics allows for the precision application of fertilizers, reducing waste and minimizing the environmental impact of excess nutrients. Studies, such as those by Weckesser et al. (2022), have shown that digital twin-based systems can enhance nutrient use efficiency, lower costs, and improve crop yields.

6 Benefits of digital twin technology for farmers

DTT offers a wide range of benefits for farmers by enhancing their ability to manage resources efficiently, increase productivity, and reduce risks. Below are the key benefits of DTT for farmers:

i. **Improved decision-making capabilities**: It provides real-time data and predictive insights, enabling farmers to make informed decisions about crop management, irrigation, fertilization, and harvesting. This leads to optimized strategies and improved efficiency in operations.
ii. **Cost reduction in inputs (water, fertilizers, pesticides)**: By optimizing resource use, digital twins help reduce water consumption, ensure the precise application of fertilizers and pesticides, and minimize waste, leading to significant cost savings on essential inputs.
iii. **Enhanced yield and productivity**: It helps track crop growth and environmental conditions, allowing farmers to identify factors affecting productivity. This data enables better management practices that boost yields and improve overall farm output.
iv. **Risk mitigation for adverse weather, pests, and diseases**: It can predict adverse weather conditions, pest invasions, and disease outbreaks, allowing farmers to take preventive actions. This reduces crop losses and helps maintain a stable harvest.
v. **Sustainability and environmental impact reduction**: By optimizing water, fertilizer, and pesticide use, this technology contributes to sustainable farming. It helps reduce environmental impacts, such as water waste and chemical runoff, while promoting soil health and overall environmental stewardship.

7 Challenges and considerations

The adoption of DTT in farming presents numerous opportunities to enhance productivity, sustainability, and decision-making. However, several challenges must be addressed for successful implementation, especially for smallholder farmers. Below are the key challenges and solutions that can enable the widespread adoption and effectiveness of digital twins in agriculture.

i. Training and adaptability

- **Challenges**: Smallholder farmers, especially in regions like India, struggle with using IoT equipment due to a lack of proper training programs.
- **Considerations**: Developing user-centered designs for diverse farming communities, conducting educational outreach programs, and providing targeted training to small and marginal farmers to enhance adaptability.
- **Solutions**: Initiatives like e-Krishi improve training accessibility, helping farmers gain the skills to use DTs.

ii. Digitization and competency gaps

- **Challenges**: Lack of digital literacy and technical competency hampers the effective use of DTT.
- **Considerations**: Provide multidisciplinary training to enhance DT system management skills, and collaborate with agricultural experts and technologists to bridge knowledge gaps.
- **Solutions**: AgTech startups in Brazil, like Solinftec, are addressing competency gaps by offering training programs that teach farmers how to use AI and digital systems effectively.

iii. Technological design and transparency

- **Challenges**: ML models and digital twin systems can sometimes be opaque or biased in their decision-making.

- **Considerations**: Prioritize assessing ML models to avoid biases, simplify interdependent models for clearer decision-making, and implement accountability measures for errors.
- **Solutions**: In Californian vineyards, DT models are used for smart irrigation; frequent assessments ensure fair and accurate water distribution, preventing biases.

iv. Use-case specificity and standardization

- **Challenges**: A lack of standardization across various farming contexts can limit the scalability of digital twin systems.
- **Considerations**: Adopting the findable, accessible, interoperable, and reusable (FAIR) framework for scalability, developing standardized criteria for DT architecture, and fostering open, inclusive systems to ensure broad applicability.
- **Solutions**: In Australia, digital twins for livestock monitoring use standardized data-sharing platforms, ensuring flexibility across different farm sizes.

v. Model reliability and validation

- **Challenges**: Ensuring the accuracy and reliability of models in diverse farming conditions.
- **Considerations**: Establish analytical standards to assess DT performance, develop protocols for validating model behavior in different scenarios, and involve domain experts for thorough testing and validation.
- **Solutions**: In European precision agriculture, DT systems for real-time pest monitoring rely on domain-specific validation processes to ensure accuracy and reliability.

vi. Cost and accessibility

- **Challenges**: High costs and limited accessibility can prevent smallholders from adopting digital twin technologies.
- **Considerations**: Create scalable, modular, and pre-validated solutions to reduce costs; explore alternative financial models, such as shared ownership or subsidies; and address connectivity issues for efficient and affordable data processing.
- **Solutions**: Low-cost modular DT solutions, such as precision seeding drones, are helping smallholders in Africa increase yields while reducing costs.

8 Case studies and real-world examples

To understand the practical impact of DTT in farming, it's important to look at real-world examples and case studies. Below are several case studies showcasing the successful application of digital twins in different agricultural contexts.

Case studies	Descriptions	Illustrations
Moghadam et al. (2020) and Ao et al. (2022)	Created an orchard digital twin with 3D LiDAR and cameras for real-time condition monitoring and	

Case studies	Descriptions	Illustrations
	decision support, lowering farmers' labor requirements.	
Skobelev et al. (2020)	A digital twin of wheat is proposed by Extended Laryukhin et al. (2019), using information databases and multiagent modeling. With a focus on climate-resilient decision support, the system is designed to identify irregularities and suggest corrective measures.	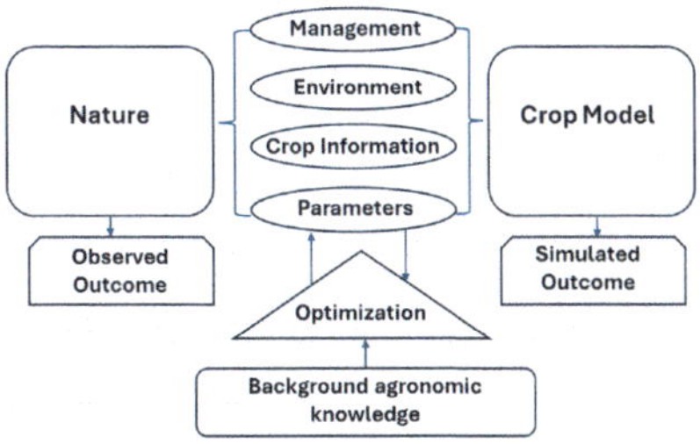
Machl et al. (2019)	Developed a spatio-temporal digital twin for cultivated landscapes to assist geo-design processes in agriculture and transportation.	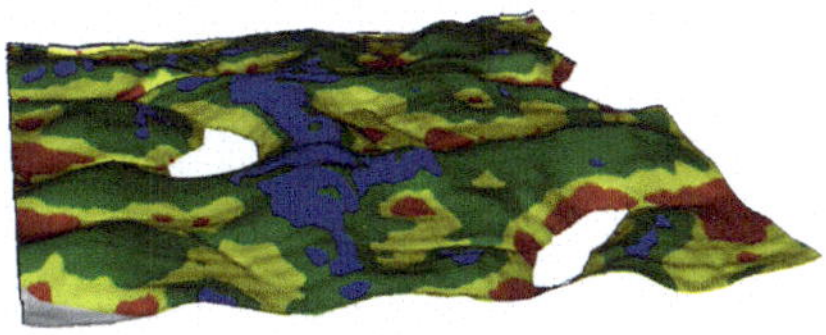
Jan et al. (2021)	Shown how IoT is being used in India to monitor water quality and fertilizers using a digital twin prototype that aims to reduce contamination.	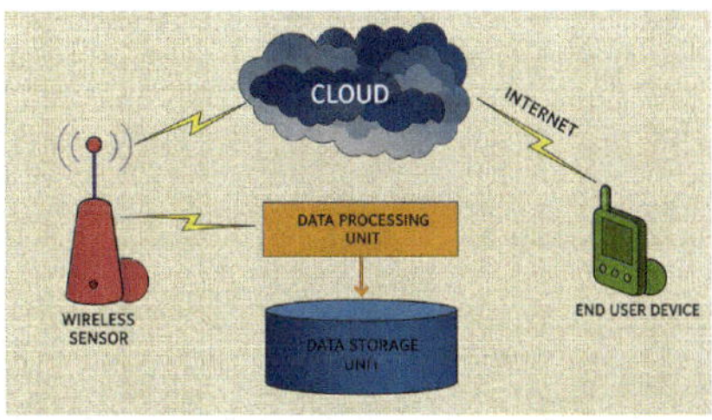

Continued

Case studies	Descriptions	Illustrations
Alves et al. (2019)	Employed a cloud-based IoT digital twin with localized sensors and remote sensing to optimize agricultural irrigation and increase production.	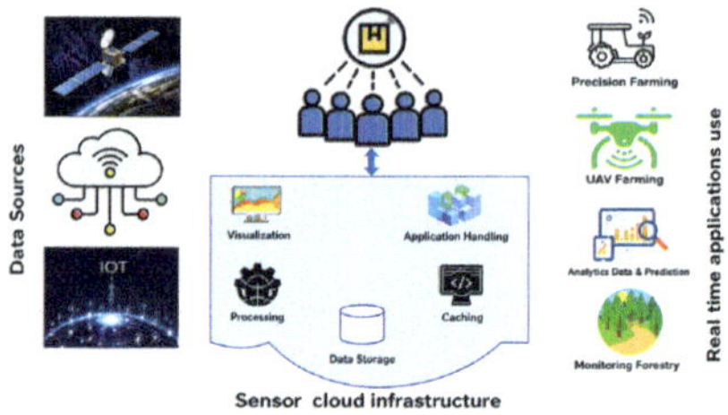
Kampker et al. (2019)	Developed a "digital potato" to maximize the calibration of potato harvesters. Utilized potato-like, sensor-equipped artifacts to determine the best harvester configuration using machine learning.	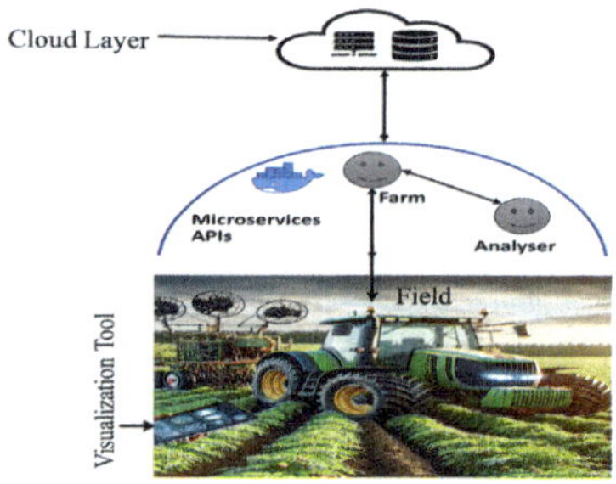
Jo et al. (2019)	Developed a digital twin that uses simulations and big data to identify scenarios and make decisions while controlling the temperature and air quality of the barn to support the growth of livestock.	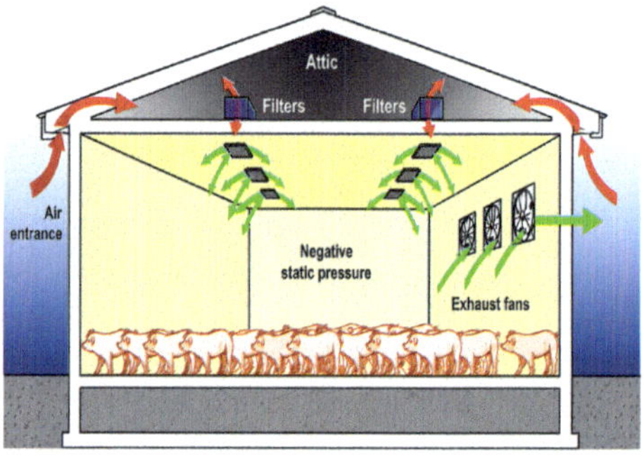 (Source: https://www.canr.msu.edu/news/basic-types-of-ventilation-in-swine-barns)

9 Prospects and emerging trends

DTT, which originated in the industrial and manufacturing sectors, is quickly expanding and finding disruptive uses in agriculture. This technology presents a viable way to optimize operations, boost efficiency, and improve decision-making as agriculture grapples with issues such as resource scarcity, environmental uncertainties, and the need for a sustainable food production system. The opportunities and new developments in this cutting-edge field are reviewed below:

- Precise monitoring of soil health, plant growth, and environmental conditions is made possible by the combination of the IoT and cutting-edge sensor technology.
- For crop growth modeling, yield prediction, and early disease and pest detection, AI augments digital twins with predictive analytics.
- Using cloud and edge computing improves precision agriculture decision-making by reducing latency and facilitating real-time data processing.
- By maximizing the use of water, fertilizers, and pesticides, digital twins assist environmentally friendly farming methods and advance sustainability.
- With the help of open-source platforms and reasonably priced IoT devices, efforts are underway to make DTT more scalable and accessible for small and medium-sized farms.
- Digital twins are easier to use and engage with because of sophisticated visualization techniques, such as virtual reality and augmented reality.
- Platforms that allow farmers, researchers, and policymakers to share data collaboratively spur innovation and increase forecast accuracy.
- Digital twins are enhanced by robotics and autonomous systems, including drones and robotic weed eaters, which carry out precise tasks based on real-time data.
- To address concerns about data privacy, ownership, and transparency in algorithmic judgments, regulatory frameworks and ethical standards are being developed.
- The goal of research and development in enhanced data fusion techniques and quantum computing is to increase the capabilities of DTT.

10 Conclusion

The DT's technology is truly a game-changer for crop management, offering farmers the tools they need to optimize their practices and make smarter, data-driven decisions. With the ability to provide real-time insights, predict crop outcomes, and reduce resource waste, digital twins are helping to create more sustainable and productive farms. From better water and fertilizer management to early warnings for pests and diseases, the technology is transforming how we approach farming.

The agricultural sector is facing increasing challenges due to environmental emergencies, resource limitations, and growing food demand, making it essential to embrace innovations such as DTT to ensure long-term success. The future of agriculture depends on our willingness to adopt new technologies that can improve efficiency, protect the environment, and increase food security. Farmers, stakeholders, and policymakers must support the adoption and integration of these technologies to unlock their full potential.

References

Alves, R.G., Souza, G., Maia, R.F., Tran, A.L.H., Kamienski, C., Soininen, J.P., et al., 2019. A digital twin for smart farming. In: 2019 IEEE Global Humanitarian Technology Conference (GHTC). IEEE, pp. 1–4. doi:10.1109/GHTC46095.2019.9033075.

Anand, R., Parray, R.A., Mani, I., Khura, T.K., Kushwaha, H., Sharma, B.B., et al., 2024a. Spectral data driven machine learning classification models for real-time leaf spot disease detection in brinjal crops. Eur. J. Agron. 161, 127384. doi:10.1016/j.eja.2024.127384.

Anand, R., Parray, R.A., Mani, I., Khura, T.K., Kushwaha, H., Sharma, B.B., et al., 2024b. A multimodal approach for enhanced disease management in cauliflower crops: integration of spectral sensors, machine learning models and targeted spraying technology. Front. Agric. Sci. Eng. 012 (2), 261–273.

Ao, Z., Wu, F., Hu, S., Sun, Y., Su, Y., Guo, Q., Xin, Q., 2022. Automatic segmentation of stem and leaf components and individual maize plants in field terrestrial LiDAR data using convolutional neural networks. Crop J. 10 (5), 1239–1250. doi:10.1016/j.cj.2022.07.005.

Benny, B., Shirley, C.P., Jingle, B.J., Vidhya, K., Elakkiya, E., 2024. Leveraging IoT and digital twins to monitor crop growth and health in agriculture. In: 2024 10th International Conference on Advanced Computing and Communication Systems (ICACCS), vol. 1. IEEE, pp. 2266–2271.

Bhalekar, D.G., Parray, R.A., Ingle, P.V., Mani, I., Khura, T.K., Sarkar, S.K., Kumbhare, N.V., 2024a. Agrochemical spray technology adoption and safety awareness assessment in crop protection in vineyard cultivation. Curr. Nat. Sci. Eng. J. 1 (3), 188–196.

Bhalekar, D.G., Parray, R.A., Mani, I., Kushwaha, H., Khura, T.K., Sarkar, S.K., et al., 2023. Ultrasonic sensor-based automatic control volume sprayer for pesticides and growth regulators application in vineyards. Smart Agric. Technol. 4, 100232. doi:10.1016/j.atech.2023.100232.

Bhalekar, D.G., Sahni, R.K., Schrader, M.J., Khot, L.R., 2024b. Pneumatic Spray Delivery-Based Fixed Spray System Configuration Optimization for Efficient Agrochemical Application in Modern Vineyards, Pest Management Science.

Bloch, V., Palosuo, T., Huitu, H., Ronkainen, A., Backman, J., Pussi, K., Suokannas, A., Pastell, M., 2022. Towards a digital twin for optimal field management. agriRxiv.

Botín-Sanabria, D.M., Mihaita, A.S., Peimbert-García, R.E., Ramírez-Moreno, M.A., Ramírez-Mendoza, R.A., Lozoya-Santos, J.D.J., 2022. Digital twin technology challenges and applications: a comprehensive review. Remote Sens. 14 (6), 1335. doi:10.3390/rs14061335.

Chowdhury, M., Khura, T.K., Parray, R.A., Upadhyay, P.K., Kushwaha, H.L., Lama, A., Kushwah, A., 2025. Multi-vegetation indices based handheld device for precise nitrogen assessment and prescription in direct seeded rice. Comput. Electron. Agric. 229, 109886.

Dai, M., Shen, Y., Li, X., Liu, J., Zhang, S., Miao, H., 2024. Digital twin system of pest management driven by data and model fusion. Agriculture 14 (7), 1099. doi:10.3390/agriculture14071099.

Dhar, T., Parray, R.A., Bashyal, B.M., Singh, A.K., Dhanger, P., Khura, T.K., et al., 2025. Optimized convolutional neural networks for real-time detection and severity assessment of early blight in tomato (Solanum lycopersicum L.). Fungal Genet. Biol. 178, 103984.

Escribà-Gelonch, M., Liang, S., van Schalkwyk, P., Fisk, I.D., Long, N.V., Hessel, V., 2024. Digital twins in agriculture: orchestration and applications. J. Agric. Food Chem. 72, 10737–10752. doi:10.1021/acs.jafc.4c01234.

García-Tejero, I.F., Durán-Zuazo, V.H., Muriel-Fernández, J.L., Rodríguez-Pleguezuelo, C.R., 2011. Water and sustainable agriculture. In: García-Tejero, I.F., Durán-Zuazo, V.H., Muriel-Fernández, J.L., Rodríguez-Pleguezuelo, C.R. (Eds.), Water and Sustainable Agriculture. Springer Netherlands, pp. 1–94. doi:10.1007/978-94-007-1868-9_1.

Ghazal, S., Munir, A., Qureshi, W.S., 2024. Computer vision in smart agriculture and precision farming: techniques and applications. Artif. Intell. Agric. 13, 64–83.

Grieves, M., Vickers, J., 2016. Digital twin: Mitigating unpredictable, undesirable emergent behavior in complex systems.In: Transdisciplinary perspectives on complex systems: New findings and approaches, Springer International Publishing, Cham, pp. 85–113.

Hijmans, R.J., Guiking-Lens, I., Van Diepen, C., 1994. WOFOST 6.0: User's Guide for the WOFOST 6.0 Crop Growth Simulation Model, Wageningen University and Research.

Jan, F., Min-Allah, N., Düştegör, D., 2021. IoT based smart water quality monitoring: recent techniques, trends and challenges for domestic applications. Water 13 (13), 1729. doi:10.3390/w13131729.

Jo, S.K., Park, D.H., Park, H., Kwak, Y., Kim, S.H., 2019. Energy planning of pigsty using digital twin. In: 2019 International Conference on Information and Communication Technology Convergence (ICTC). IEEE, pp. 723–725. doi:10.1109/ICTC46691.2019.8939836.

Kalyani, Y., Vorster, L., Whetton, R., Collier, R., 2024. Application scenarios of digital twins for smart crop farming through cloud-fog-edge infrastructure. Future Internet 16, 100. doi:10.3390/fi16060100.

Kampker, A., Stich, V., Jussen, P., Moser, B., Kuntz, J., 2019. Business models for industrial smart services–the example of a digital twin for a product-service-system for potato harvesting. Procedia. Cirp. 83, 534–540.

Kumari, K., Parray, R., Basavaraj, Y.B., Godara, S., Mani, I., Kumar, R., Khura, T., Sarkar, S., Ranjan, R., Mirzakhaninafchi, H., 2025. Spectral sensor-based device for real-time detection and severity estimation of groundnut bud necrosis virus in tomato. J. Field Robot. 42 (1), 5–19.

Laryukhin, V., Skobelev, P., Lakhin, O., Grachev, S., Yalovenko, V., Yalovenko, O., 2019. The multi-agent approach for developing a cyber-physical system for managing precise farms with digital twins of plants. Cybern. Phys. 8 (4), 257–261.

Lin, Z., Fu, R., Ren, G., Zhong, R., Ying, Y., Lin, T., 2022. Automatic monitoring of lettuce fresh weight by multi-modal fusion based deep learning. Front. Plant Sci. 13, 980581.

Machl, T., Donaubauer, A., Kolbe, T.H., 2019. Planning agricultural core road networks based on a digital twin of the cultivated landscape. J. Digit. Landsc. Archit. 4, 316–327.

Manivasagam, V.S., 2025. From bytes to farm: transferability of industrial digital twins in agricultural systems. J. Biosyst. Eng. 50 (1), 130–144.

McCown, R.L., Hammer, G.L., Hargreaves, J.N.G., Holzworth, D.P., Freebairn, D.M., 1996. APSIM: a novel software system for model development, model testing and simulation in agricultural systems research. Agric. Syst. 50 (3), 255–271. doi:10.1016/0308-521X(95)00011-Q.

Moghadam, P., Lowe, T., Edwards, E.J., 2020. Digital twin for the future of orchard production systems. Proceedings 36 (1), 92. MDPI.

Mummaneni, S., Sappa, T.S., Katakam, V.G.D., 2024. Enhancing crop health through digital twin for disease monitoring and nutrient balance. Informatyka, Automatyka, Pomiary W Gospodarce I Ochronie Środowiska 14 (1), 57–62.

Navarro, E., Costa, N., Pereira, A., 2020. A systematic review of IoT solutions for smart farming. Sensors 20 (15), 4231.

Parewai, I., Köppen, M., 2025. A digital twin approach for soil moisture measurement with physically based rendering simulations and machine learning. Electronics 14 (2), 395. doi:10.3390/electronics14020395.

Peladarinos, N., Piromalis, D., Cheimaras, V., Tserepas, E., Munteanu, R.A., Papageorgas, P., 2023. Enhancing smart agriculture by implementing digital twins: a comprehensive review. Sensors 23 (16), 7128. doi:10.3390/s23167128.

Pesantez, J.E., Alghamdi, F., Sabu, S., Mahinthakumar, G., Berglund, E.Z., 2022. Using a digital twin to explore water infrastructure impacts during the COVID-19 pandemic. Sustain. Cities Soc. 77, 103520. doi:10.1016/j.scs.2021.103520.

Piddubna, A., 2024. Digital Twins in Agriculture: A Time Machine for Your Farm. Intellias. https://intellias.com/digital-twins-in-agriculture/.

Prusinkiewicz, P., Hammel, M., Hanan, J., Mech, R., 1996. L-systems: from the theory to visual models of plants. In: Proceedings of the 2nd CSIRO Symposium on Computational Challenges in Life Sciences, vol. 3, pp. 1–32.

Rehman, A., Saba, T., Kashif, M., Fati, S.M., Bahaj, S.A., Chaudhry, H., 2022. A revisit of internet of things technologies for monitoring and control strategies in smart agriculture. Agronomy 12 (1), 127.

Ritchie, J., 1985. Description and performance of CERES wheat: a user-oriented wheat yield model. ARS Wheat Yield Proj. 38, 159–175.

Silva, L., Rodríguez-Sedano, F., Baptista, P., Coelho, J.P., 2023. The digital twin paradigm applied to soil quality assessment: a systematic literature review. Sensors 23 (2), 1007. doi:10.3390/s23021007.

Skobelev, P.O., Mayorov, I.V., Simonova, E.V., Goryanin, O.I., Zhilyaev, A.A., Tabachinskiy, A.S., Yalovenko, V.V., 2020. Development of models and methods for creating a digital twin of plant within the cyber-physical system for precision farming management. J. Phys. Conf. Ser. 1703 (1), 012022. doi:10.1088/1742-6596/1703/1/012022.

Syed, T.A., Khan, M.Y., Jan, S., Albouq, S., Alqahtany, S.S., Naqash, M.T., 2024. Integrating digital twins and artificial intelligence multi-modal transformers into water resource management: overview and advanced predictive framework. AI 5 (4), 1977–2017. doi:10.3390/ai5040098.

Tang, L., Zhu, Y., Hannaway, D., Meng, Y., Liu, L., Chen, L., Cao, W., 2009. RiceGrow: a rice growth and productivity model. NJAS Wagening. J. Life Sci. 57 (1), 83–92. doi:10.1016/j.njas.2009.12.001.

Verdouw, C., Tekinerdogan, B., Beulens, A., Wolfert, S., 2021. Digital twins in smart farming. Agric. Syst. 189, 103046. doi:10.1016/j.agsy.2021.103046.

Wang, Z., 2020. Digital twin technology. In: Industry 4.0-impact on Intelligent Logistics and Manufacturing, IntechOpen.

Weckesser, F., Beck, M., Hülsbergen, K.J., Peisl, S., 2022. A digital advisor twin for crop nitrogen management. Agriculture 12 (2), 302. doi:10.3390/agriculture12020302.

Xhimitiku, I., Bianchi, F., Proietti, M., Tocci, T., Marini, A., Menculini, L., et al., 2021. Anomaly detection in plant growth in a controlled environment using 3D scanning techniques and deep learning. In: 2021 IEEE International Workshop on Metrology for Agriculture and Forestry (Metroagrifor). IEEE, pp. 86–91. doi:10.1109/MetroAgriFor52389.2021.9628476.

Zhang, L., Chen, X., Zhou, W., Cheng, T., Chen, L., Guo, Z., et al., 2020. Digital twins for additive manufacturing: a state-of-the-art review. Appl. Sci. 10 (23), 8350. doi:10.3390/app10238350.

Zhang, R., Zhu, H., Chang, Q., Mao, Q., 2025. A comprehensive review of digital twins technology in agriculture. Agriculture 15 (9), 903.

Predicting crop yield and nutrient needs using explainable artificial intelligence

10

Oumnia Ennaji[a,b], *Leonardus Vergutz*[b], *and Achraf El Allali*[a]
[a]College of Computing, University Mohammed VI Polytechnic, Benguerir, Morocco
[b]College of Agriculture and Environmental Sciences, University Mohammed VI Polytechnic, Benguerir, Morocco

1 Introduction

Agriculture is central to food security, livelihoods, and economic development in many low and middle-income countries, particularly in Africa (Preethi et al., 2020; Tapsoba et al., 2020; Collier and Dercon, 2014; Rejeb et al., 2022). As global population growth accelerates, agricultural systems face mounting pressure to produce more food using fewer resources while coping with increasingly erratic weather (Kamilaris et al., 2017; Bharath et al., 2022). In many agrarian economies, national prosperity is intricately linked to farm productivity (Salami et al., 2010; Livingston et al., 2011). However, agriculture across the continent is marked by structural challenges, including climatic variability, water scarcity, soil heterogeneity, and limited access to data-informed decision-making tools (Aworka et al., 2022; Diao et al., 2010; Olanipekun et al., 2019; Jayne et al., 2010).

Despite significant advances in agricultural research and growing volumes of field-level data, many smallholder farmers continue to experience very low yields due to inappropriate nutrient management (Chivenge et al., 2022; Jha et al., 2019; Ennaji et al., 2023). Conventional agronomic models often fail to capture complex and nonlinear interactions between soil properties, climatic variability, and plant responses. At the same time, the increasing availability of high-resolution data from sensors, drones, and satellites has created new opportunities to transform agricultural systems through data-driven insights. This development has laid the foundation for the emergence of precision agriculture, an approach that aims to tailor farm management to the specific needs of particular locations using advanced technologies.

Among these technologies, machine learning (ML) has shown remarkable potential for predicting crop yields, diagnosing nutrient deficiencies, and optimizing input recommendations by learning patterns from historical and real-time data (Liakos et al., 2018a; Adnan et al., 2019; Patrício and Rieder, 2018; Paudel et al., 2021). However, a major factor limiting the large-scale application of ML models for

Precision Technologies for Digital Agriculture. https://doi.org/10.1016/B978-0-443-40513-6.00007-X

agriculture is the transparency of most of the best-performing models. Although these models generally make more accurate predictions than traditional methods, their "black box" nature can affect stakeholder confidence and the use of applications in low-resource environments.

Explainable artificial intelligence (XAI) is important in this context. XAI techniques, such as SHapley Additive Explanations (SHAP) and permutation feature importance, allow researchers, practitioners, and farmers to better understand the rationale behind model outputs. These methods not only improve transparency but also generate actionable insights into the key variables driving yield variation and nutrient demand. By transforming complex predictions into interpretable knowledge, XAI bridges the gap between algorithmic sophistication and practical decision-making, fostering greater trust and usability across the agricultural value chain (Sarker, 2021; Barbedo, 2019; Ennaji et al., 2025).

This chapter explores the integration of ML and XAI to improve crop yield prediction and nutrient management in precision agriculture. We provide a conceptual overview of the main algorithms and interpretation methods commonly used in agronomic applications, using two real-world case studies: yield improvement in the North China Plain (NCP) and wheat production in Morocco. We show how ML models, in conjunction with XAI, can provide accurate predictions and clear explanations that support management decisions at the field level. The first case study examines the role of climatic variability and soil nutrient availability on maize yields, while the second study highlights the effects of nutrient timing and soil pH on wheat production.

These case studies illustrate the value of interpretable ML in promoting sustainable, site-specific, and data-driven agricultural practices. By improving predictive accuracy and transparency, the integration of ML and XAI offers a promising avenue for enhancing productivity, profitability, and environmental awareness in modern agriculture.

2 Machine learning in precision agriculture

The advent of precision agriculture has inaugurated a new era of data-driven decision-making in agriculture. At the center of this change is the application of ML, which enables the analysis of complex multidimensional data sets to extract patterns and create predictive models. In agronomy, ML is increasingly being used to support critical tasks such as yield forecasting, nutrient recommendations, and real-time plant health monitoring (Ennaji et al., 2023). In the area of nutrient management and fertilizer recommendations, ML offers the potential to change traditional practices by enabling more precise and site-specific nutrient application (Fink et al., 1998). For example, ML models can predict crop nutrient requirements based on soil properties, weather patterns, and plant characteristics, helping farmers optimize fertilizer use and minimize nutrient loss. ML can also be used to diagnose nutrient deficiencies in crops, enabling timely interventions and preventing yield losses (Wang et al., 2020; de Lara et al., 2023).

Several specific ML algorithms have been used for fertilizer recommendations, each with their own strengths and weaknesses. Linear regression models offer simplicity and interpretability but can have problems with complex nonlinear relationships in the data (Ennaji et al., 2024). Support vector machines are effective in high-dimensional spaces and can deal with nonlinearity but can be computationally intensive (Ennaji et al., 2023). Random forests (RFs), an ensemble method, provide robustness and can handle missing data but can lack transparency (Paudel et al., 2021). Neural networks, inspired by the human brain and characterized by learning complex patterns, including deep learning architectures, are also on the rise, especially for image-based tasks such as weed or disease detection. However, they tend to overfit and require large data sets for training (Ennaji et al., 2023). The choice of algorithm depends on the specific problem, the availability of data, and the desired level of accuracy and interpretability. Unsupervised learning approaches, such as k-means clustering or principal component analysis, are used to reveal structures in data, e.g., when classifying soil types or identifying field zones with similar fertility patterns. Studies have investigated various ML models, including gradient boosting decision trees and neural networks, for predicting crop yields and optimizing nutrient uptake (Ennaji et al., 2023; Kaneko et al., n.d.; Ennaji et al., 2024).

While these ML algorithms form the basis for predictive modeling, optimization methods play a crucial role in fine-tuning their parameters and improving their effectiveness for tasks such as fertilizer recommendations (Ennaji et al., 2024). Gradient-based methods, such as gradient descent, iteratively adjust the model parameters by moving in the direction of steepest descent to find the minimum error. Populationbased methods, such as genetic algorithms and particle swarm optimization, investigate different solutions and evolve them over generations to achieve optimal results. Stochastic/heuristic methods, including simulated annealing and Bayesian optimization, use probabilistic and statistical approaches to guide the search for optimal solutions. Hybrid methods combine the strengths of different optimization algorithms, such as Bayesian-evolutionary hybrids and gradient-assisted evolutionary algorithms. Reinforcement learning methods, including policy-based optimization and Q-learning optimization, are important for dynamic decision problems, such as optimizing fertilizer use throughout the growing season.

ML models in agriculture use a variety of input characteristics that reflect the complex interactions between plant physiology, soil properties, climatic variability, and management practices. Soil-related variables often include pH, organic matter content, concentrations of macro and micronutrients such as nitrogen (N), phosphorus (P), and potassium (K), and physical properties such as bulk density and electrical conductivity. Climate and weather variables such as temperature, precipitation, humidity, solar radiation, and drought indices are essential for recording environmental and growing conditions throughout the crop cycle. In addition, cropping practices such as sowing dates, variety selection, fertilization, irrigation schedules, and crop rotations provide an important context for understanding how human decisions affect productivity. Finally, indicators of plant stress derived from spectral indices (e.g., normalized difference vegetation index (NDVI), Enhanced Vegetation Index (EVI)), thermal images, and field observations provide valuable clues for detecting early signs of nutrient deficiency, water stress, or disease. Together, these variables form the basis for the development of robust, powerful ML models for precision agriculture. By integrating these heterogeneous data sources, often collected via sensors, satellites, farmer surveys, and laboratory analysis, ML models can better capture yield variability and support site-specific recommendations.

Two of the best-known predictive applications of ML in precision agriculture are yield prediction and fertilizer recommendations. Yield prediction models aim to estimate future crop yields based on current and historical input data and help farmers and policy makers plan logistics, storage, and market strategies. Fertilizer recommendation systems, on the other hand, seek to optimize nutrient inputs to maximize yield while minimizing environmental impact. These systems often combine agronomic principles with data-driven models to recommend the type, timing, and amount of nutrient application. Recent studies, including (Paudel et al., 2021; Ennaji et al., 2024), have shown that ensemble methods such as RF and eXtreme Gradient Boosting (XGBoost) perform particularly well at these tasks, often outperforming traditional linear models.

The use of ML in agronomy offers several decisive advantages. ML models are characterized by their ability to capture nonlinear interactions and complex relationships between variables and provide strong predictive performance even in noisy, multidimensional data sets. They also support the integration of different types of data, including categorical, continuous, spatial, and temporal data, which is essential in precision agriculture, where data comes from multiple sources such as soil sensors, satellite imagery, and farmer records. In addition, trained models can be scaled across different regions and integrated into digital decision support systems (DSS) for broader impact. However, there are still important limitations. One major problem is interpretability: Many ML models, especially deep learning algorithms, operate as 'black boxes', making their results difficult to explain or validate. Furthermore, the performance of these models depends heavily on the quality of the data and the availability of resources, which are often limited in developing regions. Furthermore, models developed in one agroecological setting may not be transferable to other regions without significant recalibration, raising questions about generalizability. To overcome these limitations, a stronger focus on transparency and interpretability is needed. As a result, the integration of XAI techniques has become increasingly important to ensure that ML-powered tools in agriculture are trustworthy, interpretable, and actionable, which are topics we explore in more detail in the next section.

3 The role of explainable artificial intelligence

While ML has opened up new opportunities to improve agricultural productivity and sustainability, one of the biggest challenges remains: understanding how and why a model makes a particular prediction. Many powerful algorithms, especially ensemble models such as RFs and boosting methods or deep neural networks, are often considered 'black boxes.' In areas such as agriculture, where decisions impact farmers' livelihoods, the use of natural resources, and food security, this lack of transparency can be a major barrier to trust and adoption. XAI has emerged as an important solution to this problem. It refers to a set of techniques that make ML models more interpretable and help researchers, agronomists, and farmers alike understand what influences a model's decisions. In other words, XAI not only tells us what the prediction is but also helps explain why. This is especially important in agricultural science, where expert knowledge, practical experience, and environmental context all play a role (Table 10.1).

Table 10.1 Key publications on explainable AI (XAI) in agriculture.

Year	References	Title	Highlights
2016	Mohanty et al. (2016)	Using deep learning for image-based plant disease detection	Pioneering the use of CNNs in plant disease identification with high accuracy
2018	Kamilaris and Prenafeta-Boldú (2018)	Deep learning in agriculture: A survey	Comprehensive review of DL applications in agriculture
2018	Liakos et al. (2018b)	Machine learning in agriculture: A review	Overview of ML techniques in various agricultural domains
2018	Ghosal et al. (2018)	An explainable deep machine vision framework for plant stress phenotyping	Integration of explainability in deep learning for stress detection
2019	Zhang et al. (2019)	A deep learning-based approach for automated yellow rust disease detection from high-resolution hyperspectral UAV imagery	Use of hyperspectral UAV data and DL for early rust detection
2020	Arrieta et al. (2020)	Explainable Artificial Intelligence (XAI): Concepts, taxonomies, opportunities and challenges toward responsible AI	Foundational paper on XAI principles and methods
2020	van Klompenburg et al. (2020)	Crop yield prediction using machine learning: A systematic literature review	Summarizes trends and gaps in ML-based yield prediction
2022	Liang et al. (2022)	Explainable artificial intelligence for agriculture and crop yield prediction: A survey	Survey focused on XAI in yield modeling and precision agriculture
2023	Khaki and Wang (2023)	Crop yield prediction using deep neural networks: A comprehensive review and case study	Analysis of DNN performance for yield prediction with a real-case study
2023	Mohanty et al. (2023)	Explainable AI (XAI) in agriculture: Concepts, techniques, and applications	Detailed review of XAI methods with agricultural case applications

Among the most commonly used XAI methods in agricultural modeling are SHAP and permutation feature importance. SHAP is based on cooperative game theory and assigns a contribution score to each feature for each individual prediction. For example, it can show how much the timing of nitrogen application or a particular soil pH value contributes to a predicted crop yield. Permutation feature importance, on the other hand, offers a global view: It shows how model accuracy decreases when the relationship between a feature and the target is perturbed and helps us identify which variables are consistently important across the data set.

These tools are not only technically useful but also practical. In our research, SHAP and permutation methods have helped to reveal that, for example, minimum temperature and surface pressure are the most important factors in maize yield variation in the NCP, while soil pH and timing of potassium application are critical for wheat productivity in Morocco. By uncovering such insights, XAI provides a link between ML predictions and agronomic reality. It helps farmers, advisors, and policy makers make more informed decisions based on data they can understand and trust.

Beyond performance, this focus on interpretability builds trust in the technology itself. When users understand why a model recommends a certain fertilizer dose or identifies a potential yield risk, they are more likely to use, question, and improve it. Ultimately, integrating XAI into agricultural artificial intelligence (AI) tools transforms them from opaque predictive models into transparent DSS that make ML not only accurate but also actionable.

In the following case studies, we show how explainable ML has been applied in two very different agricultural contexts: Corn farming in the NCP and wheat farming in Morocco. Each case illustrates how XAI methods can uncover what really contributes to crop outcomes and how these insights can be used to improve field-level decisions.

4 Case study 1: Predicting maize yield in the North China Plain

This case study investigates predictive modeling of maize yield for the elite hybrid ZhengDan 958 in the NCP using gradient boosting and other ML algorithms (Ennaji et al., 2024). ZhengDan 958 is one of the most widely cultivated maize hybrids in the NCP due to its adaptability, high yield potential, and resilience to environmental stress. The dataset comprises over 1700 georeferenced observations collected between 2005 and 2010. It includes a wide range of variables capturing climate, soil, and management conditions, such as daily precipitation, temperature (min/max/mean), surface pressure, relative humidity, soil organic matter (SOM), available phosphorus (Olsen-P), potassium (Ava-K), and fertilizer application rates (N, P_2O_5, K_2O).

A number of algorithms were tested, including RF, gradient boosting regressor (GBR), XGBoost, multilayer perceptron (MLP), and basic models such as linear regression. All models were trained and evaluated with 5-fold cross-validation. XGBoost outperformed the others and achieved the highest prediction accuracy with $R^2=0.87$, RMSE=0.37, and MAE=0.26. GBR followed closely behind with $R^2=0.86$, RMSE=0.39, and MAE=0.27. RF, although traditionally strong, performed slightly worse in this context, highlighting the superior performance of gradient boosting algorithms for nonlinear, high-dimensional agronomic data (Figs. 10.1–10.4).

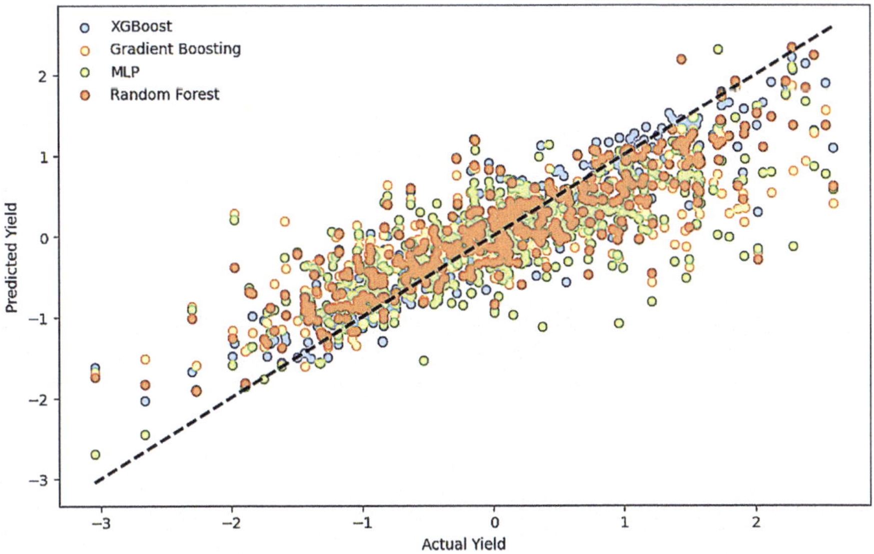

Fig. 10.1 Observed vs. predicted maize yield using XGBoost.

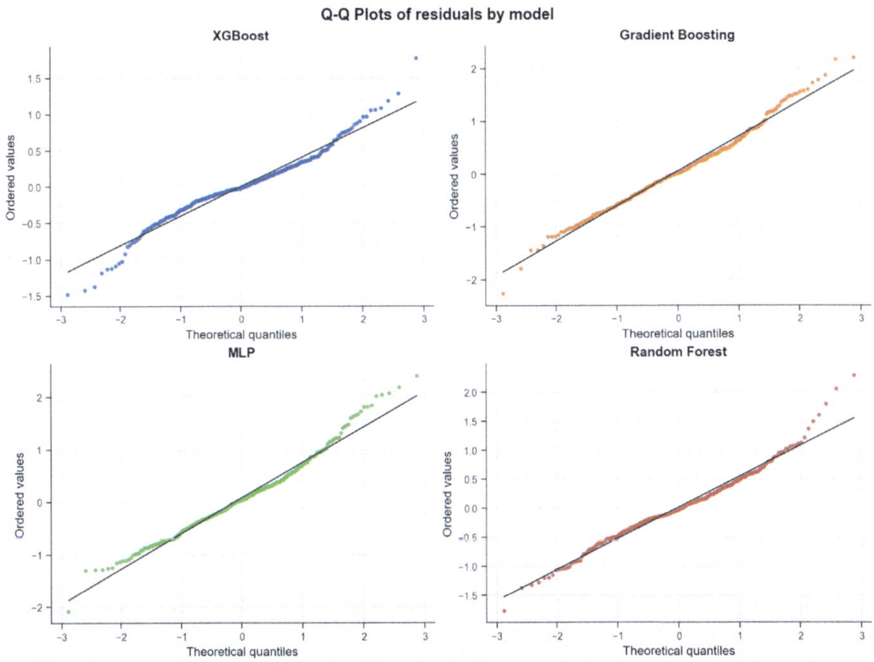

Fig. 10.2 Residual plot showing prediction errors for the XGBoost maize yield model.

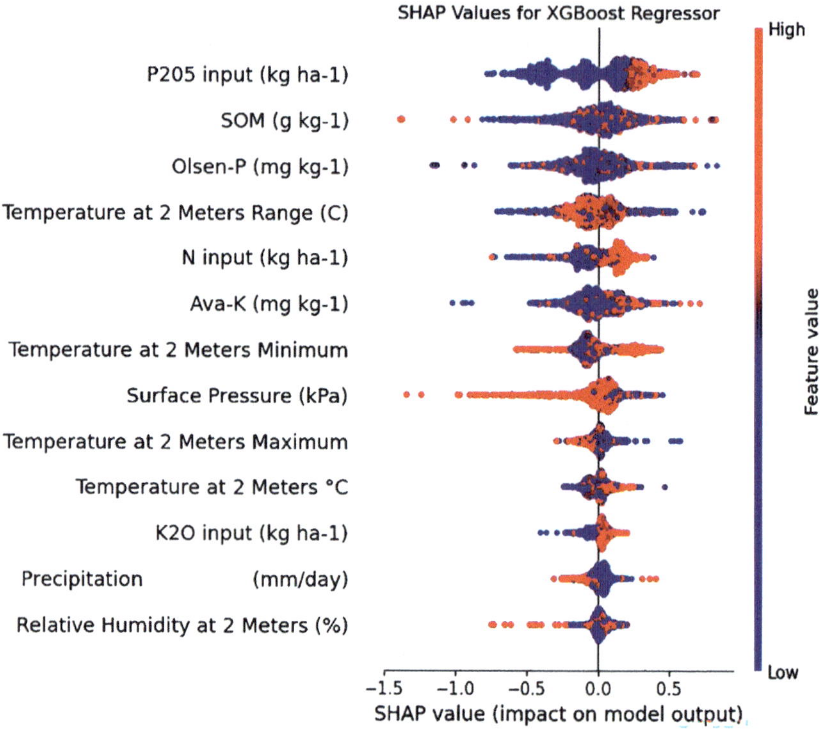

Fig. 10.3 SHAP summary plot showing the key drivers of maize yield prediction.

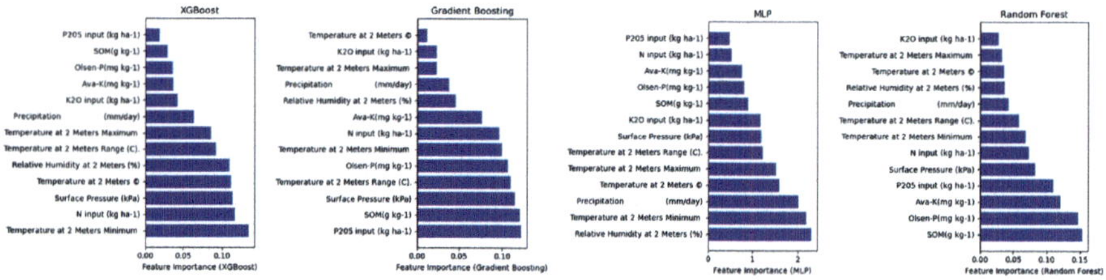

Fig. 10.4 Permutation feature importance for the maize yield prediction model.

To further investigate the model's behavior, XAI methods such as SHAP and permutation importance were used. These analyses revealed that minimum temperature, nitrogen input, surface pressure, and relative humidity were consistently among the best predictors of yield in the XGBoost and MLP models. In contrast, RF highlighted soil properties such as SOM, OlsenP, and P_2O_5 input as the most influential factors, demonstrating the variability in prioritizing traits between algorithms. The combined results suggest that weather variables, particularly temperature and humidity, play a dominant role in yield determination and that boosting methods are particularly well-suited to capture these dynamics.

In practice, these findings support more informed decisions on nutrient management and irrigation planning. By integrating historical weather data with site-specific soil parameters, gradient boosting models provide accurate, scalable, and interpretable predictions for increasing corn production in the NCP. This case illustrates the potential of ML and XAI to promote localized, data-driven agricultural management strategies.

5 Case study 2: Predicting wheat yield in Morocco

Wheat is one of the most important crops in Morocco and is grown in various agro-ecological zones under rainfed conditions. However, due to soil degradation, erratic rainfall, and suboptimal nutrient management, its production fluctuates greatly. This case study investigates the prediction of wheat yields using XGBoost, complemented by XAI techniques, to improve interpretability and support decision-making at the field level.

The dataset includes 5679 observations of wheat fields collected over three growing seasons (2018/2019, 2019/2020, 2020/2021) in eight regions in Morocco. Key characteristics include soil pH, organic matter content, phosphorus (P_2O_5) and potassium (K_2O) content, nitrogen (N) application rates (both as base and topdressing), seeding rates, crop variety, previous crops, and final wheat yields. All soil properties were collected through field sampling and laboratory analysis, while management data was collected directly from farmers as part of the Al Moutmir initiative. The selected fields cover a representative gradient of climate zones to ensure the robustness of the modeling framework.

XGBoost was trained with standardized numerical inputs and categorical variables that were one-hot encoded. Hyperparameters were tuned using grid search, and performance was evaluated using 5-fold cross-validation. The final model achieved an R^2 of 0.93 with an RMSE of 508.53 kg/ha, confirming excellent fit and generalizability.

Fig. 10.5 illustrates that the residuals are randomly distributed around zero, showing minimal prediction bias.

Fig. 10.6 confirms a strong alignment between predicted and observed wheat yields.

Partial dependency plots (Fig. 10.7) illustrate the positive effects of nitrogen and potassium fertilization on wheat yield. In particular, yields increased with nitrogen application up to a plateau, underlining the importance of correct dosing and timing. Soil pH also proved to be a key factor: yields peaked when pH was between 7.0 and 8.5; beyond that, nutrient availability likely decreased. This suggests that managing soil acidity is crucial for maximizing wheat productivity, especially on alkaline soils in Morocco.

Fig. 10.8 shows summary SHAP diagrams that quantify the influence of each trait on the model's predictions. Topdress nitrogen, the Bandera variety and potassium oxide (K_2O) at basal application were the main positive influencing factors on

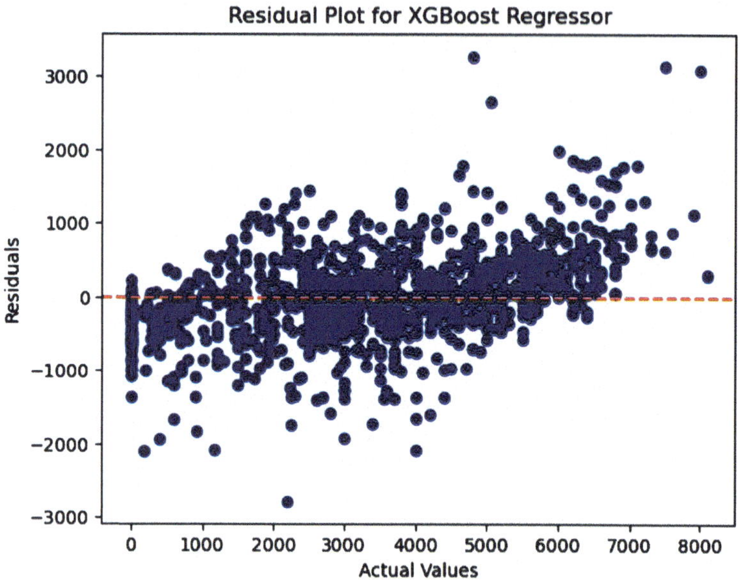

Fig. 10.5 Residual distribution plot showing prediction errors for the XGBoost model.

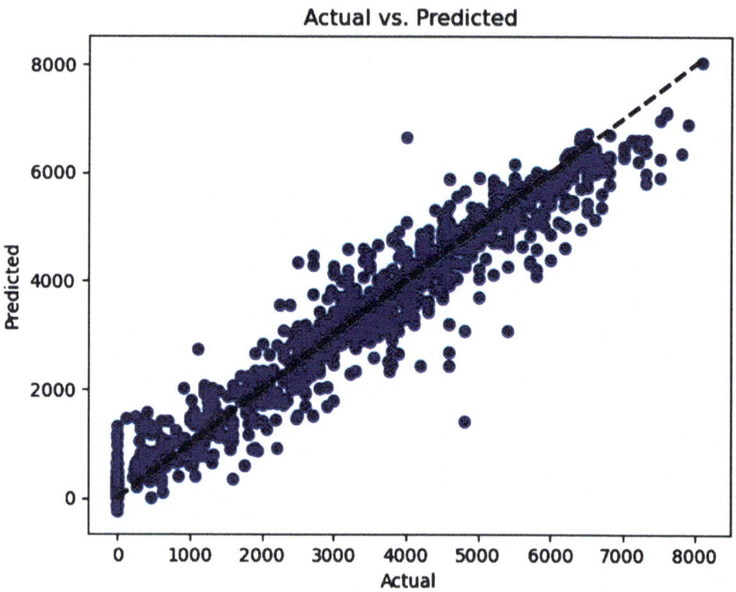

Fig. 10.6 Observed vs. predicted wheat yield using XGBoost.

predicted yield. In contrast, very high pH values showed a negative SHAP effect, confirming that nutrient availability and uptake is hindered under excessively alkaline conditions. These findings are consistent with the partial dependence plot results and improve the understanding of nutrient-soil interactions at the local scale.

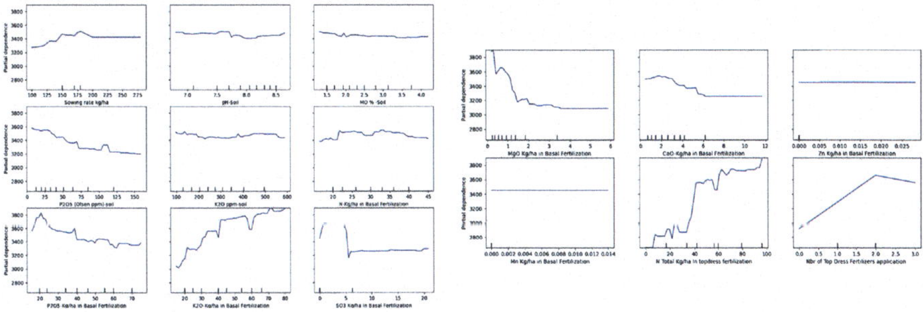

Fig. 10.7 Partial dependence plots (PDPs) showing the marginal effects of top predictors on wheat yield.

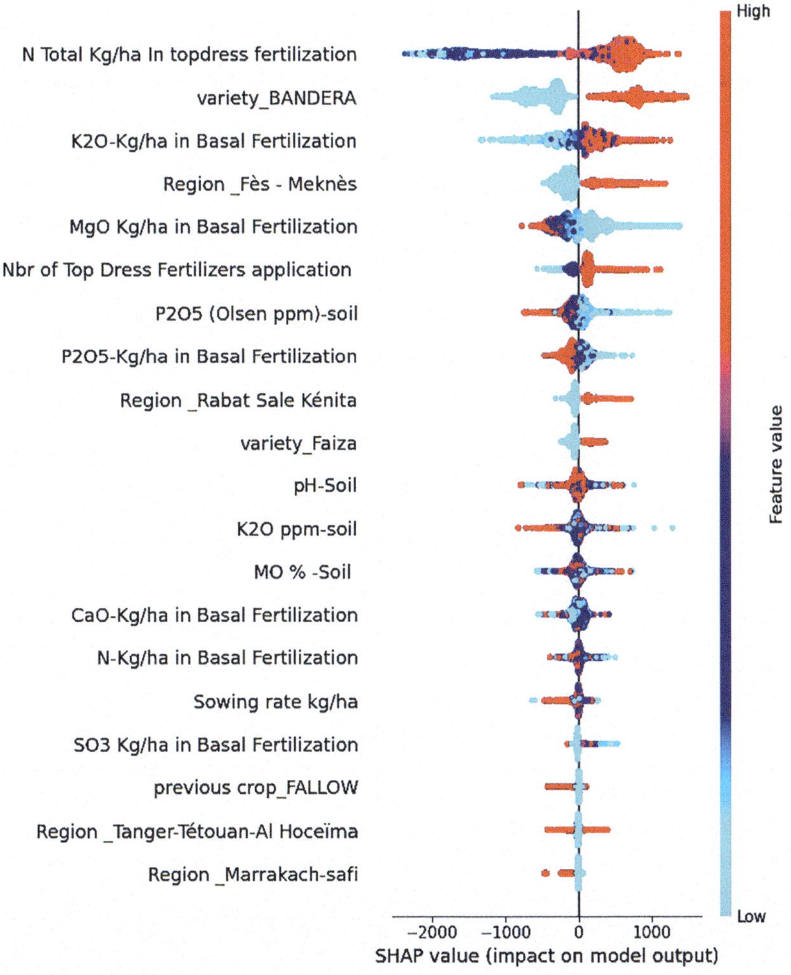

Fig. 10.8 SHAP summary plot showing feature contributions to the yield prediction model.

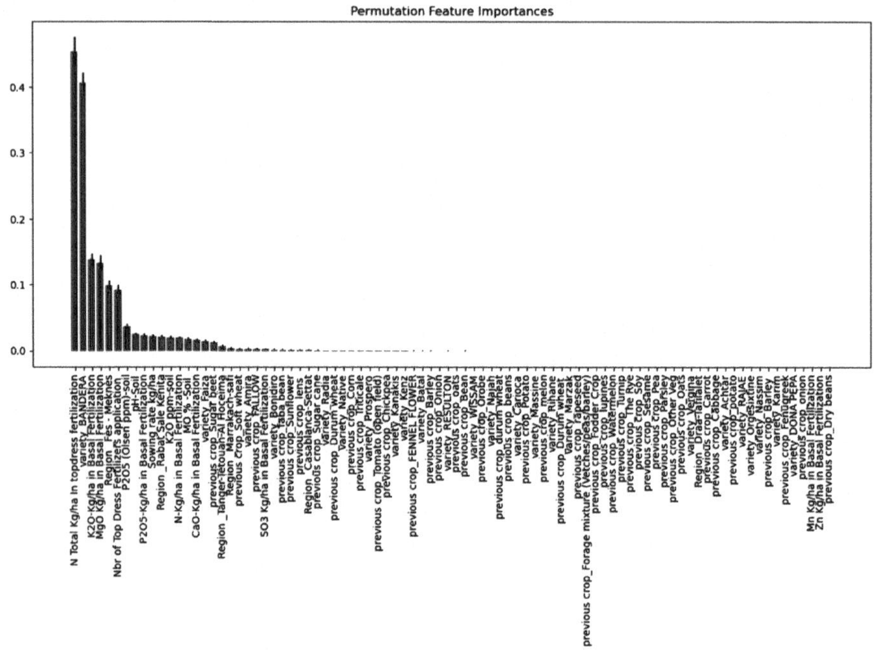

Fig. 10.9 Permutation feature importance scores from the XGBoost model.

The results of the model serve as a basis for site-specific fertilization strategies and recommendations for soil management. Prioritizing nitrogen fertilization and managing phosphorus and potassium levels can significantly improve wheat yields. In addition, the identification of fields with suboptimal pH values enables targeted interventions, such as acidification measures or the application of organic matter. These recommendations can be implemented both at the farm level and at the political level. Fig. 10.9 (permutation importance placeholder) underpins these results by ranking the features based on their overall contribution to prediction accuracy.

6 Discussion

This chapter illustrates the value of integrating ML and XAI in precision agriculture through two complementary case studies of maize in the NCP and wheat in Morocco. Despite their geographical, agronomic, and environmental differences, both studies share some key themes.

First, both modeling experiments confirm the predictive power of gradient boosting algorithms. In the case of maize, GBR and XGBoost consistently outperformed RF and other base models by achieving R^2 values above 0.85 and root mean square error (RMSE) values below 0.40. For wheat in Morocco, XGBoost also achieved high accuracy (R^2=0.93) and low prediction error (RMSE=508.53 kg/ha), indicating the ability of ML models to generalize across different cropping systems when trained on well-structured datasets.

Second, the contribution of environmental and management variables was evident in both cases. In maize, weather characteristics, especially temperature and humidity, proved to be dominant factors influencing yield prediction, while soil fertility indicators such as P_2O_5, K_2O and SOM also played a crucial role. In the wheat model, the timing of nitrogen application and soil pH were the most influential, highlighting the interaction between the timing of inputs and local edaphic conditions. These findings support a growing body of evidence that crop yields are highly sensitive to spatial and temporal variations in environmental and management conditions.

Importantly, the use of XAI tools such as SHAP values and permutation feature importance allowed these models to go beyond prediction and provide explanations. Rather than relying on black-box outputs, we were able to attribute the model's decisions to specific agronomic features highlighting, for example, the negative effects of overly alkaline soils or the yield plateau following optimal nitrogen application. This ability to interpret model behavior not only strengthens scientific knowledge but also makes AI tools more accessible and trustworthy for stakeholders such as farmers, agronomists, and policy makers.

From an implementation perspective, both case studies show that ML models can be effectively integrated into DSS for real-time agronomic planning. However, this integration must be done with user-friendliness in mind. Farmers are more likely to adopt tools that not only work well but also provide clear, actionable insights. Therefore, the inclusion of XAI is not a luxury but a necessity.

However, there are still trade-offs. While more complex models, such as gradient boosting and deep learning, often achieve higher accuracy, they demand significantly more computing power and memory compared to simpler algorithms like RF or linear regression. This raises concerns not only about computational effort but also about interpretability. Moreover, the transferability of models across regions or seasons is not guaranteed without recalibration, highlighting the need for localized data and iterative refinement. For smallholder farmers in low-resource contexts, practical deployment may rely on cloud-based platforms or lightweight pretrained versions optimized for mobile devices. Consequently, implementation must balance predictive performance with computational efficiency and accessibility.

In summary, these case studies show that the combination of ML and XAI can increase agricultural productivity, provide information on the sustainable use of inputs, and bridge the gap between algorithmic modeling and practical farm management. However, to realize this potential at scale, further investment in data infrastructure, farmer engagement, and adaptive research tailored to different agro-ecological contexts is required.

Taken together, these findings emphasize not only the predictive power and interpretability of ML and XAI approaches, but also their broader implications for sustainable agriculture, including productivity gains, environmental protection, and climate resilience.

The integration of ML and XAI into agricultural systems holds transformative potential, not only to improve productivity but also to advance sustainability goals. As the case studies on corn and wheat yield prediction show, AI-powered tools can lead to more precise use of inputs and better decision-making, which translates into tangible benefits at farm, regional, and global levels.

From a productivity and profitability perspective, ML models enable data-driven recommendations that optimize the use of resources. These tools help farmers maximize yields while minimizing unnecessary input costs. In both case studies, for example, XAI's findings enabled targeted interventions based on key yield factors such as temperature, nutrient availability, and soil pH, which supported more consistent management and more stable crop outcomes.

From an ecological point of view, conventional fertilization often leads to nutrient runoff, soil degradation, and greenhouse gas emissions. ML models with XAI can suggest site-specific nutrient schedules that are better matched to crop needs, reducing overfertilization and improving nutrient efficiency. This supports healthier soils, cleaner water, and long-term sustainability.

ML–XAI tools also strengthen climate resilience by using weather patterns to anticipate risks and guide adaptation practices. In regions facing heat stress, erratic rainfall, or shorter growing seasons, they help anticipate crop performance and intervene in time. As the XAI explains the reasons for these recommendations, farmers are more likely to trust and adopt them.

In summary, the thoughtful application of AI in agriculture aligns with the overarching goals of sustainable development: increasing productivity without compromising the environment or social equity. Combined with robust field data and participatory engagement, ML and XAI can support the transition to more resilient, resource-efficient, and climate-friendly food systems.

7 Conclusion

This chapter showed how ML and XAI can support more informed, efficient, and sustainable agricultural practices. Using diverse agronomic, environmental, and management variables, we demonstrated that ML models, such as XGBoost and RF, can accurately predict crop yields.

Beyond predictive accuracy, XAI methods such as SHAP and permutation feature importance have proven essential for interpreting results. They revealed the key features influencing yield, such as the timing of nitrogen application, soil pH, and weather variables like minimum temperature and humidity, bridging the gap between data-driven models and practical decision-making. This interpretability makes AI tools more trustworthy and actionable for farmers, agronomists, and policy makers.

Looking forward, future research should enhance model generalization across regions and seasons, refine XAI techniques for agronomic relevance, and codevelop decision tools with farming communities. By addressing challenges in data availability, model transferability, and user engagement, precision agriculture can become not only technologically advanced but also inclusive, transparent, and sustainable.

References

Adnan, A.A., Diels, J., Jibrin, J.M., Kamara, A.Y., Craufurd, P., Shaibu, A.S., Mohammed, I.B., Tonnang, Z.E.H., 2019. Options for calibrating CERES-maize genotype specific parameters under data-scarce environments. PLoS One 14 (2), 1–20.

Arrieta, A.B., Díaz-Rodríguez, N., Del Ser, J., Bennetot, A., Tabik, S., Barbado, A., et al., 2020. Explainable artificial intelligence (xai): concepts, taxonomies, opportunities and challenges toward responsible ai. Inf. Fusion 58, 82–115.

Aworka, R., Cedric, L.S., Adoni, W.Y.H., Zoueu, J.T., Mutombo, F.K., Kimpolo, C.L.M., Nahhal, T., Krichen, M., 2022. Agricultural decision system based on advanced machine learning models for yield prediction: case of east african countries. Smart Agric. Technol. 2, 100048.

Barbedo, J.G.A., 2019. Detection of nutrition deficiencies in plants using proximal images and machine learning: a review. Comput. Electron. Agric. 162 (December 2018), 482–492.

Bharath, S.M., Manoj, S., Adhappa, P., Patagar, P.L., Bhaskar, R., 2022. Crop yield prediction with efficient use of fertilizers. Lect. Notes Electr. Eng. 783, 937–943.

Chivenge, P., Zingore, S., Ezui, K.S., Njoroge, S., Bunquin, M.A., Dobermann, A., Saito, K., 2022. Progress in research on site-specific nutrient management for smallholder farmers in sub-saharan africa. Field Crops Res. 281, 108503.

Collier, P., Dercon, S., 2014. African agriculture in 50years: smallholders in a rapidly changing world? World Dev. 63, 92–101. Economic Transformation in Africa.

de Lara, A., Mieno, T., Luck, J.D., Puntel, L.A., 2023. Predicting site-specific economic optimal nitrogen rate using machine learning methods and on-farm precision experimentation. Springer Sci.+Bus. Media 24 (5), 1792–1812.

Diao, X., Hazell, P., Thurlow, J., 2010. The role of agriculture in african development. World Dev. The Future of Small Farms Including Special Section: Impact Assessment of Policy-Oriented International Agricultural Research (pp. 1453–1526). 38 (10), 1375–1383.

Ennaji, O., Baha, S., Vergutz, L., El Allali, A., 2024. Gradient boosting for yield prediction of elite maize hybrid zhengdan 958. PLOS ONE 19 (12), 1–16.

Ennaji, O., Hamma, A., Vergütz, L., El Allali, A., 2025. The assessment of soil variables relative importance for cereal yield prediction under rainfed cropping system in morocco. Smart Agric. Technol. 11, 100950.

Ennaji, O., Vergutz, L., El Allali, A., 2023. Machine learning in nutrient management: a review. Artif. Intell. Agric. 9, 1–11.

Fink, R.K., Hoskinson, R.L., Hess, J.R., 1998 From prediction to prescription: intelligent decision support for variable rate fertilization. In: 2001 ASAE Annual Meeting. American Society of Agricultural and Biological Engineers.

Ghosal, S., Blystone, D., Singh, A.K., Ganapathysubramanian, B., Singh, A., Sarkar, S., 2018. An explainable deep machine vision framework for plant stress phenotyping. Proc. Natl. Acad. Sci. USA 115 (18), 4613–4618.

Jayne, T.S., Mather, D., Mghenyi, E., 2010. Principal challenges confronting smallholder agriculture in sub-saharan africa. World Dev. 38 (10), 1384–1398.

Jha, K., Doshi, A., Patel, P., Shah, M., 2019. A comprehensive review on automation in agriculture using artificial intelligence. Artif. Intell. Agric. 2, 1–12.

Kamilaris, A., Kartakoullis, A., Prenafeta-Boldu, F.X., 2017. A review on the practice of big data analysis in agriculture. Comput. Electron. Agric. 143 (October), 23–37.

Kamilaris, A., Prenafeta-Boldú, F.X., 2018. Deep learning in agriculture: a survey. Comput. Electron. Agric. 147, 70–90.

Kaneko, A., Kennedy, T., Mei, L., Sintek, C., Burke, M., Ermon, S., Lobell, D., n.d. Deep learning for crop yield prediction in africa.

Khaki, S., Wang, L., 2023. Crop yield prediction using deep neural networks: a comprehensive review and case study. Front. Plant Sci. 14, 1085328.

Liakos, K.G., Busato, P., Moshou, D., Pearson, S., Bochtis, D., 2018a. Machine learning in agriculture: a review. Sensors 18 (8), 1–29.

Liakos, K.G., Busato, P., Moshou, D., Pearson, S., Bochtis, D., 2018b. Machine learning in agriculture: a review. Sensors 18 (8), 2674.

Liang, Q., Zhu, K., Long, J., Wang, Y., Li, C., 2022. Explainable artificial intelligence for agriculture and crop yield prediction: a survey. Agronomy 12 (8), 1887.

Livingston, G., Schonberger, S., Delaney, S., 2011. Sub-saharan africa: the state of smallholders in agriculture. In: Paper Presented at the IFAD Conference on New Directions for Smallholder Agriculture, vol. 24. Citeseer, pp. 25.

Mohanty, A., Wahi, T., Gupta, A., 2023. Explainable ai (xai) in agriculture: concepts, techniques, and applications. IEEE Access 11, 81300–81327.

Mohanty, S.P., Hughes, D.P., Salathé, M., 2016. Using deep learning for image-based plant disease detection. Front. Plant Sci. 7, 1419.

Olanipekun, I.O., Olasehinde-Williams, G.O., Alao, R.O., 2019. Agriculture and environmental degradation in africa: the role of income. Sci. Total Environ. 692, 60–67.

Patrício, D.I., Rieder, R., 2018. Computer vision and artificial intelligence in precision agriculture for grain crops: a systematic review. Comput. Electron. Agric. 153 (August), 69–81.

Paudel, D., Boogaard, H., de Wit, A., Janssen, S., Osinga, S., Pylianidis, C., Athanasiadis, I.N., 2021. Machine learning for large-scale crop yield forecasting. Agric. Syst. 187 (December 2020), 103016.

Preethi, G., Rathi Priya, V., Sanjula, S.M., Lalitha, S.D., Bindhu, B.V., 2020. Agro based crop and fertilizer recommendation system using machine learning. Eur. J. Mol. Clin. Med. 7 (4), 2043–2051.

Rejeb, A., Rejeb, K., Zailani, S., Keogh, J.G., Appolloni, A., 2022. Examining the interplay between artificial intelligence and the agrifood industry. Artif. Intell. Agric. 6, 111–128.

Salami, A., Kamara, A.B., Brixiova, Z., 2010. Smallholder Agriculture in East Africa: Trends, Constraints and Opportunities. African Development Bank Tunis, Tunisia.

Sarker, I.H., 2021. Machine learning: algorithms, real-world applications and research directions. SN Comput. Sci. 2 (3), 1–21.

Tapsoba, P.K., Aoudji, A.K.N., Kabore, M., Kestemont, M.-P., Legay, C., Achigan-Dako, E.G., 2020. Sociotechnical context and agroecological transition for smallholder farms in Benin and Burkina Faso. Agronomy 10 (9), 1447.

van Klompenburg, T., Kassahun, A., Catal, C., 2020. Crop yield prediction using machine learning: a systematic literature review. Comput. Electron. Agric. 177, 105709.

Wang, X., Miao, Y., Dong, R., Zha, H., Xia, T., Chen, Z., Kúsnierek, K., Mi, G., Sun, H., Li, M., 2020. Machine Learning-based In-season Nitrogen Status Diagnosis and Side-dress Nitrogen Recommendation for Corn, vol. 123. Elsevier BV, p. 126193.

Zhang, X., Han, L., Dong, Y., Shi, Y., Huang, W., Han, L., et al., 2019. A deep learning-based approach for automated yellow rust disease detection from high-resolution hyperspectral uav imagery. Remote Sens. 11 (13), 1554.

Section III

Deployment of smart agricultural technologies for informed decision-making

Physics-based crop modeling for climate- and water-smart irrigation management

11

Rabindra Kumar Panda[a] *and Rajiv Kumar Srivastava*[b]
[a]Centre for Climate Smart Agriculture, Siksha 'O' Anusandhan Deemed to be University, Bhubaneswar, Odisha, India
[b]Biological and Agricultural Engineering Department, Texas A&M University, College Station, TX, United States

1 Introduction

1.1 Climate change and water scarcity challenges

In the 21st century, agriculture faces a dual crisis: escalating water scarcity and intensifying climate variability. As the primary global consumer of freshwater, agriculture accounts for approximately 70% of total freshwater withdrawals, with irrigation alone consuming over 2500 km^3 annually (FAO, 2021c). Rapid population growth, changing dietary patterns, and expanding cultivation in marginal lands have further strained water resources, particularly in regions already prone to aridity or high evapotranspiration (ET). Between 2000 and 2020, the frequency of agricultural droughts worldwide rose by 29%, disproportionately affecting high water-use crop systems in vulnerable geographies (UNCCD, 2022). These trends signal a growing imbalance between water demand and natural replenishment, exacerbated by inconsistent rainfall patterns and rising temperatures.

Compounding this situation, the (IPCC Sixth Assessment Report, 2021) projects that, under current emission trajectories, rainfed crop yields may decline by 8%–15% in South Asia and sub-Saharan Africa by 2050. This is attributed to a combination of declining precipitation reliability, intensified heat stress, and increased evaporative demand, all of which undermine the productivity and stability of agro-ecosystems dependent on seasonal rainfall. Moreover, longer and more frequent dry spells, coupled with episodic flooding, disrupt sowing calendars and reduce growing season predictability, especially in tropical and subtropical zones (Contosta et al., 2024).

To illustrate global patterns of irrigation reliance, Fig. 11.1 presents irrigation intensity as a gradient map based on Food and Agriculture Organization's (FAO) Global Map of Irrigation Areas—Version 5 (GMIA v5), which quantifies the percentage of land area equipped for irrigation per 5 arc-minute grid cell (FAO

Precision Technologies for Digital Agriculture. https://doi.org/10.1016/B978-0-443-40513-6.00016-0

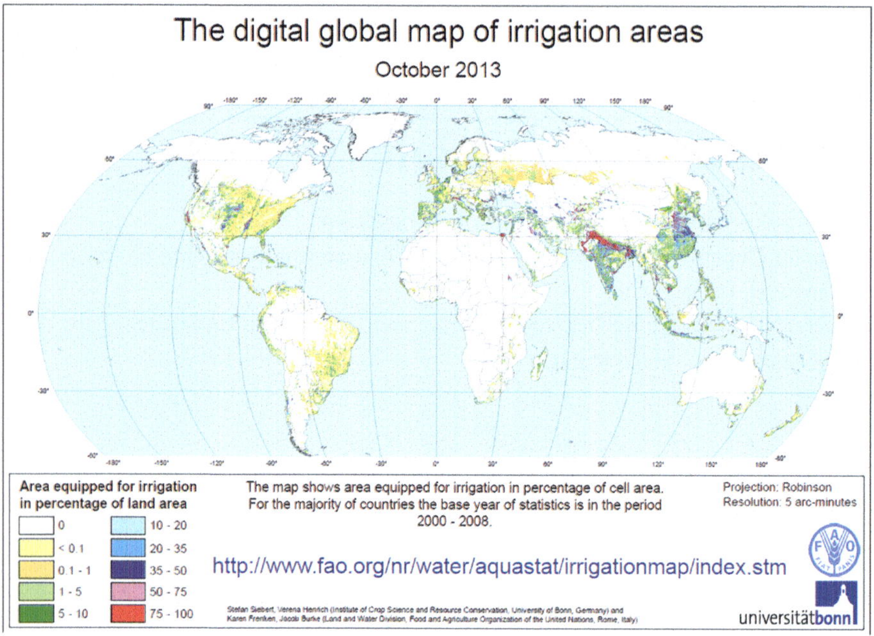

Fig. 11.1 Irrigated area spread over the globe.
Source: FAO AQUASTAT, 2013. Global map of irrigation areas v5. Retrieved on 19/07/2025. URL: https://firebasestorage.googleapis.com/v0/b/fao-aquastat.appspot.com/o/PDF%2FMAPS%2Fgmia_v5_highres.pdf?alt=media&token=0477a76c-b4a0-440e-93ec-8093db54ef0d.

AQUASTAT, 2013). Derived from subnational statistics and spatial overlays, the dataset provides a high-resolution global overview of irrigation infrastructure. Regions with intensive irrigation exceeding 50% of cropland coverage include the Indo-Gangetic Plain, North China Plain, and Nile Delta, where canal and groundwater systems are well established. Moderate irrigation zones (20%–40%) appear across southern Europe, Central Asia, and productive belts in the western United States. In contrast, large rainfed regions such as sub-Saharan Africa, the Amazon Basin, and northern Australia show irrigation coverage below 10%, reflecting uneven investment and access. The spatial distribution underscores both ecological dependence on irrigation and persistent disparities in infrastructure development across agricultural landscapes.

Altogether, this evolving landscape necessitates a paradigm shift toward adaptive, resource-efficient irrigation management. Without proactive intervention, regions already on the climatic edge risk further declines in agricultural productivity, food security, and rural livelihoods. The next sections of this chapter, therefore, explore how physics-based crop models can serve as strategic tools to guide this transition, offering scientifically grounded simulations for crop water requirements (CWRs), irrigation scheduling, and climate resilience planning.

1.2 Need for climate- and water-smart farming practices

The climate-smart agriculture (CSA) approach, developed by the FAO, promotes an integrated pathway for achieving sustainable agriculture by simultaneously enhancing productivity, building climate resilience, and lowering greenhouse gas emissions (FAO, 2021a, 2021b). Within this framework, climate and water-smart irrigation is recognized as a cornerstone intervention, essential for alleviating water stress and adapting cropping systems to increasingly erratic climatic patterns. In contrast to conventional irrigation practices, which often depend on static calendars or broad assumptions about water supply, smart irrigation strategies prioritize site-specific precision, environmental responsiveness, and resource efficiency. This paradigm shift relies on a fusion of agronomic insights and emerging technologies to:

- Quantify crop-specific water requirements at various phenological stages
- Optimize irrigation timing using weather forecasts, soil moisture sensors, and real-time ET data
- Maximize water-use efficiency (WUE) by reducing losses due to evaporation, percolation, or runoff
- Enhance decision-making through dynamic feedback loops, incorporating field observations and environmental signals

One of the most widely adopted strategies in climate- and water-smart irrigation is deficit irrigation, wherein water is purposefully withheld during less sensitive crop stages to reduce usage without significantly compromising yield (Patle et al., 2020). Studies have shown that wheat, maize, and cotton can achieve up to 20%–40% higher water productivity under carefully managed deficit conditions, particularly in semi-arid environments (Cheng et al., 2021; Shoukat Hafiza et al., 2025; Zou et al., 2021). Likewise, alternate wetting and drying (AWD) in rice systems significantly cuts irrigation water use and methane emissions while maintaining comparable yields, offering both mitigation and adaptation benefits (Al Viandari et al., 2020). These practices are increasingly supported by policy, with national programs such as India's Per Drop More Crop (Kayatz et al., 2019) and international climate adaptation funds promoting solar-powered pumps, on-farm water management, and micro-irrigation systems (Srivastava et al., 2024). In this evolving policy and agro-ecological landscape, physics-based crop models are indispensable as they offer a platform to simulate, test, and optimize these interventions under location-specific conditions (Chaurasiya et al., 2025), thereby enhancing decision-making and accelerating the transition to sustainable irrigation systems.

1.3 Role of crop models in irrigation decision-making

To operationalize smart irrigation, DSS that simulate and forecast crop–water interactions across varying scenarios are indispensable. Physical or system-based crop models such as the Decision Support System for Agrotechnology Transfer (DSSAT), AquaCrop, and Simulation of Transpiration, Interactions, and Coupling of Soil-Plant

Systems (STICS) serve this purpose by integrating scientific principles of soil-water-plant-atmosphere dynamics with empirical data and user-defined management practices (Murmu et al., 2025).

These models allow users to:

- Simulate crop water use (ET) and nutrient uptake under current and future climates
- Optimize irrigation strategies for maximum yield or water productivity
- Evaluate trade-offs between water input, growth stages, and yield outcomes
- Assess the impact of new technologies like subsurface drip irrigation or AI-powered scheduling

1.4 Scope and objectives of the chapter

This chapter synthesizes current knowledge and advances in the use of physics-based crop models for climate- and water-smart irrigation. The main objectives are to:

1. Clarify the conceptual foundations of crop water demand and irrigation scheduling within the soil-plant-atmosphere framework.
2. Introduce leading physical models and explain their structure, input requirements, and calibration techniques.
3. Present real-world applications that showcase model-driven evaluations of irrigation strategies, especially under water-limited and climatically volatile conditions.
4. Discuss integration with emerging technologies, including sensor-based monitoring, remote sensing, and artificial intelligence, to build adaptive and responsive irrigation systems.

The focus is both global and context-specific, drawing on case studies from rainfed and irrigated systems across climatic zones. Emphasis is also placed on user accessibility and the practical adoption of these models by researchers, extension officers, and farmers.

2 Precision irrigation management concepts

Precision irrigation is a modern approach to managing water in agricultural systems by tailoring irrigation to the specific needs of crops across time and space (Himanshu et al., 2023; Kumar et al., 2022a, 2022b, 2023; Lena et al., 2022; Patle et al., 2020). Unlike conventional methods, which often apply water uniformly across entire fields, precision irrigation recognizes that factors such as soil texture, elevation, plant growth stage, and weather conditions vary within a field (Kumar et al., 2022b; Bastia et al., 2025). These variations lead to different water requirements at different locations and times. The theoretical foundations of precision irrigation are rooted in systems thinking and ecological principles, aiming to optimize water use efficiency (WUE) by aligning irrigation practices with crop physiological needs and environmental variability (Lakhiar et al., 2024). By identifying management zones based on field variability, farmers can apply water where it is needed most, reducing losses from evaporation, runoff, and deep percolation. This spatial specificity enhances both yield and sustainability by reducing excess water application and improving plant health (Levidow et al., 2014).

Another key theoretical element of precision irrigation is its reliance on real-time feedback mechanisms and adaptive decision-making (Bwamb et al., 2023; Ale et al., 2023). Instead of following fixed irrigation calendars, precision irrigation systems depend on continuous monitoring of environmental and crop conditions. Tools such as soil moisture sensors, crop canopy temperature sensors, and weather forecasting models provide high-resolution data that can be used to guide irrigation timing and quantity (King et al., 2020). Decision support systems (DSS) and crop simulation models interpret this data to inform farmers when and how much to irrigate, enabling dynamic responses to changing weather patterns and plant water stress (Puig et al., 2025). These technologies are based on hydrological, crop physiological, and agronomic models that quantify crop water demand and the impact of irrigation strategies on crop performance. This shift from schedule-based to needs-based irrigation supports resilience in the face of climate change, where rainfall patterns are increasingly erratic and droughts are more frequent (Bhatnagar et al., 2024).

Precision irrigation also plays a critical role in climate- and water-smart agriculture by enhancing resource-use efficiency while maintaining or increasing crop yields (Amankwaa-Yeboah et al., 2024). From a sustainability perspective, precision irrigation reduces water withdrawals from surface and groundwater sources, helping to preserve aquatic ecosystems and buffer communities against water scarcity. The theoretical framework behind this contribution includes the principles of sustainable intensification and climate adaptation: producing more with less by making farming systems more resilient and efficient. For example, precision irrigation can be integrated with other smart farming practices, such as conservation agriculture and remote sensing-based nutrient management, to form a holistic strategy for climate-resilient food production (Lawrence et al., 2024). The adoption of such practices also supports broader policy goals related to climate mitigation and sustainable water governance. In educational settings, exposing students to the interconnections between technology, environmental systems, and agricultural productivity builds a foundation for informed decision-making and innovation in future agroecosystems (Francis et al., 2011).

2.1 Crop water requirement and evapotranspiration

Accurate estimation of CWR is central to designing efficient irrigation systems, particularly in the case of climate-smart and water-efficient farming (Bhatnagar et al., 2024). CWR refers to the amount of water needed by a crop to grow optimally without water stress (Ruan et al., 2020). This need varies depending on the type of crop, its developmental stages, and the surrounding environmental conditions such as temperature, humidity, solar radiation, and wind speed. One of the most widely accepted scientific approaches to estimate CWR is through the concept of ET, which combines two processes: evaporation of water from the soil and transpiration from plant leaves. ET essentially reflects the "thirst" of both the plant and its environment and provides a sound theoretical basis for irrigation planning under varying climatic regimes.

To quantify ET, agricultural scientists use the reference ET (ET_0), which is the ET from a hypothetical well-watered grass surface under standard conditions (Cai et al., 2007). This is calculated using climatic data and formulas such as the FAO Penman-Monteith equation (FAO 56), which accounts for solar radiation, temperature, wind, and humidity (Cai et al., 2007). ET_0 is then multiplied with the crop coefficient (Kc) that adjusts for the water needs of specific crops at different growth stages, giving the actual crop ET (ET_c) (Eq. 1):

$$ET_c = K_c \times ET_0 \tag{1}$$

Where,
ET_c = Crop evapotranspiration,
K_c = Crop coefficients,
ET_0 = Potential evapotranspiration.

This equation links meteorology with crop physiology and helps farmers and irrigation planners tailor water applications based on real biological demand rather than approximations. It forms the theoretical basis for dynamic irrigation scheduling and precision water management under climate variability (Cai et al., 2007).

Moreover, integrating ET-based estimation into irrigation planning enhances resource use efficiency and resilience. Since climate conditions, and thereby ET rates, can vary seasonally or even daily, real-time monitoring of meteorological parameters becomes essential. Remote sensing, weather stations, and mobile-based irrigation apps now allow farmers to compute irrigation volumes based on predicted ET trends. This capability to align irrigation water application with environmental fluctuations is a pillar of CSA, where both yield sustainability and water savings are targeted. ET not only reveals the crop water demand but also shapes broader decisions about crop selection, planting calendars, and water allocation strategies (Calera et al., 2017). For nonprofessional users, understanding the principles behind accurate estimation of CWR and ET is critical for managing agricultural landscapes in a changing climate.

2.2 Irrigation scheduling

Irrigation scheduling refers to the process of determining when, how much, and how often water should be applied to crops to meet their physiological needs without waste (FAO, 2021c). From a theoretical standpoint, irrigation scheduling aims to maintain soil moisture within an optimal and readily available range between field capacity and the critical level, thereby supporting growth while avoiding surplus or deficient availability of water. The decision-making process is influenced by several variables, including soil properties (like water-holding capacity), crop type and stage, climatic conditions, and system capacity. At the core of scheduling theory lies the soil water balance model, which treats soil as a reservoir: inputs include rainfall and irrigation, while outputs include ET, drainage, and runoff. Understanding this balance helps determine the irrigation interval (time between applications) and the irrigation depth (amount per event).

Modern irrigation scheduling increasingly relies on data-driven approaches rather than fixed calendars. Tools like tensiometers, time-domain reflectometry sensors (Srivastava et al., 2017, 2018), and weather-based forecasting models provide real-time or near-real-time data on soil moisture and atmospheric demand. ET-based scheduling uses climate data to estimate water loss and aligns irrigation accordingly. When ET exceeds precipitation, irrigation compensates for the deficit. Plant-based indicators such as leaf wilting, canopy temperature, or sap flow add physiological precision (Fernández, 2014). These tools are grounded in principles from plant physiology, meteorology, and hydrology, and support dynamic scheduling that adjusts to short-term weather anomalies and long-term climate trends.

In the context of climate- and water-smart agriculture, irrigation scheduling is both a risk mitigation tool and a resource optimization strategy. Effective scheduling reduces water use, energy costs, and nutrient leaching while protecting yields under uncertain rainfall patterns (Bhatnagar et al., 2024). For example, regulated deficit irrigation (RDI) applies less water than full crop ET, strategically stressing crops during noncritical stages to conserve water without compromising yield (Panda et al., 2003; Himanshu et al., 2021). Similarly, partial root-zone drying alternates irrigation between halves of the root zone, triggering drought-tolerance responses while maintaining growth. These techniques are supported by agroecological and physiological theories and are increasingly validated by simulation models. Teaching irrigation scheduling in this broader context encourages students and researchers to see it not only as a technical exercise but also as a strategic response to global water scarcity and climate variability.

2.3 Efficient irrigation water management under deficit conditions

Deficit irrigation is a strategy that deliberately underirrigates crops compared to their full ET requirement, with the goal of conserving water while minimizing yield reduction (Ale et al., 2023; Singh et al., 2022). The theoretical foundation of this approach lies in plant-water relations and the understanding that crops tolerate mild to moderate water stress at specific phenological stages without significant yield penalties. Key concepts include critical growth stages, yield response factors (K_y), and economic water productivity. The K_y concept, introduced by Doorenbos and Kassam (1979), expresses the proportional decrease in yield relative to the reduction in crop ET (Eq. 2):

$$\frac{Y_a}{Y_m} = 1 - K_y\left(1 - \frac{ET_c}{ET_0}\right) \tag{2}$$

Where,

Y_a and Y_m are actual and maximum yield,

ET_c and ET_0 are crop and potential evapotranspiration, and

K_y is the crop yield response factor.

This equation aids planners in selecting which crops and stages to stress during water scarcity, maximizing yield per unit of water applied.

Efficient water management under such conditions also involves optimizing irrigation performance through the application of various efficiency indicators. These include:

- **Application efficiency (E_a)**: Application efficiency measures the fraction of delivered water that is stored in the root zone and available to crops (Eq. 3):

$$E_a = \left(\frac{Waterstoredintheroot}{Waterdeliveredtothefield}\right) \times 100 \tag{3}$$

Where,

E_a=Application efficiency. High E_a indicates minimal losses through surface runoff or deep percolation. Drip irrigation systems may achieve E_a values above 90%, while surface systems often fall below 60%.

- **Distribution uniformity (DU)**: Assesses how evenly water is applied across a field (Eq. 4):

$$DU = \left(\frac{Averageoflowestquarterapplicationdepths}{Averageapplicationdepth}\right) \times 100 \tag{4}$$

Critical for reducing underirrigation or overirrigation in heterogeneous soils, this is highly relevant in sprinkler and micro-irrigation systems.

- **Conveyance efficiency (E_c)**: Addresses water lost during transport from source to field. Improved efficiency can be achieved through practices such as drip irrigation, surge flow in furrows, and AWD in rice. AWD, for instance, allows fields to dry intermittently, reducing water use while maintaining yields and lowering methane emissions, linking efficient irrigation to both water-smart and climate-smart objectives.
- **WUE:** WUE is a physiological agronomic indicator linking crop output to water input (Eq. 5):

$$WUE = \frac{Grainyield(kg)}{Waterused(mmorm^3)} \tag{5}$$

Reflects both transpiration and photosynthetic productivity, influenced by genotype, management practices, and climatic conditions.

- **Irrigation efficiency (IE):** A broader term that encapsulates the combined effectiveness of water delivery, storage, and usage (Eq. 6):

$$IE = \left(\frac{Waterbeneficiallyused}{Waterappliedordiverted}\right) \times 100 \tag{6}$$

Where "beneficial use" includes transpiration and improved soil moisture, it helps identify losses at the field, farm, or system level.

Note: Application efficiency measures water stored in the root zone relative to the water applied in the field, while IE accounts for the total water beneficially used from the source to the crop. Thus, application efficiency is a component of system-wide IE.

- **Efficiency fraction (f):** Defined in FAO models like CROPWAT, the efficiency fraction reflects the effectiveness of a particular irrigation method (Eq. 7):

$$f = \frac{1}{E_a} \tag{7}$$

A higher f means greater inefficiencies and, therefore, increased gross irrigation requirement.

From a systems perspective, integrating deficit irrigation with decision support tools, crop models, and economic thresholds allows for a nuanced understanding of trade-offs between water savings and yield reduction. Simulation models like AquaCrop or DSSAT can evaluate different irrigation regimes under variable climates, guiding farmers toward resilient strategies (Murmu et al., 2025). These models incorporate soil, crop, and weather parameters to forecast growth and yield outcomes. In addition, combining deficit irrigation with conservation agriculture practices, such as mulching and residue incorporation, enhances water retention and reduces surface evaporation. Educating students on these synergies prepares them to approach water management not as isolated scheduling but as a component of whole-farm planning within the broader context of environmental stewardship and sustainable intensification.

2.4 Sensor-based and Internet of Things-enabled irrigation systems

The integration of sensors and Internet of Things (IoT) technologies into irrigation systems represents a transformational shift toward real-time, precision-based water management (Puig et al., 2025). The theoretical underpinning of this approach lies in cyber-physical systems and adaptive control theory, where physical processes (soil water dynamics, crop transpiration) are continuously monitored and regulated through digital feedback loops. These systems use sensors to capture environmental and crop-specific data, such as soil moisture, canopy temperature, and atmospheric humidity, and transmit it via wireless networks to a central processing unit. The data is then processed by decision-support algorithms, which adjust irrigation volumes and timing based on real-time conditions. This creates a closed-loop control framework where water is applied not on a calendar basis but in response to actual crop water stress.

Key technologies include a range of sensors and controllers, each tailored to specific variables. Capacitance and dielectric soil moisture sensors estimate volumetric water content with high accuracy (Kashyap and Kumar, 2021). Infrared thermometers or thermal cameras detect canopy temperature as a proxy for transpiration and stress (Mertens et al., 2023). Weather stations measure solar radiation, rainfall, wind speed, and ET drivers. These components are typically connected through low-power wide-area networks or cellular IoT platforms, enabling scalability and remote access. Advanced systems may incorporate machine learning models to forecast CWRs based on historical patterns, phenology, and predicted weather (Worrall et al., 2023). Unlike traditional approaches, which are reactive and often wasteful, sensor-based systems offer predictive and site-specific irrigation control, aligning with water- and climate-smart principles.

In applied contexts, sensor-based and IoT-enabled systems significantly improve efficiency metrics, including:

- **Application efficiency**: Higher due to precise targeting of wetted zones
- **DU**: Improved by uniform emitter performance, based on sensor feedback
- **WUE**: Enhanced as water is applied when and where crops need it most. These systems are especially valuable under deficit irrigation regimes, where fine-tuning water delivery ensures yield stability while minimizing inputs. From a sustainability standpoint, IoT-enabled irrigation reduces overirrigation, energy use, and nutrient leaching, while enabling data-driven farm management. For undergraduates, this section highlights how interdisciplinary knowledge spanning crop science, electronics, data analytics, and hydrology converges to develop next-generation agricultural solutions. Exposure to such systems prepares future practitioners to implement smart farming approaches that address both resource scarcity and climate resilience.

3 Biophysical basis of climate and water interactions in agriculture

3.1 Climate-resilient cropping systems

Climate-resilient cropping systems address challenges from erratic rainfall, rising temperatures, and extreme events by integrating drought- and heat-tolerant cultivars with diversified rotations and conservation practices (Srivastava et al., 2025). Cereal-legume intercropping and zone-specific cultivar selection enhance system stability. For example, resilient rice varieties in South Asia, such as *Sahbhagi Dhan*, *Swarna Shreya*, and *Binadhan 67,* and drought-tolerant maize hybrids in sub-Saharan Africa, developed through Drought Tolerant Maize for Africa (DTMA), Stress Tolerant Maize for Africa (STMA), and TEGO, have maintained stable yields under water stress, reducing crop failure risks by 30%–50% (Mottaleb et al., 2017; Muralikrishnan et al., 2021; TAAT, 2025; CIMMYT, 2008). Precision technologies like sensor-guided irrigation and weather forecasting further enhance adaptive capacity. A study in Eastern India showed that combining resilient rice varieties with tensiometric and Chameleon sensors reduced water use by 25% and increased yields by 18% (ICAR-NRRI, 2021). These systems also support ecosystem services: agroforestry buffers microclimates and sequesters carbon, while cover cropping improves soil health and moisture retention. Scaling such innovations requires investment in seed systems, localized germplasm banks, and farmer-led trials (Srivastava et al., 2025).

3.2 The soil-plant-atmosphere continuum

The soil-plant-atmosphere continuum (SPAC) represents the integrated pathway through which water moves from soil, through plant tissues, and into the atmosphere. It is driven by gradients in water potential and mediated by root hydraulic conductivity, xylem transport, and stomatal responses. Environmental factors such as temperature, relative humidity, and vapor pressure deficit significantly influence transpiration rates and plant water demand, which must be carefully modeled in irrigation systems (Murmu et al., 2025). Efficient irrigation scheduling hinges on a thorough understanding of SPAC dynamics. Real-time monitoring using soil moisture sensors and canopy temperature detectors provides actionable data to align irrigation with plant physiological needs. Integration of SPAC into crop models enables predictive analytics that guide smart irrigation interventions. Recent advancements include remote sensing technologies and machine learning algorithms that interpret SPAC-related variables at scale (Shi, 2024). These tools allow adaptive irrigation decisions based on spatial variability and crop growth stages.

3.3 Dynamics of water availability under varying climate and water management

Shifting climate patterns increasingly disrupt hydrological regimes, constraining irrigation through delayed monsoons, intensified ET, and erratic rainfall, which are linked to annual groundwater declines of 0.5–1.2 m in regions across South Asia and sub-Saharan Africa (Taylor et al., 2013). Climate-smart water management (CSWM) mitigates these impacts by deploying efficient techniques such as AWD, rainwater harvesting, and RDI, yielding water savings of up to 40% and reducing methane emissions by 30% in rice cultivation (Richard and Sander, 2014). Integrative CSWM strategies involve coordinated upstream and downstream infrastructure, supported by participatory water governance, digital monitoring tools, and AI-enabled decision systems. Embedding climate projections and hydrological models into planning frameworks enhances adaptive capacity and reduces vulnerability to droughts, floods, and crop failure.

3.4 Climate-resilient decision support systems for the of water use efficiency and income

Improving water productivity and farmers' income amid climate stress demands responsive, data-driven decision frameworks. Climate-smart advisory platforms, such as mobile apps and sensor-integrated systems, guide farmers in irrigation scheduling, crop selection, and input use, with field studies in Burkina Faso showing a 52% boost in water productivity and a 35% rise in net income through smart scheduling and deficit irrigation (Morris and Barron, 2014). Simulation models such as DSSAT, Agricultural Production Systems sIMulator (APSIM), and AquaCrop support scenario planning by evaluating yield responses to water inputs under variable climate conditions, helping farmers adapt proactively. Integration of water-efficient practices with market incentives, such as sustainability certifications, contract farming, and institutional support—further enhances income potential. DSS that merge socio-economic data with biophysical modeling enable tailored, context-specific solutions, converting climate risks into pathways for resilience and agronomic prosperity (FAO, 2021a, 2021b).

4 Physics/system-based crop modeling frameworks

4.1 Structure and logic

System-based crop models simulate plant growth by integrating biophysical processes across the soil–plant–atmosphere continuum. Modular platforms such as DSSAT, APSIM, and STICS represent key subsystems, including phenology, photosynthesis, root development, and stress response, driven by site-specific inputs such as climate data, soil characteristics, and cultivar traits (Murmu et al., 2025). These frameworks support decision-making across scales and are compatible with GIS, remote sensing, and AI-based controllers. When linked with economic or environmental modules, they enable comprehensive sustainability assessments. Fig. 11.2 shows the structure of DSSAT.

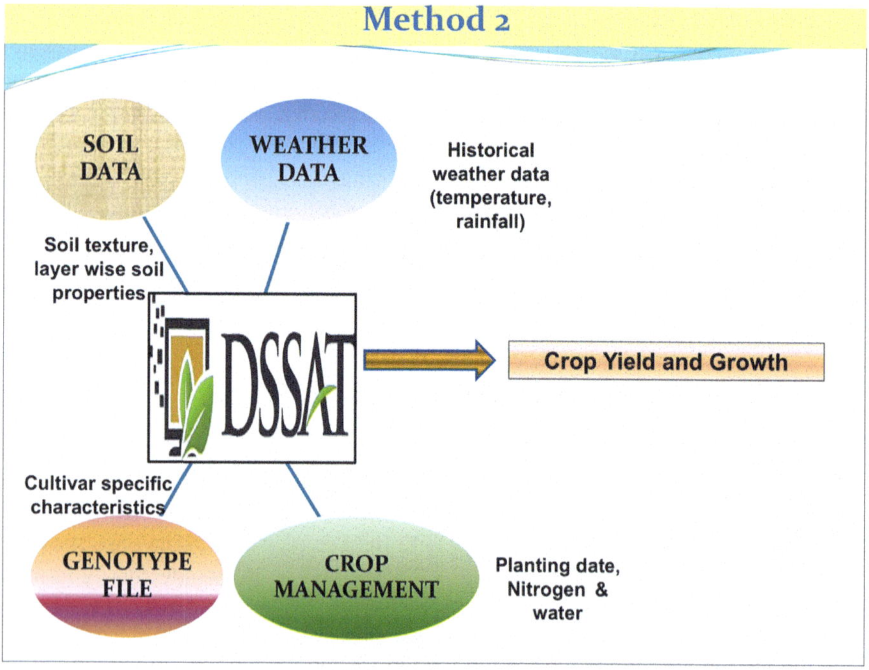

Fig. 11.2 Functional components of DSSAT.

4.2 Hydrological and energy dynamics

While the detailed mathematical formulations are embedded within each model, platforms like DSSAT and APSIM internally simulate water balance, energy fluxes, and crop–soil interactions. These models use embedded routines to estimate ET, soil moisture movement, and root water uptake based on site conditions and crop physiology. Integration with plant growth modules allows dynamic feedback between water availability and biomass production, supporting irrigation planning and stress analysis under variable climate scenarios.

4.3 Temporal and spatial resolution

These models typically operate at daily time steps, with subdaily options in controlled environments. They scale spatially from plot-level homogeneity to region-wide heterogeneity, using grid-based layers or GIS integration (Calera et al., 2017; Lawrence et al., 2024). Model compartments, such as soil horizons, canopy layers, and root zones, are parameterized independently, enabling modular calibration and sensitivity testing. The choice of resolution affects simulation accuracy, model scalability, and responsiveness to climatic or agronomic fluctuations (Thirumeninathan et al., 2024).

4.4 Climate variability and extremes

Crop models increasingly incorporate stochastic and deterministic representations of climate variability through historical data, climate projections, and synthetic weather series (White et al., 2011). Platforms such as DSSAT, AquaCrop, and APSIM include built-in routines for simulating responses to heatwaves, droughts, and flooding based on stress thresholds and probabilistic climate scenarios. Real-time forecasting tools and DSS enhance adaptability, allowing users to simulate future scenarios and adjust irrigation, nutrient, and harvest strategies accordingly (Lawrence et al., 2024).

5 Data requirements and parameterization

5.1 Inputs: Weather, soil, crop phenology, and management practices

Accurate crop modeling hinges on the quality and resolution of input datasets, as shown in Fig. 11.2. These inputs define the environmental and biological context in which crops grow and respond to climate and water variability.

- **Weather data** include daily or hourly records of precipitation, temperature, solar radiation, humidity, and wind speed. These are typically sourced from meteorological stations, reanalysis datasets, or satellite-derived products. For example, the NASA POWER and ERA5 datasets offer global coverage with high temporal resolution (Mutlu, 2025).

- **Soil parameters** encompass texture, depth, bulk density, hydraulic conductivity, organic matter content, and water retention characteristics. These influence infiltration, root development, and nutrient cycling. Soil databases like ISRIC and SSURGO provide standardized profiles for modeling (Rossiter, 2016).
- **Crop phenology** involves cultivar-specific parameters such as thermal time requirements, leaf area development, flowering dates, and maturity duration. These are often derived from field trials or literature and are critical for simulating growth stages and stress responses.
- **Management practices** include sowing dates, irrigation schedules, fertilizer applications, tillage operations, and residue management. These inputs are scenario-dependent and must reflect local agronomic norms or experimental treatments.

Parameterization of these inputs requires harmonization across spatial and temporal scales to ensure consistency and model reliability.

5.2 Handling uncertainty: Calibration, validation, and sensitivity analysis

Uncertainty is inherent in crop modeling due to data limitations, model structure, and environmental variability. Robust parameterization involves iterative refinement through:

- **Calibration**: Adjusting model parameters to align simulated outputs with observed data. Techniques include manual tuning, optimization algorithms (e.g., genetic algorithms), and Bayesian inference (Tunio et al., 2024). Calibration targets may include yield, biomass, ET, or soil moisture.
- **Validation**: Testing model performance against independent datasets to assess predictive accuracy. Metrics such as RMSE, Nash–Sutcliffe efficiency, and bias are used to quantify agreement.
- **Sensitivity analysis**: Identifying parameters that most influence model outputs. Methods range from one-at-a-time perturbations to global sensitivity techniques like Sobol indices or Morris screening (Wang et al., 2025). This informs the prioritization of data collection and model simplification.

Together, these processes enhance model credibility and guide decision-making under uncertainty.

5.3 Integration of remote sensing and observed climate datasets

Remote sensing and climate datasets offer scalable, spatially explicit inputs for crop modeling, especially in data-scarce regions.

- **Satellite-derived vegetation indices** (e.g., Normalized difference vegetation index (NDVI), Enhanced Vegetation Index (EVI)) inform crop growth dynamics, canopy cover, and stress detection. These can be assimilated into models to adjust phenology or biomass estimates (Gao and Zhang, 2021).
- **Land surface temperature and soil moisture** from sensors like Moderate Resolution Imaging Spectroradiometer (MODIS), Soil Moisture Active Passive (SMAP), and

Sentinel-1 provide real-time indicators of crop water status and thermal stress (Dilip et al., 2023).

- **Climate datasets** such as Climate Hazards Group InfraRed Precipitation with Station data (CHIRPS), Tropical Rainfall Measuring Mission (TRMM), and Asian Precipitation – Highly Resolved Observational Data Integration Towards Evaluation (APHRODITE) offer gridded precipitation and temperature records, enabling spatial modeling across heterogeneous landscapes (Santra Mitra et al., 2021).
- **Data fusion techniques** combine remote sensing with ground observations using machine learning or geostatistical methods to improve accuracy and resolution (Singh and Gaurav, 2023).

These integrations support dynamic model updating, early-warning systems, and regional scenario analysis.

5.4 Scaling considerations: Plot to watershed

Crop models must be scalable to address both field-level management and watershed-level resource planning.

- **Plot-scale models** focus on detailed physiological processes and management interventions. They are ideal for experimental validation and precision agriculture applications.
- **Watershed-scale models** incorporate spatial heterogeneity, hydrological connectivity, and aggregated land-use impacts. Tools like SWAT and APEX simulate interactions between crop growth, water flow, and nutrient transport.
- **Upscaling techniques** include parameter aggregation, representative elementary areas, and nested modeling frameworks. These techniques ensure consistency across scales while preserving key dynamics.
- **Downscaling approaches** use high-resolution inputs or statistical methods to refine coarse-scale outputs for local relevance.

Effective scaling enables integrated assessments of climate-smart irrigation, water productivity, and sustainability across spatial domains.

6 Case studies

6.1 Case study 1: Adaptive irrigation frameworks under climate stress

Climate-resilient irrigation management involves precise scheduling tailored to meet crop-specific water demand under changing climatic conditions (Vilavan et al., 2024). In subhumid and subtropical regions of India, field experiments on maize, wheat, okra, and potato revealed that irrigating at 30%–45% maximum allowable depletion (MAD) of available soil water consistently maximized yield and WUE. In the case of maize, irrigation at 45% MAD yielded 4.2 t ha^{-1} of grain and maximized WUE at 1.87 kg m^{-3}, with a major portion of soil moisture extraction concentrated in the top 45 cm (Panda et al., 2004). Wheat experiments revealed that irrigating at 45% MAD yields 3.8 t ha^{-1} of grain as well as improved leaf area index (LAI) and aboveground dry matter (ADM)

(Panda et al., 2003). In the case of okra, sprinkler irrigation at 30% MAD resulted in the highest yield of 12.4 t ha^{-1} and nitrogen use efficiency of 38.2%, while minimizing deep percolation losses considerably compared to furrow and basin methods (Home et al., 2002). For potato, stage-wise crop coefficients (K_c) were determined through lysimetric experiments and were found to be 0.42 at the initial stage, 0.85 at the development stage, 1.27 at the reproductive stage, and 0.57 at the maturity stage. The Penman-Monteith method was used for accurate estimation of ET_0 with root mean square error (RMSE) = 0.080 mm/day and R^2 = 0.91 (Kashyap and Panda, 2001, 2003). These findings underscore the importance of stage-specific irrigation thresholds and accurate crop ET estimation to sustain productivity under increased evaporative demand due to global warming as well as erratic rainfall conditions.

Integrating irrigation scheduling with balanced fertilization further enhances resource efficiency and environmental sustainability. In wheat, a fertilization regime of $N:P_2O_5:K_2O$ at 80:40:40 kg ha^{-1} combined with 40% MAD irrigation, emerged as the threshold for optimal nitrogen and phosphorus use efficiency. This treatment minimized nitrogen leaching losses ($\leq$12.5 kg ha^{-1}) and maintained a high grain yield (3.6 t ha^{-1}), while phosphorus leaching remained negligible due to soil fixation (Behera and Panda, 2009). The study also revealed that low-frequency, high-volume irrigation reduced deep percolation losses compared to high-frequency, low-volume irrigation, which increased water and nutrient loss. Nitrogen uptake was not significantly affected by the irrigation schedule, but leaching was sensitive to both water and fertilizer levels. Phosphorus transport remained stable across treatments, indicating its low mobility in the soil solution. These findings underscore the importance of synchronizing irrigation depth and timing with moderate fertilization to reduce environmental risks and improve the water-nutrient balance. Such integrated field-based strategies are essential for CSA, particularly in regions facing erratic rainfall, rising ET, and declining groundwater reserves.

6.2 Case study 2: Calibration of crop models with limited data

In situations where the availability of comprehensive datasets is scarce, physically based crop models offer a pragmatic solution for simulating crop responses to climate variability and guiding irrigation strategies. This case study synthesizes findings from the grain legume crop growth model (CROPGRO)-Peanut, Subroutine for storage organ growth (SUBSTOR)-Potato, and crop environment resource synthesis (CERES)-Maize models, each calibrated using limited field experimental data from eastern India (Halder et al., 2014, 2017; Srivastava et al., 2019, 2023). Calibration of the models was performed using the trial-and-error method to estimate the genetic coefficients, utilizing the minimum available experimental data on phenology, biomass, and yield. Despite data constraints, model validation showed strong performance across all systems, with R^2 values exceeding 0.72, normalized RMSE below 8%, and the index of agreement (d-stat) consistently above 0.94. These results affirm the reliability of crop models even under limited calibration conditions. Sensitivity analysis played a pivotal role in identifying optimal sowing windows and input levels.

For instance, CROPGRO-Peanut simulations revealed that sowing between January 14 and February 21, combined with moderate phosphorus fertilization (30–50 kg P_2O_5 ha^{-1}), maximized yield and minimized irrigation demand (Halder et al., 2017). Similarly, SUBSTOR-Potato simulations indicated that November 20 sowing dates offered resilience against warming scenarios (Srivastava et al., 2019), while CERES-Maize identified June 10 (rainfed) and January 15–25 (irrigated) as optimal sowing periods to reduce water stress and yield variability (Srivastava et al., 2023).

Climate change impact assessments using these models demonstrated that rising temperatures, particularly during reproductive stages—significantly reduce crop yields. Simulations under +1°C to +4°C warming scenarios showed yield declines of up to 18% in maize, with elevated CO_2 (760 ppm) offering partial mitigation. However, the timing of sowing remained the most influential factor in maintaining yield stability and reducing irrigation requirements. ET-based irrigation scheduling using the FAO-56 Penman-Monteith method further enhanced crop WUE and irrigation WUE, especially under moderate soil moisture depletion thresholds (50%–70% MAD) (Halder et al., 2014). Sensitivity analyses confirmed that early sowing dates aligned better with cooler temperatures and lower ET rates, thereby reducing irrigation demand. The integration of appropriate sowing dates with optimum nitrogen and water management emerged as a low-cost, scalable adaptation strategy. These case studies underscore the utility of crop models calibrated with limited data for designing climate-resilient irrigation systems and guiding agronomic decisions. They highlight the potential of simulation-based approaches to support sustainable agriculture under increasing climatic uncertainty, especially in data-constrained regions.

6.3 Case study 3: Sowing date adjustments for water-smart-irrigation

A robust application of physical-based crop models such as CERES-Maize, CERES-Rice, and CROPGRO-Peanut across Eastern India with a subhumid tropical climate reveals that the strategic adaptation of sowing dates can significantly mitigate climate-induced yield losses while optimizing water demand. Under future climate scenarios (Representative Concentration Pathway (RCPs) 2.6 to 8.5 for the 2050s and 2080s), maize yield simulations indicate that early sowing under irrigated conditions (e.g., 25 December to 5 January) reduces ET and avoids peak heat stress during tasseling, thereby sustaining yield with lower irrigation input (Srivastava et al., 2022). Conversely, rainfed maize benefits from slightly earlier sowing (e.g., 30 May to 30 June), which aligns crop phenology with anticipated rainfall shifts, reducing water stress and enhancing grain filling (Srivastava et al., 2022). For peanut and rice, early sowing and transplanting, respectively (14 to 28 January for peanut and 30 June to 30 July for rice), were found to be optimal under elevated CO_2 and warming conditions, improving WUE and yield stability (Halder et al., 2020). Elevated CO_2 concentrations further enhance photosynthetic rates and reduce stomatal conductance, contributing to lower CWRs, especially under irrigated regimes. Sensitivity analyses underscore the importance of genotype-specific parameters (e.g., P1, Soil Lower Limit of plant-extractable water (SLLL)) and soil moisture thresholds in determining ET and biomass accumulation

(Srivastava et al., 2021). Collectively, these findings advocate for site-specific, model-guided sowing date adjustments as a low-cost, scalable strategy to achieve climate-smart irrigation, ensuring sustainable yield while minimizing water inputs under increasingly variable climatic conditions.

From a policy and implementation standpoint, the integration of crop simulation models into regional agricultural planning offers a pragmatic pathway toward climate-resilient irrigation management. The high predictive accuracy of models like DSSAT-Cropping system model (CSM) demonstrates their utility in guiding localized agronomic decisions for maize, peanut, and rice crops. By leveraging genotype-specific calibration and global sensitivity analysis, policymakers and extension services can tailor sowing calendars and irrigation schedules to anticipated climate trajectories (Srivastava et al., 2022). Importantly, the economic viability of sowing date adjustments, which require minimal infrastructural investment, makes them particularly attractive for resource-constrained farming systems. When embedded within digital advisory platforms or DSS, these models can empower farmers with real-time recommendations, enhancing adaptive capacity at scale. Furthermore, the observed interaction between elevated CO_2 and temperature suggests that future irrigation strategies must account not only for water availability but also for physiological crop responses under changing climatic conditions (Halder et al., 2020). Thus, a coordinated approach combining model-driven agronomic planning, farmer awareness, and climate-sensitive policy frameworks is essential to operationalize water-smart irrigation across diverse agroecological zones.

7 Summary

Field-level data play a critical role in managing climate-resilient agricultural systems, especially as global warming intensifies ET and disrupts rainfall patterns. Empirical observations such as soil moisture dynamics, crop phenology, and yield responses enable targeted interventions that improve water-use efficiency and nutrient uptake. When calibrated with such field data, system-based crop models, i.e., DSSAT, simulate soil–plant–atmosphere interactions with high fidelity, allowing for optimized input management even under data-scarce conditions. These models can extrapolate localized insights to broader scales, supporting crop-specific strategies across diverse agroecological zones. Once robustly validated, they become powerful tools for regional-to-global climate change assessments, enabling evaluation of environmental impacts, adaptation strategies, and policy interventions tailored to specific cropping systems. The integration of field data, calibrated models, and scalable analytics is essential for designing sustainable, climate-smart agricultural frameworks.

Acknowledgment

Support provided by all concerned, including the authorities of Siksha 'O' Anusandhan, is sincerely acknowledged.

References

Al Viandari, N., Adriany, T.A., Pramono, A., 2020. Alternate wetting and drying system (AWD) combined with farmyard manure to increase rice yield and reduce methane emission and water use. IOP Conf. Ser. Mater. Sci. Eng. 980 (1), 012066.

Ale, S., Su, Q., Singh, J., Himanshu, S., Fan, Y., Stoker, B., Gonzalez, E., Sapkota, B.R., Adams, C., Biggers, K., Kimura, E., Wall, J., 2023. Development and evaluation of a decision support mobile application for cotton irrigation management. Smart Agric. Technol. 27723755. 5, 100270. doi:10.1016/j.atech.2023.100270.

Amankwaa-Yeboah, P., Yeboah, S., Oppong-Sekyere, D., 2024. Water-smart crop farming: a holistic approach to its practice as a climate change adaptation strategy in Sub-Saharan Africa. Front. Sustain. Food Syst. 8, 1378191.

Bastia, D.K., Behera, S.K., Tripathy, S., Nanda, A., Rath, S., Srivastava, R.K., 2025. Building resilience: climate adaptation practices in Indian rainfed farming. In: Mitigation and Adaptation Strategies against Climate Change in Natural Systems. Springer Nature Switzerland, Cham, pp. 109–147.

Behera, S.K., Panda, R.K., 2009. Integrated management of irrigation water and fertilizers for wheat crop using field experiments and simulation modeling. Agric. Water Manag. 03783774. 96 (11), 1532–1540. doi:10.1016/j.agwat.2009.06.016.

Bhatnagar, S., Chaudhary, R., Sharma, S., Janjhua, Y., Thakur, P., Sharma, P., Keprate, A., 2024. Exploring the dynamics of climate-smart agricultural practices for sustainable resilience in a changing climate. Environ. Sustain. Indic. 26659727. 24, 100535. doi:10.1016/j.indic.2024.100535.

Bwambale, E., Abagale, F.K., Anornu, G.K., 2023. Data-driven model predictive control for precision irrigation management. Smart Agric. Technol. 3, 100074.

Cai, J., Liu, Y., Lei, T., Pereira, L.S., 2007. Estimating reference evapotranspiration with the FAO Penman-Monteith equation using daily weather forecast messages. Agric. For. Meteorol. 01681923. 145 (1–2), 22–35. doi:10.1016/j.agrformet.2007.04.012.

Calera, A., Campos, I., Osann, A., D'Urso, G., Menenti, M., 2017. Remote sensing for crop water management: from ET modelling to services for the end users. Sensors. 1424-8220. 17 (5), 1104. doi:10.3390/s17051104.

Chaurasiya, A., Singh, G., Mohan, M., Kumar, D., Kumari, P., Hussain, M.Z., Dutta, S.K., Kumar, R., 2025. Harnessing Conservation Agriculture for Climate Resilience Transition to Regenerative Agriculture: Principles and Indicators of Soil Health Management. Springer Nature, India, pp. 27–54. Available from: https://link.springer.com/book/9789819614219. doi:10.1007/978-981-96-1421-9_2.

Cheng, M., Wang, H., Fan, J., Zhang, S., Wang, Y., Li, Y., Sun, X., Yang, L., Zhang, F., 2021. Water productivity and seed cotton yield in response to deficit irrigation: a global meta-analysis. Agric. Water Manag. 255, 107027.

CIMMYT, 2008. Drought Tolerant Maize Varieties and Hybrids. Retrieved on 18/07/2025. URL: https://stma.cimmyt.org/drought-tolerant-maize-varieties-and-hybrids/.

Contosta, A.R., Arndt, K.A., Baulch, H.M., Casson, N.J., Harpold, A., Morelli, T.L., Sirén, A.P.K., Templer, P.H., 2024. Threshold changes in winter temperature and precipitation drive threshold responses across nine global climate zones and associated biomes. Annu. Rev. Ecol. Evol. Syst. 15452069. 55 (1), 271–300. Available from: http://www.annualreviews.org/journal/ecolsys. doi:10.1146/annurev-ecolsys-110421-102101.

Dilip, T., Kumari, M., Murthy, C.S., Neelima, T.L., Chakraborty, A., Devi, M.U., 2023. Monitoring early-season agricultural drought using temporal Sentinel-1 SAR-based combined drought index. Environ. Monit. Assess. 15732959. 195 (8). Available from: https://www.springer.com/journal/10661. doi:10.1007/s10661-023-11524-y.

Doorenbos, J., Kassam, A.H., 1979. Yield response to water. FAO Irrigation and Drainage, Paper 33, 193.

FAO AQUASTAT, 2013. Global map of irrigation areas v5. Retrieved on 19/07/2025. URL: https://firebasestorage.googleapis.com/v0/b/fao-aquastat.appspot.com/o/PDF%2FMAPS%2Fgmia_v5_highres.pdf?alt=media&token=0477a76c-b4a0-440e-93ec-8093db54ef0d

FAO, 2021a. AQUASTAT—FAO's Global Information System on Water and Agriculture. URL: https://www.fao.org/aquastat/en/overview/methodology/water-use

FAO, 2021b. Strategic Framework 2021-31. Retrieved on 19/07/2025. URL: https://openjicareport.jica.go.jp/pdf/1000054456_02.pdf

FAO, 2021c. Climate-smart agriculture case studies 2021—projects from around the world. Rome. doi:10.4060/cb5359en

Fernández, J.E., 2014. Plant-based sensing to monitor water stress: applicability to commercial orchards. Agric. Water Manag. 142, 99–109.

Francis, C.A., Jordan, N., Porter, P., Breland, T.A., Lieblein, G., Salomonsson, L., Langer, V., 2011. Innovative education in agroecology: experiential learning for a sustainable agriculture. Crit. Rev. Plant Sci. 30 (1–2), 226–237.

Gao, F., Zhang, X., 2021. Mapping crop phenology in near real-time using satellite remote sensing: challenges and opportunities. J. Remote Sens. doi:10.34133/2021/8379391.

Halder, D., Kheroar, S., Srivastava, R.K., Panda, R.K., 2020. Assessment of future climate variability and potential adaptation strategies on yield of peanut and Kharif rice in eastern India. Theor. Appl. Climatol. 140 (3), 823–838.

Halder, D., Panda, R.K., Srivastava, R.K., 2014. Evapotranspiration based irrigation water management strategy for peanut under water scarcity and climate variability conditions. J. Agrometeorol. 1 (16), 72–78.

Halder, D., Panda, R.K., Srivastava, R.K., Kheroar, S., 2017. Evaluation of the CROPGRO-Peanut model in simulating appropriate sowing date and phosphorus fertilizer application rate for peanut in a subtropical region of eastern India. Crop J. 5 (4), 317–325.

Himanshu, S.K., Ale, S., Bell, J., Fan, Y., Samanta, S., Bordovsky, J.P., Brauer, D.K., 2023. Evaluation of growth-stage-based variable deficit irrigation strategies for cotton production in the Texas High Plains. Agric. Water Manag. 280, 108222.

Himanshu, S.K., Fan, Y., Ale, S., Bordovsky, J., 2021. Simulated efficient growth-stage-based deficit irrigation strategies for maximizing cotton yield, crop water productivity and net returns. Agric. Water Manag. 250, 106840.

Home, P.G., Panda, R.K., Kar, S., 2002. Effect of method and scheduling of irrigation on water and nitrogen use efficiencies of Okra (Abelmoschus esculentus). Agric. Water Manag. 55 (2), 159–170.

ICAR-NRRI, 2021. Farmers' Field School on Climate Smart Water Management Practices in Rice. Retrieved on 18/07/2025. URL: https://www.icar-nrri.in/wp-content/uploads/2022/01/Technology-Bulletin-178.pdf

IPCC Sixth Assessment Report (2021). IPCC AR6—climate Change 2021: the Physical Science Basis. https://www.ipcc.ch/assessment-report/ar6/

Kashyap, B., Kumar, R., 2021. Sensing methodologies in agriculture for soil moisture and nutrient monitoring. IEEE Access 9, 14095–14121.

Kashyap, P.S., Panda, R.K., 2001. Evaluation of evapotranspiration estimation methods and development of crop-coefficients for potato crop in a sub-humid region. Agric. Water Manag. 50 (1), 9–25.

Kashyap, P.S., Panda, R.K., 2003. Effect of irrigation scheduling on potato crop parameters under water stressed conditions. Agric. Water Manag. 59 (1), 49–66.

Kayatz, B., Harris, F., Hillier, J., Adhya, T., Dalin, C., Nayak, D., Green, R.F., Smith, P., Dangour, A.D., 2019. "More crop per drop": exploring India's cereal water use since 2005. Sci. Total Environ. 673, 207–217.

King, B.A., Shellie, K.C., Tarkalson, D.D., Levin, A.D., Sharma, V., Bjorneberg, D.L., 2020. Data-driven models for canopy temperature-based irrigation scheduling. Trans. ASABE 63 (5), 1579–1592.

Kumar, H., Srivastava, P., Lamba, J., Diamantopoulos, E., Ortiz, B., Morata, G., Bondesan, L., 2022b. Site-specific irrigation scheduling using one-layer soil hydraulic properties and inverse modeling. Agric. Water Manag. 273, 107877.

Kumar, H., Srivastava, P., Lamba, J., Lena, B., Diamantopoulos, E., Ortiz, B., Bondesan, L., 2023. A methodology to optimize site-specific field capacity and irrigation thresholds. Agric. Water Manag. 286, 108385.

Kumar, H., Srivastava, P., Lamba, J., Ortiz, B.V., Way, T.R., Sangha, L., Molinari, R., 2022a. Within-field variability in nutrients for site-specific agricultural management in irrigated cornfield. J. ASABE 65 (4), 865–880.

Lakhiar, I.A., Yan, H., Zhang, C., Wang, G., He, B., Hao, B., Rakibuzzaman, M., 2024. A review of precision irrigation water-saving technology under changing climate for enhancing water use efficiency, crop yield, and environmental footprints. Agriculture 14 (7), 1141.

Lawrence, M., Nuthammachot, N., Som-ard, J., 2024. Nexus between conservation agriculture and remote sensing: a review. Agric. Nat. Resour. 58 (5), 665–678.

Lena, B.P., Bondesan, L., Pinheiro, E.A.R., Ortiz, B.V., Morata, G.T., Kumar, H., 2022. Determination of irrigation scheduling thresholds based on HYDRUS-1D simulations of field capacity for multilayered agronomic soils in Alabama, USA. Agric. Water Manag. 259, 107234.

Levidow, L., Zaccaria, D., Maia, R., Vivas, E., Todorovic, M., Scardigno, A., 2014. Improving water-efficient irrigation: prospects and difficulties of innovative practices. Agric. Water Manag. 146, 84–94.

Mertens, S., Verbraeken, L., Sprenger, H., De Meyer, S., Demuynck, K., Cannoot, B., Wuyts, N., 2023. Monitoring of drought stress and transpiration rate using proximal thermal and hyperspectral imaging in an indoor automated plant phenotyping platform. Plant Methods 19 (1), 132.

Morris, J., Barron, J., 2014. Agricultural water management technology expansion and impact on crop yields in northern Burkina Faso (1980-2010): a review. URL: https://cgspace.cgiar.org/items/bc8d07f7-a11c-4860-a36c-5069581ef15f

Mottaleb, K.A., Rejesus, R.M., Murty, M.V.R., Mohanty, S., Li, T., 2017. Benefits of the development and dissemination of climate-smart rice: ex ante impact assessment of drought-tolerant rice in South Asia. Mitig. Adapt. Strateg. Glob. Change 22 (6), 879–901.

Muralikrishnan, L., Padaria, R.N., Dass, A., Choudhary, A.K., Kakade, B., Shokralla, S., Elansary, H.O., 2021. Elucidating traditional rice varieties for consilient biotic and abiotic stress management under changing climate with landscape-level rice biodiversity. Land 10 (10), 1058.

Murmu, K., Sarkar, A., Sarma, S.S., Srivastava, R.K., 2025. An Overview of Crop Simulation Models: A Decision Support System for Evaluation of Agronomic Managements in Climate Change Context. Mitigation and Adaptation Strategies Against Climate Change in Natural Systems, 399–426.

Mutlu, A., 2025. Comparative evaluation of NASA, ERA5, and observational data for accuracy and reliability. Theor. Appl. Climatol. 156 (7), 367.

Panda, R.K., Behera, S.K., Kashyap, P.S., 2003. Effective management of irrigation water for wheat under stressed conditions. Agric. Water Manag. 63 (1), 37–56.

Panda, R.K., Behera, S.K., Kashyap, P.S., 2004. Effective management of irrigation water for maize under stressed conditions. Agric. Water Manag. 66 (3), 181–203.

Patle, G.T., Kumar, M., Khanna, M., 2020. Climate-smart water technologies for sustainable agriculture: a review. J. Water Clim. Change 11 (4), 1455–1466.

Puig, F., García-Vila, M., Soriano, M.A., Rodríguez-Díaz, J.A., 2025. AquaCrop-IoT: a smart irrigation platform integrating real-time images and weather forecasting. Comput. Electron. Agric. 235, 110372.

Richard, Sander, 2014. Alternate wetting and drying in irrigated rice. Retrieved on 19/07/2025. URL: https://cgspace.cgiar.org/rest/bitstreams/34363/retrieve

Rossiter, D.G., 2016. Digital soil resource inventories: status and prospects in 2015. In: Zhang, G.L., Brus, D., Liu, F., Song, X.D., Lagacherie, P. (Eds.), Digital Soil Mapping Across Paradigms, Scales and Boundaries. Springer Environmental Science and Engineering. Springer, Singapore, pp. 275–286. doi:10.1007/978-981-10-0415-5_22.

Ruan, H., Yu, J., Wang, P., Wang, T., 2020. Increased crop water requirements have exacerbated water stress in the arid transboundary rivers of Central Asia. Sci. Total Environ. 713, 136585.

Santra Mitra, S., Santra, A., Kumar, A., 2021. Catchment specific evaluation of Aphrodite's and TRMM derived gridded precipitation data products for predicting runoff in a semi gauged watershed of Tropical India. Geocarto Int. 36 (11), 1292–1308.

Shi, H., 2024. Potential of constructing all-encompassing soil–plant-atmosphere-continuum stations and datasets from meteorological, flux, soil moisture station networks and plant-relevant observations. Hydrol. Process. 38 (10), e15284.

Shoukat Hafiza, B., Ishaque, W., Ahmad, S., Ali, S., El-Sheikh, M.A., 2025. Optimizing wheat productivity and water productivity through deficit irrigation strategies in semi-arid environments. Sci. Rep. 15 (1), 20630.

Singh, A., Gaurav, K., 2023. Deep learning and data fusion to estimate surface soil moisture from multi-sensor satellite images. Sci. Rep. 13 (1), 2251.

Singh, J., Ale, S., DeLaune, P.B., Himanshu, S.K., Barnes, E.M., 2022. Modeling the impacts of cover crops and no-tillage on soil health and cotton yield in an irrigated cropping system of the Texas Rolling Plains. Field Crops Res. 287, 108661.

Srivastava, R.K., Mequanint, F., Chakraborty, A., Panda, R.K., Halder, D., 2022. Augmentation of maize yield by strategic adaptation to cope with climate change for a future period in Eastern India. J. Clean. Prod. 339, 130599.

Srivastava, R.K., Panda, R.K., Chakraborty, A., 2021. Assessment of climate change impact on maize yield and yield attributes under different climate change scenarios in eastern India. Ecol. Indic. 120, 106881.

Srivastava, R.K., Panda, R.K., Chakraborty, A., 2025. Sustainable agricultural engineering: integrating science, technology, and practical applications. In: Mitigation and Adaptation Strategies against Climate Change in Natural Systems. Springer Nature Switzerland, Cham, pp. 3–27.

Srivastava, R.K., Panda, R.K., Chakraborty, A., Halder, D., 2018. Enhancing grain yield, biomass and nitrogen use efficiency of maize by varying sowing dates and nitrogen rate under rainfed and irrigated conditions. Field Crops Res. 221, 339–349.

Srivastava, R.K., Panda, R.K., Chakraborty, A., Singh, S.P., 2023. Drought tolerance agronomic management strategies for rainfed and irrigated maize crop in Eastern India. In: Integrated Drought Management, vol. 2. CRC Press, pp. 593–606.
Srivastava, R.K., Panda, R.K., Halder, D., 2017. Effective crop evapotranspiration measurement using time-domain reflectometry technique in a sub-humid region. Theor. Appl. Climatol. 129 (3), 1211–1225.
Srivastava, R.K., Talla, A., Swain, D.K., Panda, R.K., 2019. Quantitative approaches in adaptation strategies to cope with increased temperatures following climate change in potato crop. Potato Res. 62 (2), 175–191.
Srivastava, S.K., Kishore, P., Birthal, P.S., Shirsath, P.B. (2024). Harnessing the Potential of Solar-Powered Micro-Irrigation for Sustainable Intensification of Agriculture. URL: https://www.researchgate.net/publication/388350700_Harnessing_the_Potential_of_Solar_Powered_Micro-Irrigation_for_Sustainable_Intensification_of_Agriculture
TAAT, 2025. DroughtTEGO: drought tolerant and high yield maize varieties. Retrieved on 18/07/2025. URL: https://e-catalogs.taat-africa.org/com/technologies/droughttego-drought-tolerant-and-high-yield-maize-varieties
Taylor, R.G., Scanlon, B., Döll, P., Rodell, M., Van Beek, R., Wada, Y., Longuevergne, L., Leblanc, M., Famiglietti, J.S., Edmunds, M., Konikow, L., 2013. Ground water and climate change. Nat. Clim. Change 3 (4), 322–329.
Thirumeninathan, S., Pazhanivelan, S., Mohan, R., Pouchepparadjou, A., Sudarmanian, N.S., Ragunath, K., Satheesh, S., 2024. Integrating S1A microwave remote sensing and DSSAT CROPGRO simulation model for groundnut area and yield estimation. Eur. J. Agron. 161, 127348.
Tunio, M.H., Li, J.P., Zeng, X., Akhtar, F., Shah, S.A., Ahmed, A., Heyat, M.B.B., 2024. Meta-knowledge guided Bayesian optimization framework for robust crop yield estimation. J. King Saud Univ. Comput. Inf. Sci. 36 (1), 101895.
UNCCD, 2022. World "at a crossroads" in drought management, up 29% in a generation and worsening, says UN. URL: https://www.unccd.int/news-stories/press-releases/world-crossroads-drought-management-29-generation-and-worsening-says-un
Vilavan, S., Das, D., Ullah, H., Gade, S.A., Ahmed, S.F., Cha-Um, S., Himanshu, S.K., 2024. Exploring the impacts of climate change and identifying potential adaptation strategies for sustainable rice production in Thailand's Lower Chao Phraya Basin through crop simulation modeling. Environ. Monit. Assess. 196 (12), 1192.
Wang, C., Harrison, M.T., Brown, H., Qiao, Y., Yin, X., Yang, R., Liu, K., 2025. Global sensitivity analyses prior to parameter calibration significantly improves prediction quality. Comput. Electron. Agric. 238, 110750.
White, J.W., Hoogenboom, G., Kimball, B.A., Wall, G.W., 2011. Methodologies for simulating impacts of climate change on crop production. Field Crops Res. 124 (3), 357–368.
Worrall, G., Judge, J., Boote, K., Rangarajan, A., 2023. In-season crop phenology using remote sensing and model-guided machine learning. Agron. J. 115 (3), 1214–1236.
Zou, Y., Saddique, Q., Ali, A., Xu, J., Khan, M.I., Qing, M., Azmat, M., Cai, H., Siddique, K.H., 2021. Deficit irrigation improves maize yield and water use efficiency in a semi-arid environment. Agric. Water Manag. 243, 106483.

Metagenomics: A next-generation tool for monitoring soil microbial communities and managing soil health

12

Kamlesh Thakur[a], *Karan Saini*[b], *Sibiraj Murugesan*[c], *Prachi Bhatnagar*[d], *Navneet Kumar*[a,e], *Rahul Yadav*[f], *and Gaurav Saxena*[g]

[a]School of Biotechnology, Faculty of Applied Sciences and Biotechnology, Shoolini University, Solan, Himachal Pradesh, India
[b]Applied Phycology and Food Technology Laboratory, CSIR-Institute of Himalayan Bioresource Technology (IHBT), Palampur, Himachal Pradesh, India
[c]School of Bioengineering and Food Technology, Faculty of Applied Sciences and Biotechnology, Shoolini University, Solan, Himachal Pradesh, India
[d]Department of Biosciences and Bioengineering, Indian Institute of Technology (IIT), Roorkee, Uttarakhand, India
[e]Microbiology Department, Shoolini Life Sciences Private Limited, Solan, Himachal Pradesh, India
[f]National Accreditation Board for Testing and Calibration, Laboratories (NABL), Quality Council of India (QCI), Gurugram, Haryana, India
[g]Department of Life Science, Faculty of Life Sciences, Mandsaur University, Mandsaur, Madhya Pradesh, India

1 Introduction

Soil is the planet's most biodiverse environment, with an estimated 1000 Gbp of microbial DNA sequences per gram. Microorganisms are essential to many ecological processes, including the biogeochemical cycle, biogenesis, soil formation, and bioremediation. They also help maintain soil structure and fertility. Therefore, for the sustainable management of soil ecosystem function and the prediction of the effects of environmental change, it is crucial to understand the factors that affect the growth, activity, and diversity of soil microbial communities, how soil physicochemical characteristics influence them, and how they interact with other inhabitants of the soil. However, the majority of soil bacteria are unknown in terms of their roles and remain unidentified. While metagenomic sequencing provides information on

Precision Technologies for Digital Agriculture. https://doi.org/10.1016/B978-0-443-40513-6.00015-9

functional genes and microbial identities, it also includes DNA from a wide range of physiological states in bacteria. Therefore, metagenomics only forecasts a community's ability to regain ecosystem function—a function that is now in jeopardy due to changing climatic and precipitation patterns, degraded soil, and inefficient land management practices.

The drawbacks of culture-based approaches, whose use has grown significantly in recent years, can now be remedied through metagenomics. Metagenomics is a technique for studying genomes from a population of uncultured environmental bacteria (Ghosh et al., 2019). Instead of using laboratory culture, metagenomics extracts DNA directly from environmental sources (Neelakanta and Sultana, 2013). DNA analysis of microbial diversity produces accurate and comprehensive results (Tripathi et al., 2018). Metagenomics can be used to explore the benefits of natural gene resources and learn more about the interdependencies of microbial populations involved in the natural biogeochemical cycle (Bastida et al., 2019). Complex microbial communities found in soil are essential for several ecological activities, including plant and animal life (Wilpiszeski et al., 2019). It is crucial to comprehend the structure and function of these ecosystems since they are sensitive to anthropogenic pressure and climate change (Jansson and Hofmockel, 2020).

The understanding of soil's sensitivity to perturbations and functional resilience, such as changes in land use or climate, is limited, despite its importance in various aspects of life, including soil biogeochemistry, geographic complexity, and how soil processes like carbon cycling impact global warming. Although microorganisms are crucial to soil health, very few (<0.5%) have been cultivated in the lab, and only a small number have genome sequences accessible. The application of metagenomics in research is expected to lead to discoveries about microbial community data and natural enzymes (Ferrer et al., 2013). However, further evaluation of metagenomics research is required to deepen insights into soil metagenomic investigations. As a result, this chapter discusses the benefits and drawbacks of soil metagenomics and how it can be used to monitor the composition and activities of microbial communities in the soil. This data can enhance soil quality, health, fertility, and aid in the bioremediation of contaminants for a healthier environment.

2 Overview of metagenomics

Metagenomics is a method that examines the genetic makeup of entire microbial populations in a specific environment, independent of culture (Gudiña et al., 2020). Soil metagenomics is one molecular method that may be used to discover and exploit the wide variety of soil microbial communities without requiring cultivation. Utilizing this approach, clone libraries are created and screened, and soil DNA is isolated. High-quality DNA isolation from soil samples involves recovering cells and lysing them, or lysing them directly before purifying the DNA. Generous amounts of soil DNA are needed to create small-insert libraries in plasmid vectors or large-insert libraries in cosmid, fosmid, or bacterial artificial chromosome (BAC) vectors. While small-insert libraries identify specific genes or operons with distinct metabolic activities, large-insert libraries help analyze gene clusters and genomic regions of uncultured soil microorganisms. The activities of library clones or sequence similarities with known genes are used as screening criteria for metagenomic soil libraries. Novel biomolecules, such as antibiotics and enzymes of economic significance, have been discovered using this technique. Before library development, procedures for enriching soil microorganisms carrying the necessary features can be utilized to boost gene detection frequencies. Several novel enzymes with potential catalytic activity have already been discovered using soil metagenomics, and many of them have been patented and commercialized (Berini et al., 2017). The success of soil metagenomics depends on a number of factors, including open system data management and sharing, cloning, appropriate DNA extraction methods, screening tactics, and careful sample selection (Vogel et al., 2009).

3 Metagenomics in understanding soil microbiome

Soil ecosystems are shaped by soil microorganisms like bacteria, archaea, and fungi, which promote the cycling of essential elements like N, C, and P. This cycling not only shapes the structure and function of ecosystems but also provides soil with capabilities that offer various services to people (Aislabie et al., 2013). The productivity of agroecosystems is determined by the soil microbiome, which is the collective function of the soil's microbes (Van Der Heijden et al., 2008). Research focuses on regulating the soil microbiome for sustainable agriculture, as it depends on its diversity and health. Researchers aim to enhance microbiome functions amidst anthropogenic and natural activities like pest and pathogen strains, climate change, pollution, and chemical fertilizer (CF) use, which threaten reliable agricultural production (Callaway, 2016). For sustainable soil and agricultural management, the complex interactions among plant genotypes, microbiome architectures, and diverse environmental circumstances are essential.

In this context,(Ahmed et al., 2018) a study on soil microbes used metagenomic profiling to identify salt stress-resistant genes. The study found that Actinobacteria, Proteobacteria, Gemmatimonadetes, Firmicutes, Bacteroidetes, and Acidobacteria are the most prevalent halophilic and halotolerant phylotypes. Several salt-tolerant genes, such as duf3445, GSDH, STK PknB, and BCAA ABCtp, were discovered using a functional metagenomics approach. These genes can be used to improve soils, crops, and bioactive compound production in high-salt environments, reducing contamination risks.

The soil microbial populations and their functional characteristics have been analyzed using cross-biome metagenomic techniques by Fierer et al. (2012). Metagenomic sequencing was used in research to examine the functional properties and composition of 16 soil microbial communities from a variety of habitats, including forests, grasslands, tundra, cold deserts, and hot deserts. Desert microbial communities often have the lowest levels of phylogenetic, taxonomic, and functional diversity, despite being distinct from nondesert communities. In all soil types, there is a high correlation between taxonomic and phylogenetic beta diversity and functional beta diversity. The genes linked to osmoregulation and dormancy are more abundant in desert ecosystems, whereas the genes linked to nutrient cycling and the breakdown of organic compounds originating from plants are less abundant. Antibiotic resistance genes (ARGs) are three times less abundant in desert soils than in nondesert soils, suggesting that abiotic factors are more significant in determining the composition of desert microbial communities. Through the use of metagenomic approaches, this work demonstrates how it is possible to anticipate changes in microbial function and diversity across terrestrial biomes.

Muñoz-Arenas et al. (2020) conducted metagenomic surveys in a high-mountain temperate forest to investigate the decrease in soil microbial diversity due to land-use alteration. In high-altitude temperate oak and pine-oak forests, as well as in arable soils in Mexico, the taxonomic and functional diversity of bacteria, archaea, and DNA viruses were investigated using shotgun metagenomics. Arable soils exhibited higher viral diversity than forest soils. Converting these forests to arable land increased bacterial diversity but did not affect archaeal or viral diversity. In the Loxahatchee Nature Preserve in the Florida Everglades,(Abraham et al., 2020) carried out a shotgun metagenomic analysis on microbial communities. They used 17 soil samples from four locations and found Actinobacteria, Acidobacteria, and Proteobacteria as the most prevalent phyla, but also observed heterogeneity in relative phylum composition. Euryarchaeota was the most frequent phylum across all Archean sites. According to alpha and beta diversity analyses, most samples from Sites 1, 3, and 4 showed significant taxonomic diversity congruence; however, Site 2 exhibited very low congruence with the other sites. Shotgun metagenomic sequencing identified biogeochemical biomarkers in Everglades soil samples, including mrcA, nifH, and dsrB. Normalized gene abundances showed a negative association with organic carbon and a positive correlation with nitrogen and water levels. This study found connections between organic content and bacterial gene frequency, providing insights into wetland ecological processes. The taxonomic profile also offers a foundation for understanding prokaryotic microbial life diversity.

The soil microbiome in organic and conventionally farmed coffee plantations in tropical agroecosystems was studied by Cabrera-Rodríguez et al. (2020) using Illumina metagenomic sequencing. In tropical eastern Mexico, they employed organic and conventional agronomic methods to evaluate the soil microbiome. With Shannon's diversity index of 9.7 and Simpson's index of 0.99, they discovered a varied array of phyla, classes, orders, families, and genera in both treatments. Proteobacteria, which are defined as copiotrophic, were more prevalent in CAM soils, while Acidobacteria, which are classified as oligotrophic, were more prevalent in OAM soils. The study indicates that bio-fertilizers have led to an increase in Acidobacteria, a phylum involved in carbon sequestration and positively linked to organic matter. Uncovering the hidden features of the soil microbiome with metagenomics is undoubtedly an exciting and modern genetic tool.

4 Metagenomics in understanding soil resistome

Soil worldwide, including Antarctic surface soils, is the most likely source of ARGs due to the extensive variety found in these genes (Braga et al., 2017; Wang et al., 2016b; Xiao et al., 2016). Through horizontal gene transfer, several medicinally significant ARGs were extracted from the soil resistome (Dantas et al., 2008; Pearlman, 2012). Additional research shows that various human activities dramatically increase the number of native ARGs in soils (Reece et al., 2017). It is essential to comprehend soil resistomes in all major ecosystems, particularly in native soils that have not been subjected to human activity, in order to precisely identify background levels and types of ARGs. This is necessary to evaluate the risk associated with new human activities.

In order to study the kinds and quantities of ARGs in the soil DNA of three natural ecosystems—the Amazon rainforest, the US Midwest grassland, and the Alaskan tundra—Qian et al. (2021) used a metagenomic technique with datasets that had well-described metadata. They also examined the effects of converting the latter two ecosystems to pasture and agriculture, respectively. The quantity of ARGs and the diversity of their subtypes were both high across all habitats, with efflux pumping and multidrug resistance being the most common classes and mechanisms. Ten regulatory genes were identified, making up 13%–35% of soil resistome abundances; of these, the most abundant and ubiquitous were arlR, vanR, ompR, cpxR, and vanS. Around 81% of the nonregulatory resistome abundance was explained by the 55 nonregulatory ARGs that were found in all 26 soil metagenomes from the three habitats. Seven of the ten most abundant ARGs were discovered in all three of the principal ARG hosts: Proteobacteria, Firmicutes, and Actinobacteria. Between adjoining long-term developed agricultural soil and natural prairie soil, the research observed no discernible differences in ARG abundance or diversity. At the sequence level, they selected 12 clinically important ARGs to examine and found that they differed from those found in human infections. When combined, they were even more dissimilar. The study found a significant association between the resistome profile and the structure of the bacterial community, indicating that the composition of the bacterial community significantly influences resistome profile variance. Their research identifies candidate background ARGs, counts resistance classes, categorizes candidate background hosts (common to all 26 soils), and discloses probable future resistance gene variants.

Liu et al. (2023) used metagenomic analysis to investigate the potential deterministic impacts of pH on the formation of soil resistomes, which are the compositions of ARGs. Utilizing a structured ARG database, they analyzed the distribution of ARGs across 41 metagenomes constructed from soil and 36 soil metagenomes. The multidrug efflux pump genes were found to be more abundant in the acidic soil resistome, and they were shown to be significantly pH-dependent, indicating that they had undergone natural selection due to the high soil proton activity, particularly for the PMF-driven inner membrane transferase. In addition, they assessed the consistency between the neutrally built microbiome and the resistome, which was assembled deterministically, and identified the likely microbial contributors involved in pH shaping soil resistance. They proposed a new idea to explain the impact of edaphic variables on soil resistance mechanisms. This theory suggests that, independent of the microbial environment, bottom-up selection on ARGs might result from the natural selection of fitting resistance mechanisms. If we want to understand how future human actions will affect the spread of ARGs, we need to account for these kinds of natural alterations to the resistome.

In South Chinese paddy soils, Xiao et al. (2016) investigated the metagenomic profiles of genes resistant to antibiotics. They identified sixteen distinct ARG categories, with 110 distinct ARG subtypes assigned to each. While the distribution patterns and abundances of the sediment from human-impacted estuaries were similar to those of activated sludge and pure deep ocean sediment, the sediment from paddy soil was different. Genes for multidrug resistance were the most prevalent across all samples, accounting for 37.5%–47.5% of ARGs and covering all three primary resistance mechanisms, with efflux pump extrusion being the most common. Redundancy analysis revealed a substantial ($P < 0.05$) correlation between pH and the distribution of ARG subtypes. Torres-Cortés et al. (2011) conducted a study on novel antibiotic-resistance genes in South China's paddy soil using functional metagenomics. They discovered 11 novel ARGs, including three that confer resistance to ampicillin, gentamicin, chloramphenicol, and trimethoprim. A unique trimethoprim resistance gene that encodes a 26.8 kDa protein similar to unassigned dihydrofolate reductases was found in one of the clones. With dihydrofolate reductase activity and kinetic constants similar to those of other type I and type II reductases, the gene produces an enzyme. The study also provided the first description of a novel reductase type that imparts trimethoprim resistance. The findings suggest this method as a promising approach for discovering novel enzymes with unpredictable actions due to soil bacteria's high genetic diversity and ARGs.

Willms et al. (2019) discovered eight unidentified ARGs in soils from grasslands and forests. These ARGs, originating from diverse soil bacteria, encode proteins with a minimum of 46% identity to other antibiotic resistance determinants. They also discovered nonmobile dihydropteroate synthase (DHPS) genes granting resistance to sulfonamides in forest soil for the first time, even without prior exposure to artificial medications. Three distinct sulfonamide-resistant DHPS genes were found in the soil of beech or pine forests, each with a different taxonomic origin, proving that sulfonamide resistance is a natural occurrence in soil bacterial communities inhabited by trees. They found sulfonamide resistance proteins in forest soil and DHPS genes connected to Chloroflexi in grassland soil; the former provides less susceptibility to sulfamethazine, which is often used in animal production for food. Newly identified members of the major facilitator superfamily include four efflux proteins associated with tetracycline resistance. It is worth mentioning that one of these proteins also reduced lincomycin susceptibility.

5 Metagenomics in understanding biogeochemical cycling in soil

Soil metagenomic research aims to understand the C, N, P, S, and other elemental cycles in soil. Using high-throughput sequencing, White et al. (2023) examined the microbial population in the switchgrass rhizosphere on marginal land both before and after nitrogen fertilization. Researchers resolved 29 individual bacterial genomes from various soil taxa, including Verrucomicrobia, Candidate Phyla UBA10199, Dormibacterota, and Acidobacteria. They discovered a decrease in the bulk metabolic capacity of carbohydrate-active enzymes but no discernible changes in community structure or diversity between the prefertilization and postfertilization phases. Dormibacterota AD3 was identified as potentially affecting carbon partitioning and storage. The unusual diazotroph Janthinobacterium might be responsible for the fertilizer-resistant switchgrass development.

In order to assess the microbial organic carbon transformation capabilities in various habitats,(Dai et al., 2021) used metagenomics to investigate soil microbial communities engaged in carbon cycling over elevational climatic sequences, particularly at Mount Kilimanjaro. According to the research, there was a hump-shaped pattern at mid-elevations where the largest gene abundances for compounds that are highly mineralizable, such as chitin, pectin, and carbohydrate esters, were discovered. On the other hand, due to the high pH of the soil, the genes for chemicals that are readily mineralized and resistant, like lignin, exhibited a U-shaped pattern. These genes showed positive associations with relevant C compounds, indicating consistent reactions of microbial functional profiles and metabolites to higher climatic sequences. This understanding could help in controlling terrestrial carbon storage.

A grazing experiment was conducted in a typical steppe grassland in Inner Mongolia by Zhang et al. (2022) to investigate the effects of three different grazing intensities (none, moderate, and heavy) on the microbial communities in the bulk soils and rhizosphere of a grassland dominated by Stipa grandis. It was discovered that grazing significantly increased the following: soil alkaline phosphatase, soil available P concentration, the relative abundance of genes responsible for microbial P-uptake and transport, inorganic P solubilization and organic P mineralization in rhizosphere soil (but not in bulk soil), the node degree of the genes encoding the C-P lyase multienzyme complex in rho, the abundance of rhizosphere bacteria that promote P-cycling potential (but not fungal communities), and the network of genes involved in P-transformation,which is reduced. The changes in microbial functional gene abundance caused by grazing enhance P-solubilization and mineralization. According to the data, grazing has distinct effects on soil microbial communities and functional genes involved in P-transformation in the rhizosphere compared to bulk soils. These results provide important new information on the microbial processes that control P cycling in grassland ecosystems. At the aggregate level of soil,(Wu et al., 2021) used metagenomic technology to study how organic material additions impact phosphorus and nitrogen cycling. There was a strong correlation between soil NO_3 and soil organic carbon (SOC) and the richness of taxonomic and functional alpha bacteria.

Macroscopic aggregates had the highest levels of denitrification and ammonia oxidizer genes, namely nasB, nirK, and amoA. In treatments that included manure and peat, the enzymes napA and nrfA, which are involved in the dissimilatory nitrate reduction to ammonium (DNRA), were shown to be drastically diminished. Macroscopic and large macroaggregates had much higher relative abundances of Gcd, making it a more reliable measure of soil inorganic P solubilization. Seven bacterial genomes and three almost complete genomes from the Nitrososphaeraceae (AOA) family included genes that were engaged in nitrification and P solubilization. These findings provide information on microbial genetic potential for DNRA and P-solubilizing activity across soil aggregate fractions and fertilization regimes.

In the black soil area of northeast China, Hu et al. (2023) used metagenomics to determine whether soil-accessible phosphorus was enhanced by organic fertilization in Mollisols. Results demonstrated a favorable and statistically significant relationship between manure amendments and soil total phosphorus (TP), accessible phosphorus (AP), and functional gene patterns linked to the microbial P cycle (AP). Soil AP concentrations were increased both directly and indirectly as a result of manure addition due to changes in microbial functional genes involved in soil P cycling. Soil microbial organic P-mineralization, absorption, and transport may be inhibited, since manure inputs consistently increased phnP, opd, and phoN gene abundances while decreasing phnC gene abundances across three sites. Incorporating manure improved microbial inorganic P-solubilization through enrichment of the ppa, gcd, and pqqC genes at two of the three locations, while adding CF barely increased the functional gene abundances involved in regulating microbial P-starvation response and P-uptake and transport. An association between soil AP concentration and gene abundances was found to be positive for opd and phoN genes and negative for phnC genes. The phnC gene was also detected in all 23 Metagenome-Assembled Genomes (MAGs), which included soil samples from various parts of the globe. In each of these genomes, the number of copies of this gene varied between 1 and 19. Regardless, the only bin with a negative correlation with soil AP (r=0.361, P=0.030) was bin44, suggesting it may be a signal for modulating P-uptake and transport by microbes. In sustainable intensive agriculture, manure inputs greatly improved microbial P-transformations, which in turn aided in the development of efficient P management systems.

6 Metagenomics in understanding rehabilitation of disturbed land

To prevent net losses of biodiversity and ecosystem functioning, environmental rehabilitation of disturbed areas must be carried out as closely as feasible to their pre-disturbance levels. To satisfy the objectives of environmental licensing authorities, it is necessary to monitor rehabilitation sites. As of now, there is no consensus on the optimum scientific indicators for assessing the monitoring process. Recently, multidisciplinary strategies aimed at providing such metrics have been proposed

(Perring et al., 2015; Gastauer et al., 2018). Along with assessments of flora and fauna (Audino et al., 2014), rapid environmental change detection is possible through the examination of microbial populations (Valentini et al., 2016), offering insights into whether rehabilitation goals have been achieved (Hamonts et al., 2017; Eaton et al., 2017). Due to their ability to identify the diversity of microbes present in various environments, metagenomic methods provide valuable information about environmental changes (Garris et al., 2016). The composition of microbial communities in damaged settings and reference areas can be compared to gain a better understanding of the effectiveness of restoration.

In a study conducted by Catania et al. (2022), the soil microbial biomass and diversity were examined using metagenomic analysis in areas affected by desertification in southern Europe. They found that bacterial diversity was negatively linked to SOC, while soil microbial biomass was positively correlated with SOC. The study also found that pH had the greatest impact on soil alpha-diversity. The study used 16S rRNA gene metagenomics to examine the core microbiota of semiarid soils, which include Firmicutes, Acidobacteria, Proteobacteria, and Actinobacteria. The study establishes a baseline for monitoring semiarid European soils affected by climate change and evaluating adaptation methods.

Wiryawan et al. (2022) used a metagenomic method to study soil microbial community organization. They found that forest soil had higher concentrations of micronutrients like iron, zinc, and magnesium compared to tilled land, as well as higher diversity indices of soil bacteria. Forest soil contained more unique taxa compared to tilled land, including less common species such as *Rhodoplanes* sp., *Candidatus Koribacter versatilis*, *Luteitalea pratensis*, *Solibacter usitatus*, and *Betaproteobacteria*. *Janthinobacterium* sp. and *Anseongella ginsenosidimutans* were identified as the most prevalent species in forest soil and tilled land, respectively. Soil management approaches influence the microbial community structure and the levels of micro- and macronutrients, with critical nutrient content and bacterial community structure being shaped by management practices. To maintain soil quality in areas close to forests, more effective management approaches are required.

Soil microbial populations in Brazil's Urucum Massif were studied in a metagenomic study by Gastauer et al. (2019a) and (2019b) following rehabilitation efforts after iron ore extraction. They examined three soil samples for chemical, physical, and metagenomic analyses. NextSeq 500 Illumina's paired-end library sequencing method was employed, and MEGAHIT was used to assemble the reads. Kaiju was utilized to identify coding DNA sequences (CDS) in combination with nonredundant NCBI BLAST reference sequences from archaea, bacteria, and viruses. In addition, EMG v2.3.2 carried out a functional classification. This study revealed that the taxonomic makeup of the microorganism communities from the rehabilitation and reference sites is strikingly comparable, demonstrating that Urucum Massif's rehabilitation efforts may successfully restore soil microorganisms following iron ore extraction.

Carbonetto et al. (2014) studied soil microbiomes in the Pampas of Argentina, focusing on their structure, content, and metagenomic profile. They found that cultivated soils had more prevalent phyla Gemmatimonadetes, Nitrospirae, and

WS3, while noncultivated soils had more Verrucomicrobia, Plactomycetes, Actinobacteria, and Chloroflexi. Cultivated soils had higher gene abundances related to transcription, protein modification, nucleotide transport, metabolism, and secretion. In addition, they found that both conventional and no-till systems significantly altered the taxonomic composition and organisation of the microbiome. The study reveals that microorganisms' life-history strategies have evolved due to agronomical land usage and tillage systems. Noncultivated soils have oligotrophic microbiomes, unlike fertilized soils with increased nutrient amendment. Although no-tillage methods are less likely to encourage copiotrophy, conventional tillage methods, which decrease the stability of soil organic matter, may be more successful.

7 Metagenomics in understanding soil pollution

The nature of the microbial community can be utilized to analyze the soil's properties, assess potential dangers of polluted soils, and inform prospective methods for remediating the soil using naturally occurring microorganisms. The use of certain rhizosphere bacteria for bio/phytoremediation is a well-established practice (Navarro-Noya et al., 2010; Lundberg et al., 2012). The presence of microbes, the most dynamic part of the soil ecosystem, may serve as a good indicator of how well the soil is holding up against human activities (Kaschuk et al., 2010; Preston-Mafham et al., 2002). Research conducted by Li et al. (2015) discovered that when soil contains excessive amounts of cadmium, it alters the taxonomic diversity and function of the local microbial species. The dominant genera shifted due to increasing bacterial alpha diversity caused by fluctuating heavy metal concentrations near an iron mining location, according to research by Hong et al. (2015). Tonin et al. (2001) found that arbuscular mycorrhizal fungi and plants, which are known to be resistant to heavy metals, can significantly increase the accumulation of heavy metals in plant roots.

The clear identification of the contaminant's cleanup procedure is crucial to effectively tackling the soil issue. The majority of our present understanding of this, however, comes from cultured microbes, which makes it harder to identify the factors that determine metal resistance. Traditional microbiological methods only successfully identify and cultivate a tiny amount (0.1%–1%) of the microorganisms found in soil samples (Amann et al., 1995); hence, they provide only partial information about soil bacteria. Due to their potential role in bioremediation, some uncultivable bacteria might nevertheless transmit metal resistance, which is crucial for the environment. The ambient materials are too close to known genes to be useful for identifying genes using culture-independent approaches; hence, they cannot be used to create new mechanisms to resist metals.

In order to understand the microbial population and their functions in soil poisoned with cadmium, Feng et al. (2018) used a metagenomic approach. The findings indicated that these soils contained a significant number of microorganisms, many of which were capable of performing a wide range of physiological tasks. Cadmium poisoning may lead to a decline in soil microbial diversity, potentially altering community structure. The study of vital microorganisms, such as Proteobacteria, Sulfuricella, and Thiobacillus, revealed that these bacteria are crucial for dealing with Cd pollution. The coding DNA sequences (COG) annotation revealed that the microbial metabolism cluster was dominant in both soil samples, with higher metabolic gene abundance in Cd-contaminated soil. Metabolic pathways involving microbial communities were shown to include fewer genes, pathways, modules, orthologies, and enzymes, according to the Kyoto Encyclopedia of Genes and Genomes (KEGG) annotation. However, Cd-contaminated soil showed an increase in dominant KEGG pathways, primarily enriched for amino acid metabolism, biosynthesis, and degradation, which are connected to microorganisms' Cd tolerance. The taxonomic species of soil microorganisms can decline due to cadmium exposure, and the microbial makeup can change. The microbial structure can alter the functional pathways in the soil. Many of these pathways are connected to the tolerance of soil microorganisms to heavy metals. Exploring bacteria that are cadmium-resistant or tolerant could benefit from the Cd-contaminated soil microbiome.

Yu et al. (2023)utilized metagenomics to assess the impact of residual chlorine on soil microbial diversity and composition. They found no significant long-term effects on the microbial community after 14 days of continuous treatment. However, the rhizosphere microbial community experienced a slight negative impact but returned to normal. The abundances of some resistance genes fluctuated for 7 days before returning to normal. The study suggests that chlorine residue left after daily disinfection may have minimal long-term effects on soil micro-ecology and plant growth, supporting the use of chlorine-containing disinfectants and reducing environmental risks associated with disease control, such as COVID-19.

Li et al. (2023) found that mycorrhiza-mediated recruitment of fully denitrifying Pseudomonas lowers soil N_2O emissions. Arbuscular mycorrhizal fungi (AMF) hyphae decreased denitrification-derived N_2O emissions in carbon- and nitrogen-rich residue patches and increased the quantity and expression of the clade I nosZ gene. The N_2O-reducing Pseudomonas selectively enriched by AMF was associated with decreased N_2O emissions in the hyphosphere and increased abundance of important genes for the bacterial citrate cycle. Upregulated nosZ expression in Pseudomonas fluorescens diminished net N_2O emission ability. The collaboration between AMF and the Pseudomonas that reduce N_2O emissions and are found on hyphae dramatically lowers N_2O emissions in the microsites. *P. fluorescens* is drawn in by the carboxylates that hyphae release, and the *nosZ* gene is stimulated by these carboxylates. The study suggests that enhancing the interactions between AMF and the hyphosphere microbiome could potentially increase N_2O intake in nutrient-rich microsites, thereby reducing soil N_2O emissions and providing new opportunities for sustainable agriculture and climate change mitigation.

Su et al. (2017) utilized metagenomic assembly to study microbial responses to redox variation in acid sulphate soil. They found significant changes in microbial populations in both topsoil and parent material due to aerobic incubation. After incubation, archaea numbers decreased, and multiple sulphide oxidation genes increased, with higher sulphur cycling genes in the parent material layer. Eight important draught genomes were created from ASS thanks to metagenomic assembly. *Thermoplasmatales, Acidothermales* (Acidothermus), and *Acidimicrobiales* new strains make up three of them (GS3, GS6, and GT3). These population genomes' functional gene annotations showed that genes for sulphur cycling and acid tolerance predominated. These results show how microbes respond to environmental change and reveal how they adapt ecologically and how they survive in acid sulphate soils.

Cai et al. (2020) studied the soil microbial community under perfluorooctanoic acid (PFOA) and perfluorooctane sulfonate (PFOS) stress using metagenomic analysis. They found Acidobacteria, Chloroflexi, Actinobacteria, Proteobacteria, Firmicutes, and Gemmatimonadetes to be the most prevalent bacterial phyla, while Sphingomonas and Bacillus reacted poorly. They utilized the KEGG and National Center for Biotechnology Information (NCBI) databases and carbohydrate-active enZYmes (CAZy) to understand soil microbial responses to PFC pollution. PFC stress was found to affect membrane transport, global maps, and glucose metabolism in soil microorganisms. According to a detailed investigation, glycoside hydrolases were significantly inhibited, whereas glycosyl transferases were more active in soils contaminated with pollutants.

8 Metagenomics in understanding bioremediation in soil

Metagenomic techniques are instrumental in identifying uncultivable microorganisms and exploring their potential for bioremediation of soil contaminants. The study of the microbial load in contaminated or polluted soil, as well as its function in bioremediation, can be conducted effectively using metagenomics. The metagenomic study of pollutants bioremediated by soil microbial communities has been based on several research studies. Functional evaluation of crude oil-contaminated soils in Bodo West Community, Ogoni, was carried out by Chikere et al. (2019) using Illumina MiSeq technology. The study compared these soils to other polluted areas. Predictive functional and biomarker studies reveal that bacteria in oil-contaminated environments are involved in various metabolic pathways for breaking down xenobiotic aromatic chemicals. Most of the hydrocarbon degraders found in the oil-polluted soils were long-established members of the groups *Alcanivoracaceae* and *Oceanospirillaceae*. The biomarker taxa included *Oceanicaulis, Sneathiella, Parvibaculum*, and *Sphingobium*. At the Bodo West spill site, compared to the other comparison sites, the bacterial groups *Desulfobacteraceae* and *Acidithiobacillaceae* were differentially more common. The study revealed that the bacterial population in Ogoniland is highly dependent on hydrocarbons, and the acid-tolerant bacterial family significantly contributes to biodegradation at sites with acidic pH, according to the findings.

According to Wang et al. (2021), sulfur cycling and cadmium mobilization under sodium sulfate stress activate soil microorganisms, genes, and metabolites. Their findings point to sulfur-oxidizing bacteria and metal-resistant bacteria. Gene clusters encoding znt, czc, and cad were carried by both Cd and dsr and sqr, respectively. Soil metabolites changed after being exposed to sodium sulfate. The tricarboxylic acid (TCA) cycle was specifically activated by the intensification of sulfur metabolism, which encouraged soil bacteria to secrete carboxylic acids and their precursors. A potential explanation for the observed increase in Cd mobility is the formation of organic acids brought on by high sodium sulfate dosages. Novel insights into microbial processes under salinity stress have been revealed by the improved adaptability of Pseudomonas and Erythrobacter spp. to heavy metal or sulfur-induced stress, because of increased genes and metabolites for sulfur cycling and Cd resistance.

Eyice et al. (2015) discovered uncultivated Methylophilaceae as dimethylsulfide-degrading bacteria in soil and lake sediment using SIP metagenomics. In terrestrial environments, such as agricultural soil and lake silt, they identified bacteria that used dimethyl sulfide (DMS) as a carbon and energy source. The Methylophilaceae family was found to be the predominant DMS-degrading population in soil and lake sediment. Incubations with SIP also yielded Thiobacillus species in 13C-DNA. While gammaproteobacterial communities were detected in the marine environment, the research indicates that betaproteobacterial populations were not recognized as DMS degraders by SIP. *Methylophilaceae* might play a significant role in the cycling of DMS in terrestrial ecosystems.

A model organophosphorus xenobiotic's rate of breakdown in pesticide-contaminated soils is predicted by metagenomic functional potential, according to Jeffries et al. (2018) . The study identified two soil groups with distinct functional and taxonomic profiles, indicating the potential for organophosphorus degradation. The fastest-degrading community showed increased nutrient cycling, transport, and enzymes involved in phosphorus metabolism. This was accompanied by taxonomic changes linked to pollution adaptation and exposure effects. The study suggests the use of holistic, system-wide metagenomic techniques to predict microbial degradation in contaminated habitats.

Wang et al. (2016b) found Actinobacteria in compost-enriched microbial consortiums degrade lignocellulose using metagenomic analysis. Firmicutes, Chloroflexi, Proteobacteria, and Bacteroidetes were the most common phylum among bacteria in the RSA consortium, following Actinobacteria. These bacterial phyla's CAZyme genes were found to be broadly dispersed according to the CAZymes profile. Unexpectedly, the gene repertoires for *pectin lyase, arabinofuranosidase, cellobiohydrolase, acetyl xylan esterase, glucosidase,* and *ligninase* were greatly enlarged in actinomycetal populations. The study reveals that thermophilic Actinobacteria play a crucial role in the biodegradation of lignocellulose in compost habitats, disproving previous theories about their ecological and functional diversity. One new standard for microbial enzymatic deconstruction of lignocelluloses and possible candidates for effective enzyme mining in the biofuel business are actinomycetes found in compost ecosystems.

In order to determine the variety of microbes and their capacity to break down hydrocarbons,(Goma-Tchimbakala et al., 2023) used shotgun metagenomics on soils polluted with diesel and gasoline, as well as soils from auto-mechanic workshops. Over 50 families, including Gordoniaceae and Pseudomonadaceae, were identified in the microbial diversity. There were also 16 classes, with Actinobacteria and Gammaproteobacteria dominating these groups. Gordonia and Pseudomonas were the most prevalent bacterial genera in the two soils, with 57.9% containing alkanes and aromatic compounds. Researchers used HUMANn2 to identify genes and pathways responsible for hydrocarbon degradation. The microbial inhabitants of the two soils may be suitable for bioaugmentation of oil-contaminated soils due to their diverse range of microorganisms with a genetic package for hydrocarbon breakdown.

9 Strength and limitations (biases) in soil metagenomics technologies

Metagenomics has revolutionized microbial ecology by enabling the study of microorganisms directly from environmental samples without the need for culturing. It provides insights into the diversity, structure, and functional potential of microbial communities inhabiting complex ecosystems such as soil. However, soil metagenomics is prone to several methodological biases that arise at different stages of sample processing, DNA extraction, amplification, sequencing, and bioinformatic analysis. These biases can distort the representation of microbial taxa and functions, leading to misleading interpretations of microbial community composition and dynamics.

One of the primary sources of bias originates from the sampling stage. Soil is a highly heterogeneous environment, and microbial populations vary significantly with depth, texture, pH, moisture, and organic matter content. Small sample volumes or inconsistent sampling techniques may fail to capture this variability. Moreover, improper handling and storage, such as temperature fluctuations, delays before DNA extraction, or repeated freeze–thaw cycles, can alter microbial community structures. To minimize sampling bias, it is essential to use composite samples, maintain consistent sampling depths, employ sterile tools, and preserve samples immediately at low temperatures. Another major contributor to methodological bias is DNA extraction. Efficient recovery of high-quality DNA from soil is challenging due to the complex matrix and the presence of inhibitory substances such as humic acids and clay particles. Some microorganisms, such as

Gram-positive bacteria and fungal spores, possess robust cell walls that resist lysis, leading to their underrepresentation in metagenomic data. Variations in commercial extraction kits and protocols further influence the yield and microbial diversity profiles obtained. Optimized extraction methods combining chemical, enzymatic, and mechanical lysis, along with inhibitor removal steps, can improve DNA recovery and reduce this bias.

Polymerase chain reaction (PCR) amplification introduces another layer of bias in metagenomic workflows. Universal primers targeting conserved genes, such as 16S rRNA, may not perfectly match all microbial taxa, resulting in the preferential amplification of certain groups. Differences in Guanine–Cytosine (GC) content, the number of amplification cycles, and the formation of chimeric sequences during PCR can also skew abundance estimates and artificially inflate diversity. These biases can be mitigated by using multiple primer sets, minimizing the number of amplification cycles, employing high-fidelity polymerases, and screening for chimeras using bioinformatic tools. Biases also occur during sequencing due to platform-specific limitations. Different sequencing technologies, such as Illumina, PacBio, and Oxford Nanopore, vary in read length, accuracy, and error rates, influencing the detection of low-abundance taxa. Library preparation steps, like fragmentation and adapter ligation, may selectively include or exclude certain fragments, while uneven sequencing depth may cause dominant taxa to overshadow rare species. Normalizing DNA libraries and ensuring adequate sequencing depth can help achieve a more balanced representation of microbial communities.

The final stage of potential bias lies in bioinformatic analysis. Data processing steps such as quality filtering, sequence trimming, clustering, and taxonomic assignment can significantly alter community composition profiles. Overly stringent filtering may remove valid rare reads, while incomplete or biased reference databases can lead to incorrect taxonomic or functional annotations. The choice between clustering into operational taxonomic units or using amplicon sequence variants also affects diversity estimates. To reduce these biases, researchers should use curated databases, validate pipelines with mock communities, and apply appropriate normalization and statistical methods for comparative analyses. Overall, the cumulative effect of these methodological biases can distort ecological interpretations, especially when comparing datasets generated using different protocols or sequencing platforms. Recognizing and minimizing these biases through standardized protocols, validation using mock microbial communities, and transparent reporting of methods is crucial for improving reproducibility and reliability in soil metagenomics research. In addition, integrating metagenomic data with complementary omics approaches, such as metatranscriptomics or metaproteomics, can help validate functional inferences and provide a more comprehensive understanding of soil microbial ecosystems. Conclusively, careful experimental design, methodological consistency, and bioinformatic rigor are essential to ensure that metagenomic data accurately reflect the true complexity of microbial life in soil environments.

10 Challenges in soil metagenomics and future prospects

The vast variety of soils offers an endless exploration of various industrial, agricultural, and environmental functions. The highest potential for resolving numerous open questions pertaining to soil habitats, changes in microbial abundances with changing edaphic conditions, microbial interactions, diversity of soil microorganisms, and the study of soil microbial communities, including their dynamics over time, climate change effects, resistance, and resilience, is crucial for understanding soil ecology. Despite advancements in metagenomics and molecular techniques, the vast diversity of soil microbes hinders scientists from fully addressing the scope and scale of relevant microbiology questions about this complex ecosystem. In reality, the majority of microorganisms in a soil community still have unknown metabolic roles. It is more common to analyze several samples at a shallow depth using high-throughput sequencing technologies than to fully examine all genomes from a single sample. This has made it easier to define and measure soil diversity.

Methodological biases persist as a significant obstacle in characterizing microbial communities, alongside their high diversity. Biases in molecular surveys and metagenomics include contributions from extracellular DNA, soil extraction methods, nucleic acid adsorption, sample preparation, sequencing techniques, sequence analysis, and functional annotation. Microbial ecologists may struggle to understand the contributions of soils and microbes due to the millions of reads generated by modern sequencing technology. Molecular surveys and metagenomics rely on specific DNA extraction procedures, amplification tactics, sequencing processes, bioinformatic analyses, and databases to provide their quantitative and qualitative results (Nam et al., 2023). Variations in soil composition at the micro-scale level hinder the creation of accurate replicas, as one gram of soil differs from another, and determining the number of species in a single soil sample is also challenging.

The ability to characterize the soil microbiome using soil metagenomics has been demonstrated, but like any technology, it has drawbacks and biases. Molecular approaches, including cell lysis, effective nucleic acid extraction, and sequencing errors, share common drawbacks and biases that hinder accurate gene annotation and environmental gene sequence quantification. In addition, it might be challenging, and perhaps impossible, to attribute detected sequences to species (Prosser, 2015). Data analysis pipelines that can selectively delete or repair incorrect sequences, such as MEGAN, Mothur, QIIME, DADA2, and Deblur, are available to address some of these issues (Callahan et al., 2016; Caporaso et al., 2010; Huson et al., 2007; Schloss et al., 2009; Amir et al., 2017). High-throughput sequencing techniques face the issue of compositional data, where the sequencing tool affects the overall number of sequences created, limiting the analysis and interpretation of the data (Gloor et al., 2017). Working with a physically diverse material is another drawback of using metagenomics in soil research. To effectively extract DNA, thorough dispersion of soil aggregates and cell separation from various soil types and textures are necessary, requiring multiple extractions of a specific soil sample to maximize cell output (Williamson et al., 2011). Repeated DNA extractions can produce cells with unique microorganism assemblages, which would be missed by single-pass extractions. Despite homogenizing soil samples, DNA extractions are typically performed on small subsamples, causing potential bias even with technical replicates.

Soil metagenomics lacks a magic solution, but experimental methods can clarify ecological theories, quantify methodological bias, and provide a reliable research framework. Creating a thorough record of every member of the microbial community and all of their roles for at least one reference soil is a crucial first step in resolving some of the problems that soil microbiologists confront. A comprehensive reference dataset could reveal the unidentified form of soil microbial species frequency distribution and serve as a benchmark for analyzing changes in community composition across soil landscapes, thereby enhancing beta diversity analysis. Through complementary datasets, the reverse engineering of a reference soil can also lead to new findings. Single-cell genomics can help identify unique microbial "dark matter" in aquatic samples by collecting and characterizing cells, as demonstrated by Rinke et al. (2013). To better manage soil resources and employ informed strategies to preserve its functionality, extensive metagenomic study of soil can provide valuable information. To understand soil diversity, it is crucial to define its range and depth, understand strain and species variation, and identify the active portion of the soil community. This includes identifying dominant and rare members, determining their contributions to ecosystem functions, and assessing the variation in the microbial community and soil's physical and chemical composition over time. Anthropogenic inputs, such as fertilizers and pollutants, as well as changes in land use and climate, all have an impact on the soil community's composition, activity, and function. These data are helpful for evaluating and designing strategies to minimize these effects.

11 Conclusions and recommendation

Sequencing the soil metagenome will have significant positive effects on the economy and ecology. Soil metagenomics is a promising approach to improve the understanding of the soil microbial community, which contains numerous genes and pathways that encode novel biocatalysts for biosynthetic or biodegradation activities. This technology can provide crucial knowledge for the preservation of the terrestrial environment and offer insights into the ecology of microorganisms that ensure ecosystem services and crop production. However, substantial sequencing efforts are needed to understand the diversity and function of the soil microbial community and to unravel its complexity. Successful soil metagenomics requires selecting suitable samples, using efficient DNA extraction methods, and performing cloning, screening, and sequencing operations. We can increase our understanding by investigating and assessing the biological diversity of soil in a coordinated manner.

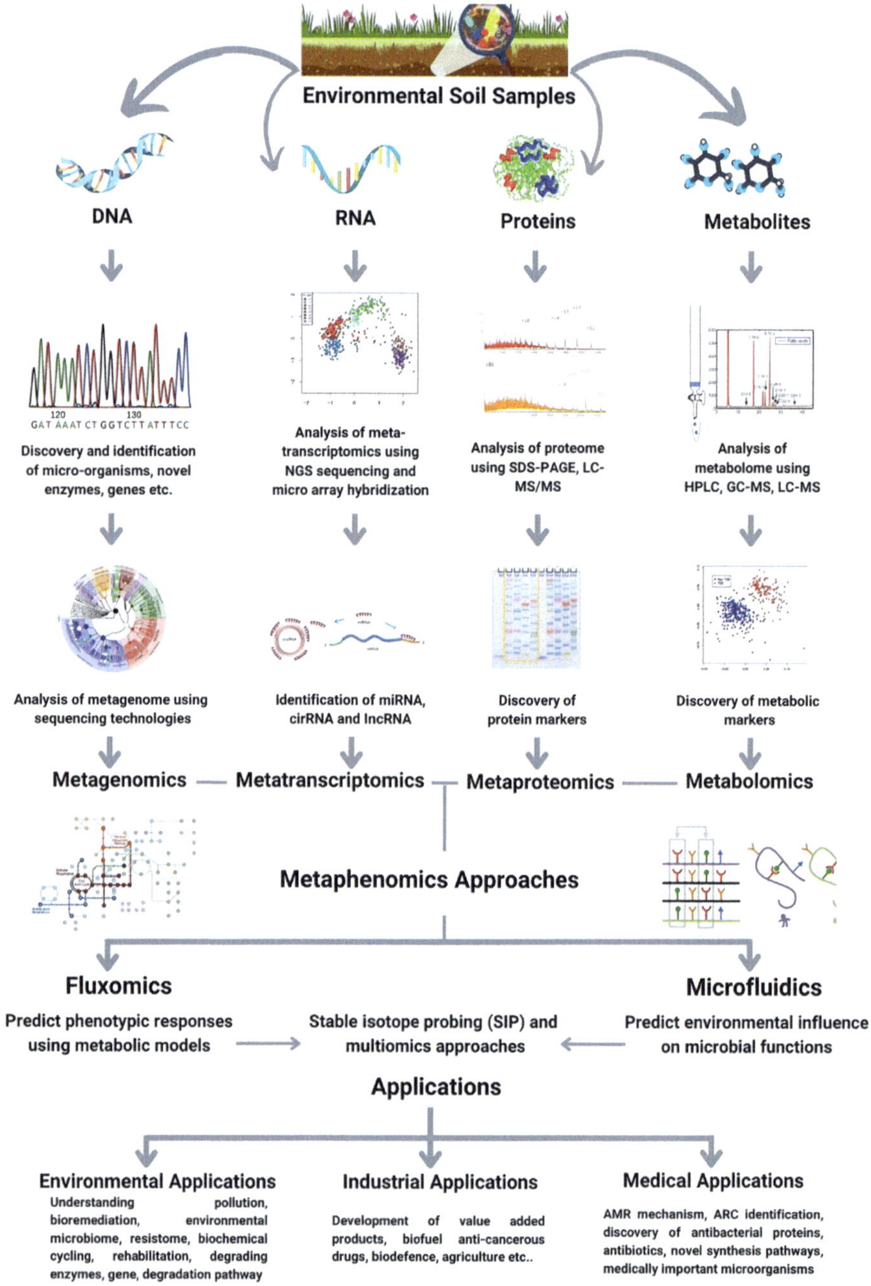

Fig. 12.1 Flow of soil metagenomics.

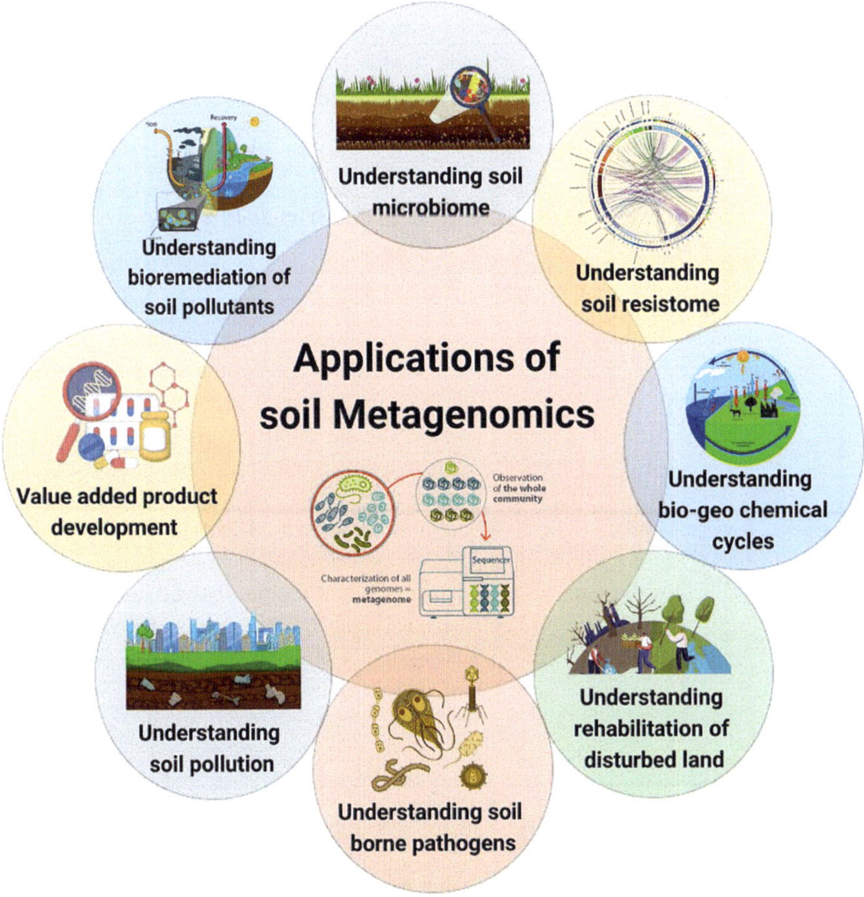

Fig. 12.2 Applications of soil metagenomics.

To better understand microbial dynamics, investigate interactions, and comprehend diversity, researchers need reductionist strategies and extensive research. This includes identifying applications in industry, environment, and agriculture, as well as clarifying microbial adaptation and evolution in soil ecosystems. Microbial ecologists have long relied on insufficient data, such as microbial satellite imaging. Currently, there is no complete list of functional genes and soil microbes for each soil type, and the degree of methodological bias remains uncertain. We still do not fully understand the extent of our ignorance. The increasing number of soil bacteria on Earth, surpassing the number of stars in the universe, suggests that it is time for humans to address soil microbial diversity (Figs. 12.1 and 12.2, and Table 12.1).

Table 12.1 Key applications of metagenomics in managing soil health.

Metagenomic technique	Objective	Microbial communities	Results	Significance	Country	References
Illumina NovaSeq 6000 platform	To learn how soil antibiotic resistance genes (ARGs) have changed due to human-caused shifts in land usage	There were 27 clinical ARGs connected to humans that were carried by plasmids, with most of them being associated with Pseudomonadota. The majority of these genes were found in Pseudomonadota and conferred resistance to tetracycline, aminoglycosides, carbapenems, phenicols, sulfonamides, fluoroquinolones, and MLS.	Areas with a high concentration of human-related ARGs and viruses, as well as those with a favorable correlation to clinical ARGs, were identified in the results	Stress the significance of monitoring the fate of clinical ARGs for a more accurate assessment of how human activities affect the soil resistome	China	(Liao et al., 2022)
Shotgun Metagenome Sequencing	To learn how sewage sludge affects soil resistance gene acquisition and the natural selection of microorganisms resistant to antibiotics	Soils supplemented with sludge showed a high enrichment of read counts for one phylum (Elusimicrobia) and one class (Chloroflexi lineage TK10). When comparing cumulative ARGs across soil treatments, no statistically significant changes emerged. Among the soil and sludge samples, 335 distinct ARGs were identified. Six genes—oqxB, otrA, tetV, aadA2, ant3″-Ia, and another otrA variation (annotated as "tet" in ResFinder)—exceeded the prevalence and abundance threshold values; however, the majority of these genes had few or even single reads.	Soil microbiome composition was unaffected by sewage sludge application, and there was no increase in phenotypic resistance, enrichment of resistance genes, or antibiotic accumulation	Swedish sewage sludge applied to farmland over an extended period does not alter the soil's bacterial population	Sweden	Rutgersson et al. (2020)
Illumina Hiseq Xten Platform	To investigate how the conversion of cropland to plantations on China's Loess Plateau has affected soil respiration	*Actinobacteria*, *Proteobacteria*, *Acidobacteria*, *Chloroflexi*, *Firmicutes*, *Cyanobacteria*, *Gemmatimonadetes*, *Planctomycetes*, and *Verrucomicrobia*	Cropland conversion to plantations altered microbial abundance, which, in turn, affected the abundance of genes involved in carbon metabolism. These changes increased organic matter storage, which encouraged microbial catabolism to produce a significant amount of CO_2 in plantation soil.	Draws attention to the effects of land-use change in this area	China	(Zhang and Lv, 2020)

NextSeq550 (Illumina)	To investigate thermotolerant bacteria capable of degrading aromatic compounds derived from lignin in thermal swamp sediments	In the sphingomonads Altererythrobacter spp. and Novosphingobium spp., as well as in the Burkholderiales species Rubrivivax, metatranscriptomic data showed activation of des/lig genes anticipated to define the metabolism of syringate, vanillate, and phenolic oligomers	A total of 351 unique genomes were recovered from hot-spring (30°C–45°C) and mesophilic (17°C) sediments by assembling shotgun metagenome libraries. Results from the annotation of 39 complete draft genomes showed oligotrophy-consistent metabolism, which includes pathways for the breakdown of aromatic chemicals, including phenol, syringate, vanillate, and p-hydroxybenzoate. A wide variety of biogeochemical and aromatic transformations have been linked to thermotolerant Burkholderiales, such as Rubivivax spp., demonstrating their impressive metabolic versatility.	Emphasize the role of temperature in shaping populations involved in biogeochemical cycles within a specific environment. Altererythrobacter spp. varied aromatic catabolic pathways might also provide thermotolerant enzymes useful for lignin valorization	Canada	Levy-Booth et al. (2021)
Illumina NovaSeq 6000 Platform	In order to learn what makes manured soils' ARG profiles change over time	The bacterial phyla Proteobacteria and Actinobacteria constituted the vast majority of the genera strongly correlated with ARG subtypes, according to the co-occurrence network	Beta-lactam ARGs predominated in all manure-amended soils; however, acidic and near-neutral soils were found to have a greater enrichment of ARGs than alkaline soils. The bacterial community played a crucial role in determining ARG profiles.	Long-term manured soil's antibiotic resistance profiles are shaped by the type of soil. This knowledge could enable initiatives to regulate the dispersion of ARGs in different types of soil	China	Zhang et al. (2021)
6S rRNA Gene Targeted Metagenomics	The goal of this study is to use the cbbL gene to investigate the diversity and abundance of CO_2-fixing bacteria in the Arachis hypogaea rhizosphere	Rhizobium leguminosarum, Bradyrhizobium sp., Sinorhizobium meliloti, Ochrobactrum anthropi, and a number of other cbbL-containing bacteria were found in the majority of the 108 clones isolated from the rhizospheric soil	Firmicutes (34.4%), Proteobacteria (18.3%), Actinobacteria (17.2%), and Bacteroidetes (17.2%) were the most abundant groups in the 16S rRNA gene clone library (16.1%). A little over half of the 16S rRNA gene clone library's nucleotide sequences belonged to new taxa that showed <95% sequence identity with previously reported sequences. According to quantitative real-time PCR (qRT-PCR), the gene copy number of the cbbL gene was 9.38 ±0.75 × 10(7), while the gene copy number of the 16S rRNA gene was 5.43±0.79 × 10(8), both expressed per gram of dry soil.	Further investigation into the role of these CO(2)-fixing bacteria in the carbon cycle, environmentally responsible farming, and environmental management might be undertaken in light of the results, which showed a bacterial community structure characterized by a high degree of bacterial diversity	India	(Yousuf et al., 2012)

Continued

Table 12.1 Key applications of metagenomics in managing soil health—cont'd

Metagenomic technique	Objective	Microbial communities	Results	Significance	Country	References
Shotgun Metagenomic Analysis	Examine the impact of chemical and organic fertilization on soil microbial C and N cycling over time at three locations in Northeast China's black soil region: southern, middle, and northern	In contrast to oligotrophic taxa like Verrucomicrobia, which are inhibited by organic fertilization, the relative abundances of nitrogen- and carbon-cycling Proteobacteria and Planctomycetes are enhanced	Soil carbon breakdown and fixation were unaffected by chemical fertilization; however, the number of genes implicated in methane oxidation increased. In contrast, the use of manure as fertilizer, particularly in conjunction with chemical and organic fertilizers (CFM), significantly decreased the number of genes engaged in carbon fixation (icd) and the reductive acetyl-CoA pathway, while increasing the number of genes involved in the rTCA cycle. Assimilatory and dissimilatory nitrate reductions were enhanced in black soils by manure fertilization, whereas soil nitrogen degradation, nitrification, and anammox genes were overrepresented in chemical fertilization.	The implications of chemical and organic fertilization on soil nutrient cycling in agroecosystems in the black soil area may be foreseen, and our study sheds light on the functional potentials of diverse microbes in this context	China	(Hu et al., 2022)
Shotgun Metagenomics (NextSeq)	To investigate how disturbances affect soil biota	Bacteria accounted for 83.1% of the average metagenomic sequences in disturbance incubation samples, with Eukaryota at 2.7%, Archaea at 0.9%, and viruses at 1.0%. (0.1%)	The functional metagenome and community structure show consistent, sequential changes across time relative to baseline circumstances, which suggests that different soil biotas have different successional niches. One would assume that genes associated with motility would predominate in early successional systems, whereas genes involved in DNA/RNA/protein metabolism, cell division, and the cell cycle would be less common. There were no notable variations in genes related to cell signaling, pathogenicity, or defense, which was unexpected. Furthermore, stress-related genes seemed to be less prevalent in succeeding generations. The early successional system manifested itself via a decrease in taxonomic variety and an increase in functional gene diversity. While the community's traits did evolve during the course of the experiment, they never reverted to their "original" state prior to the disruption	Similar to how large creatures undergo ecosystem succession, this research found that gene functions and taxa follow a predictable pattern after disruption. The time it would take for the system to recover to its condition prior to the disruption is unknown.	Korea	(Choi et al., 2017)

Illumina-MiSeq Platform	The composition and function of eukaryotic microbial communities in soils polluted with heavy metals	Filamoeba, Chrysophyte, Polycostidae, Eimeriidae, and Balantidion were the active genera of soil markers	NGS datasets processed by the Mothur pipeline and analyzed using the Parallel-Meta 3 pipeline illustrated the presence of all the major eukaryotic phyla. A functional metatranscriptomics study revealed the identification of various expressed eukaryotic genes involved in metal tolerance. These metal-tolerant gene families were phylogenetically related to the different eukaryotic lineages reported in the metagenomic analysis.	To better understand the role of active microbes in these types of environments, this approach might find widespread use in microbial ecology	India	(Thakur et al., 2022)
Shotgun Metagenomics	To identify the nitrogen metabolic pathway in the bioremediation of soil contaminated by diesel	Nitrogen from different sources has the potential to significantly change the soil microbiome and the abundance of enzyme genes. The two phyla that were found to be most common in cleaning up diesel-contaminated soil were Actinobacteria and Proteobacteria	The results showed that ammonium-based nitrogen sources significantly accelerated the breakdown of total petroleum hydrocarbons (TPH) (79%–81%) compared to the control treatment (38%) and other nitrogen additions that did not include ammonium (43%–57%). The nitrite reductase enzyme was in limited supply in the nitrate-based amendment, which slowed the decomposition of TPH and restricted the conversion of nitrite to ammonium. An examination of genetic data revealed that the nitrogen metabolism pathway most favorably utilized glutamine as a mediator between ammonium and glutamate. Glutamate synthetase and glutamine synthetase were the enzymes responsible for this transformation.	When applied to diesel-contaminated soil, nitrogen—and especially organic nitrogen sources—significantly enhanced bioremediation	China	Gao et al. (2021)

References

Abraham, B.S., Caglayan, D., Carrillo, N.V., Chapman, M.C., Hagan, C.T., Hansen, S.T., Jeanty, R.O., Klimczak, A.A., Klingler, M.J., Kutcher, T.P., Levy, S.H., Millard-Bruzos, A.A., Moore, T.B., Prentice, D.J., Prescott, M.E., Roehm, R., Rose, J.A., Yin, M., Hyodo, A., Lail, K., Daum, C., Clum, A., Copeland, A., Seshadri, R., Rio, T.G.D., Eloe-Fadrosh, E.A., Benskin, J.B., 2020. Shotgun metagenomic analysis of microbial communities from the Loxahatchee nature preserve in the Florida Everglades. Environ. Microbiomes. 25246372. 15 (1). Available from: https://standardsingenomics.biomedcentral.com/. doi:10.1186/s40793-019-0352-4.

Ahmed, V., Verma, M.K., Gupta, S., Mandhan, V., Chauhan, N.S., 2018. Metagenomic profiling of soil microbes to mine salt stress tolerance genes. Front. Microbiol. 1664302X. 9. Available from: https://www.frontiersin.org/articles/10.3389/fmicb.2018.00159/full. doi:10.3389/fmicb.2018.00159.

Aislabie, J., Deslippe, J.R., Dymond, J., 2013. Soil microbes and their contribution to soil services. In: Ecosystem Services in New Zealand–Conditions and Trends, Vol. 1(12). Manaaki Whenua Press, Lincoln, New Zealand, pp. 143–161.

Amann, R.I., Ludwig, W., Schleifer, K.H., 1995. Phylogenetic identification and in situ detection of individual microbial cells without cultivation. Microbiol. Rev. 59 (1), 143–169.

Amir, A., McDonald, D., Navas-Molina, J.A., Kopylova, E., Morton, J.T., Xu, Z.Z., Kightley, E.P., Thompson, L.R., Hyde, E.R., Gonzalez, A., Knight, R., 2017. Deblur rapidly resolves single-nucleotide community sequence patterns. mSystems. 23795077. 2 (2). http://msystems.asm.org/content/msys/2/2/e00191-16.full.pdf. doi:10.1128/mSystems.00191-16.

Audino, L.D., Louzada, J., Comita, L., 2014. Dung beetles as indicators of tropical forest restoration success: is it possible to recover species and functional diversity? Biol. Conserv. 169, 248–257.

Bastida, F., Jehmlich, N., Martínez-Navarro, J., Bayona, V., García, C., Moreno, J.L., 2019. The effects of struvite and sewage sludge on plant yield and the microbial community of a semiarid Mediterranean soil. Geoderma. 00167061. 337, 1051–1057. doi:10.1016/j.geoderma.2018.10.046.

Berendsen, R.L., Pieterse, C.M., Bakker, P.A., 2012. The rhizosphere microbiome and plant health. Trends Plant Sci. 17 (8), 478–486.

Berini, F., Casciello, C., Marcone, G.L., Marinelli, F., 2017. Metagenomics: novel enzymes from non-culturable microbes. FEMS Microbiol. Lett. 15746968. 364 (21). Available from: http://femsle.oxfordjournals.org/. doi:10.1093/femsle/fnx211.

Braga, L.P.P., Alves, R.F., Dellias, M.T.F., Navarrete, A.A., Basso, T.O., Tsai, S.M., 2017. Vinasse fertirrigation alters soil resistome dynamics: an analysis based on metagenomic profiles. BioData Min. 17560381. 10 (1). Available from: http://www.biodatamining.org/. doi:10.1186/s13040-017-0138-4.

Brierley, A.S., Fernandes, P.G., Brandon, M.A., Armstrong, F., Millard, N.W., McPhail, S.D., Stevenson, P., Pebody, M., Perrett, J., Squires, M., Bone, D.G., 2002. Antarctic krill under sea ice: elevated abundance in a narrow band just south of ice edge. Science 295 (5561), 1890–1892.

Cabrera-Rodríguez, A., Trejo-Calzada, R., García-de la Peña, C., Arreola-Ávila, J.G., Nava-Reyna, E., Vaca-Paniagua, F., Díaz-Velásquez, C., Meza-Herrera, C.A., 2020. A metagenomic approach in the evaluation of the soil microbiome in coffee plantations under organic and conventional production in tropical agroecosystems. Emir. J. Food Agric. 32 (4) , 263–270.

Cai, Y., Chen, H., Yuan, R., Wang, F., Chen, Z., Zhou, B., 2020. Metagenomic analysis of soil microbial community under PFOA and PFOS stress. Environ. Res. 00139351. 188, 109838. doi:10.1016/j.envres.2020.109838.

Callahan, B.J., McMurdie, P.J., Rosen, M.J., Han, A.W., Johnson, A.J.A., Holmes, S.P., 2016. DADA2: high-resolution sample inference from Illumina amplicon data. Nat. Methods. 15487105. 13 (7), 581–583. Available from: http://www.nature.com/nmeth/. doi:10.1038/nmeth.3869.

Callaway, E., 2016. Devastating wheat fungus appears in Asia for first time. Nature. 0028-0836. 532 (7600), 421–422. doi:10.1038/532421a.

Caporaso, J.G., Kuczynski, J., Stombaugh, J., Bittinger, K., Bushman, F.D., Costello, E.K., Fierer, N., Pẽa, A.G., Goodrich, J.K., Gordon, J.I., Huttley, G.A., Kelley, S.T., Knights, D., Koenig, J.E., Ley, R.E., Lozupone, C.A., McDonald, D., Muegge, B.D., Pirrung, M., Reeder, J., Sevinsky, J.R., Turnbaugh, P.J., Walters, W.A., Widmann, J., Yatsunenko, T., Zaneveld, J., Knight, R., 2010. QIIME allows analysis of high-throughput community sequencing data. Nat. Methods. 15487105. 7 (5), 335–336. Available from: https://www.nature.com/nmeth/. doi:10.1038/nmeth.f.303.

Carbonetto, B., Rascovan, N., Álvarez, R., Mentaberry, A., Vázquez, M.P., Structure, K.T., 2014. Composition and metagenomic profile of soil microbiomes associated to agricultural land use and tillage systems in argentine pampas. PLoS ONE. 1932-6203. 9 (6), e99949. doi:10.1371/journal.pone.0099949.

Catania, V., Bueno, R.S., Alduina, R., Grilli, E., La Mantia, T., Castaldi, S., Quatrini, P., 2022. Soil microbial biomass and bacterial diversity in southern European regions vulnerable to desertification. Ecol. Indic. 145, 109725.

Chikere, C.B., Mordi, I.J., Chikere, B.O., Selvarajan, R., Ashafa, T.O., Obieze, C.C., 2019. Comparative metagenomics and functional profiling of crude oil-polluted soils in Bodo West Community, Ogoni, with other sites of varying pollution history. Ann. Microbiol. 18692044. 69 (5), 495–513. Available from: http://springerlink.com/content/121707/. doi:10.1007/s13213-019-1438-3.

Choi, S., Song, H., Tripathi, B.M., Kerfahi, D., Kim, H., Adams, J.M., 2017. Effect of experimental soil disturbance and recovery on structure and function of soil community: a metagenomic and metagenetic approach. Sci. Rep. 20452322. 7 (1). Available from: www.nature.com/srep/index.html. doi:10.1038/s41598-017-02262-6.

Dai, Z., Zang, H., Chen, J., Fu, Y., Wang, X., Liu, H., Shen, C., Wang, J., Kuzyakov, Y., Becker, J.N., Hemp, A., Barberán, A., Gunina, A., Chen, H., Luo, Y., Xu, J., 2021. Metagenomic insights into soil microbial communities involved in carbon cycling along an elevation climosequences. Environ. Microbiol. 14622920. 23 (8), 4631–4645. Available from: http://onlinelibrary.wiley.com/journal/10.1111/(ISSN)1462-2920. doi:10.1111/1462-2920.15655.

Dantas, G., Sommer, M.O.A., Oluwasegun, R.D., Church, G.M., 2008. Bacteria subsisting on antibiotics. Science. 10959203. 320 (5872), 100–103. doi:10.1126/science.1155157.

Eaton, W.D., Shokralla, S., McGee, K.M., Hajibabaei, M., Chain, F.J.J., 2017. Using metagenomics to show the efficacy of forest restoration in the New Jersey Pine Barrens. Genome. 08312796. 60 (10), 825–836. Available from: http://pubs.nrc-cnrc.gc.ca. doi:10.1139/gen-2015-0199.

Eyiçe, Ö., Namura, M., Chen, Y., Mead, A., Samavedam, S., Schäfer, H., 2015. SIP metagenomics identifies uncultivated Methylophilaceae as dimethylsulphide degrading bacteria in soil and lake sediment. ISME J. 1751-7362. 9 (11), 2336–2348. doi:10.1038/ismej.2015.37.

Feng, G., Xie, T., Wang, X., Bai, J., Tang, L., Zhao, H., Wei, W., Wang, M., Zhao, Y., 2018. Metagenomic analysis of microbial community and function involved in cd-contaminated soil. BMC microbiology, 18 (1), 11.

Ferrer, M., Nechitaylo, T.Y., Golyshin, P.N., 2013. Terrestrial invertebrate animal metagenomics, earthworms. In: Encyclopedia of Metagenomics: Animal Metagenomics. Springer. pp. 1–9.

Fierer, N., Leff, J.W., Adams, B.J., Nielsen, U.N., Bates, S.T., Lauber, C.L., Owens, S., Gilbert, J.A., Wall, D.H., Gregory Caporaso, J., 2012. Cross-biome metagenomic analyses of soil microbial communities and their functional attributes. Proc. Natl. Acad. Sci. USA. 0027-8424. 109 (52), 21390–21395. doi:10.1073/pnas.1215210110.

Gao, Y., Du, J., Bahar, M.M., Wang, H., Subashchandrabose, S., Duan, L., Yang, X., Megharaj, M., Zhao, Q., Zhang, W., Liu, Y., 2021. Metagenomics analysis identifies nitrogen metabolic pathway in bioremediation of diesel contaminated soil. Chemosphere, 271, 129566.

Garris, H.W., Baldwin, S.A., Van Hamme, J.D., Gardner, W.C., Fraser, L.H., 2016. Genomics to assist mine reclamation: a review. Restor. Ecol. 1526100X. 24 (2), 165–173. Available from: http://www.blacksci.co.uk/~cgilib/bsinc.bin?Journal=ecology. doi:10.1111/rec.12322.

Gastauer, M., Silva, J.R., Junior, C.F.C., Ramos, S.J., Filho, P.W.M.S., Neto, A.E.F., Siqueira, J.O., 2018. Mine land rehabilitation: modern ecological approaches for more sustainable mining. J. Clean. Prod. 09596526. 172, 1409–1422. doi:10.1016/j.jclepro.2017.10.223.

Gastauer, M., Souza Filho, P.W.M., Ramos, S.J., Caldeira, C.F., Silva, J.R., Siqueira, J.O., Neto, A.E.F., 2019b. Mine land rehabilitation in Brazil: goals and techniques in the context of legal requirements. Ambio. 16547209. 48 (1), 74–88. Available from: http://www.bioone.org/loi/ambi. doi:10.1007/s13280-018-1053-8.

Gastauer, M., Vera, M.P.O., De Souza, K.P., Pires, E.S., Alves, R., Caldeira, C.F., Ramos, S.J., Oliveira, G., 2019a. A metagenomic survey of soil microbial communities along a rehabilitation chronosequence after iron ore mining. Sci. Data 6 (1), 1–10.

Ghosh, A., Mehta, A., Khan, A.M., 2019. Metagenomic analysis and its applications.

Gloor, G.B., Macklaim, J.M., Pawlowsky-Glahn, V., Egozcue, J.J., 2017. Microbiome datasets are compositional: and this is not optional. Front. Microbiol. 1664302X. 8. Available from: https://www.frontiersin.org/articles/10.3389/fmicb.2017.02224/full. doi:10.3389/fmicb.2017.02224.

Goma-Tchimbakala, E.J.C.D., Pietrini, I., Goma-Tchimbakala, J., Corgnati, S.P., 2023. Use of shotgun metagenomics to assess the microbial diversity and hydrocarbons degrading functions of auto-mechanic workshops soils polluted with gasoline and diesel fuel. Microorganisms. 20762607. 11 (3). Available from: www.mdpi.com/journal/microorganisms. doi:10.3390/microorganisms11030722.

Gudiña, E.J., Amorim, C., Braga, A., Costa, Â., Rodrigues, J.L., Silvério, S., Rodrigues, L.R., 2020. Biotech green approaches to unravel the potential of residues into valuable products. In: Sustainable Green Chemical Processes and Their Allied Applications. Springer International Publishing, Cham, pp. 97–150.

Hamonts, K., Bissett, A., Macdonald, B.C.T., Barton, P.S., Manning, A.D., Young, A., 2017. Effects of ecological restoration on soil microbial diversity in a temperate grassy woodland. Appl. Soil Ecol. 09291393. 117–118, 117–128. Available from: www.elsevier.com/inca/publications/store/5/2/4/5/1/8/index.htt. doi:10.1016/j.apsoil.2017.04.005.

Hong, C., Si, Y., Xing, Y., Li, Y., 2015. Illumina MiSeq sequencing investigation on the contrasting soil bacterial community structures in different iron mining areas. Environ. Sci. Pollut. Res. 22 (14), 10788–10799.

Hu, X., Gu, H., Liu, J., Wei, D., Zhu, P., Cui, X.A., Zhou, B., Chen, X., Jin, J., Liu, X., Wang, G., 2023. Metagenomic strategies uncover the soil bioavailable phosphorus improved by organic fertilization in Mollisols. Agric. Ecosyst. Environ. 349, 108462.

Hu, X., Gu, H., Liu, J., Wei, D., Zhu, P., Cui, X., Zhou, B., Chen, X., Jin, J., Liu, X., Wang, G., 2022. Metagenomics reveals divergent functional profiles of soil carbon and nitrogen cycling under long-term addition of chemical and organic fertilizers in the black soil region. Geoderma. 00167061. 418, 115846. doi:10.1016/j.geoderma.2022.115846.

Huson, D.H., Auch, A.F., Qi, J., Schuster, S.C., 2007. MEGAN analysis of metagenomic data. Genome Res. 15495469. 17 (3), 377–386. Available from: http://www.genome.org/cgi/reprint/17/3/377. doi:10.1101/gr.5969107Germany.

Jansson, J.K., Hofmockel, K.S., 2020. Soil microbiomes and climate change. Nat. Rev. Microbiol. 17401534. 18 (1), 35–46. Available from: http://www.nature.com/nrmicro/index.html. doi:10.1038/s41579-019-0265-7.

Jeffries, T.C., Rayu, S., Nielsen, U.N., Lai, K., Ijaz, A., Nazaries, L., Singh, B.K., 2018. Metagenomic functional potential predicts degradation rates of a model organophosphorus xenobiotic in pesticide contaminated soils. Front. Microbiol. 1664302X. 9. Available from: https://www.frontiersin.org/articles/10.3389/fmicb.2018.00147/full. doi:10.3389/fmicb.2018.00147.

Kaschuk, G., Alberton, O., Hungria, M., 2010. Three decades of soil microbial biomass studies in Brazilian ecosystems: lessons learned about soil quality and indications for improving sustainability. Soil Biol. Biochem. 42 (1), 1–13.

Levy-Booth, D.J., Hashimi, A., Roccor, R., Liu, L.Y., Renneckar, S., Eltis, L.D., Mohn, W.W., 2021. Genomics and metatranscriptomics of biogeochemical cycling and degradation of lignin-derived aromatic compounds in thermal swamp sediment. ISME J. 15 (3), 879–893.

Li, Y., Chen, L., Wen, H., 2015. Changes in the composition and diversity of bacterial communities 13 years after soil reclamation of abandoned mine land in eastern China. Ecol. Res. 30, 357–366.

Liao, H., Li, H., Duan, C.S., Zhou, X.Y., An, X.L., Zhu, Y.G., Su, J.Q., 2022. Metagenomic and viromic analysis reveal the anthropogenic impacts on the plasmid and phage borne transferable resistome in soil. Environ. Int. 18736750. 170. Available from: www.elsevier.com/locate/envint. doi:10.1016/j.envint.2022.107595.

Liu, Z., Zhao, Y., Zhang, B., Wang, J., Zhu, L., Hu, B., 2023. Deterministic effect of pH on shaping soil resistome revealed by metagenomic analysis. Environ. Sci. Technol. 57 (2), 985–996.

Li, Zhao, Li, Wang, Bei, Ju, An, Li, Kuyper, Christie, Bender, 2023. Mycorrhiza-mediated recruitment of complete denitrifying Pseudomonas reduces N2O emissions from soil. Microbiome 11 (1), 45.

Lundberg, D.S., Lebeis, S.L., Paredes, S.H., Yourstone, S., Gehring, J., Malfatti, S., Tremblay, J., Engelbrektson, A., Kunin, V., Rio, T.G.D., Edgar, R.C., Eickhorst, T., Ley, R.E., Hugenholtz, P., Tringe, S.G., Dangl, J.L., 2012. Defining the core Arabidopsis thaliana root microbiome. Nature. 14764687. 488 (7409), 86–90. Available from: http://www.nature.com/nature/index.html. doi:10.1038/nature11237.

Muñoz-Arenas, L.C., Fusaro, C., Hernández-Guzmán, M., Dendooven, L., Estrada-Torres, A., Navarro-Noya, Y.E., 2020. Soil microbial diversity drops with land-use change in a high mountain temperate forest: a metagenomics survey. Environ. Microbiol. Rep. 12 (2), 185–194.

Nam, N.N., Do, H.D.K., Loan Trinh, K.T., Lee, N.Y., 2023. Metagenomics: an effective approach for exploring microbial diversity and functions. Foods, 12 (11), 2140.

Navarro-Noya, Y.E., Jan-Roblero, J., Del Carmen González-chávez, M., Hernández-Gama, R., Hernández-Rodríguez, C., 2010. Bacterial communities associated with the rhizosphere of pioneer plants (Bahia xylopoda and Viguiera linearis) growing on heavy metals-contaminated soils. Antonie van Leeuwenhoek. 0003-6072. 97 (4), 335–349. doi:10.1007/s10482-010-9413-9.

Nazir, A., 2016. Review on metagenomics and its applications. Imp J Intersd Res 2 (10).

Neelakanta, G., Sultana, H., 2013. The use of metagenomic approaches to analyze changes in microbial communities. Microbiol. Insights. 1178-6361. 6. doi:10.4137/mbi.s10819.

Pearlman, B.L., 2012. Protease inhibitors for the treatment of chronic hepatitis C genotype-1 infection: the new standard of care. Lancet Infect. Dis. 14744457. 12 (9), 717–728. doi:10.1016/S1473-3099(12)70060-9.

Perring, M.P., Standish, R.J., Price, J.N., Craig, M.D., Erickson, T.E., Ruthrof, K.X., Whiteley, A.S., Valentine, L.E., Hobbs, R.J., 2015. Advances in restoration ecology: rising to the challenges of the coming decades. Ecosphere. 21508925. 6 (8). Available from: http://www.esajournals.org/loi/ecsp. doi:10.1890/ES15-00121.1.

Preston-Mafham, J., Boddy, L., Randerson, P.F., 2002. Analysis of microbial community functional diversity using sole-carbon-source utilisation profiles—a critique. FEMS Microbiol. Ecol. 42 (1), 1–14.

Prosser, J.I., 2015. Dispersing misconceptions and identifying opportunities for the use of 'omics' in soil microbial ecology. Nat. Rev. Microbiol. 17401534. 13 (7), 439–446. Available from: http://www.nature.com/nrmicro/index.html. doi:10.1038/nrmicro3468.

Qian, X., Gunturu, S., Guo, J., Chai, B., Cole, J.R., Gu, J., Tiedje, J.M., 2021. Metagenomic analysis reveals the shared and distinct features of the soil resistome across tundra, temperate prairie, and tropical ecosystems. Microbiome, 9 (1), 108.

Reece, S.M., Sinha, A., Grieshop, A.P., 2017. Primary and photochemically aged aerosol emissions from biomass cookstoves: chemical and physical characterization. Environ. Sci. Technol. 51 (16), 9379–9390.

Rinke, C., Schwientek, P., Sczyrba, A., Ivanova, N.N., Anderson, I.J., Cheng, J.F., Darling, A., Malfatti, S., Swan, B.K., Gies, E.A., Dodsworth, J.A., 2013. Insights into the phylogeny and coding potential of microbial dark matter. Nature, 499 (7459), 431-437.

Rutgersson, C., Ebmeyer, S., Lassen, S.B., Karkman, A., Fick, J., Kristiansson, E., Larsson, D.J., 2020. Long-term application of Swedish sewage sludge on farmland does not cause clear changes in the soil bacterial resistome. Environ. Int. 137, 105339.

Schloss, P.D., Westcott, S.L., Ryabin, T., Hall, J.R., Hartmann, M., Hollister, E.B., Lesniewski, R.A., Oakley, B.B., Parks, D.H., Robinson, C.J., Sahl, J.W., Stres, B., Thallinger, G.G., Van Horn, D.J., Weber, C.F., 2009. Introducing mothur: open-source, platform-independent, community-supported software for describing and comparing microbial communities. Appl. Environ. Microbiol. 10985336. 75 (23), 7537–7541. Available from: http://aem.asm.org/cgi/reprint/75/23/7537. doi:10.1128/AEM.01541-09.

Sousa, J., Silvério, S.C., Costa, A.M., Rodrigues, L.R., 2022. Metagenomic approaches as a tool to unravel promising biocatalysts from natural resources: soil and water. Catalysts 12 (4), 385.

Starke, R., Jehmlich, N., Bastida, F., 2019. Using proteins to study how microbes contribute to soil ecosystem services: the current state and future perspectives of soil metaproteomics. J. Proteomics 198, 50–58.

Su, J.Q., Xia, Y., Yao, H.Y., Li, Y.Y., An, X.L., Singh, B.K., Zhang, T., Zhu, Y.G., 2017. Metagenomic assembly unravel microbial response to redox fluctuation in acid sulfate soil. Soil Biol. Biochem. 00380717. 105, 244–252. Available from: www.elsevier.com/inca/publications/store/3/3/2. doi:10.1016/j.soilbio.2016.11.027.

Thakur, B., Yadav, R.K., Marmeisse, R., Prashanth, S., Krishnamohan, M., Tachct, L.F., Reddy, M.S., 2022. Metagenomic analysis of heavy metal-contaminated soils reveals distinct clades with adaptive features. Int. J. Environ. Sci. Technol. 20 (11) , 12155-12166.

Tonin, C., Vandenkoornhuyse, P., Joner, E.J., Straczek, J., Leyval, C., 2001. Assessment of arbuscular mycorrhizal fungi diversity in the rhizosphere of Viola calaminaria and effect of these fungi on heavy metal uptake by clover. Mycorrhiza. 0940-6360. 10 (4), 161–168. doi:10.1007/s005720000072.

Torres-Cortés, G., Millán, V., Ramírez-Saad, H.C., Nisa-Martínez, R., Toro, N., Martínez-Abarca, F., 2011. Characterization of novel antibiotic resistance genes identified by functional metagenomics on soil samples. Environ. Microbiol. 14622920. 13 (4), 1101–1114. doi:10.1111/j.1462-2920.2010.02422.x.

Tripathi, M., Singh, D., Vikram, S., Singh, V., Kumar, S., 2018. Metagenomic approach towards bioprospection of novel biomolecule(s) and environmental bioremediation. Annu. Res. Rev. Biol. 2347565X. 22 (2), 1–12. doi:10.9734/arrb/2018/38385.

Valentini, A., Taberlet, P., Miaud, C., Civade, R., Herder, J., Thomsen, P.F., Bellemain, E., Besnard, A., Coissac, E., Boyer, F., Gaboriaud, C., 2016. Next-generation monitoring of aquatic biodiversity using environmental DNA metabarcoding. Mol. Ecol. 25 (4), 929–942.

Van Der Heijden, M.G., Bardgett, R.D., Van Straalen, N.M., 2008. The unseen majority: soil microbes as drivers of plant diversity and productivity in terrestrial ecosystems. Ecol. Lett. 11 (3), 296–310.

Vogel, T.M., Simonet, P., Jansson, J.K., Hirsch, P.R., Tiedje, J.M., van Elsas, J.D., Bailey, M.J., Nalin, R., Philippot, L., 2009. TerraGenome: a consortium for the sequencing of a soil metagenome. Nat. Rev. Microbiol. 17401526. 7 (4), 252. doi:10.1038/nrmicro2119.

Wang, F., Stedtfeld, R.D., Kim, O.S., Chai, B., Yang, L., Stedtfeld, T.M., Hong, S.G., Kim, D., Lim, H.S., Hashsham, S.A., Tiedje, J.M., 2016b. Influence of soil characteristics and proximity to Antarctic research stations on abundance of antibiotic resistance genes in soils. Environ. Sci. Technol. 50 (23), 12621–12629.

Wang, M., Wang, L., Shi, H., Liu, Y., Chen, S., 2021. Soil bacteria, genes, and metabolites stimulated during sulfur cycling and cadmium mobilization under sodium sulfate stress. Environ. Res. 00139351. 201, 111599. doi:10.1016/j.envres.2021.111599.

White III, R.A., Garoutte, A., Mclachlan, E.E., Tiemann, L.K., Evans, S., Friesen, M.L., 2023. Genome-Resolved Metagenomics of Nitrogen Transformations in the Switchgrass Rhizosphere Microbiome on Marginal Lands.

Williamson, K.E., Kan, J., Polson, S.W., Williamson, S.J., 2011. Optimizing the indirect extraction of prokaryotic DNA from soils. Soil Biol. Biochem. 00380717. 43 (4), 736–748. doi:10.1016/j.soilbio.2010.04.017.

Willms, I.M., Kamran, A., Aßmann, N.F., Krone, D., Bolz, S.H., Fiedler, F., Nacke, H., 2019. Discovery of novel antibiotic resistance determinants in forest and grassland soil metagenomes. Front. Microbiol. 10, 460.

Wilmes, P., Heintz-Buschart, A., Bond, P.L., 2015. A decade of metaproteomics: where we stand and what the future holds. Proteomics 15 (20), 3409–3417.

Wilpiszeski, R.L., Aufrecht, J.A., Retterer, S.T., Sullivan, M.B., Graham, D.E., Pierce, E.M., Zablocki, O.D., Palumbo, A.V., Elias, D.A., 2019. Soil aggregate microbial communities: towards understanding microbiome interactions at biologically relevant scales. Appl. Environ. Microbiol. 10985336. 85 (14). Available from: https://journals.asm.org/doi/pdf/10.1128/aem.00324-19. doi:10.1128/AEM.00324-19.

Wiryawan, A., Eginarta, W.S., Hermanto, F.E., Ustiatik, R., Dinira, L., Mustafa, I., 2022. Changes in essential soil nutrients and soil disturbance directly affected soil microbial community structure: a metagenomic approach. Polskie Towarzystwo Inzynierii Ekologicznej (PTIE) Indones. J. Ecol. Eng. 22998993. 23 (7), 238–245. Available from: http://www.jeeng.net. doi:10.12911/22998993/149972.

Wu, X., Peng, J., Liu, P., Bei, Q., Rensing, C., Yong, L., Yuan, H., Liesack, W., Zhang, F., Cui, Z., 2021. Metagenomic insights into nitrogen and phosphorus cycling at the soil aggregate scale driven by organic material amendments. Sci. Total Environ. 00489697. 785, 147329. doi:10.1016/j.scitotenv.2021.147329.

Xiao, K.Q., Li, B., Ma, L., Bao, P., Zhou, X., Zhang, T., Zhu, Y.G., 2016. Metagenomic profiles of antibiotic resistance genes in paddy soils from South China. FEMS Microbiol. Ecol. 92 (3), fiw023.

Yousuf, B., Keshri, J., Mishra, A., Jha, B., 2012. Application of targeted metagenomics to explore abundance and diversity of CO2-fixing bacterial community using cbbL gene from the rhizosphere of Arachis hypogaea. Gene. 03781119. 506 (1), 18–24. doi:10.1016/j.gene.2012.06.083.

Yu, Y., Zhang, Q., Zhang, Z., Xu, N., Li, Y., Jin, M., Feng, G., Qian, H., Lu, T., 2023. Assessment of residual chlorine in soil microbial community using metagenomics. Soil Ecol. Lett. 5 (1), 66–78.

Zhang, L., Lv, J., 2020. Metagenomic analysis of microbial community and function reveals the response of soil respiration to the conversion of cropland to plantations in the Loess Plateau of China. Glob. Ecol. Conserv. 23519894. 23, e01067. doi:10.1016/j.gecco.2020.e01067.

Zhang, M., Li, X., Li, M., Yin, P., 2022. Effects of Grazing on Soil Organic Carbon in the Rhizosphere of Stipa Grandis in a Typical Steppe of Inner Mongolia, China. Sustainability, 14 (19), 11866.

Zhang, Y., Cheng, D., Zhang, Y., Xie, J., Xiong, H., Wan, Y., Shi, X., 2021. Soil type shapes the antibiotic resistome profiles of long-term manured soil. Sci. Total Environ. 786, 147361.

Next-gen data-driven precision irrigation: Application of artificial intelligence with in-situ and remote sensing measurements for management zones and variable-rate irrigation technologies

13

Ronald Singh[a], *Sushil Kumar Himanshu*[d], *and Hemendra Kumar*[a,b,c]
[a]Digital and Precision Agriculture Lab, University of Maryland, College Park, MD, United States
[b]College of Agriculture and Natural Resources, University of Maryland, College Park, MD, United States
[c]Central Maryland Research and Education Center, University of Maryland Extension, Upper Marlboro, MD, United States
[d]Agricultural Systems and Engineering, Faculty of Food, Agriculture and Natural Resources, Asian Institute of Technology, Pathum Thani, Thailand

1 Introduction

Water is one of the most important agricultural inputs for crop growth, productivity, and sustainability. However, increasing water scarcity, exacerbated by climate change, has made efficient irrigation practices central to ensuring food security, making effective water management a necessity for modern agriculture (Kumar et al., 2021, 2022a). Optimized irrigation through precision water management can significantly improve water-use efficiency and enhance yield outcomes.

Conventional irrigation systems, which assume homogeneous field conditions, often lead to inefficiencies such as overwatering or under-irrigation. Some regions experience water shortages that affect agricultural yields, while other areas receive excess irrigation, causing surface runoff, waterlogging, and nutrient leaching. New evidence suggests that variable rate irrigation (VRI) and other precision irrigation frameworks are vital for more efficient and precise irrigation water management (Kumar et al., 2022b; 2023a). As such, one precision agriculture technology that

Precision Technologies for Digital Agriculture. https://doi.org/10.1016/B978-0-443-40513-6.00014-7

has been developed to overcome these issues is VRI, which enables site-specific water application based on field variability (O'Shaughnessy et al., 2016; Kumar et al., 2022b; 2023a).

Management zones (MZs) define land units within a field that exhibit homogeneous soil texture, topography, moisture levels, and crop response characteristics. These zones allow for precise VRI applications by isolating treated subunits, ensuring efficient water use and optimized crop performance. Their implementation is crucial for sustainable and resource-efficient agricultural management practices. Traditional methods, like soil sampling alone, may lack the capacity to capture spatiotemporal variability (Shan et al., 2021a). Recent advancements in remote sensing and geospatial data fusion technologies have enhanced the delineation of MZs, enabled more systematic and data-driven irrigation scheduling, and improved water-use efficiency. Inefficient zone segmentation can lead to suboptimal irrigation practices, adversely affecting agricultural productivity and the sustainable use of vital agricultural resources such as water, soil nutrients, and energy inputs (labor and technology). Historically, approaches like remote sensing, electrical conductivity (EC) mapping, and soil sampling have been used to characterize MZs (Li et al., 2020a, 2020b). However, these approaches frequently require manual interpretation, and they might not adequately reflect the dynamic interplay of crop physiology, soil, and climate (Shan et al., 2021b). The accuracy of MZ delineation has been significantly refined by consolidating machine learning (ML) algorithms with geostatistical techniques and high-resolution remote sensing data, specifically through the combined use of variables such as the normalized difference vegetation index (NDVI) for vegetation vigor, soil EC for identifying soil moisture and texture variability (Morari et al., 2021), evapotranspiration (ET) rates derived from thermal remote sensing (Kustas et al., 2022), and terrain metrics like slope, aspect, and topographic wetness index (TWI) extracted from digital elevation models (DEMs) (Kumar et al., 2022a, 2023b; Yousaf et al., 2024).

Advancements in artificial intelligence (AI) and ML enable scientists and farmers to analyze complex agricultural data more effectively, enhancing irrigation management based on creating MZs (Liu et al., 2018). ML approaches, such as clustering algorithms (e.g., k-means, fuzzy c-means), decision trees, deep learning (DL), and ensemble methods, enable precise categorization of field conditions based on real-time and historical data (Sun et al., 2022). To produce data-driven insights for adaptive irrigation decision-making, these algorithms combine data from various sources, such as weather stations, soil moisture sensors, unmanned aerial vehicle (UAV)-based hyperspectral data, and satellite imagery (Feng et al., 2019; Ale et al., 2023; Gade et al., 2025).

ML algorithms improve irrigation scheduling and enhance MZs delineation by forecasting crop water demand based on ET rates, soil moisture levels, and weather projections (Ojha, 2022). DL-based recurrent neural networks have proven effective in capturing the temporal dynamics of soil moisture, ET, and crop water stress, enabling more accurate simulation of water stress fluctuations and optimized irrigation scheduling, thereby enhancing water-use efficiency and maintaining or increasing crop yields (Rahman, 2021). Researchers have shown that self-adaptive irrigation systems utilize reinforcement learning methods, which enable AI to refine irrigation strategies based on environmental input and ancillary data (Kamilaris and Prenafeta-Boldú, 2018). ML-driven VRI systems still face multiple challenges, including data quality problems and model interpretability issues, which need to be resolved before these systems can be implemented on a larger scale. The key barriers include data quality, model interpretability, computing expenses, and challenges with real-time system integration. Although advancements have been made, these barriers still exist ahead of the large-scale implementation of ML-based VRI systems. The full potential of AI in precision irrigation requires the creation of high-resolution training datasets and robust machine-learning frameworks that work across different agro-climatic zones (Shan et al., 2021a).

This chapter investigates the role of AI technologies, particularly ML models, in contributing to MZ-based VRI systems. It presents a detailed overview of ML techniques applied in zone delineation, decision support systems (DSS), and adaptive irrigation scheduling. Furthermore, it explores emerging directions for integrating AI into sustainable irrigation practices, while critically addressing the technical and practical challenges hindering large-scale implementation.

2 Concept of management zones in precision agriculture

MZs are spatial units within a field that exhibit analogous combinations of yield-limiting factors such as soil texture, organic matter content, topography, and crop vigor. These zones are designed to guide site-specific management practices, including irrigation, fertilizer application, and seeding, to help farmers maximize crop yield. Notably, MZs do not have to be spatially adjacent; zones with similar agronomic characteristics may be distributed across different parts of the field. The underlying principle is not proximity but rather the homogeneity of conditions that influence crop response to inputs. Farmers can optimize resource use and enhance productivity by individually treating these zones based on their specific requirements rather than using a uniform strategy across the field (Morari et al., 2021; Franzen et al., 2023).

MZs originated in precision agriculture as a significant sector that seeks to match crop demands with agricultural inputs at the appropriate time and location (Schimmelpfennig, 2024). A more realistic and economical way to apply site-specific management techniques is through the MZs approach. On the other hand, early precision agriculture focused on grid-sampling-based variable-rate applications. When regions with similar characteristics are clustered, farmers can capture the most significant geographic variability in their fields and make easier decisions.

2.1 Methods for delineating MZs

The integration of many data sources is necessary to represent geographic variability when identifying MZs. The main techniques consist of:

Soil properties-based zoning

In addition to these physical characteristics, soil EC has long been recognized as a powerful proxy variable in delineating MZs. EC reflects a combination of important soil attributes, including salinity, moisture content, clay content, and cation exchange capacity. Due to its ability to capture spatial variability in subsurface conditions, EC has been widely employed over the past two decades using proximal sensors such as electromagnetic induction (e.g., EM38) or contact-type sensors (e.g., Veris) for rapid field-scale mapping (Morari et al., 2021; Franzen et al., 2023). When integrated with other data layers—such as elevation models, yield maps, or vegetation indices—EC significantly enhances the reliability of zone delineation, especially in heterogeneous or poorly characterized soils. A recent study by Ben-Gal et al. (2025) demonstrated that including soil texture variability in MZs delineation improved irrigation water use efficiency by up to 25% when compared to uniform application strategies. The amount of organic matter in the soil is also significant for water availability and retention; better soil structure and increased water-holding capacity are often associated with higher organic matter levels. Geography, natural soil formation processes, and management practices significantly influence the amount of organic matter in a field (Diacono et al., 2022). Recent advances have highlighted the importance of optimizing field capacity and irrigation thresholds using zone-specific soil hydraulic properties and inverse modeling approaches such as HYDRUS-1D, which significantly improves irrigation scheduling accuracy and water-use efficiency (Kumar et al., 2022a, 2023b).

Topography-based zoning

Microclimatic conditions, soil depth, and water circulation are influenced by field topography. Elevated regions could have more drainage and evaporative losses, whereas low-lying places might receive more water from surface runoff and subsurface flow. ET rates are influenced by the slope aspect, which also affects solar radiation interception (Gonzalez-Dugo et al., 2022). DEMs derived from Veris, Light Detection and Ranging (LiDAR), or drone surveys can provide high-resolution topographic data that can be used to determine terrain characteristics (Kumar et al., 2021, 2022a) such as slope, aspect, and TWI. When creating MZs, these derivatives might be useful instruments. By considering natural soil moisture, Kumar et al. (2022b, 2023a, 2024a) and Yousaf et al. (2024) discovered that adding the TWI to zone delineation increased the accuracy of irrigation scheduling in undulating landscapes.

Crop characteristics-based zoning

Crop varieties vary in their water requirements, rooting depths, and capacity to withstand drought. However, within the same crop variety, spatial variability in crop growth, density, and health can result in differences in water consumption throughout a field. According to Khorsand et al. (2023), the phenological stage of plants has a substantial influence on water requirements. Plants in critical development periods typically require precision irrigation control. Crop vigor and water stress spatial patterns can be captured by vegetation indices generated from satellite or UAV data, such as the NDVI, enhanced vegetation index, and normalized difference water index (NDWI). Areas with persistent variations in crop production that can benefit from zone-based management can be identified by analyzing these indicators throughout the growing season (Sishodia et al., 2024).

Remote sensing and geographic information system-based zoning

Delineating MZs for precision irrigation can also be achieved using remote sensing and geographic information system (GIS) technology. These technologies make large-scale, high-resolution data gathering possible for crop health, soil characteristics, and water stress situations. Spatial heterogeneity in agricultural fields may be efficiently studied and controlled by combining multispectral and hyperspectral imaging, UAV-based sensing, and spatial clustering algorithms (). Satellite-based remote sensing offers a large-scale and cost-effective approach for field condition monitoring, providing frequent observations at lower per-hectare costs. Crop health evaluation, soil moisture estimation, and vegetation stress detection are made possible by the data collected by various sensors at various spectral resolutions. The following are a few popular satellite platforms.

2.2 Satellite-based monitoring

a. The European Space Agency's (ESA) Sentinel-2, which is outfitted with multispectral sensors that function in 13 spectral bands, offers high-resolution data (10–60 m) for soil characterization, water content calculation, and vegetation monitoring (Drusch et al., 2012).
b. Data from the NASA/USGS Landsat 8 and 9 satellites, Operational Land Imager, and Thermal Infrared Sensor provide both multispectral and thermal data that are useful in classifying land use and evaluating water stress (Roy et al., 2014).

2.3 Unmanned Aerial Vehicle-based monitoring and decision support

UAVs offer extremely high-resolution data suited for field-scale precision irrigation. Equipped with multispectral or thermal cameras, UAVs can detect intra-field variability in crop health, canopy temperature, and water stress with fine granularity.

Irrigation scheduling

Ballester et al. (2018) demonstrated UAV-based monitoring to schedule irrigation in greenhouse cucumber systems using soil water potential and ET data.

Real-time decision support

Near real-time UAV imagery enables prompt identification of vegetative stress and supports dynamic irrigation control strategies (Hunt et al., 2018; Praseartkul et al., 2023).

Parameter	Satellite Platforms	UAV Platforms
Spatial resolution	10–60 m (e.g., Sentinel-2), 30 m (Landsat) (Drusch et al., 2012; Roy et al., 2014)	2–20 cm, depending on the sensor and altitude (Ballester et al., 2018)
Temporal resolution	5–16 days (Sentinel-2:5 days; Landsat: 16 days) (Roy et al., 2014)	On-demand, can be deployed anytime (Hunt et al., 2018)
Coverage scale	Regional to global (Chatterjee et al., 2022)	Field-to-farm scale (Ballester et al., 2018)
Use cases	ET mapping, crop health, NDVI, water stress (Chatterjee et al., 2022)	Precision irrigation, intra-field variability, and canopy temperature mapping (Hunt et al., 2018)
Relative cost	Low cost per hectare; many datasets are free (Copernicus, USGS) (Drusch et al., 2012)	Higher operational costs due to hardware and deployment (Ballester et al., 2018)
Weather dependency	Affected by clouds, radar like Sentinel-1 is all-weather	Weather-sensitive; limited by wind and rain (Hunt et al., 2018)
Sensor types	Multispectral, thermal, SAR (Drusch et al., 2012; Roy et al., 2014)	Multispectral, thermal, hyperspectral (custom payloads) (Ballester et al., 2018)

3 Variable rate irrigation technology

VRI is a highly improved irrigation method that optimizes water application by managing irrigation rates based on spatial variability within a field. VRI enables precision irrigation by considering different parameters such as soil type, topography, crop water requirements, and environmental conditions, unlike conventional irrigation methods, which apply a uniform amount of water over a field (Kumar et al., 2022a, 2023a, 2023b). This technology is quite useful in diverse fields; for example, various zones need varied quantities of water to enhance agricultural productivity and efficiency. A VRI system typically consists of the following key components.

3.1 Components of a VRI system

Sensors

Sensors are critical for monitoring various environmental and crop parameters. These include:

a. **Soil moisture sensors**—Water availability in various zones can be evaluated using soil moisture sensors, such as capacitance, Time-Domain Reflectometry (TDR), or gypsum block sensors.
b. **Meteorological sensors**—Weather sensors detect rainfall, wind speed, temperature, and humidity, so that irrigation can be adjusted in real time.
c. **Crop Sensors**—Crop health sensors that measure crop water stress and identify variations in vegetation indices (NDVI, NDWI), including multispectral and thermal cameras.

3.2 GIS

GIS technology helps map and manage spatial data related to crop type and health, soil texture and moisture, topographic variation (e.g., slope and elevation), and agronomic variables such as planting density, fertilizer application rates, and historical yield performance.

Automation and control systems

VRI systems use automation to control irrigation based on real-time data. These systems include:

a. **Internet of Things (IoT)-enabled irrigation systems**: VRI systems commonly incorporate IoT technology to remotely control and monitor irrigation using mobile apps or web dashboards. These systems integrate field sensors (e.g., soil moisture probes, temperature sensors) and actuators (e.g., solenoid valves), allowing real-time decision-making based on dynamic soil and weather conditions (Fossa, 2024).
b. **Global Positioning System (GPS)-guided application controllers**: GPS-enabled systems allow irrigation equipment to apply variable water volumes across different zones based on geospatial coordinates. GPS precision enables alignment with MZs, ensuring that application maps are executed with minimal positional error, especially in center-pivot or lateral-move irrigation setups (Sadler et al., 2005).
c. **Edge computing and AI controllers**: Modern VRI systems increasingly integrate edge computing devices with AI models that process sensor inputs locally, reducing latency and bandwidth dependency. These AI-powered controllers predict crop water needs based on historical trends and current stress conditions, enabling proactive irrigation control even in remote areas with limited internet connectivity .

3.3 Advantages of a well-integrated VRI system

By integrating sensors, automation, and GIS-based Decision Support System (DSS), VRI technology can maintain agricultural output while reducing water consumption by 20%–30% (Evett et al., 2012; Hedley and Yule, 2009). It improves crop health and production by directing water to regions based on crop demands and increases irrigation efficiency by applying precision watering based on real-time data.

4 Role of machine learning in irrigation scheduling

The recent advancements in the field of AI, such as ML and DL algorithms, have proven to be powerful tools for managing various challenges in the fields of agriculture, crop management, soil analysis, and precision irrigation. Due to the combination of sensors and ML technology, traditional farming has now been transformed into data-driven precision agriculture.

4.1 Key ML techniques in agriculture for variable-rate irrigation

Supervised learning

It is used for crop yield prediction, pest detection, and disease classification. The algorithms are: -

a. **Random forest (RF)**—These ensemble methods combine multiple decision trees to improve prediction accuracy and robustness. Sadler (2022) used RFs to predict zone-specific irrigation requirements based on soil moisture trends, weather forecasts, and crop phenology, achieving 15%–20% water savings compared to conventional scheduling.
b. **Support vector machine (SVM)**—SVMs find optimal decision boundaries in high-dimensional feature spaces, making them practical for predicting irrigation needs based on multiple variables. Kustas et al. (2022) applied SVMs to predict irrigation requirements based on soil moisture, meteorological inputs, and remote sensing data, achieving crop yields comparable to those under traditional full irrigation regimes while significantly reducing water use.
c. **Artificial neural networks (ANNs)**—ANNs can model complex, nonlinear relationships between environmental variables and crop water requirements. Tiwari et al. (2024) developed a multilayer perceptron neural network that integrates soil moisture sensor data, weather parameters, and crop coefficients to predict zone-specific irrigation needs accurately.
d. **Gradient boosting machines**—Algorithms such as eXtreme Gradient Boosting (XGBoost) and Light Gradient Boosting Machine (LightGBM) combine multiple weak learners to create strong predictive models. Xiao et al. (2023) used gradient boosting to predict zone-specific ET rates from weather data and vegetation indices, achieving more precise irrigation scheduling than traditional methods.

Unsupervised learning

This method helps cluster soil types, segment land-use patterns, and group similar crop stress responses. Some examples are as follows:

a. **K-means clustering**—Within-cluster sum of squares is minimized in this approach, which divides the data into a fixed number of clusters. Despite its processing efficiency, K-means necessitates prespecifying the number of clusters. To maximize the number of zones, cluster validity indices, like the elbow method and the Silhouette coefficient, are usually evaluated (Albornoz et al., 2022).
b. **Fuzzy C-means**—Fuzzy C-means permits points to belong to many clusters with different membership levels, unlike K-means, which allocates each point to exactly one cluster. This method can more accurately capture the ongoing character of crops and soil variability. In contrast to complex clustering techniques, Morari et al. (2021) showed that Fuzzy C-means clustering of soil EC and topography data generated more realistic MZs. Kumar et al. (2022a; Kumar et al., 2022b, 2023a, 2023b, 2024b) have used Fuzzy C-means for creating the MZs based on soil texture, elevation, and 10 years of historical yield data, and developed the VRI scheduling method and studied the impact on nutrient dynamics in an irrigated field.

c. **Hierarchical clustering**—Creating a dendrogram that depicts nested clusters makes it possible to choose the correct number of zones depending on the data's structure. In addition to revealing linkages between zones at various dimensions, hierarchical clustering eliminates the need to predetermine the number of clusters. Guo et al. (2023) defined irrigation MZs based on soil characteristics and past yield data using agglomerative hierarchical clustering in conjunction with Ward's technique.
d. **Density-based spatial clustering of applications with noise (DBSCAN)**—This technique does not require a fixed number of clusters. Instead, it identifies clusters as dense regions separated by regions of lower density. Outliers that may not fit well into any zone can be detected using DBSCAN, which can also identify zones with irregular shapes. DBSCAN was utilized by Khanal et al. (2022) to define MZs using topography and soil EC data and found that it effectively captured spatial patterns in distinct fields.

Hybrid approaches

This process combines ML with remote sensing, GIS, and IoT to enhance precision agriculture. For instance, integrating DL with UAV imagery has improved soil nutrient classification and irrigation efficiency.

5 Data sources for irrigation scheduling

To maintain the efficiency of irrigation scheduling, timely and precise data on crop water requirements, soil moisture, and environmental factors are important inputs for most models and algorithms. Advancements in satellite technology have equipped digital agriculture with multiple data sources, including in-situ sensor networks, satellites, and UAVs. The most-used data sources that support modern irrigation techniques are described in this section, along with how integrating these data helps in decision-making in variable-rate irrigation systems.

5.1 Satellite remote sensing

Satellites are integral to irrigation monitoring at the regional and national levels, often referred to as large-scale applications, typically covering areas greater than 10,000 ha or multiple administrative units such as districts or agroecological zones. One of their key advantages is the ability to deliver multitemporal observations that capture crop phenology, seasonal growth trends, and anomalies in vegetation health or soil moisture.

a. Sentinel-2: Operated by the ESA, Sentinel-2 is widely adopted in agricultural remote sensing due to its 10-meter spatial resolution in key spectral bands (visible and near-infrared [NIR]), 5-day revisit time, and 13 spectral bands spanning the visible, NIR, and shortwave infrared regions. These spectral properties enable the calculation of vegetation indices such as NDVI, NDWI, and Soil Adjusted Vegetation Index (SAVI), which are crucial for monitoring crop vigor, water stress, and biomass accumulation.

b. Landsat 8 and 9—These sensors are ubiquitous among researchers, as they consist of thermal and multispectral imaging sensors that help estimate ET and monitor land surface temperature.
c. Moderate Resolution Imaging Spectroradiomete (MODIS)—Because of its high temporal frequency, MODIS is favored for regional-scale monitoring, aiding in tracking agricultural phenology and water requirements.

5.2 UAV-based multispectral and thermal imaging

UAVs offer extremely high-resolution imagery that is perfect for accuracy at the field scale. Several types of cameras can be attached to UAVs for field-scale analysis, and some of the commonly used ones are:

a. **Multispectral cameras**—UAVs equipped with multispectral cameras can provide accurate NDVI mapping to pinpoint areas under stress and those that receive sufficient irrigation (Mekonnen et al., 2023).
b. **Thermal cameras**—Thermal imaging is a reliable predictor of drought (crop water condition), and changes in canopy temperature are relatively easy to detect using thermal cameras (Patel et al., 2024).

By detecting intra-field variability, these techniques facilitate the adoption of VRI and enable the real-time identification of irrigation inefficiencies (Rahman et al., 2023a; 2023b).

5.3 In situ sensor networks

Ground-based sensors are highly important as they validate the remote sensing data by emphasizing real-time measurements. Some of the widely used sensors are:

a. ***Soil moisture sensors***—The availability of optimal irrigation timing and root-zone water is determined using soil moisture sensors, such as capacitance and TDR.
b. ***Weather stations***—To calculate reference ET (ET_0), which is essential for irrigation modeling, weather stations supply inputs such as solar radiation, wind, and humidity.
c. ***Plant-based sensors***—Plant water uptake and stress are directly measured by plant-based sensors, such as sap flow meters and dendrometers.
d. ***IoT-enabled networks***—This allows irrigation systems to be controlled automatically by wirelessly transmitting sensor data.

5.4 Integration of multisource data

Irrigation choices can be improved by combining in situ data, UAVs, and remote sensing:

a. **Data fusion**—Multiresolution data may be combined using data fusion techniques to produce solid insights.
b. **ML Models**—Predicting irrigation requirements from integrated datasets can utilize ML models like RF and Long short-term memory (LSTM).
c. **DSS**—This integration is used by DSS to provide irrigation advisories that are geographically aware and delivered in real-time.

6 Developing an ML-based irrigation model

Irrigation scheduling has been transformed by ML, which makes site-specific, adaptive, and data-driven decision-making possible. Data collection and preprocessing, feature selection, model training, and validation are the usual systematic processes in creating an ML-based irrigation model. These algorithms use large and varied information to predict irrigation needs, manage water consumption, and boost agricultural output.

6.1 Data collection and preprocessing

The effectiveness of ML models depends on the caliber and variety of the input data. Familiar sources of data are mentioned in Section 5 of this chapter. Once the data is collected, it requires preprocessing before being used for further analysis. Normalization, gap-filling for missing values, noise filtering, and temporal alignment of data from several sources are all part of preprocessing (Patel et al., 2024). In addition, feature engineering enhances model inputs through the creation of rolling averages and change detection (Ghosh et al., 2023).

6.2 Feature selection and model training

Selecting the most relevant features from large agricultural datasets is essential for improving model performance and interpretability. According to Ali et al. (2023), techniques such as principal component analysis, recursive feature elimination, and mutual information ranking are widely used to identify key predictors, such as soil moisture, vegetation indices, and temperature, that mostly influence irrigation demand. These methods help eliminate redundant or irrelevant variables, thereby reducing the model's complexity without compromising accuracy.

Once the most informative features are selected, the dataset is typically divided into training, validation, and testing subsets. Ali et al. (2023) emphasize the use of k-fold cross-validation during training to ensure that the model's performance is consistent across different data partitions. This process also aids in hyperparameter tuning, which involves optimizing parameters like learning rate, tree depth (in ensemble models), or the number of neurons (in neural networks). These practices collectively reduce the risk of overfitting and enhance the model's generalizability to unseen data.

6.3 Popular ML algorithms for irrigation scheduling

RF is quite effective for tabular and nonlinear data and provides excellent accuracy with little adjustment (Liu et al., 2024). Complex input-output interactions between environmental factors and irrigation requirements can be modeled using ANNs (Das and Pandey, 2023). Convolutional neural networks (CNNs) are beneficial for immediately recognizing spatial patterns in image data, such as satellite or UAV imagery. For time-series forecasting, such as projecting future irrigation demands based on past patterns, LSTM is a good fit (Chatterjee et al., 2022). The kind of incoming data, real-time demands, and available computing power influence the method used.

6.4 Validation and performance assessment

Model validation guarantees robustness and dependability in field applications. To confirm model performance under various circumstances, K-fold cross-validation and external validation with separate field datasets are advised. Typical metrics consist of:

a. **Root mean square error (RMSE)** and **mean absolute error (MAE)**—For regression accuracy, use MAE and RMSE.
b. **coefficient of determination (R^2)**—To evaluate model fit, use the coefficient of determination, or R^2.
c. **Confusion matrix**, **precision**, **recall**, and **F1 score**—F1 score, accuracy, recall, and confusion matrix for irrigation triggers of the classification type.

7 Case studies and applications

The combination of VRI with ML has shown great potential in tackling the world's problems with food security and water management. Applications in the real world across many crops and countries have shown how successful these technologies are in maximizing agricultural output, reducing environmental effects, and improving water usage efficiency.

7.1 Real-world applications of ML and VRI in different crops and agriculture sectors

Different studies indicate the use of ML and VRI across the globe, some of which are as follows:

a. **Maize in the US Corn Belt**—In the US Corn Belt, farmers have been able to cut water consumption by 25% without sacrificing output because of VRI devices combined with ML-based soil moisture and NDVI mapping (Singh et al., 2024).
b. **Rice in India**—Water productivity increased by 18% when ML algorithms (RF and LSTM) were utilized to forecast irrigation requirements using MODIS and in-situ meteorological data (Kumar et al., 2023a).
c. **Cotton in Australia**—Field-level VRI was made possible using ANN models and UAV-based NDVI mapping, which increased irrigation schedule accuracy and resulted in a 22% reduction in input expenditures (Jackson et al., 2022).
d. **Grapevines in Spain**—Using Sentinel-2 and climate data, LSTM models forecast short-term irrigation requirements, resulting in 30% water savings and enhanced grape quality (Torres-Sánchez, 2023).
e. **Israel's agri-tech sector**—Water use efficiency and export-grade crop production have increased nationally because of precision irrigation powered by sensor networks and ML decision-support systems (Ben-Ari and Alon, 2022).
f. **China's smart irrigation in wheat farming**—According to Wang et al. (2023), automated irrigation scheduling, made possible by CNN models that analyze soil-water balance and UAV footage, has increased wheat output by 15% while conserving 20% of water.

7.2 Economic and environmental benefits of precision irrigation

It helps to reduce energy consumption and water pumping expenses by 15%–30% (Ali et al., 2023). The automation also helps to reduce labor costs and operating expenses. The use of VRI significantly reduces runoff and nutrient leaching by avoiding excessive irrigation (Ghosh et al., 2023). It decreases the carbon footprint due to using less power and gasoline. It also improves soil health, resulting in decreased erosion and better moisture management (Liu et al., 2024).

8 Challenges and future perspectives

VRI and ML-based irrigation scheduling have great potential, but there are several barriers, such as weak policy support, limited extension infrastructure, and poor interagency coordination, along with financial and technological obstacles hindering real-world adoption. Simultaneously, new technologies are expanding the capabilities of intelligent irrigation systems. These issues are examined in this section, along with potential future paths for scalable and sustainable adoption.

8.1 Technical and practical challenges in implementing VRI and ML-based irrigation

a. **Data availability and quality**—High-resolution, continuous data from sensors, satellites, and field monitoring devices is essential for accurate irrigation scheduling. Nonetheless, gaps caused by cloud cover, malfunctioning sensors, or uneven calibration are frequent.
b. **Integration complexity**—In heterogeneous fields with different crop varieties and soil profiles, it can be challenging to integrate remote sensing, meteorological, and in-situ data with ML models.
c. **Model interpretability**—Deep neural networks and other black-box models frequently lack transparency, which makes it difficult for technicians and farmers to accept and trust their advice.
d. **Hardware and infrastructure**—Smallholder adoption of VRI-compatible irrigation systems is constrained by the high upfront costs and the technical expertise required for installation and maintenance.

8.2 Future trends in smart irrigation

IoT-based climate stations and soil moisture sensors allow for continuous data collection, facilitating irrigation choices in real-time. New AI models improve forecast accuracy and allow for flexible irrigation scheduling in the face of changing weather patterns. In remote agricultural regions, on-site data processing using edge devices decreases latency, improves real-time responses, and reduces the requirement for cloud infrastructure. Farmers can view water status, receive warnings, and model irrigation scenarios with integrated DSS platforms that use cloud-based dashboards.

Chapter summary

This chapter presented a comprehensive overview of how ML techniques and VRI technologies are being integrated to support precision irrigation. It began by outlining the importance of delineating MZs based on spatial variability in soil properties, crop response, and topographic features. The role of remote sensing data from satellites, UAVs, and in-situ sensors was examined in the context of spatial analysis and real-time monitoring for irrigation scheduling.

Various ML algorithms, such as RF, ANN, CNN, and LSTM networks, were discussed for their ability to model crop water demand and forecast irrigation needs using multisource inputs. Emphasis was placed on the significance of feature selection, model training practices, and the integration of explainable AI for decision support.

Case studies were reviewed to demonstrate successful field-level applications of VRI and ML across different crop types and agroclimatic regions. The chapter also identified technical and institutional challenges, including data integration, digital infrastructure, and farmer adoption. Finally, emerging tools such as Edge AI, IoT networks, and automation platforms were highlighted as enablers for more responsive and scalable irrigation systems.

Ronald Singh: Writing – review & editing, Writing – original draft, Visualization, Software, Methodology, Conceptualization, Investigation, Data Curation, Formal analysis. **Hemendra Kumar**: Writing – review & editing, Writing – original draft, Visualization, Software, Methodology, Conceptualization, Validation, Supervision, Resources, Project administration, Investigation, Funding Acquisition, Data Curation, Formal analysis. **Sushil Kumar Himanshu**: Writing – review & editing, Visualization, Methodology, Validation.

References

Albornoz, M.R., 2022. Evaluation of clustering validity indices for agricultural zone delineation using geospatial data. Comput. Electron. Agric. 198, 107041.

Ale, S., Su, Q., Singh, J., Himanshu, S., Fan, Y., Stoker, B., et al., 2023. Development and evaluation of a decision support mobile application for cotton irrigation management. Smart Agric. Technol. 5, 100270.

Ali, A., Sharma, R., Singh, P., 2023. Integration of multi-source data for decision support in precision irrigation. Comput. Electron. Agric. 212, 108123.

Ballester, C., Perez-Ruiz, M., Agüera, J., Carballido, J., 2018. Scheduling drip irrigation for greenhouse cucumber using soil water potential sensors and evapotranspiration data. Agric. Water Manag. 210, 68–78.

Ben-Ari, A., Alon, D., 2022. National strategies in precision irrigation: the Israeli model. Irrig. Sci. 40 (5), 987–1001. doi:10.1007/s00271-022-00772-5.

Ben-Gal, A., Barski, A., Bukris, O., Yasuor, H., O'Shaughnessy, S.A., Hansen, N.C., Peeters, A., Cohen, Y, 2025. Dynamic approaches to precision irrigation of cotton. Precision Agric. 26 (4), doi:10.1007/s11119-025-10261-1.

Chatterjee, S., Verma, A., Das, M., 2022. Regional-scale irrigation planning using MODIS time-series data. Remote Sens. Environ. 276, 113125. doi:10.1016/j.rse.2022.113125.

Das, P., Pandey, R., 2023. Real-time weather station data for irrigation management in semi-arid regions. Agric. Water Manag. 278, 108127.

Diacono, M., Montemurro, F., Fiore, A., 2022. Influence of soil formation processes and management on organic matter distribution in Mediterranean agricultural systems. Soil Tillage Res. 217, 105277. doi:10.1016/j.still.2022.105277.

Drusch, M., Del Bello, U., Carlier, S., Colin, O., Fernandez, V., Gascon, F., Bargellini, P., 2012. Sentinel-2: ESA's optical high-resolution mission for GMES operational services. Remote Sens. Environ. 120, 25–36. doi:10.1016/j.rse.2011.11.026.

Evett, S.R., Schwartz, R.C., Casanova, J.J., Heng, L.K., 2012. Soil water sensing for water balance, ET and WUE. Agric. Water Manag. 104, 1–9. doi:10.1016/j.agwat.2011.12.002.

Feng, X., Li, J., Zhou, Y., 2019. Integration of UAV hyperspectral imagery and soil moisture sensors for real-time irrigation scheduling. Remote Sens. Environ. 232, 111307.

Fossa, M., 2024. Advances in soil moisture sensing for precision irrigation: technologies and field applications. J. Agric. Sens. Syst. 12 (1), 34–49.

Franzen, D.W., Khan, M., Nagle, M., 2023. Irrigation zone delineation by coupling neural networks with spatial statistics. NEJSDS: Northeast J. Spat. Data Sci. 5 (1), 25–42. doi:10.32629/NEJSDS.2023.050104.

Gade, S.A., Madolli, M.J., García-Caparrós, P., Ullah, H., Cha-um, S., Datta, A., Himanshu, S.K., 2025. Advancements in UAV remote sensing for agricultural yield estimation: a systematic comprehensive review of platforms, sensors, and data analytics. Remote Sens. Appl. Soc. Environ. 37, 101418.

Ghosh, N., Mishra, D., Bose, R., 2023. IoT-based smart irrigation: a field-level implementation. Sensors 23 (3), 1203.

Gonzalez-Dugo, M.P., Zarco-Tejada, P.J., Fereres, E., 2022. Evaluating the influence of slope and aspect on evapotranspiration and canopy temperature using high-resolution imagery. Agric. Water Manag. 263, 107416. doi:10.1016/j.agwat.2022.107416.

Guo, L., Zhao, W., Han, J., 2023. Delineation of irrigation management zones using agglomerative hierarchical clustering and Ward's method. Precis. Agric. 24 (2), 145–162.

Hedley, C.B., Yule, I.J., 2009. Soil water status mapping and two variable-rate irrigation scenarios. Precision Agric. 10 (4), 342–355. doi:10.1007/s11119-009-9119-z.

Hunt, E.R., Daughtry, C.S.T., McMurtrey, J.E., 2018. Remote sensing of crop leaf area index and canopy chlorophyll content using unmanned aerial vehicles. Remote Sens. 10 (3), 320. doi:10.3390/rs10030320.

Jackson, T., McCarthy, M., Wilson, D., 2022. VRI applications in cotton: insights from UAV and neural network modeling. Precis. Agric. 23 (4), 515–530. doi:10.1007/s11119-022-09864-4.

Kamilaris, A., Prenafeta-Boldú, F.X., 2018. Deep learning in agriculture: a survey. Comput. Electron. Agric. 147, 70–90. doi:10.1016/j.compag.2018.02.016.

Khanal, S., Fulton, J.P., Shearer, S.A., 2022. Delineation of management zones using DBSCAN clustering of soil electrical conductivity and topographic attributes. Precis. Agric. 23 (2), 335–352.

Khorsand, A., Zhang, L., Morales, J., 2023. Influence of crop phenology on dynamic irrigation scheduling in heterogeneous agroecosystems. Agric. Water Manag. 278, 108190.

Kumar, H., Ortiz, B.V., Srivastava, P., Lamba, J., 2024a. Assessing nutrient variability across irrigation management zones using unsupervised learning and mixed models. In: Agri-Tech Approaches for Nutrients and Irrigation Water Management. CRC Press, pp. 68–82.

Kumar, H., Srivastava, P., Lamba, J., Diamantopoulos, E., Ortiz, B., Morata, G., Takhellambam, B., Bondesan, L., 2022a. Site-specific irrigation scheduling using one-layer soil hydraulic properties and inverse modeling. Agric. Water Manag. 273, 107877.

Kumar, H., Srivastava, P., Lamba, J., Lena, B., Diamantopoulos, E., Ortiz, B., Takhellambam, B., Morata, G., Bondesan, L., 2023a. A methodology to optimize site-specific field capacity and irrigation thresholds. Agric. Water Manag. 286, 108385. doi:10.1016/j.agwat.2023.108385.

Kumar, H., Srivastava, P., Lamba, J., Ortiz, B.V., Lyon, S.W., Takhellambam, B.S., Morata, G., Bondesan, L., 2023b. Integral approach of unsupervised learning and temporal stability in irrigation management. Am. Soc. Agric. Biol. Eng. (ASABE) 2301023. doi:10.13031/aim.20231023.

Kumar, H., Srivastava, P., Lamba, J., Ortiz, B.V., Way, T.R., Sangha, L., Takhellambam, B.S., Morata, G., Molinari, R., 2022b. Within-field variability in nutrients for site-specific agricultural management in irrigated cornfield. J. ASABE 65 (4), 865–880.

Kumar, H., Srivastava, P., Ortiz, B.V., Morata, G., Takhellambam, B.S., Lamba, J., Bondesan, L., 2021. Field-scale spatial and temporal soil water variability in irrigated croplands. Trans. ASABE 64 (4), 1277–1294.

Kumar, S., Roy, S., Verma, H., 2024b. Machine learning approaches for irrigation water requirement prediction using fused datasets. Agric. Syst. 220, 103011.

Kustas, W.P., Anderson, M.C., Alfieri, J.G., Colaizzi, P.D., 2022. Predicting irrigation water needs using support vector machines with soil moisture and remote sensing data. Agric. Water Manag. 262, 107372. doi:10.1016/j.agwat.2022.107372.

Li, Y., Simunek, J., Jing, L., Zhang, Z., Ni, L., 2020a. Evaluation of water movement and water losses in a direct-seeded-rice field experiment using HYDRUS-1D. Agric. Water Manag. 142, 38–46. doi:10.1016/j.agwat.2014.04.021.

Li, Z., Liu, X., Chen, Y., 2020b. Delineation of management zones using soil and remote sensing data in precision agriculture. Precis. Agric. 21 (4), 765–782.

Liu, H., Zhang, L., Wang, Q., 2018. Applications of artificial intelligence in agricultural irrigation systems: a review. J. Precis. Agric. 19 (3), 210–225.

Liu, Y., Zhang, M., Sun, Q., 2024. Plant-based sensing for real-time water stress monitoring. Precis. Agric. 25 (2), 339–353.

Mekonnen, B., Tadesse, M., Hagos, H., 2023. UAV-based NDVI for irrigation management in maize. Agric. Remote Sens. 8 (1), 45–55.

Morari, F., Castrignanò, A., Giordano, A., 2021. Delineating management zones using fuzzy clustering of soil electrical conductivity and topographic attributes. Precis. Agric. 22 (3), 456–470.

O'Shaughnessy, S.A., Evett, S.R., Colaizzi, P.D., Howell, T.A., 2016. A crop water stress index and time threshold for automatic irrigation scheduling of grain sorghum. Agric. Water Manag. 174, 1–11.

Ojha, R.S., 2022. Machine learning-based approaches for forecasting crop water demand using remote sensing and sensor data. J. Agric. Inform. 13 (2), 87–98.

Patel, K., Joshi, R., Bhatt, R., 2024. Thermal UAV imaging for real-time irrigation stress mapping. J. Precis. Agric. 25 (1), 18–27.

Praseartkul, P., Taota, K., Pipatsitee, P., Tisarum, R., Sakulleerungroj, K., Sotesaritkul, T., et al., 2023. Unmanned aerial vehicle-based vegetation monitoring of aboveground and below-ground traits of the turmeric plant (Curcuma longa L.). Int. J. Environ. Sci. Technol. 20 (8), 8673–8686.

Rahman, M.Z., 2021. Deep learning-based recurrent neural networks for irrigation optimization under water stress conditions. Comput. Electron. Agric. 186, 106170.

Rahman, S., Gupta, D., Chaudhary, A., 2023a. DSS-enabled smart irrigation: design and deployment experiences. Agric. Water Manag. 266, 107654.

Rahman, S., Gupta, D., Chaudhary, A., 2023b. Field-scale irrigation zoning using UAV-derived imagery. Agric. Water Manag. 265, 107614.

Roy, D.P., Wulder, M.A., Loveland, T.R., Woodcock, C.E., Allen, R.G., Anderson, M.C., Zhu, 2014. Landsat-8: science and product vision for terrestrial global change research. Remote Sens. Environ. 145, 154–172. doi:10.1016/j.rse.2014.02.001.

Sadler, E., Evans, R., Stone, K., Camp, C., 2005. Opportunities for conservation with precision irrigation. J. Soil Water Conserv. 60 (6), 371–379. doi:10.1080/00224561.2005.12435829.

Sadler, E.J., 2022. Predicting site-specific irrigation needs using random forests and soil-crop -climate dynamics. Agric. Water Manag. 263, 107385. doi:10.1016/j.agwat.2022.107385.

Schimmelpfennig, D., 2024. Advancements in precision agriculture and management zone development. J. Agric. Econ. Technol. 36 (1), 1–15.

Shan, X., Zhang, W., Liu, J., 2021a. Integrating soil, crop, and climate dynamics for smart irrigation management. Agric. Syst. 193, 103233.

Shan, X., Zhang, W., Liu, J., 2021b. Advanced sensor-based analysis for crop-soil-climate interaction in precision agriculture. J. Precis. Agric. Res. 19 (2), 112–125.

Singh, R., Yadav, A., Chauhan, S., 2024. VRI and NDVI-based irrigation optimization in maize: a case study from the U.S. Midwest. J. Precis. Agric. 24 (3), 211–222. doi:10.1007/s11119-024-09941-0.

Sishodia, R.P., Ray, R.L., Giri, S., 2024. Remote sensing-based identification of persistent yield gaps for precision zone management in irrigated agriculture. Precis. Agric. 25 (1), 88–104.

Sun, Y., Chen, J., Wu, D., 2022. Machine learning techniques for precision irrigation scheduling: a review. Comput. Electron. Agric. 198, 106993.

Tiwari, R., Sharma, V., Gupta, N., 2024. Predicting zone-specific irrigation needs using multilayer perceptron neural networks and integrated sensor data. Comput. Electron. Agric. 216, 108237.

Torres-Sánchez, J.P.-G., 2023. Precision irrigation in viticulture using LSTM modeling and satellite data. Remote Sens. 15 (8), 1812. doi:10.3390/rs15081812.

Wang, X., Zhao, H., Li, Y., 2023. CNN-based irrigation scheduling in Chinese wheat systems using UAVs and weather data. Agric. Syst. 218, 103035.

Xiao, Y., Zhang, H., Liu, J., 2023. Gradient boosting models for evapotranspiration prediction using vegetation indices and weather data. Comput. Electron. Agric. 205, 107475.

Yousaf, M., Rehman, T., Iqbal, M., 2024. Integration of topographic wetness index and soil moisture gradients for precision irrigation zone delineation. Precis. Agric. 25 (1), 99–114.

Optimizing unmanned aerial vehicle spray application for pesticide deposition in watermelon production systems in the Mid-Atlantic

14

Abubakar Palli[a,b], *Haley Sater*[b], *and Hemendra Kumar*[a,b,c]
[a]Digital and Precision Agriculture Lab, University of Maryland, College Park, MD, United States
[b]College of Agriculture and Natural Resources, University of Maryland, College Park, MD, United States
[c]Central Maryland Research and Education Center, University of Maryland Extension, Upper Marlboro, MD, United States

1 Introduction

As producers strive to sustain high yields and crop health while minimizing environmental impact, the precise application of crop protection products has become essential in modern agriculture (Srinivasarao et al., 2021). Conventional methods, such as manual spraying or large-scale ground machinery, have long been criticized for their application inefficiencies, often leading to excessive chemical input and heightened risks to nontarget organisms and ecosystems. Thus, precise application using drones reduces both chemical usage and the time required for their application. In addition, a ground sprayer is more expensive than a drone and requires additional drive rows, which cannot be planted with crops, leading to a reduction in overall yield. These limitations underline the necessity for more refined, data-driven application strategies, such as drones, that can deliver agrochemicals where they are needed the most, improving efficiency while reducing excessive chemical use and environmental impact.

In this context, unmanned aerial application systems (UAAS) have emerged as a promising technology for advancing precision agriculture practices (Praseartkul et al., 2023; Gade et al., 2025). Byers et al. (2024) conducted a comprehensive evaluation of UAAS performance under varying operational conditions, focusing on key parameters like nozzle configurations, application heights, and flight speeds. Their findings suggested a nuanced interaction among these factors, demonstrating

Precision Technologies for Digital Agriculture. https://doi.org/10.1016/B978-0-443-40513-6.00003-2

that increasing the application height enhanced the coverage area for one unmanned aerial vehicle (UAV) platform while producing negligible changes for another, which indicates the need for more UAAS studies in different climates and flight parameters to improve understanding and develop a standard protocol. Moreover, adjustments to flight speed appeared to change coverage uniformity, highlighting a complex balance of trade-offs when deriving application protocols. Subsequent research has built on these findings, analyzing the environmental and crop-specific factors affecting the efficacy of aerial application systems. For example, Yan et al. (2023) investigated the effect of increasing spray volume from 60 to 120 L/ha in citrus orchards and found that this adjustment substantially boosted droplet density and canopy coverage, particularly in the more challenging lower canopy zones. Similarly, Zhang et al. (2020) explored the relationship between spray parameters and droplet deposition within sugarcane canopies. They identified a combination of spray volume, flight height, and velocity that significantly improved droplet penetration and distribution, underscoring the importance of parameter optimization for ensuring effective coverage within complex canopy architectures.

Further refinement of UAAS techniques has focused on understanding how flight altitude influences droplet behavior and off-target movement or drift. Lou et al. (2018) reported that reducing flight height could enhance droplet uniformity and deposition. Building on the importance of droplet characteristics, Chen et al. (2020) emphasized that larger droplet sizes were instrumental in increasing deposition efficiency while curbing drift. By carefully selecting droplet size distributions, practitioners can achieve higher precision in targeting pests and diseases, minimizing chemical wastage and potential contamination of surrounding areas. Overall, based on the findings, researchers have begun to develop integrated models that account for the interdependent factors affecting aerial spray outcomes. Such comprehensive models are invaluable for practitioners and researchers alike, as they offer predictive insights for optimizing application strategies and ensure that emerging technologies can be adapted to a range of cropping systems, canopy types, and pest pressures.

The studies mentioned thus far underscore the multivariable nature of UAAS-based pesticide applications. By examining how different UAV platforms respond to modifications in operational parameters, adjusting droplet characteristics, and integrating environmental variables into these models, we gain a nuanced understanding of how best to deploy these systems. Ultimately, incorporating flight height, speed, and nozzle diameter into broader precision agriculture practices can enable producers to achieve more consistent deposition, enhanced target coverage, and reduced environmental risks, thereby strengthening the sustainability and efficiency of modern crop protection practices.

However, achieving truly optimized and precise chemical delivery relies on a thorough understanding of how operational parameters such as flight speed, application height, and nozzle diameter interact to influence coverage patterns under diverse environmental conditions, such as wind speed and humidity. As new drone models and nozzle configurations emerge, and as environmental variables fluctuate, it becomes increasingly challenging to apply insights gained from one scenario to another without a systematic approach. To advance this understanding, we focused on exploring the spray deposition characteristics of a drone across varying application heights and cruise speeds. Specifically, this study aimed to evaluate and compare the coverage area achieved within a consistent swath and assess coverage area changes across different spraying swaths, all while varying critical operational parameters. This study aimed to gain insight into how operational factors such as application height, flight speed, and nozzle diameter affect on-target coverage and help mitigate unintended chemical distribution. This approach provided a comprehensive framework for refining drone practices to achieve more precise, responsible, and adequate coverage of pesticide application. This study is also unique since it was carried out in an actual watermelon cropping system instead of on artificial surfaces like wooden planks and sheets with water-sensitive paper (WSP).

2 Materials and methodologies

2.1 Drone description

A commercially available DJI Agras T40 UAAS was utilized in this study, hereafter referred to as the "T40". The T40 is configured in a quadcopter arrangement and is equipped with a 40 L spray tank. Instead of conventional spray nozzles, the T40 employs two dual-disk centrifugal atomizers positioned directly beneath the rear rotors along the centerline of rotation. These atomizers operate without a prescribed spray angle, offering a variable droplet distribution compared to traditional nozzle systems.

The T40 was controlled using a DJI RC Plus flight controller operating on dual frequency bands (2.4–2.4835 GHz and 5.725–5.850 GHz) and managed through the DJI Smart Farm application software. To enhance positional accuracy, the UAS was integrated with an Emlid Reach RS3 real-time kinematics (RTK) base station during all trials. According to the manufacturer's specifications, this setup of drone plus RTK station provides a horizontal and vertical positioning accuracy of ±10 cm, ensuring precise navigation and application consistency.

The T40 platform itself weighs approximately 50 kg (drone and battery) and can reach a maximum takeoff weight of 90 kg when fully loaded. It measures 2.8 m × 3.15 m × 0.8 m and is powered by a 30,000 mAh, 52.22 V battery. Under no-load conditions, the T40 can hover for up to 18 minutes, providing ample operational flexibility for field applications. Flight planning software was employed to preprogram flight paths, allowing for consistent swath coverage and repeatable application conditions across different operational scenarios.

Multiple campaigns were conducted for spraying using various application parameters. The variations in application height, speed, and nozzle diameter allowed for an evaluation of how operational parameters influence coverage and deposition characteristics. Specific treatments, including various agricultural chemicals and fertilizers, were selected to meet management objectives. The details are summarized in Table 14.1.

Table 14.1 Specifications for the DJI Agras T40 UAS.

Platform	DJI T40 (Stock configuration)
Platform weight (empty) (kg)	50
Operating payload (kg)	40
Dimensions (unfolded) (m)	2.8 m × 3.15 m × 0.8 m
Number of Nozzles	2
Nozzle type	Dual disk centrifugal atomizers
Nozzle location	Centerline of rear rotors underside
Hovering time (full payload) (min)	18 (without payload)
Battery capacity (mAh, V)	30,000, 52.22 V
Positional accuracy (cm)	10 cm (RTK-based)
Flight application software	DJI Smart Farm
Controller type	DJI RC Plus
Frequency band	2.4–2.4835 GHz, 5.725–5.850 GHz
Other modules	Emlid Reach RS3 RTK Base Station

2.2 Experimental setup

Field trials were conducted in a farmer's field located in Delmar, Sussex County, Delaware, USA. We had six swaths of watermelon beds, each 270 m in length and 14.5 m in width. The watermelon swath was divided into 67 m long and 7.3 m wide spray swaths for spraying, with three rows of watermelon beds in each block. This experimental design was adopted based on the common practice of farmers growing watermelons in the region. To enhance the robustness of the findings, each treatment was replicated four times at different locations within the field, ensuring that variations in coverage were not confined to a single location. These treatments were randomized to capture the maximum variability of deposition and coverage during spraying with respect to height and flight speed. Within the designated application area, WSPs were placed along each watermelon row to capture droplet distribution for each treatment scenario. Additional WSPs were positioned outside the spray application area to document off-target drift potentially influenced by environmental factors and rotor-induced airflow patterns.

2.3 Data collection

Data collection focused on evaluating spray deposition patterns under different operational conditions using Syngenta WSP to record droplet distribution directly on the crop canopy. Over multiple test days, experiments were conducted in the watermelon field, where WSPs were affixed to the leaves of the watermelon plants via clips, as shown in Fig. 14.1, ensuring that each sample captured a realistic representation of spray deposition on watermelon leaves in the field setting. This approach differed from previous studies that relied on artificial surfaces like wooden planks and sheets with WSPs pinned on them, which did not represent the actual crop and canopy conditions while spraying. In this study, the WSPs' placement on plant leaves allowed for more realistic insights into how droplets interact with an actual crop canopy.

Fig. 14.1 Placement of the WSP onto the leaves.

Two application heights (2.44 and 3 m) were tested to understand how the vertical height of the UAV affected droplet coverage within the canopy. Two nozzle diameters (140 and 300 μm) were also compared to examine how droplet size influenced deposition patterns. Two flight speeds of 2.6 and 5.8 m/s were tested for spraying. These changes in flight speed allowed for assessing how coverage and droplet distribution responded to different UAV movement rates. In each aerial treatment, the UAV followed a preprogrammed flight path designed in the DJI Smart Farm application. Spray applications were conducted in collaboration with a professional pilot (Scout Applicators LLC, MD, USA). The DJI Agras T40 was operated in autonomous waypoint mode with a fixed application rate prescribed in the mission planning software. Pump flow was automatically adjusted to maintain this rate, while preprogrammed flight speeds and spray heights were controlled by the autopilot to ensure consistent application. After completing each pass, we immediately collected the WSPs and stored them in sealed and labeled envelopes to prevent contamination by ambient moisture. This careful handling ensured that any observed differences in droplet density or size were attributable to the tested parameters rather than environmental artifacts. The combination of different application heights and flight speeds generated a wide range of conditions for evaluating spray deposition.

2.4 Data analysis

All collected samples were processed using the DropScope instrument (SprayX, São Paulo, Brazil), which provided a precise measurement of spray coverage in percentage by detecting and quantifying droplet marks on each card. The SprayX scanner utilizes an app for a phone or tablet for analysis. Once the WSP card has been scanned, the image data is communicated to the application, where the app detects the droplets using an AI-based vision model. The attributes of the card, like the covered area and the quantity of drops, are calculated and stored appropriately. This automated image analysis tool enabled the consistent and reliable characterization of deposition patterns under varying operational conditions. After DropScope analysis, the resulting data was exported into a structured format suitable for further investigation.

Subsequent data handling and statistical analyses were conducted using Python's Pandas library. This approach facilitated efficient filtering, grouping, and summarizing of the dataset, as well as the generation of descriptive statistics like mean spray coverage and standard deviations for each treatment and experiment day. Treatment in this study includes application height, flight speed, and nozzle diameter. A T-test was conducted for pair-wise comparison of individual treatments using Python's Scipy library. Data visualization and graphical representations of the results were created using Python's Matplotlib library, allowing for clear comparisons across treatments and highlighting trends in droplet deposition patterns. By leveraging a flexible, script-based analytical workflow, this data analysis process supports repeatable, transparent, and customizable evaluations of how application parameters affected on-target coverage in different treatments.

3 Results and discussions

3.1 Effect of nozzle diameter on spray deposition

Nozzle diameters were evaluated using two distinct sizes, 140 and 320 μm, to determine how spray characteristics shift under the same flight speed and application heights (2.44 and 3 m). Overall, the smaller nozzle diameter (140 μm) produced slightly greater coverage than the larger diameter (300 μm) at 3 m height, but at 2.44 m height, the larger diameter produced slightly better coverage (Fig. 14.2). However, these differences were not statistically significant due to considerable variation among samples (Fig. 14.2). While finer droplets from the smaller nozzle can improve coverage, they may also be more susceptible to drift and evaporation (Sinha et al., 2022), highlighting the practical trade-offs between spray coverage and potential off-target movement. Despite the lack of statistical significance, the 140 μm nozzle diameter demonstrated slightly better performance overall and was therefore selected and utilized for the rest of the study.

In the end, environmental conditions, particularly wind speed and direction, significantly influenced spray outcomes; for example, when wind speed was higher in a particular direction, the deposition toward the swath in that direction was higher. Adverse conditions, such as high humidity and strong wind gusts during several test days, rendered the data unusable, emphasizing the importance of multiday trials to capture variability and ensure reliable conclusions. These findings underscore the necessity for precision in UAV operations, as even slight changes in environmental factors can drastically affect performance. In addition to wind speed, challenges arose when relative humidity was high (>80%) in July. On these spray trial days, we

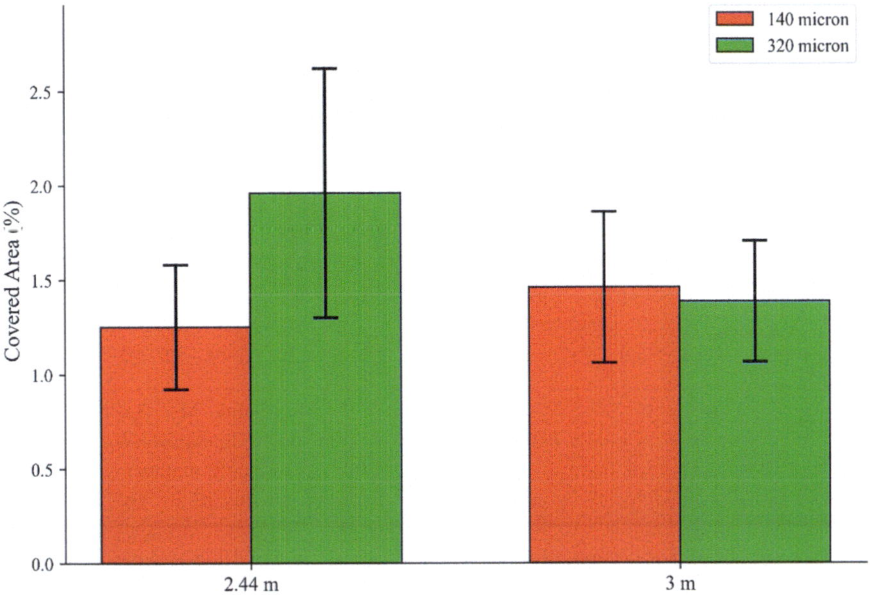

Fig. 14.2 Coverage area at the central swath for different nozzle diameters at 2.44 m and 3 m.

attempted to spray in the morning or in the evening around dusk, when wind speeds were lower than during mid-day. However, because of the high relative humidity, WSPs began to absorb atmospheric moisture after clipping the WSP cards to the watermelon leaf surface. Therefore, we recommend that researchers conduct spray studies during stable weather with lower wind speeds and lower relative humidity conditions.

3.2 Effect of application height on spray deposition

Application height remained a key factor influencing both the covered area and the quantity of drops observed across the swath (Fig. 14.3). In these tests, the 2.44 m application height consistently produced a higher concentration of droplets at the center of the swath compared to the 3 m height. At the flight speed of 5.8 m/s, this difference was evident in the density of droplets measured at the center, where the 2.44 m height peaked at around 300 drops/cm^2 with a total number of 35,000 droplets, while the 3 m height peaked closer to 200 drops/cm^2 with a total number of droplets closer to 20,000 (Fig. 14.3). A similar trend was noted when the flight speed was 2.6 m/s, but with an even greater gap in the number of droplets (Fig. 14.3A). The 2.44 m application height surpassed 60,000 droplets at its highest point, whereas the 3 m height reached approximately 40,000. At a flight speed of 5.8 m/s, the average droplet coverage at the center of the swath differed significantly between the two altitudes. As shown in Fig. 14.2, the 2.44 m application height resulted in a mean coverage of 3.43% (SE=0.83%), while the 3 m height yielded a mean coverage of 2.10% (SE=0.55%). This difference was statistically significant (P=0.0085), indicating that flying at a lower altitude (2.44 m) produced significantly greater central

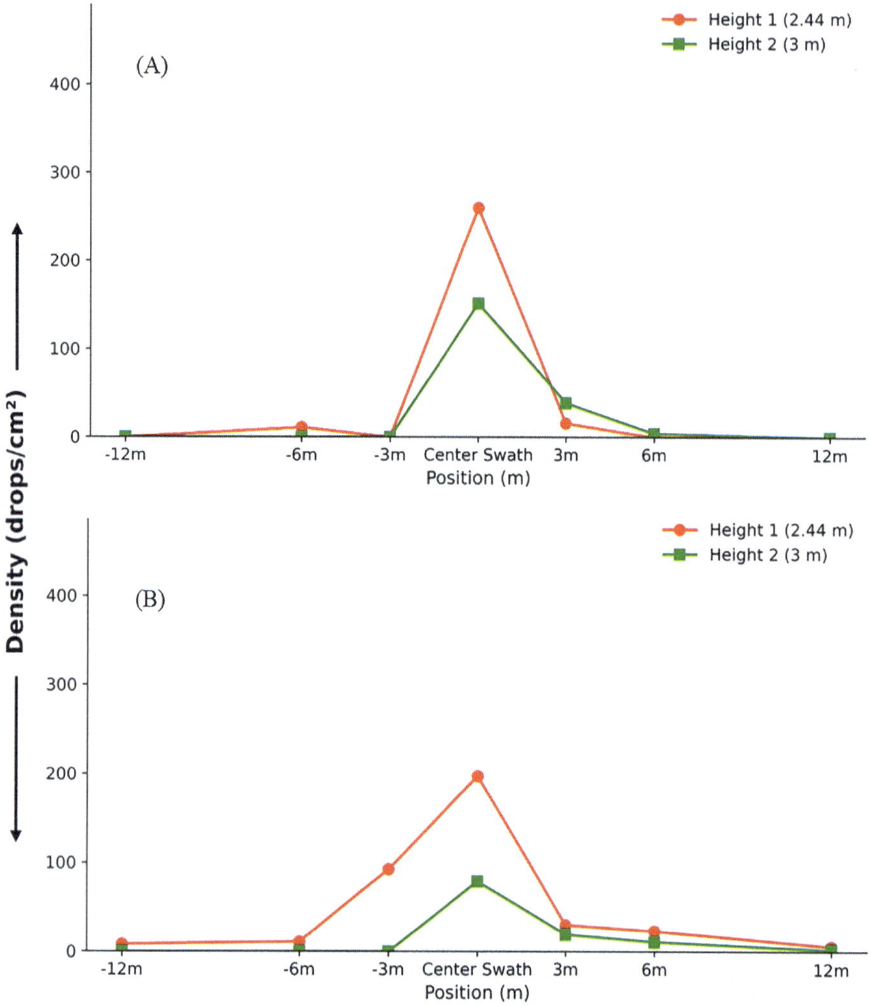

Fig. 14.3 Quantity of drops by position for different heights: (A) 2.6 m/s and (B) 5.8 m/s.

swath coverage than the higher altitude (3 m). This improvement is likely due to shorter droplet travel distances, which reduce the potential for drift and evaporation.

These findings align with the observed coverage area patterns, where the lower application height yielded a sharper peak in spray coverage near the center of the swath (Fig. 14.4). Our results are consistent with expectations, as operating the drone at a lower altitude (2.44 m) reduces the distance droplets must travel before contacting the plant canopy, thereby improving deposition efficiency. Limiting the distance between the sprayer applicator and the crop helps to minimize off-target losses, including drift and evaporation, which become more pronounced at higher flight altitudes (Younis, 1973; Bird et al., 1996; Martin et al., 2019).

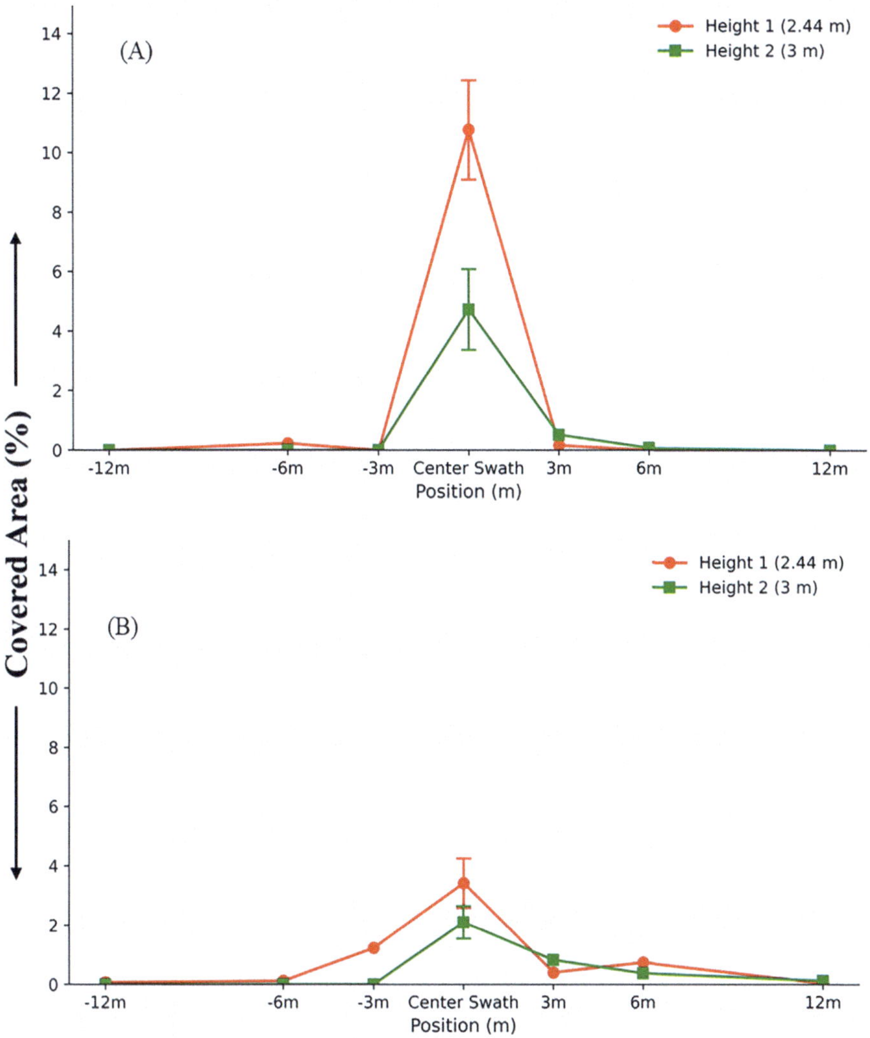

Fig. 14.4 Total coverage area average on WSP by position for different heights: (A) 2.6 m/s, and (B) 5.8 m/s.

Downwash from the UAV rotors plays a significant role in directing droplets downward and affects how they disperse in the canopy (Teske et al., 2018). At lower application altitudes, rotor-induced airflow is more concentrated, facilitating more efficient droplet transport and deposition within the target canopy zone. In contrast, at 3 m, droplets experience a longer airborne trajectory, increasing their exposure to environmental variables such as crosswinds and evaporation. Despite differences observed at the center of the swath, both application heights exhibited a characteristic bell-shaped deposition pattern, with steep declines in droplet density and count (Fig. 14.3) and coverage (Fig. 14.4) toward the swath edges. This distribution reflects a general concentration of droplets directly beneath the flight path. Under more stable atmospheric conditions, the central peak in droplet deposition becomes even more pronounced. However, flying at lower altitudes may also introduce challenges, such as increased rotor-induced turbulence, which can negatively affect deposition uniformity at the swath's periphery. These findings illustrate the importance of optimizing application height based on site wind conditions and crop canopy structure to balance deposition efficiency and uniformity.

3.3 Effect of flight speed on spray deposition

Flight speed had an even greater impact on spray coverage, as demonstrated by the comparison between 5.8 m/s and 2.6 m/s. Higher speeds at the same height reduce total coverage (Fig. 14.5), which is consistent with several previous studies (Woldt et al., 2018; Sinha et al., 2022; Martin et al., 2019). The slower speed (2.6 m/s) in this study produced a noticeably higher coverage peak (Fig. 14.5). This finding suggests that the extended dwell time associated with the slower flight enables droplets to remain more concentrated under the UAV's flight path, thereby improving on-target deposition. At 2.44 m height, when the flight speed was 5.8 m/s, the average coverage was 3.43% with a standard error (SE) of 0.83%, whereas 2.6 m/s had a significantly higher coverage of 10.77% with an SE of 1.67%, which is roughly three times the former coverage. The same trend can be observed at the 3 m height, with 2.6 m/s achieving 4.73% coverage, while 5.8 m/s only achieved 2.10%. The p-value from the t-tests for this data was calculated as 0.0008, which shows that a slower flight speed performs significantly better (Fig. 14.5).

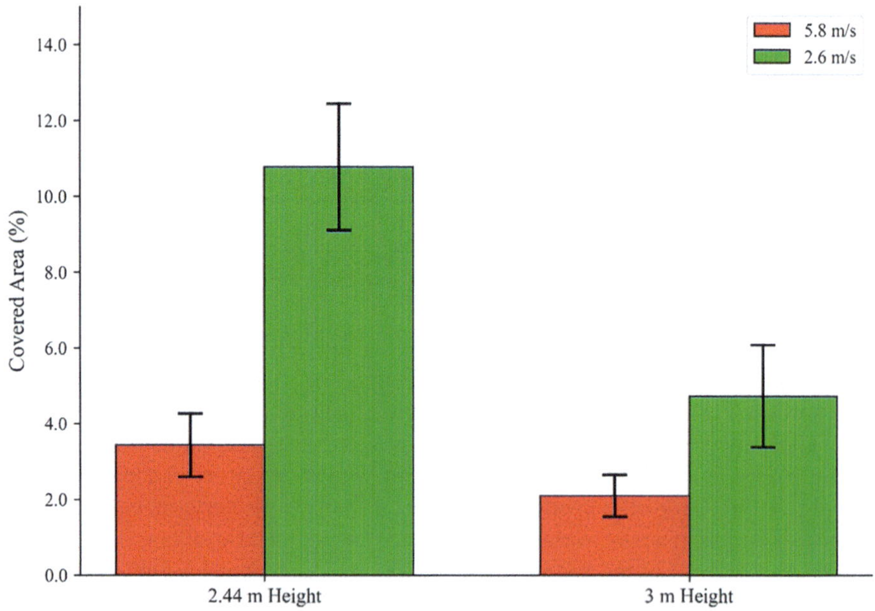

Fig. 14.5 Coverage area at central swaths for different flight speeds at 2.44 m and 3 m.

Both speeds followed a general "bell-shaped" coverage curve, indicating that the greatest droplet density occurred near the center of the swath (Fig. 14.6). Although rotor downwash is concentrated at the center of the swath, the data show that reducing flight speed can enhance droplet penetration in the central region. Nevertheless, some standard deviations at certain measurement points highlight the role of wind and other variables in shaping UAV-based spray outcomes. Previous studies examining the effects of flight speed on UAAS spray coverage have reported both consistent and divergent results, reflecting the variability and complexity of outcomes in UAAS research. For example, researchers using DJI MG-1 and HSE V6A platforms without rate control found that the slowest tested flight speed (1.0 m/s) yielded the highest coverage when compared to speeds ranging from 1.0 to 7.0 m/s (Woldt et al., 2018; Hunter et al., 2020; Sinha et al., 2022; Martin et al., 2019). In contrast, Martin et al. (2019) observed no significant difference among flight speeds of 3.0, 5.0, and 7.0 m/s for these same UAAS, whereas (Woldt et al., 2018) reported a decline in deposition above 3.0 m/s for the V6A but not for the MG-1. (Lv et al., 2019) further demonstrated how, in wind tunnel tests, raising flight speed (0.3–1.0 m/s) diminished overall coverage and droplet uniformity.

Overall, our findings show that applying pesticides at a slower flight speed can substantially improve coverage, especially around the center swath, likely due to longer exposure time. This is an advantage when targeting specific crop areas rather than the entire field. This trade-off suggests that slower speeds are better at achieving precise pesticide delivery or are better suited for crop systems with dense canopies or multiple layers of crop foliage to penetrate, whereas higher speeds may be more appropriate for achieving broader coverage uniformity in less sensitive areas. However, flight speed should be strategically decided relative to other parameters, such as application height, nozzle type, and local wind conditions, to achieve optimal coverage for each unique field scenario.

3.4 Effect of application height and flight speed on droplet density

Droplet density patterns were significantly affected by both application height and flight speed, as illustrated in Fig. 14.7. At 2.44 m above the canopy, the central swath consistently recorded higher droplet densities than at 3 m, regardless of the flight speed. For instance, at a 5.8 m/s flight speed, the peak density at the center reached approximately 200 drops/cm^2 for 2.44 m, whereas it was closer to 100 drops/cm^2 at 3 m (Fig. 14.7B). This difference became even more pronounced when the flight speed decreased to 2.6 m/s, with the lower altitude achieving around 250 drops/cm^2 at the center, compared to roughly 200 drops/cm^2 at 3 m (Fig. 14.7A). Statistical analysis confirmed that changes in application height were significant, with a p-value of 0.0008.

At 5.8 m/s, droplets had less time to settle directly beneath the UAV's flight path, resulting in lower peak densities across both altitudes (Fig. 14.7B). In contrast, flying at 2.6 m/s produced sharper peaks in droplet density at the center swath for both

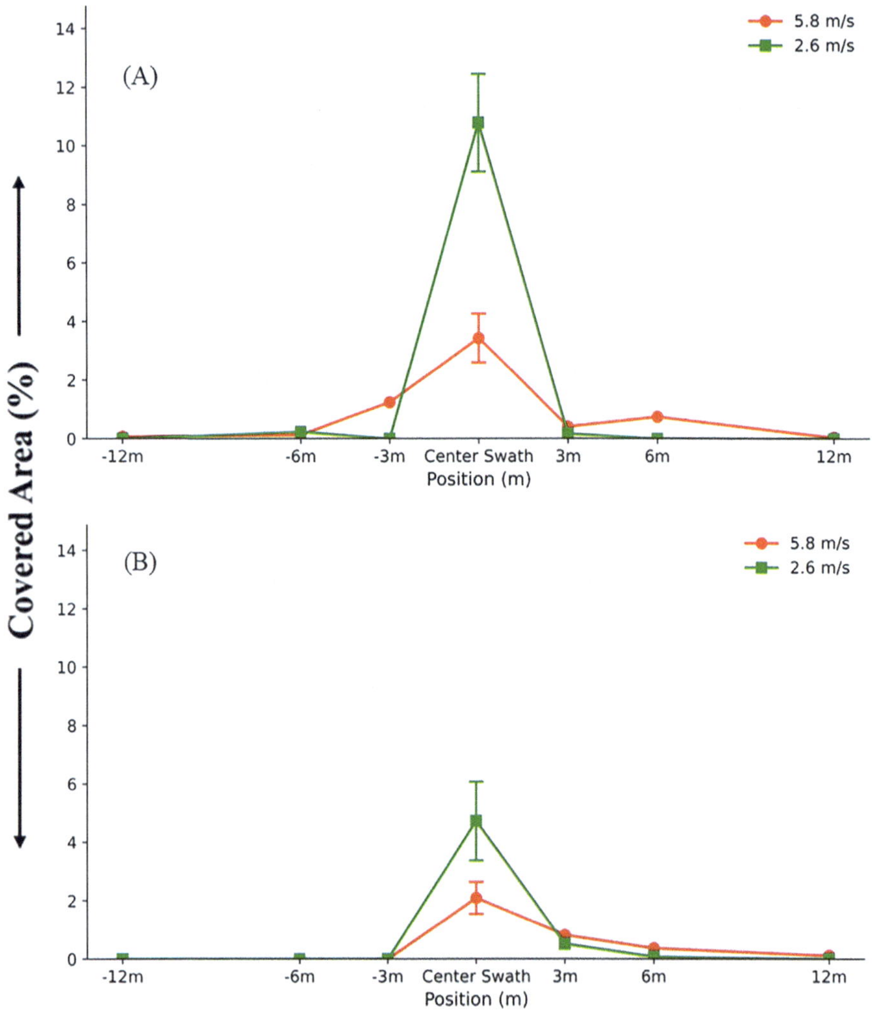

Fig. 14.6 Coverage area by position for different flight speeds: (A) at 2.44 m, and (B) at 3 m.

heights (Fig. 14.7A). This aligns with the notion that slower speeds extend the dwell time, allowing droplets to cluster more effectively before dispersal forces, such as wind or rotor downwash, carry them away. The difference in droplet density between 5.8 and 2.6 m/s was statistically significant, with a p-value of 0.0007.

Overall, these results mirror a characteristic "bell-shaped" deposition curve, where the highest droplet densities occur directly below the flight path and taper off toward the edges. Flying at a lower altitude (2.44 m) and slower speeds (2.6 m/s) both enhance on-target droplet density, likely due to reduced droplet travel distances and extended settling time. While these settings increase deposition in the central swath, they may also introduce challenges like rotor-induced turbulence or longer application

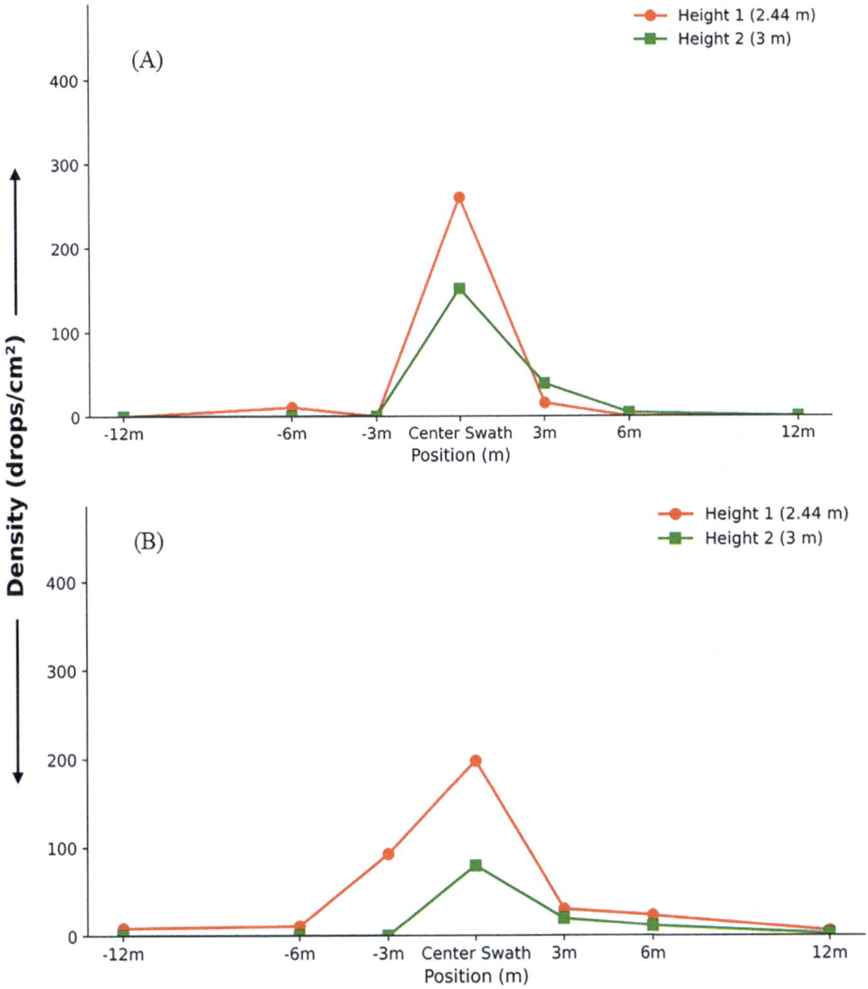

Fig. 14.7 Density of droplets by position for different heights: (A) at 2.6 m/s and (B) at 5.8 m/s.

times. Consequently, a balanced approach that accounts for site-specific conditions, such as canopy structure and wind speed, is key to optimizing droplet distribution and minimizing off-target drift.

4 Future directions

Building on this study, future research should explore integrating UAV operations with advanced technologies such as autonomous navigation, AI-based optimization, and high-resolution mapping. For instance, companies like DJI Agriculture have already implemented autonomous drone systems using AI-driven spraying strategies that optimize chemical distribution according to crop health data captured through high-resolution imaging. Examining additional variables, like spray formulations and canopy architectures, will provide a more comprehensive understanding of UAV applications. Predictive models that simulate diverse field scenarios will offer invaluable tools for researchers aiming to maximize UAV efficiency and efficacy.

5 Conclusion

This study highlights the critical role of UAV operational parameter optimization in modern agriculture. By systematically evaluating the effects of application height, flight speed, and nozzle diameter, our research identifies strategies to enhance spray deposition and coverage uniformity while minimizing off-target impacts. Lower altitudes and slower speeds were shown to significantly improve on-target deposition, while nozzle selection demonstrated the importance of adapting equipment to environmental conditions.

The spray performance of the DJI Agras T40 UAV was evaluated for coverage uniformity under varying operational parameters, including application height, flight speed, and nozzle diameter. The following key conclusions can be drawn from this study:

1. Lower heights (2.44 m) provided higher droplet concentrations at the swath's center compared to 3 m, minimizing drift and evaporation while maintaining a characteristic bell-shaped deposition pattern.
2. Slower speeds (2.6 m/s) improved central coverage and droplet penetration, while faster speeds (5.8 m/s) enhanced uniformity across the swath but reduced deposition at the swath's center.
3. Smaller nozzles (140 μm) generally provide better coverage but are more prone to drift and evaporation compared to larger nozzles (300 μm), which perform better under windy conditions.
4. Weather conditions, particularly wind, significantly influenced droplet trajectory and uniformity, emphasizing the need for multiday trials to account for real-world variability.

These findings highlight the importance of optimizing UAV parameters to achieve efficient and precise pesticide applications. Tailoring operational settings to specific field and crop conditions can improve spray efficacy, reduce chemical waste, and enhance environmental sustainability. The results of this study provide actionable insights into advancing UAV-based precision agriculture, offering a foundation for future innovations in sustainable farming practices.

Acknowledgement

The authors would like to thank the Mar-Del Watermelon Watermelon Association for providing funding to support this research. Authors would also like to thank the farmer and intern students involved in this study.

Abubakar Palli: Writing—review editing, Writing—original draft, Visualization, Software, Methodology, Conceptualization. Hemendra Kumar: Writing—review editing, Writing—original draft, Visualization, Software, Methodology, Conceptualization, Validation, Supervision, Resources, Project administration, Investigation, Funding Acquisition, Data Curation, Formal analysis. Haley Sater: Writing—review editing, Resources, Methodology, Data Curation, Conceptualization, Visualization, Funding acquisition.

References

Bird, S.L., Esterly, D.M., Perry, S.G., 1996. Off-target deposition of pesticides from agricultural aerial spray applications. J. Environ. Qual. 25, 1095–1104. doi:10.2134/jeq1996.00472425002500050024x.

Byers, C., Virk, S., Rains, G., Li, S., 2024. Spray deposition and uniformity assessment of unmanned aerial application systems (UAAS) at varying operational parameters. Front. Agron. 26733218. 6. Available from: www.frontiersin.org/journals/agronomy#. doi:10.3389/fagro.2024.1418623.

Chen, S., Lan, Y., Zhou, Z., Ouyang, F., Wang, G., Huang, X., Deng, X., Cheng, S., 2020. Effect of droplet size parameters on droplet deposition and drift of aerial spraying by using plant protection UAV. Agronomy. 2073-4395. 10 (2), 195. doi:10.3390/agronomy10020195.

Gade, S.A., Madolli, M.J., García-Caparrós, P., Ullah, H., Cha-um, S., Datta, A., Himanshu, S.K., 2025. Advancements in UAV remote sensing for agricultural yield estimation: a systematic comprehensive review of platforms, sensors, and data analytics. Remote Sens. Appl. Soc. Environ. 23529385. 37. Available from: https://www.sciencedirect.com/science/journal/23529385. doi:10.1016/j.rsase.2024.101418.

Hunter, J.E., Gannon, T.W., Richardson, R.J., Yelverton, F.H., Leon, R.G., 2020. Coverage and drift potential associated with nozzle and speed selection for herbicide applications using an unmanned aerial sprayer. Weed Technol. 15502740. 34 (2), 235–240. Available from: https://www.cambridge.org/core/journals/weed-technology/latest-issue. doi:10.1017/wet.2019.101.

Lou, Z., Xin, F., Han, X., Lan, Y., Duan, T., Wei, F., 2018. Effect of unmanned aerial vehicle flight height on droplet distribution, drift and control of cotton aphids and spider mites. Agronomy 2073-4395. 8 (9), 187. doi:10.3390/agronomy8090187.

Lv, M., Xiao, S., Tang, Y., He, Y., Jinhua Polytechnic College and College of Biosystems Engineering and Food Science, et al., 2019. Influence of UAV flight speed on droplet deposition characteristics with the application of infrared thermal imaging. Int. J. Agric. Biol. Eng 12, 10–17. doi:10.25165/j.ijabe.20191203.4868.

Martin, D.E., Woldt, W.E., Latheef, M.A., 2019. Effect of application height and ground speed on spray pattern and droplet spectra from remotely piloted aerial application systems. Drones 3, 83. doi:10.3390/drones3040083.

Praseartkul, P., Taota, K., Pipatsitee, P., Tisarum, R., Sakulleerungroj, K., Sotesaritkul, T., Himanshu, S.K., Datta, A., Cha-um, S., 2023. Unmanned aerial vehicle-based vegetation monitoring of aboveground and belowground traits of the turmeric plant (Curcuma longa L.). Int. J. Environ. Sci. Technol. 1735-1472. 20 (8), 8673–8686. doi:10.1007/s13762-022-04545-6.

Sinha, R., Johnson, J., Power, K., Moodie, A., Warhurst, E., Barbosa, R., 2022. Understanding spray attributes of commercial UAAS as impacted by operational and design parameters. Drones. 2504-446X. 6 (10), 281. doi:10.3390/drones6100281.

Srinivasarao, A., Khura, T.K., Parray, R.A., Kushwaha, H.L., Mani, I., 2021. Spray droplet deposition, collection, and analysis techniques: a review. Pharma Innov. J. 10 (SP-6), 156–165.

Teske, M.E., Wachspress, D.A., Thistle, H.W., 2018. Prediction of aerial spray release from UAVs. Trans. ASABE 61, 909–918. doi:10.13031/trans.12701.

Woldt, W., Martin, D., Latheef, M., Kruger, G., Wright, R., McMechan, J., Proctor, C., Jackson-Ziems, T., 2018. Field evaluation of commercially available small unmanned aircraft crop spray systems. In: ASABE 2018 Annual International Meeting. American Society of Agricultural and Biological Engineers, United States. doi:10.13031/aim.201801143.

Yan, Y., Lan, Y., Wang, G., Hussain, M., Wang, H., Yu, X., Shan, C., Wang, B., Song, C., 2023. Evaluation of the deposition and distribution of spray droplets in citrus orchards by plant protection drones. Front. Plant Sci. 1664462X. 14. Available from: https://www.frontiersin.org/journals/plant-science. doi:10.3389/fpls.2023.1303669.

Younis, S., 1973. Effect of Thickener and Lowered Air Density on Spray Drop Formation and Evaporation Rate. University of Illinois at Urbana-Champaign, Champaign.

Zhang, X.Q., Song, X.P., Liang, Y.J., Qin, Z.Q., Zhang, B.Q., Wei, J.J., Li, Y.R., Wu, J.M., 2020. Effects of spray parameters of drone on the droplet deposition in sugarcane canopy. Sugar Tech. 09740740. 22 (4), 583–588. Available from: http://www.springer.com/life+sci/agriculture/journal/12355. doi:10.1007/s12355-019-00792-z.

Real-world applications of smart technologies in protected farming

15

Yibin Wang, Al Bashir, Partho Ghose, Mike Ojo, and Azlan Zahid
Department of Biological and Agricultural Engineering, Texas A&M AgriLife Research, Texas A&M University, Dallas, TX, United States

Abbreviations

AI	artificial intelligence
ANN	artificial neural network
CEA	controlled environment agriculture
CNN	convolutional neural network
CO_2	carbon dioxide
DL	deep learning
FPN	feature pyramid network
FPS	frames per second
GAN	generative adversarial network
IoT	internet of things
IoU	intersection over union
LLM	large language model
mAP	mean average precision
MAPE	mean absolute percentage error
PF	protected farming
PID	proportional-integral-derivative
R^2	coefficient of determination
RGB-D	red green blue-depth
RMSE	root mean square error
YOLO	you only look once

1 Introduction

Protected farming (PF) or controlled environment agriculture (CEA) employs smart technologies to actively regulate environmental conditions to optimize plant growth and overall productivity (Kumar et al., 2021). The expansion of CEA offers increased food production within and near urban areas, as these systems do not depend on traditional arable farmland. CEA increasingly relies on cutting-edge technologies to enhance the production of specialty crops, such as berries, herbs, microgreens, and high-value vegetables. The integration of smart technologies, including artificial intelligence (AI), the Internet of Things (IoT), sensors, and robotics, has revolutionized how farmers

Precision Technologies for Digital Agriculture. https://doi.org/10.1016/B978-0-443-40513-6.00008-1

monitor and manage these crops, leading to more sustainable and profitable operations (Subeesh and Mehta, 2021). One of the key benefits of smart technologies in specialty crop farming is precision control over environmental factors such as temperature, humidity, light intensity, and CO_2 levels. Smart climate control systems equipped with sensors continuously monitor these variables and automatically adjust heating, cooling, shading, and ventilation to maintain optimal growing conditions. For instance, AI-driven climate control in greenhouses can predict weather changes and make real-time adjustments, reducing energy consumption (Lee et al., 2024). Another critical area of focus is automated irrigation and nutrient management. Specialty crops often have specific water and nutrient requirements, and traditional irrigation methods can lead to resource overuse or underuse. Smart irrigation systems use soil moisture sensors and robotic systems incorporated with AI algorithms to deliver precise amounts of water and fertilizers based on the crop's real-time needs (Maraveas, 2022). This strategy conserves water and prevents nutrient runoff, enhancing both sustainability and crop productivity. Pest and disease management has also been transformed through smart technologies in controlled environments. Early indicators of pests or plant diseases can be identified with advanced imaging and machine learning, allowing farmers to act immediately and limit the spread (Chakraborty et al., 2022). Automated sprayers and drones equipped with smart imaging can apply targeted treatments, reducing pesticide use while maintaining crop health. Therefore, integrating smart technologies into crop farming within protected environments has significant potential to improve efficiency, sustainability, and crop quality.

Robotics and automation further enhance efficiency in protected crop farming. Automated harvesting robots, for instance, are increasingly adopted for labor-intensive crops in greenhouses, such as strawberries, cucumbers, and tomatoes (Tangarife and Díaz, 2017). These robots are equipped with computer vision techniques to precisely identify ripe produce and harvest it, reducing dependency on manual labor and improving overall efficiency. Major robotics-assisted activities include sowing, transplanting, harvesting, and resource management, such as water usage and the application of chemical inputs like fertilizers, herbicides, and fungicides. These agricultural processes are highly suitable for automation through advanced technologies, including image processing for disease and weed detection, fruit identification during harvesting, precision water management, and localized spraying. In addition, automated mechanical systems, such as robotic arms, autonomous tractors, and conveyor-based harvesting units, are increasingly integrated into both PF and general agriculture to streamline labor-intensive tasks, minimize human intervention, and optimize productivity (Cheein and Carelli, 2013). As related technology continues to evolve and global demand for specialty crops continues to grow, the combination of robotics and AI is expected to further revolutionize precision farming, enabling real-time decision-making and adaptive management of agricultural operations. The integration of smart technologies in CEA is becoming essential, as it provides higher yields, superior quality, and promotes resource efficiency. Accordingly, the main objectives of this book chapter are summarized as:

1. Clarify the scope and operational requirements of PF, as well as the roles of AI, IoT, automation, and robotics in sensing, control, and decision support within controlled environments.
2. Review real-world applications and system architecture in PF, including sensing, AI algorithms, and in-field operation pipelines.
3. Illustrate end-to-end integration through two greenhouse case studies regarding AI-enabled production management and robotic operations.
4. Conclude future research directions and practical guidelines for resilient and resource efficient PF systems.

2 Review of recent advancements

2.1 Artificial intelligence applications in protected farming

With nearly 10 billion people anticipated by 2050, the demand for food production is increasing exponentially (FAO, 2018). AI, leveraging its ability to analyze vast datasets and offer practical insights, is transforming the agricultural sector into a data-driven and precision-oriented industry. From disease detection to irrigation management and yield prediction, AI offers various applications capable of reshaping modern farming practices. Some of the AI applications in PF are listed in Table 15.1.

Disease detection: Timely detection of crop diseases remains a significant challenge in agriculture, with delayed diagnosis often resulting in substantial yield loss. Recent developments in AI have demonstrated its efficacy in addressing this issue. AI systems can identify early-stage disease symptoms through subtle changes in leaf coloration, texture, and morphology by analyzing high-resolution red-green-blue (RGB), thermal, multispectral, and hyperspectral images. For example, an improved Faster Region-based Convolutional Neural Network (R-CNN) framework was introduced for strawberry disease detection, utilizing a custom dataset of leaves, flowers, and fruits. The model obtained a mAP score of 92.18% with an inference time of 229 ms, outperforming Mask R-CNN and YOLOv3 in terms of both accuracy and computational efficiency (Zhao et al., 2022a).

Pest control and management: Pests are another major threat to agricultural productivity, even persisting within CEA systems, and often lead to significant economic losses. AI-driven pest management solutions offer more precise and sustainable alternatives suited to CEA. Imaging and optical-acoustic sensors enable real-time detection of pest infestations by analyzing movement patterns, thermal emissions, and acoustic signatures (Akbar et al., 2024). For instance, a DL algorithm was implemented to recognize pests and diseases in Brassica chinensis, integrate image tiling and deblurring algorithms, and enhance the CenterNet DL architecture. The system attained an average precision (AP50) of 87.2% and a coefficient of determination (R^2) of 94.7%, indicating robust detection performance and highlighting its potential to advance precision pest management and crop health monitoring in CEA systems (Zhao and Shi, 2022b).

Table 15.1 Application of AI for different operations in protected farming.

Agricultural practices	System description	Method and performance	References
Disease detection	Disease detection under natural field conditions with a complex background for strawberries.	Modified Faster R-CNN, incorporated with ResNet, Feature Pyramid Network (FPN), and Convolutional Block Attention Module (CBAM), achieved a mean Average Precision (mAP) of 92.181%, recording a mean inference time frame of 229 ms.	(Zhao et al., 2022a)
Pest control and management	An AI-based enhanced model was introduced for pest and disease identification in Brassica chinensis. Unmanned aerial vehicles (UAVs) captured aerial imagery to train the model.	An enhanced model, called CenterNet, was developed to improve localization accuracy. It demonstrated strong performance, achieving an average precision (AP50) of 87.2% and an R^2 value of 94.7% for pest and disease detection.	(Zhao and Shi, 2022b)
Microclimate prediction and management	A two-stage control system was developed for hydroponic lettuce cultivation in CEA. The system optimizes microclimate regulation to minimize energy use while maintaining ideal growth parameters under changing weather conditions.	A multifactor optimization applied the nondominated sorting genetic algorithm to identify ideal temperature and humidity setpoints, and a radial basis function neural network predicted them in real time under varying weather conditions for real-time control adaptability.	(Abbood et al., 2023)
Nutrient solution management	An AI-based control system was implemented to optimize nutrient concentration, pH, and water temperature in hydroponic basil cultivation.	The system provided optimal conditions with EC levels, a pH between 5.8 and 6.3, and a nutrient solution temperature of 28°C–30°C, resulting in basil plants reaching an average height of 14 cm.	(Tuan et al., 2020)

Nutrient stress prediction	Deep learning (DL) network was applied to assess the nutrient insufficiencies in hydroponic basil.	DenseNet201, ResNet101V2, MobileNet, and VGG16 were evaluated to detect plant nutrient deficiencies. After fine-tuning, VGG16 achieved the utmost accuracy of 93.82%, effectively identifying key visual symptoms associated with nutrient stress.	(Gul and Bora, 2023)
Yield prediction	AI-based yield prediction system using YOLOv3 model for yield prediction of tomatoes in the plant factory with artificial lighting (PFAL).	Demonstrated strong predictive capabilities with an improved YOLOv3 integrated with the loss function derived from the generalized intersection over union (GIoU). The AI system achieved a 99.3% mAP value for yield prediction.	(Wang et al., 2022)
3D reconstruction and phenotyping	An unmanned greenhouse robot was used for image acquisition, and the Neural Radiation Fields (NeRF) algorithm was applied for 3D reconstruction of tomato crops in greenhouse conditions.	Internode length estimation achieved a 97.30% R^2 score and a Mean Absolute Percentage Error (MAPE) of 0.089, while leaf area and fruit volume estimations showed 95.30% and 96.00% R^2 scores with MAPE values of 0.090 and 0.135, respectively.	(Choi et al., 2024)

Microclimate prediction management: Microclimate variability is a major and often unpredictable factor affecting productivity in CEA. Maintaining stable and optimal indoor conditions in dynamic environments such as greenhouses is essential for maximizing plant growth and productivity. AI-assisted intelligent microclimate control systems offer a promising solution by leveraging predictive modeling to manage complex environmental parameters autonomously (Rasheed et al., 2025). As an illustration, Abbood et al. (2023) introduced a two-stage framework that combines optimization techniques with machine learning algorithms, achieving 25.88% energy savings (Abbood et al., 2023). These outcomes demonstrate the efficacy of AI-driven approaches for improving environmental control and energy efficiency in agricultural systems.

Nutrient solution management: Accurate nutrient management is essential in CEA, as imbalances can affect plant growth, reduce yield, and compromise crop quality. AI enables precise nutrient monitoring by integrating data from imaging sensors, pH and EC probes, and environmental control systems (Catota-Ocapana et al., 2025). AI-based nutrient detection systems can provide timely insights, reduce reliance on manual sampling, and optimize nutrient delivery for healthier, more productive crops. For instance, an AI-based control system was developed in a hydroponic setup to optimize parameters such as nutrient concentration, pH, and temperature, leading to healthy basil growth (Tuan et al., 2020).

Nutrient stress prediction: Predicting nutrient stress in CEA helps maintain healthy plant development and prevents potential yield losses by enabling early interventions. In this regard, AI facilitates the early detection of nutrient stress by processing data from imaging modalities such as RGB, multispectral, and thermal sensors, enabling timely interventions and improved input use efficiency (Jawad et al., 2025). Advancements in DL architecture, including CNNs, have been utilized to predict and regulate macronutrient concentrations in plants accurately (Akbar et al., 2024). For example, Z. Gul et al. explored neural networks such as VGG16, DenseNet201, and ResNet101V2 to identify nutrient deficiencies in hydroponic basil with high accuracy by effectively recognizing key plant stress symptoms (Gul and Bora, 2023).

Yield prediction: In CEA, accurate yield prediction is essential for optimizing resource use, guiding environmental control strategies, and maximizing productivity and economic returns. AI-based models provide considerably accurate forecasts of crop yields, enabling timely decision-making and promoting optimized resource management (Akbar et al., 2024). The development and deployment of advanced AI, such as artificial neural networks, generative adversarial networks, and You Only Look Once (YOLO), have further strengthened the precision and robustness of yield estimation systems. For example, Wang et al. introduced an improvised YOLOv3 architecture for estimating tomato yields in lit plant factories and achieved an mAP value of 99.3% with a processing time of 15 ms (Wang et al., 2022).

3D reconstruction and phenotyping: The advent of three-dimensional (3D) reconstruction technologies has significantly advanced precision farming by enabling multiangle visualization and dynamic growth analysis. These models support noninvasive phenotyping in plant breeding programs, allowing traits like plant height and canopy structure to be measured efficiently. Accurate virtual representations of crops also enhance decision-making while saving time and resources (Akbar et al., 2024). For instance, Choi et al. (2024) employed an unmanned greenhouse robot and the Neural Radiance Fields algorithm to reconstruct 3D models of tomato plants, achieving high accuracy in estimating internode length, leaf area, and fruit volume (Choi et al., 2024). These outcomes underscore the prospect of AI to automate and improve plant trait analysis in CEA systems.

In summary, AI is transforming CEA by enhancing productivity, resource efficiency, and sustainability. By integrating advanced sensors, machine learning, and predictive analytics, AI enables early disease detection, precise pest control, and accurate crop growth and environmental condition forecasting. Beyond monitoring, AI actively optimizes irrigation, nutrient delivery, and yield prediction processes. Emerging tools like 3D reconstruction and phenotyping further advance crop analysis and breeding. Collectively, AI-driven solutions strengthen daily farm management and play a role in the long-term resilience and food security of agricultural systems.

2.2 Automation and robotics

Greenhouse automation has significantly impacted agricultural operations, with robotics emerging as one of the most influential advancements. Robotics enhances efficiency in performing farming tasks and contributes to reducing environmental impact. This section provides an overview of recent advancements in automation technologies within a controlled environment, categorized based on different practices. Table 15.2 summarizes automation systems adopted for major agricultural applications.

Harvesting: Automation in harvesting addresses labor shortages and increases efficiency by employing robotic systems embedded with advanced sensors and machine learning algorithms. These robots can navigate fields, identify ripe produce, and perform delicate picking operations, reducing reliance on manual labor and minimizing crop damage. For instance, autonomous machines capable of visual detection and navigating orchards to harvest crops such as cucumbers and sweet peppers have been developed to enhance productivity and address labor challenges (Ting et al., 2009; Sa et al., 2017).

Phenotyping: Automated phenotyping employs advanced sensors and imaging technologies to noninvasively measure and analyze plant traits such as height, fresh weight, dry weight, leaf number, and leaf area. These systems operate continuously, providing precise data without damaging plants, accelerating research outcomes, and reducing manual labor. For example, the PlantEye F600 H-Portal scanner enables precise, automated measurements, significantly reducing the workload of plant scientists (Manavalan et al., 2021).

Seeding and transplanting: Seeding and transplanting automation systems utilize precision machinery and robotic systems to ensure accurate placement and spacing of seeds and seedlings. These technologies enhance crop establishment by maintaining uniform planting depths and densities, thereby improving yields. Autonomous seeders and robotic transplanters can adjust to varying field conditions, optimizing the planting

Table 15.2 Recent advancements in automation and robotic systems in CEA.

Agricultural practices	Description of associated automation systems	Methods and performance	References
Harvesting	Robotic arms and autonomous harvesters, equipped with image processing technologies, identify and pick ripe produce	Supervised learning on RGB-D–derived color and geometric features enabled accurate detection of sweet pepper peduncles	(Ting et al., 2009; Sa et al., 2017)
Phenotyping	Sensors, imaging technologies, and AI-driven analytics are utilized to noninvasively measure and analyze plant traits such as height, biomass, and leaf area	3D vision combined with multispectral imaging nondestructively acquires plant data, providing precise, objective morphological and physiological parameters	(Manavalan et al., 2021)
Seeding and transplanting	Automated seeders and robotic transplanters ensure uniform planting depth and spacing	Transplanting machines are synchronized in motion and positioning, reasonable in assembling the control system, and stable in operation	(Yang et al., 2020)
Microclimate control	AI-powered systems monitor and optimize environmental parameters, including temperature, humidity, and CO_2 to maintain ideal growing conditions	The model generated 24-hour and 7-day forecasts of microclimate variables, including ambient temperature and light radiation	(Ma et al., 2019; Petrakis et al., 2022)
Growth monitoring	UAVs and IoT sensors stream up-to-the-minute measurements of crop health, growth patterns, and the local environment	A wireless sensor network was implemented for temperature, humidity, light collection, and data-driven growth prediction	(Rodríguez et al., 2017; Zhang et al., 2022)
Pest management	Advanced imaging systems detect early signs of pest infestations, enabling targeted pesticide application and reducing chemical use	Insect activity was effectively characterized in space and time, with the counting algorithm achieving 93% average temporal accuracy	(Rustia et al., 2020)
Irrigation nutrient management	Smart irrigation and automated nutrient delivery systems analyze soil and plant data to administer precise water and fertilizer dosages	Decision support systems were developed as smartphone apps using web-based software to provide site-wise and crop-specific irrigation and nutrient recommendations	(Gallardo et al., 2020)

process and reducing labor requirements. A recent study achieved success at various planting frequencies, with overall success rates exceeding 96.5% (Yang et al., 2020).

Microclimate control: Microclimate control systems in greenhouses often integrate IoT devices, sensors, and AI algorithms to monitor and regulate environmental parameters such as temperature, humidity, light, and CO_2 concentration. These systems provide real-time adjustments to guide and maintain optimal growing conditions, which enhance crop yields and resource efficiency. The simulation models can accurately predict temperatures and light radiation over different times and positions in a greenhouse (Ma et al., 2019). Integrating AI approaches such as neural networks facilitates precision agriculture by offering detailed, real-time information to farming practitioners (Petrakis et al., 2022).

Growth monitoring: Continuous plant monitoring enables timely interventions, optimizing resource use and improving overall crop performance. A wireless sensor network was designed and implemented to collect agricultural environment data such as temperature, humidity, and light (Rodríguez et al., 2017). Each sensor node is capable of sending monitoring data to the cloud. In addition, a convolutional neural network (CNN) was adopted and demonstrated superior estimation performance for the flat-type cultivars of Flandria and Tiberius (Zhang et al., 2022).

Pest management: Automated pest management systems integrate sensors, computer vision techniques, and robotics to detect, identify, and reduce pest populations at an early stage. These systems can deploy targeted interventions, such as precision spraying of pesticides, thereby reducing chemical use and environmental impact. An automated insect-counting and environmental-monitoring system was developed using camera-equipped embedded nodes within a wireless sensor network, providing an efficient tool for long-term pest behavior studies and applied Integrated Pest Management (IPM) in protected environments (Rustia et al., 2020).

Irrigation and nutrient management: Advanced irrigation systems employ soil moisture sensors, climate data, and machine learning algorithms to optimize water usage. These systems deliver precise amounts of water based on real-time assessments, conserving resources and ensuring adequate hydration for crops. Smart irrigation technologies are particularly beneficial in regions facing water scarcity, enhancing sustainability in agricultural practices. Adopting decision support systems leads to improved water efficiency and crop yields (Gallardo et al., 2020). Automation in nutrient management involves using advanced sensors to monitor soil nutrient levels and plant health indicators. Fertilization schedules and compositions can be adjusted in real-time, ensuring proper nutrient delivery tailored to specific crop needs.

In summary, automation and robotics are transforming agriculture by significantly improving efficiency, precision, and sustainability across various farming practices. In CEA, these advanced technologies optimize critical processes such as harvesting, seeding, transplanting, weeding, pest control, nutrient management, and irrigation, reducing manual labor dependency and increasing productivity. Automated microclimate control systems regulate temperature, humidity, CO_2, and lighting, maintaining optimal growing conditions with minimal human input. The continued integration of automation and robotics in CEA will lead to a more precise, resilient, and productive agricultural system, ensuring food security, reducing waste, and improving overall sustainability.

3 Case studies

3.1 Case study 1: AI-enabled real-time lettuce growth monitoring

Introduction

Lettuce exhibits a strong responsiveness to regulated environmental circumstances, facilitating the precise optimization of light, temperature, and nutrient levels. This adaptivity ensures consistent quality and yield, rendering lettuce a favored choice for urban agriculture and greenhouse farming systems (Goh et al., 2023). For growth data analytics, it is essential to gather crop performance data, such as fresh weight, dry weight, leaf area, plant diameter, and plant height. Conventional growth monitoring methods necessitate the destructive sampling of plant samples, incurring labor costs, which are the primary expense in agriculture (Levy and Jarvis, 1999; Zhang et al., 2018). Image-based techniques have been routinely utilized to derive crop performance metrics. Images provide a noninvasive, efficient, precise, and cost-effective method for assessing crop growth.

The data collected from sensors such as RGB cameras, RGB-depth (RGB-D) cameras (Regazzoni et al., 2014), or Light Detection and Ranging (LiDAR) (Maksymova et al., 2018) undergoes segmentation to distinguish plant pixels from the background using either traditional methods or machine learning approaches. However, there has been limited research utilizing both RGB and depth images for crop growth estimation, and the few existing studies are inadequate for real-time applications in resource-constrained agricultural settings. Addressing this gap, this study aims to develop a lightweight semantic segmentation framework based on DeepLabv3+ combined with MobileNetv2, tailored for lettuce plant analysis. The model extracts essential phenotypic traits—including fresh and dry weight, leaf area, plant height, and plant diameter—through multimodal fusion of RGB-D data and deep learning-based regression. This lightweight model is designed for integration into an edge-AI system, enabling real-time crop growth monitoring.

Materials and methods

Image data collection and processing

To acquire image datasets consisting of both color and depth images, an Intel® RealSense™ Depth Camera D455 was utilized. Image collection took place before each harvesting event, followed by destructive measurements. This study contains 302 RGB images and 302 corresponding depth images. The 2D image segmentation dataset was annotated using Computer Vision Annotation Tool (CVAT), a computer vision annotation tool. Each lettuce plant instance in the raw RGB images was outlined with a polygon, which served as the class label. The dataset was then divided into 302 images, with corresponding masks split into 232 image-mask pairs for training and 70 for testing.

To enhance the dataset, the Python-based image augmentation library, Augmentations (Buslaev et al., 2020) was employed. Various transformations—including cropping, rotations (-90° and +90°), horizontal and vertical mirroring, deformation, and brightness and contrast adjustments—were applied to the training dataset. These augmentations increased the dataset size to a total of 2718 images.

Leaf segmentation network

A DeepLabv3+-based network was employed to segment lettuce leaves from RGB images (Fig. 15.1). While DeepLabv3+ offers robust segmentation capabilities, its high parameter count and complex structure present challenges for deployment on embedded devices. The default backbone feature extractor, Xception (Chollet, 2017), was replaced with MobileNetv2 (Sandler et al., 2018) to address these limitations. MobileNetv2 optimizes computational efficiency while maintaining accuracy by integrating depth-wise convolution, inverted residuals, and a linear bottleneck. Each pixel in the input images is classified, with leaf pixels labeled as 1 and background pixels as 0. To evaluate segmentation performance, the similarity between the predicted mask and the ground truth was assessed using the Dice loss function (Milletari et al., 2016) which was used to optimize the model.

Multimodal regression network

The assessment of fresh weight, dry weight, and leaf area is essential for obtaining comprehensive insights into crop performance, facilitating resource management for enhanced yield and quality. This work utilized image segmentation results to extract geometric features, which are then applied in a regression model to predict fresh weight, dry weight, and leaf area. The projected leaf area, computed with the OpenCV library, represents the two-dimensional surface area of a leaf when flattened on a surface. The plant diameter was also calculated using OpenCV-based

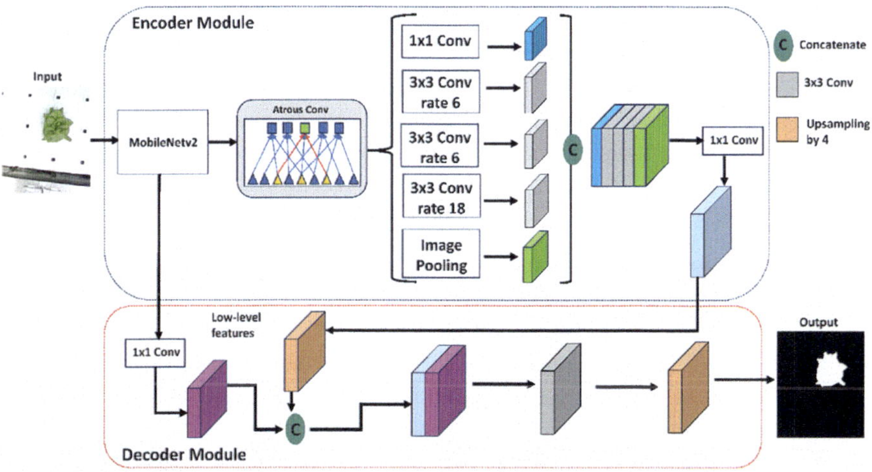

Fig. 15.1 DeepLabv3+ with MobileNetv2 network architecture for leaf segmentation. Conv: convolutional layer.

edge and contour detection techniques by locating the pair of points farthest apart in pixels from the edge pixels. To accurately estimate lettuce growth traits, this study employs a multibranch regression network consisting of two essential blocks: (1) the feature extraction block and (2) the regression block. This work utilized ResNet-50 as the feature extraction block for multimodal input, comprising RGB and depth images together with growth attributes, efficiently utilizing multiscale feature maps. Removing the 1000 × 1 fully connected layer from each branch enables the extraction of RGB and depth characteristics from the flattened layer. Meanwhile, a multilayer perception layer aids in extracting traits from the segmentation network. Adam optimizer, which iteratively modifies the weights to minimize the estimated loss, is used to optimize the model's performance. The model continuously adjusts its internal weights and biases during training by analyzing the disparity, also known as loss or error, between its predictions and the actual ground truth values. This discrepancy is quantified through a loss function, with mean squared error (MSE, Eq. 1).

$$Loss_{MSE} = \frac{1}{N}\sum_{I}^{N}(z_i - \hat{z}_i)^2 \tag{1}$$

The models were implemented in Python using the PyTorch framework. To optimize the network's weights and biases, the learning rate and weight decay were both set to 0.0001. Training was conducted on an NVIDIA RTX A6000 GPU with a batch size of 32, and the process continued until the model reached convergence. The model's accuracy was measured using the coefficient of determination (R^2), root mean square error, and mean absolute percentage error.

Results and discussion

Segmentation model performance

DeepLabv3+ with MobileNetv2 (Fig. 15.2) outperforms other state-of-the-art segmentation networks, including U-Net, PSPNet, and FPN models, achieving a mean IoU of 0.9979, an accuracy of 0.9985, and a mean dice loss of 0.0045. Furthermore, the algorithm delivers the highest frames per second (FPS), highlighting its notably

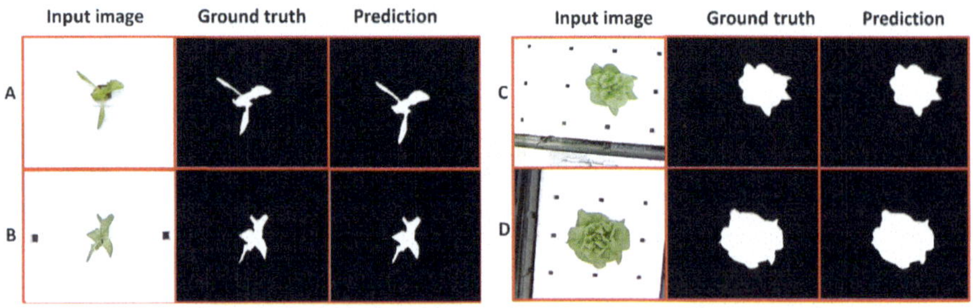

Fig. 15.2 Segmentation results obtained by DeepLabv3+ with MobileNetv2.

shorter prediction times compared to other segmentation models. Interestingly, when using MobileNetv2 instead of Xception as the backbone, the FPS decreases by 0.059. These findings emphasize that DeepLabv3+ with MobileNetv2 offers a strong balance of segmentation accuracy and processing speed, making it a highly effective choice for growth monitoring (Table 15.3).

Phenotypic prediction performance

The RGB-D model was assessed using the test dataset, showcasing strong predictive performance in estimating fresh weight, leaf area, dry weight, plant diameter, and plant height, with R^2 values of 0.968, 0.953, 0.943, 0.906, and 0.965, respectively (Fig. 15.3). The results demonstrated that combining the RGB-D model with phenotypic traits led to higher estimation accuracy, surpassing models that relied only on RGB with phenotypic traits, as shown in Fig. 15.3.

Table 15.3 Performance comparison of different segmentation models.

Model	**Backbone**	**mIoU**	**Accuracy**	**IoU**		**Mean dice loss**	**FPS**
				Leaf	***Background***		
DeepLabv3+	Xception	0.9962	0.9976	0.9948	0.9976	0.0115	0.134
	VGG16	0.9835	0.9972	0.9722	0.9977	0.0934	0.239
	VGG19	0.9829	0.9964	0.968	0.9978	0.0972	0.218
	MobileNetv2	**0.9979**	**0.9985**	**0.9978**	**0.998**	**0.0045**	**0.075**
FPN	ResNext50	0.9796	0.9952	0.9682	0.991	0.0651	0.164
U-Net	–	0.9951	0.9968	0.9925	0.9977	0.0093	0.186
PSPNet	MobileNetv2	0.9775	0.9942	0.9596	0.9954	0.1003	0.134
	ResNet101	0.9689	0.9938	0.943	0.9948	0.1523	0.364

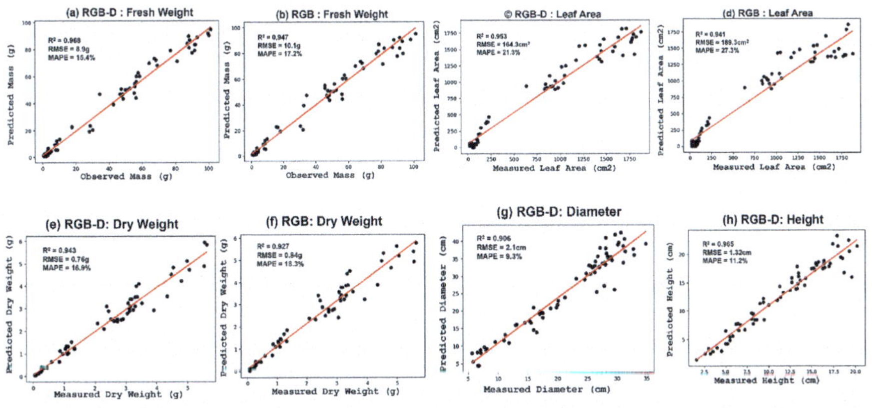

Fig. 15.3 Comparison of predicted and observed values on the test dataset using the developed model: (A) RGB-D fresh weight, (B) RGB fresh weight, (C) RGB-D leaf area, (D) RGB leaf area, (E) RGB-D dry weight, (F) RGB dry weight, (G) RGB-D diameter, and (H) RGB-D height.

Future directions

In greenhouse agriculture, the use of computer vision combined with AI-driven predictive analytics is becoming increasingly prominent, enabling real-time assessment of plant development, environmental conditions, and resource utilization. This study is a foundational step in crop monitoring, forming a crucial part of an integrated system that analyzes data to support growers in making informed decisions. By leveraging AI and edge computing—Edge-AI—this approach enables greenhouse operators to optimize resource allocation, reduce costs, and improve efficiency. Specifically, it contributes to developing an autonomous farming system for lettuce growth monitoring, where localized data processing minimizes network latency, lowers energy consumption, and reduces financial overhead, ultimately enhancing overall system performance. Despite the potential of AI and data analytics to transform greenhouse production, obstacles remain, including difficulties in integrating diverse datasets, ensuring reliable decision-making, and addressing privacy-related issues.

3.2 Case study 2: Robotics for greenhouse lettuce harvesting

Introduction

Field production of lettuce is highly vulnerable to extreme weather, such as droughts and extreme temperatures (Kerr et al., 2018; Kistner et al., 2018). As a promising alternative, CEA offers enhanced and sustainable productivity and excellent crop quality (Lages et al., 2015). Despite these benefits, CEA lettuce production is still emerging and costs 2–3 times more than conventional methods (Nicholson et al., 2020). Operations such as transplanting, scouting, spraying, and harvesting demand skilled labor, which accounts for nearly one-third of total expenses in greenhouse production (Calvin, 2010; Castillo et al., 2021; Stemmler and Meemken, 2023). Labor shortages have further increased these costs (Biswas et al., 2018). Therefore, integrating AI-based automated systems for greenhouse production is a promising solution to reduce labor dependency and ensure year-round production.

In intelligent lettuce harvesters, the end effector plays a vital role. While many end effectors have been designed for harvesting spherical fruits such as tomatoes, strawberries, apples, and mangoes (Goulart et al., 2023; Zhang et al., 2023), these designs are not suitable for leafy greens like lettuce. Birrell et al. (2020) developed an end effector mounted on a UR10 robotic arm equipped with a mobile platform for the cultivation of iceberg lettuce in field production. However, this top-down approach achieved only a 52% success rate. Researchers also proposed a pull-from-top method using four pneumatic flexible fingers that achieved nearly 90% success in controlled tests (Yidong et al., 2022). However, alternative harvesting methods are needed as hydroponic lettuces are grown closely, and their leaves overlap. Limited research has focused on robotic harvesting of lettuce, particularly in greenhouse settings, leaving a significant research gap. Considering this knowledge gap, the main objective of this study was to design and evaluate a robotic end-effector for harvesting hydroponic lettuce.

Robotic end-effector development

End-effector design

Harvesting lettuce is challenging due to its complex structure, delicate leaves, and nonuniform growth, as well as the constraints of limited workspace. To prevent damage to the target lettuce and adjacent plants, a prototype of a sideways-approach robotic end-effector was designed using SolidWorks (v.2022, Dassault Systèmes, Vélizy, France), as illustrated in Fig. 15.4A. This end-effector is engineered to perform two simultaneous functions: cutting the stem and gently holding the lettuce.

The design incorporates a shear cutter at the bottom for stem cutting and two gripping fingers at the top to gently secure the lettuce. A utility linear actuator (FA-U-110-12-2, FIRGELLI® Automations, Ferndale, WA, USA) synchronously operates both the cutter and the grippers, allowing simultaneous opening and closing operations. When open, the end-effector easily accommodates the lettuce, and when closed, it holds the plant gently, and the stem is cut. The cutter and grippers operate on a simple lever mechanism, ensuring smooth movement and precise operation in tight spaces. The spacing between the components was carefully designed to avoid damage to the hydroponic system and to ensure that the stem is cut very close (<5 mm) to the base.

End-effector prototype assembly

Following the end-effector design, an XY linear manipulator was developed, featuring four legs with a fixed Z-axis, which was adjusted to the NFT channels. The system was modeled in SolidWorks with the integration of the end-effector, as shown in Fig. 15.4B. For side-way (cut-and-hold) harvesting, the end-effector moves along the

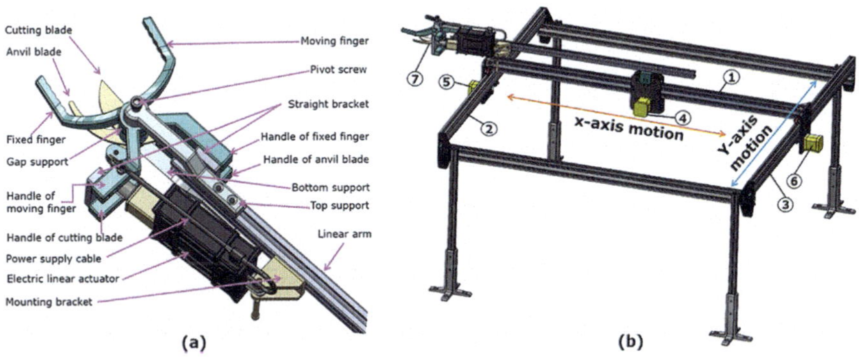

Fig. 15.4 Conceptual design of a lateral approach end-effector mounted to a manipulator (Bashir et al., 2025). (A) Lettuce harvesting end-effector designed in SolidWorks. Manipulator with XY motion: (1) linear arm on the X-axis, (2 & 3) two linear arms on the Y-axis, (4) motor (M1) for X-axis movement, (5 & 6) motors (M2, M3) for Y-axis movement, (7) mounted end-effector.

X and Y axes. This movement is driven by a stepper motor (M1) on the X-axis, utilizing a V-slotted linear arm with a gantry-belt mechanism. On the Y-axis, two synchronized stepper motors (M2 and M3) operate the motion using a similar setup. This configuration allows for precise positioning of the end-effector during lettuce harvesting in a fixed-height NFT system. Motion control of the manipulator and the opening and closing operations of the end-effector are managed by an ATmega328p microcontroller. The system first initializes by moving to the home position (0,0). The positions for each lettuce (8 inches apart from one plant to another) and the harvest bucket are set manually through the control panel. At each target, a single actuator closing action simultaneously cuts the stem and holds the lettuce head. Then, the end-effector moves to the bucket, where the actuator opens to release the lettuce, and finally, the system is ready for the next cycle.

The prototype of the end-effector was assessed using hydroponic lettuce grown in a greenhouse setup. A total of 288 plants, consisting of 144 Bergam's Green and 144 Rouxai, were cultivated in the Nutrient Film Technique (NFT) system for this evaluation. The position of each target was set via the controller interface, and the actuator of the end-effector operated at full duty cycle to maximize cutting and holding torque. The test recorded successful harvests (cutting and holding), as well as uncut or dropped leaves. In addition, reasons for failures were noted for future improvements.

Performance matrices

In lettuce harvesting tests (Table 15.4), the end-effector successfully cut 133 and held 131 out of 144 Bergam's Green, while it achieved 141 successful cuts and holds for Rouxai out of 144 plants. Most failures were due to manual misalignment and leaf obscuration at the stem cutting point. Across all 288 plants, an average of one remnant leaf per two plants and dropped leaves averaging 0.075 per plant were noted. Fig. 15.5 illustrates the end-effector prototype harvesting test.

Limitations and future directions

In this study, a prototype end-effector for greenhouse-grown hydroponic lettuce harvesting was developed and tested. However, target positioning in the current system is done manually, and leaf damage was not assessed due to the lack of an

Table 15.4 End-effector performance in greenhouse tests for cutting the lettuce stem and holding the lettuce head.

Cultivar	Sample size	Cut success	Hold success	Remnant leaves per plant (avg. count)	Fallen leaves per plant (avg. count)
Bergam's green	144	133	131	0.59	0.19
Rouxai	144	141	141	0.50	0.05
Overall	288	274	272	0.54	0.075

Fig. 15.5 Harvesting process illustration: (A) Cut and hold, (B) cut and hold with fallen leaves, (C) cut and hold with remnant leaves, and (D) cut but not hold (Bashir et al., 2025).

integrated vision system. Future improvements should focus on automating dynamic tests, upgrading the end-effector with soft, sensor-equipped fingers and proportional-integral-derivative control, and integrating a vision system with Z-axis movement for accurate and adaptable targeting across various greenhouse conditions.

4 Future directions

Adopting advanced technologies in CEA is set to further enhance precision, efficiency, and sustainability. Emerging innovations such as edge AI, digital twins, and large language models (LLMs) utilization can revolutionize how crops are grown and cultivated in controlled environments, enabling self-regulation, data-driven, and highly optimized automated agricultural systems. These cutting-edge tools will allow CEA to become more responsive to environmental fluctuations, improve resource management, and enhance yield forecasting, ultimately making food production more resilient, sustainable, and scalable.

One of the most promising advancements in CEA is the implementation of Edge AI, which enables real-time, on-site data processing for faster and more efficient decision-making (Ojo and Zahid, 2022). Unlike traditional cloud-based AI models, which require data to be transmitted to remote servers for processing, Edge AI operates directly on local devices such as cameras, IoT sensors, and robotic systems within the protected environment. This approach significantly reduces latency and ensures critical climate control adjustments, such as modifying temperature, humidity, or CO_2 concentration, are executed instantaneously based on sensor readings. In automated pest and disease management, Edge AI can power on-site computer vision systems to detect early-stage infections (Santhosh and Balarengadurai, 2024), triggering automated interventions such as targeted biocontrol applications or robotic removal of infected plants. On the other hand, low-latency operation enhances the efficiency of climate control, pest detection, and automated irrigation systems. It also reduces bandwidth usage and dependence on cloud connectivity, making it ideal for remote agricultural settings. In addition, enhanced security and data privacy are achieved by processing information locally, minimizing risks associated with cyber threats. Edge AI also contributes to energy efficiency by lowering the need for continuous data transmission to cloud servers, making it a sustainable solution for precision agriculture (Qazi et al., 2022). By decentralizing AI processing, Edge AI has been demonstrated to enhance the speed, reliability, and security of smart farming, reducing dependence on cloud infrastructure while improving operational efficiency in CEA.

Another revolutionary technology shaping the future of CEA is the digital twin concept, which involves creating virtual replicas of physical farming environments to simulate and predict plant growth under varying conditions (Purcell et al., 2023). A digital twin system is a dynamic simulation of an entity in virtual space, continuously updated through data collection, models, and what-if simulations (VanDerHorn and Mahadevan, 2021). Digital twins of greenhouses, vertical farms, and hydroponic systems are expected to integrate real-time data from IoT sensors, AI-driven climate models, and historical crop performance to simulate different growing scenarios before implementing changes in the actual farm. Digital twin systems have great potential to serve as a fundamental tool for real-time monitoring, management, and control to optimize crop productivity. For instance, a digital twin-powered AI system can analyze how different irrigation schedules, lighting intensities, or nutrient compositions impact plant development and yield, allowing growers to refine their cultivation strategies before making real-world adjustments. Moreover, integrating digital twins with autonomous farming robots will enable self-optimizing agricultural systems, where robots continuously adapt and refine operations based on real-time virtual simulations, leading to increased efficiency, reduced resource waste, and enhanced sustainability in CEA. Nevertheless, the development of validated and scalable digital twin applications relies on frameworks that facilitate model validation and improve accessibility. Collaborative efforts between researchers and agricultural stakeholders will be crucial in refining scalable digital twin solutions, ensuring effectiveness and widespread adoption in precision farming.

LLMs represent another significant advancement in knowledge-driven decision-making for precision agriculture. Agricultural extension services have long faced challenges in effectively reaching and supporting farmers, primarily due to limited institutional capacity, geographic constraints, and accessibility issues. Traditional models often struggle to provide timely, location-specific, personalized agricultural recommendations, leaving many smallholder farmers without crucial technical guidance. In this context, LLMs present a transformative opportunity to bridge the knowledge gap and enhance agricultural advisory services (Tzachor et al., 2023). LLMs have the potential to simplify complex scientific knowledge and deliver real-time, data-driven recommendations tailored to specific crops, climates, and soil conditions. By learning from extensive datasets, these AI-driven tools can provide accurate, localized, and adaptive solutions for farmers, empowering them to make informed decisions about cultivation, pest control, and resource management. These AI-powered models can analyze vast scientific literature, historical farm data, and sensor inputs to provide actionable insights for optimizing crop production (Chen and Huang, 2025). In addition, LLMs can enhance human-machine interactions in CEA by enabling voice-controlled climate management systems, AI-driven chatbots for real-time troubleshooting, and predictive analytics platforms that assist farmers in anticipating potential risks such as disease outbreaks, equipment failures, or resource shortages. Integrating LLMs with robotic automation can also create self-learning farming systems, where AI continuously refines growing protocols based on real-time data, improving precision and productivity in controlled agriculture. However, despite their promise, LLM-based agricultural advisory systems also present significant shortcomings and limitations. Challenges can arise due to accuracy, contextual relevance, and the need for human oversight in AI-generated recommendations. While LLMs hold great potential to enhance the scalability and reach of agricultural practices, their successful implementation requires continuous refinement, domain-specific fine-tuning, and collaboration between AI developers, agronomists, and extension agents to ensure the delivery of reliable, actionable, and farmer-friendly guidance.

Looking ahead, automated systems involving Edge AI, digital twins, and LLMs will lead to the development of fully autonomous, AI-driven CEA systems that require minimal human intervention. AI-powered robotic arms, autonomous drones, and smart irrigation networks will synchronize with digital twins and Edge AI-powered analytics to ensure optimal environmental conditions in real-time. These innovations will facilitate resource efficiency by minimizing energy waste, optimizing space utilization, and reducing dependency on chemical inputs, making CEA a more sustainable and scalable solution for global food production. Moreover, as concerns regarding climate change and urban food demand increase, advanced PF technologies will be crucial in enhancing food security, reducing supply chain vulnerability, and supporting the year-round production of high-quality crops. In conclusion, the future of CEA is inseparable from the advancement of intelligent technologies, where Edge AI, digital twins, LLMs, and automation will create more responsive, self-optimizing, and highly efficient agricultural ecosystems. The integration of these state-of-the-art innovations will not only enhance productivity but also contribute to more resilient, data-driven, and sustainable approaches to food production.

5 Conclusions

Integrating smart technologies into PF systems marks a significant milestone in the evolution of modern agriculture. By leveraging AI, IoT, sensors, robotics, and edge computing, CEA offers precise control over growing conditions, enabling higher productivity, superior crop quality, and resource efficiency. Given the case studies introduced in this chapter, real-world implementations demonstrate that smart systems can automate harvesting and growth monitoring, reducing resource usage and operational costs. Robotics and automation further streamline labor-intensive tasks such as harvesting, planting, and resource application, reducing dependency on manual labor and increasing scalability. These advancements make PF more sustainable and economically viable, positioning it as a critical solution to global challenges like food security, urbanization, and environmental degradation.

References

Abbood, H.M., Nouri, N.M., Riahi, M., Alagheband, S.H., 2023. An intelligent monitoring model for greenhouse microclimate based on RBF neural network for optimal setpoint detection. J. Process Control 129, 103037.

Akbar, J.U.M., Kamarulzaman, S.F., Muzahid, A.J.M., Rahman, M.A., Uddin, M., 2024. A comprehensive review on deep learning assisted computer vision techniques for smart greenhouse agriculture. IEEE Access 12, 4485–4522.

Bashir, A., Majeed, Y., Zahid, A., 2025. Development of an end-effector for robotic harvesting of hydroponic lettuce. J. ASABE 68 (4), 645–657.

Birrell, S., Hughes, J., Cai, J.Y., Iida, F., 2020. A field-tested robotic harvesting system for iceberg lettuce. J. Field Robot. 37 (2), 225–245.

Biswas, T., Wu, F., Guan, Z., 2018. Labor shortages in the Florida strawberry industry. FE1041, EDIS, 5.

Buslaev, A., Iglovikov, V.I., Khvedchenya, E., Parinov, A., Druzhinin, M., Kalinin, A.A., 2020. Albumentations: fast and flexible image augmentations. Information 11 (2), 125.

Calvin, L., 2010. The US Produce Industry and Labor: Facing the Future in a Global Economy, DIANE Publishing.

Castillo, M., Simnitt, S., Astill, G., Minor, T., 2021. Examining the Growth in Seasonal Agricultural H-2A Labor.

Catota-Ocapana, P., Minaya-Andino, C., Astudillo, P., Pichoasamin, D., 2025. Smart control models used for nutrient management in hydroponic crops: a systematic review. IEEE Access 13, 13070–13087. doi:10.1109/ACCESS.2025.3526171.

Chakraborty, A., Das, S., Mondal, B., 2022. Integrating neural network for pest detection in controlled environment vertical farm. Indian J. Sci. Technol. 15 (17), 829–838.

Cheein, F.A.A., Carelli, R., 2013. Agricultural robotics: unmanned robotic service units in agricultural tasks. IEEE Ind. Electron. Mag. 7 (3), 48–58.

Chen, D., Huang, Y., 2025. Integrating reinforcement learning and large language models for crop production process management optimization and control through a new knowledge-based deep learning paradigm. Comput. Electron. Agric. 232, 110028.

Choi, H.B., Park, J.K., Park, S.H., Lee, T.S., 2024. NeRF-based 3D reconstruction pipeline for acquisition and analysis of tomato crop morphology. Front. Plant Sci. 15, 1439086.

Chollet, F., 2017. Xception: deep learning with depthwise separable convolutions. In: Proceedings of the 2017 IEEE Conference on Computer Vision and Pattern Recognition (CVPR). IEEE, pp. 1800–1807.

FAO, 2018. The State of Food Security and Nutrition in the World 2018: Building Climate Resilience for Food Security and Nutrition, Italy, Rome.

Gallardo, M., Elia, A., Thompson, R.B., 2020. Decision support systems and models for aiding irrigation and nutrient management of vegetable crops. Agric. Water Manag. 240, 106209.

Goh, Y.S., Hum, Y.C., Lee, Y.L., Lai, K.W., Yap, W.S., Tee, Y.K., 2023. A meta-analysis: food production and vegetable crop yields of hydroponics. Sci. Hortic. 321, 112339.

Goulart, R., Jarvis, D., Walsh, K.B., 2023. Evaluation of end effectors for robotic harvesting of mango fruit. Sustainability 15, 6769.

Gul, Z., Bora, S., 2023. Exploiting pre-trained convolutional neural networks for the detection of nutrient deficiencies in hydroponic basil. Sensors 23 (12), 5407.

Jawad, M., Wahid, F., Ali, S., Ma, Y., Alkhyyat, A., Khan, J., Lee, Y., 2025. Energy optimization and plant comfort management in smart greenhouses using the artificial bee colony algorithm. Sci. Rep. 15 (1), 1752.

Kerr, A., Dialesandro, J., Steenwerth, K., Lopez-Brody, N., Elias, E., 2018. Vulnerability of California specialty crops to projected mid-century temperature changes. Clim. Change 148, 419–436.

Kistner, E., Kellner, O., Andresen, J., Todey, D., Morton, L.W., 2018. Vulnerability of specialty crops to short-term climatic variability and adaptation strategies in the Midwestern USA. Clim. Change 146, 145–158.

Kumar, S., Saravaiya, S.N., Pandey, A.K., 2021. Precision Farming and Protected Cultivation: Concepts and Applications, CRC Press.

Lages, B.G., Almeida Gadelha, F.D., Kublik, N., Proctor, A., Reichelm, L., Weissinger, E., Wohlleb, G.M., Halden, R.U., 2015. Comparison of land, water, and energy requirements of lettuce grown using hydroponic vs. conventional agricultural methods. Int. J. Environ. Res. Public Health 12, 6879–6891.

Lee, M.H., Yao, M.H., Kow, P.Y., Kuo, B.J., Chang, F.J., 2024. An artificial intelligence-powered environmental control system for resilient and efficient greenhouse farming. Sustainability 16 (24), 10958.

Levy, P.E., Jarvis, P.G., 1999. Direct and indirect measurements of LAI in millet and fallow vegetation in HAPEX-Sahel. Agric. For. Meteorol. 97 (3), 199–212.

Ma, D., Carpenter, N., Maki, H., Rehman, T.U., Tuinstra, M.R., Jin, J., 2019. Greenhouse environment modeling and simulation for microclimate control. Comput. Electron. Agric. 162, 134–142.

Maksymova, I., Steger, C., Druml, N., 2018. Review of LiDAR sensor data acquisition and compression for automotive applications. In: Proceedings, vol. 2 (13). MDPI, pp. 852.

Manavalan, L.P., Cui, I., Ambrose, K.V., Panjwani, S., DeLong, S., Mleczko, M., Spetsieris, K., 2021. Systematic approach to validate and implement digital phenotyping tool for soybean: a case study with PlantEye. Plant Phenom. J. 4 (1), e20025.

Maraveas, C., 2022. Incorporating artificial intelligence technology in smart greenhouses: current State of the Art. Appl. Sci. 13 (1), 14.

Milletari, F., Navab, N., Ahmadi, S.-A., 2016. V-Net: fully convolutional neural networks for volumetric medical image segmentation. In: Proceedings of the 2016 Fourth International Conference on 3D Vision (3DV). IEEE, pp. 565–571.

Nicholson, C.F., Harbick, K., Gómez, M.I., Mattson, N.S., 2020. An Economic and Environmental Comparison of Conventional and Controlled Environment Agriculture (CEA) Supply Chains for Leaf Lettuce to US Cities. In: Aktas, E., Bourlakis, M. (Eds.), Food Supply Chains in Cities. Palgrave Macmillan, Cham, doi:10.1007/978-3-030-34065-0_2.

Ojo, M.O., Zahid, A., 2022. Deep learning in controlled environment agriculture: a review of recent advancements, challenges and prospects. Sensors 22 (20), 7965.

Petrakis, T., Kavga, A., Thomopoulos, V., Argiriou, A.A., 2022. Neural network model for greenhouse microclimate predictions. Agriculture 12 (6), 780.

Purcell, W., Neubauer, T., Mallinger, K., 2023. Digital Twins in agriculture: challenges and opportunities for environmental sustainability. Curr. Opin. Environ. Sustain. 61, 101252.

Qazi, S., Khawaja, B.A., Farooq, Q.U., 2022. IoT-equipped and AI-enabled next generation smart agriculture: a critical review, current challenges and future trends. IEEE Access 10, 21219–21235.

Rasheed, A., Yang, B., Mei, Q., Chen, Y., 2025. Advancements in Greenhouse Energy Management: a Review of Machine Learning Approaches for Microclimate Prediction and System Control. Available at SSRN 5142250.

Regazzoni, D., De Vecchi, G., Rizzi, C., 2014. RGB cams vs RGB-D sensors: low cost motion capture technologies performances and limitations. J. Manuf. Syst. 33 (4), 719–728.

Rodríguez, S., Gualotuña, T., Grilo, C., 2017. A system for the monitoring and predicting of data in precision agriculture in a rose greenhouse based on wireless sensor networks. Procedia Comput. Sci. 121, 306–313.

Rustia, D.J.A., Lin, C.E., Chung, J.Y., Zhuang, Y.J., Hsu, J.C., Lin, T.T., 2020. Application of an image and environmental sensor network for automated greenhouse insect pest monitoring. J. Asia-Pac. Entomol. 23 (1), 17–28.

Sa, I., Lehnert, C., English, A., McCool, C., Dayoub, F., Upcroft, B., Perez, T., 2017. Peduncle detection of sweet pepper for autonomous crop harvesting—combined color and 3-D information. IEEE Robot. Autom. Lett. 2 (2), 765–772.

Sandler, M., Howard, A., Zhu, M., Zhmoginov, A., Chen, L.-C., 2018. MobileNetV2: inverted residuals and linear bottlenecks. In: Proceedings of the 2018 IEEE/CVF Conference on Computer Vision and Pattern Recognition. IEEE, pp. 4510–4520.

Santhosh, B.J., Balarengadurai, C., 2024. Edge-AI based plant leaf disease identification and prevention for smart agriculture. Grenze Int. J. Eng. Technol. 10, 6556.

Stemmler, H., Meemken, E.M., 2023. Greenhouse farming and employment: evidence from Ecuador. Food Policy 117. doi:10.1016/j.foodpol.2023.102443.

Subeesh, A., Mehta, C.R., 2021. Automation and digitization of agriculture using artificial intelligence and internet of things. Artif. Intell. Agric. 5, 278–291.

Tangarife, H.I., Díaz, A.E. (2017). Robotic applications in the automation of agricultural production under greenhouse: a review. In: 2017 IEEE 3rd Colombian Conference on Automatic Control (CCAC). IEEE, pp. 1–6.

Ting, Y., Chao, J., Wei, L., Junxiong, Z., 2009. Information acquisition of cucumber fruit in greenhouse environment based on nearing infrared image. Nongye Jixie Xuebao/Trans. Chin. Soc. Agric. Mach. 29 (8), 2054–2058.

Tuan, V.N., Khattak, A.M., Zhu, H., Gao, W., Wang, M., 2020. Combination of multivariate standard addition technique and deep kernel learning model for determining multi-ion in hydroponic nutrient solution. Sensors 20 (18), 5314.

Tzachor, A., Devare, M., Richards, C., Pypers, P., Ghosh, A., Koo, J., et al., 2023. Large language models and agricultural extension services. Nat. Food 4 (11), 941–948.

VanDerHorn, E., Mahadevan, S., 2021. Digital Twin: generalization, characterization and implementation. Decis. Support Syst. 145, 113524.

Wang, X., Vladislav, Z., Viktor, O., Wu, Z., Zhao, M., 2022. Online recognition and yield estimation of tomato in plant factory based on YOLOv3. Sci. Rep. 12 (1), 8686.

Yang, Q., Huang, G., Shi, X., He, M., Ahmad, I., Zhao, X., Addy, M., 2020. Design of a control system for a mini-automatic transplanting machine of plug seedling. Comput. Electron. Agric. 169, 105226.

Yidong, M., Pengzhan, H., Xin, J., Xinping, L., Chao, Z., Yin, Z., 2022. Design and experiment of low damage flexible harvesting device for hydroponic lettuce. Trans. Chin. Soc. Agric. Mach. 53, 175–183.

Zhang, F., Chen, Z., Wang, Y., Bao, R., Chen, X., Fu, S., Tian, M., Zhang, Y., 2023. Research on flexible end-effectors with humanoid grasp function for small spherical fruit picking. Agriculture 13, 123.

Zhang, L., Verma, B., Stockwell, D., Chowdhury, S., 2018. Density Weighted Connectivity of Grass Pixels in image frames for biomass estimation. Expert Syst. Appl. 101, 213–227.

Zhao, R., Shi, F., 2022b. A novel strategy for pest disease detection of Brassica chinensis based on UAV imagery and deep learning. Int. J. Remote Sens. 43 (19–24), 7083–7103.

Zhao, S., Liu, J., Wu, S., 2022a. Multiple disease detection method for greenhouse-cultivated strawberry based on multiscale feature fusion Faster R_CNN. Comput. Electron. Agric. 199, 107176.

Section IV

Case studies and economic analysis of smart agricultural solutions

Integrating climate-informed crop modeling to optimize planting dates for sustainable productivity of mustard (*Brassica juncea*)

16

Anamika[a], *Md. Rahat Bari Rupok*[a], *Ridwan Ahmed*[c], *Md. Rashedur Rahman*[c], *Mallappa Jadiyappa Madolli*[d,e], *Sushil Kumar Himanshu*[e], *and Sheikh Faruk Ahmed*[b]
[a]Department of Agronomy, Gazipur Agricultural University, Gazipur, Bangladesh
[b]Department of Crop Botany, Gazipur Agricultural University, Gazipur, Bangladesh
[c]Department of Agronomy, Bangladesh Agricultural University, Mymensingh, Bangladesh
[d]Department of Agricultural Engineering, University of Agricultural Sciences, Bangalore, India
[e]Agricultural Systems and Engineering, Faculty of Food, Agriculture and Natural Resources, Asian Institute of Technology, Pathum Thani, Thailand

1 Introduction

Over the years, climate change has adversely affected agricultural productivity through a number of factors such as altered precipitation patterns, temperature extremes, and an increased frequency of climate-related disasters (Yuan et al., 2024). Yield reductions and crop failures are becoming more common due to extreme weather events. For instance, droughts and heatwaves are bringing forth food shortages, threatening global food security (Lesk et al., 2016). Bangladesh, the seventh most climate-vulnerable country, has already faced 185 extreme weather events in the last two decades (Eckstein et al., 2021). Predictions of future climate indicate even more alarming scenarios, particularly concerning atmospheric temperature, precipitation, and crop water availability due to global warming as well as greenhouse gas emissions (Ahmed et al., 2023; Vilavan et al., 2024). This destabilization not only hampers immediate food availability but also threatens long-term food security (Masipa, 2017; Wheeler and Von Braun, 2013). Abrupt changes in climatic events interfere with existing cultivation practices and the patterns of most crops cultivated in Bangladesh (Ghaffar et al., 2022; Hossain et al., 2012). Hence, rethinking the temporal and spatial adjustment of crops has become a priority for re-optimizing crop productivity and ensuring long-term food security.

Precision Technologies for Digital Agriculture. https://doi.org/10.1016/B978-0-443-40513-6.00009-3

Mustard (*Brassica juncea* L.) is one of the top-ranked oilseed crops in Bangladesh (BBS, 2024). It holds both traditional and economic importance for the country due to its contribution to food security and the livelihoods of rural farmers. The oil from the crop is used for cooking and other culinary applications to enhance taste (Man and Weir, 1988). The tender mustard plant is also widely consumed as a vegetable because of its antimutagenic, anticarcinogenic, and antioxidative health benefits (Kapusta-Duch et al., 2012). In Bangladesh, mustard is normally sown between late October and early November (Chowhan et al., 2023). Because of its short growing period and low input requirements, it is an ideal intermediary rainfed crop in many agroecosystems (Alberti et al., 2019). As a rainfed crop, mustard is highly dependent on natural precipitation for its water requirements. Existing soil moisture from the immediately preceding rainy season, coupled with occasional rainfall and dew droplets, supports its growth, resulting in a considerably good yield of mustard (Singh et al., 2021). However, due to the consequences of climate change, precipitation patterns have abruptly changed in both quantity and occurrence (Boomiraj et al., 2010). Moreover, increasing temperatures are further worsening the situation through frequent major and minor droughts (Gill et al., 2012). As a result, the suitable natural agro-climatic conditions for mustard cultivation are gradually losing their viability over time (Boomiraj et al., 2010). Given the challenges posed by climate change on mustard, it is crucial to explore alternative approaches for strategic management and better adaptability.

Understanding the spatial and temporal changes, as well as predicting the possible future climatic trends of both temperature and precipitation, remains fundamental for planning optimal crop management (Bhandari, 2021; Bandara and Cai, 2014). Global climate models (GCMs) could potentially be an effective tool for this purpose. The development of cropping plans utilizing GCM-based predictions should help identify a suitable window within the current climatic trends where mustard can be optimally grown without any yield penalty. GCMs are sophisticated tools used to simulate the Earth's climate system (Tanimu et al., 2024). These models are essential for projecting future climate scenarios, aiding agricultural planning by providing insights into temperature trends, precipitation variability, and extreme weather events (Abel et al., 2024). Recent research has revealed that integrating GCMs can effectively aid in developing climate-resilient cropping strategies, including optimal sowing dates, crop selection, and irrigation management (Abel et al., 2024; Wang et al., 2024). According to Hofmann et al. (2022), regional downscaling of GCMs further enhances projections, making them applicable for site-specific agricultural decision-making. By utilizing GCM-based forecasts, climate-smart agriculture strategies can be formulated, and several risks associated with climate change can be mitigated (Sylla et al., 2016). As climate change continues to challenge mustard production, integrating GCM data into planning could ensure better resilience and adaptability, contributing to long-term sustainability. In rabi crops like mustard, inappropriate timing in sowing can expose the crop to various adverse weather events, such as higher temperatures and inadequate rainfall during critical phases, causing unexpected yield losses (Rahman et al., 2020). Very recently, (Langangmeilu et al., 2023) showed that thermal abnormalities can decrease biomass and lower oil content in rapeseed. Therefore, it is essential to understand climate change trends, particularly temperature and precipitation patterns, and thereby plan a suitable sowing date to avoid unintentional damage to the crop while ensuring optimal productivity. Thus, the current study was conducted to evaluate the productivity restoration of mustard via optimization of the sowing date based on GCM-based climate predictions.

2 Materials and methods

2.1 Experimental location and conditions

The study was conducted at the experimental field of the Department of Agronomy, Bangladesh Agricultural University, Mymensingh, Bangladesh (24.75° N and 90.50° E). The local popular mustard cultivar "BARI Sharisha 18," released by the Bangladesh Agricultural Research Institute (BARI), was used as the plant material. The study initially involved the collection, visualization, and interpretation of long-term climatic trends of Mymensingh, as well as GCM-based forecasting of future climatic scenarios. In the later stage, based on the GCM-based climate trend, the placement of mustard by changing its sowing date was executed, and the productivity performance was evaluated via an on-field trial.

2.2 Historical climate evaluation and future climate projection

Both annual and monthly data of three main climatic parameters, namely maximum temperature, minimum temperature, and precipitation, were considered for the climatic variability study in this experiment. Climatic data on the above-mentioned parameters from Mymensingh, spanning the years 1950 to 2020, available at the Climate Change Knowledge Portal (CCKP) of the World Bank, were utilized to extract both annual and monthly trends of climatic patterns. Three different trends (long-term: 1950–2020, mid-term: 1971–2020, and short-term: 1991–2020) were visualized in both annual and monthly climatic trends for a better understanding of the nature and degree of change in the climatic parameters. In the case of future climate projections, a multimodal ensemble approach, as depicted by the CCKP of the World Bank, was adopted because of its higher accuracy and reliability. Projections were visualized up to the year 2100 under four different shared socioeconomic pathways (SSPs) (SSP1-2.6, SSP2-4.5, SSP3-7.0, and SSP5-8.5) as per GCMs archived in Phase 6 of the Coupled Model Intercomparison Project (CMIP6).

2.3 Experimentation details

Based on the interpretation of the climatic parameters, an on-field trial of "BARI Sharisha 18" was designed under a randomized complete block design with three replications. For treatments, five sowing dates were selected (Table 16.1). Plots (4 m × 2.5 m) were prepared on well-tilled, friable silty loam soil (20% sand, 67% silt,

Table 16.1 Experimental treatments on differential sowing dates of mustard.

Treatment code	Treatment details
P_1	Seed sowing at 5th October
P_2	Seed sowing at 12th October
P_3	Seed sowing at 19th October
P_4	Seed sowing at 25th October
C	Seed sowing on 1st November [control] (common practice).

13% clay, 0.86% organic carbon, and a pH of 6.7). Seeds were sown in lines, maintaining 30 cm row-to-row and 15 cm plant-to-plant spacing. Initially, three seeds were sown per spot, and after seedling establishment, only one plant among the three was allowed to grow. Standard fertilization was applied to the crop as per the guidelines provided by BARI. Weeding was done manually on the 18th and 34th days after sowing (DAS).

2.4 Data collection and statistical analysis

Plant height was manually measured using a centimeter-marked meter scale from the soil level to the tip of the plant at 80 DAS. Ten plants from each plot were randomly selected, and their individual heights were averaged to record the presented plant height. Branches per plant were also manually counted from ten randomly selected plants, and the average value was recorded. Upon harvesting the crop, siliquae per plant were also manually counted from ten randomly selected plants, and the average value was recorded. To calculate seeds per siliqua, three plants were randomly selected, all the siliquae were counted, properly air-dried, and opened up to count the number of seeds. Finally, the number of seeds was divided by the number of siliquae to extract the value of seeds per siliqua. The harvested plants from individual crops were threshed using a mechanical thresher. Seeds were properly dried to obtain 12% moisture content. Randomly, a thousand seeds were counted and weighed on an electronic balance to get the 1000-seed weight. Seed yield was calculated by weighing the total seeds harvested from each plot and expressed in t/ha.

The obtained data were managed in a Microsoft Excel spreadsheet. Later on, the data were analyzed using the Statistix 10 software program for one-way analysis of variance, and the means of significantly different treatments were separated by post-hoc analysis based on Tukey's honest significant difference test. Graphs and tables were constructed for data visualization using the Microsoft Excel software program.

3 Results

3.1 Trends in current climatic variability

The analysis of climate trends from 1950 to 2020 revealed a steady change in both seasonal and annual minimum temperature, maximum temperature, and precipitation patterns (Fig. 16.1). Both the annual minimum and maximum temperatures have

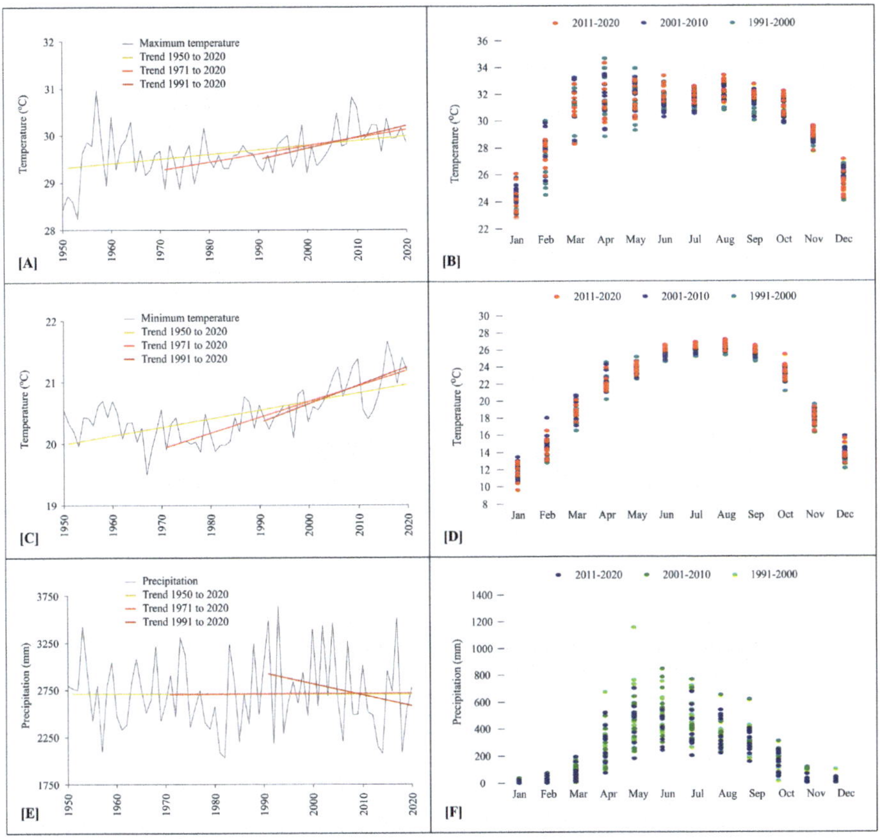

Fig. 16.1 The trend of variability in historical (A) annual maximum temperature, (B) seasonal maximum temperature, (C) annual minimum temperature, (D) seasonal minimum temperature, (E) annual precipitation, and (F) seasonal precipitation of Mymensingh over the time period of 1950–2020.

Figures were extrapolated from the open-source platform Climate Change Knowledge Portal (CCKP) of the World Bank. Available at: https://climateknowledgeportal.worldbank.org/country/bangladesh (Accessed on: 5th April 2025).

shown an increasing trend over the decades. The sharpest rise was observed after 1990, indicating that the warming trend has accelerated in recent years, with the period from 1991 to 2020 experiencing the most pronounced increase (Fig. 16.1A and C). A seasonal analysis of minimum and maximum temperatures showed that the hottest months are March to June, which also exhibited comparatively higher temperatures in the most recent decade (2011–2020) compared to earlier periods (2001–2010 and 1991–2000) (Fig. 16.1B and D). Likewise, from both temperature trends, it can also be understood that there has been a rise in temperature during the cold season (November to February). This phenomenon suggests that winters are becoming warmer. Although the minimum temperature remained quite unchanged, the maximum temperature contributed the most to warming during this period, especially within the recent decade (Fig. 16.1B and D). In contrast to temperature trends, precipitation patterns exhibited variability with no clear long-term increase; however, a steep declining trend has been observed since 1991. The trend lines indicated that the most recent period (1991–2020) experienced reduced precipitation compared to earlier decades (Fig. 16.1E). Seasonal precipitation data reveal that rainfall, which remained concentrated in the pre-monsoon and monsoon months (April to September), has changed considerably. In recent years (2011–2020), we found a monthly decline in overall precipitation compared to previous decades. During the winter season (November to February), precipitation declined significantly, exhibiting water stress exposure for rainfed agricultural crops during this period (Fig. 16.1F).

3.2 Trends of future climatic variability

The projection analysis of future climatic patterns based on historical data under four different SSP scenarios revealed alarming results (Fig. 16.2). Until 2075, projections under all SSPs indicated a rise in minimum and maximum temperatures. Both temperatures continued to rise until 2100, except for SSP1-2.6, which exhibited a modest decrease (Fig. 16.2A and B). According to the projections of SSP1-2.6, maximum temperature might stay below 34°C, whereas SSP5-8.5 projected a sharp rise, reaching nearly 40°C by the end of the century (Fig. 16.2A). Minimum temperature projections followed a similar trend to maximum temperature, with a pronounced increase under high-emission scenarios. The SSP1-2.6 scenario predicted a minor increase (remaining below 22°C), while SSP5-8.5 projected minimum temperatures exceeding 28°C by 2100 (Fig. 16.2B). Precipitation projections, on the other hand, displayed considerable variability. A slight increasing trend was observed under all SSP scenarios. However, higher-emission scenarios like SSP3-7.0 and SSP5-8.5 exhibited more extreme fluctuations. Thus, an increased likelihood of heavy rainfall events and prolonged dry spells could be suggested (Fig. 16.2C). The projected increase in precipitation and frequent aberrations under high-emission pathways might lead to more frequent extreme weather conditions, including flooding and shifts in monsoon patterns (Fig. 16.2C).

3.3 Level of significance in studied parameters

The statistical analysis of tested agronomic parameters in the current study revealed highly significant variations among the measured traits due to differential sowing dates. Plant height, branches per plant, siliquae per plant, and seed yield exhibited extremely significant differences ($P < 0.001$), indicating strong environmental influences on these traits (Table 16.2). The number of seeds per siliqua and 1000-seed weight also showed significant variation ($P < 0.01$), suggesting that these parameters were influenced by the differential sowing dates but with slightly lower levels of significance compared to the other traits (Table 16.2).

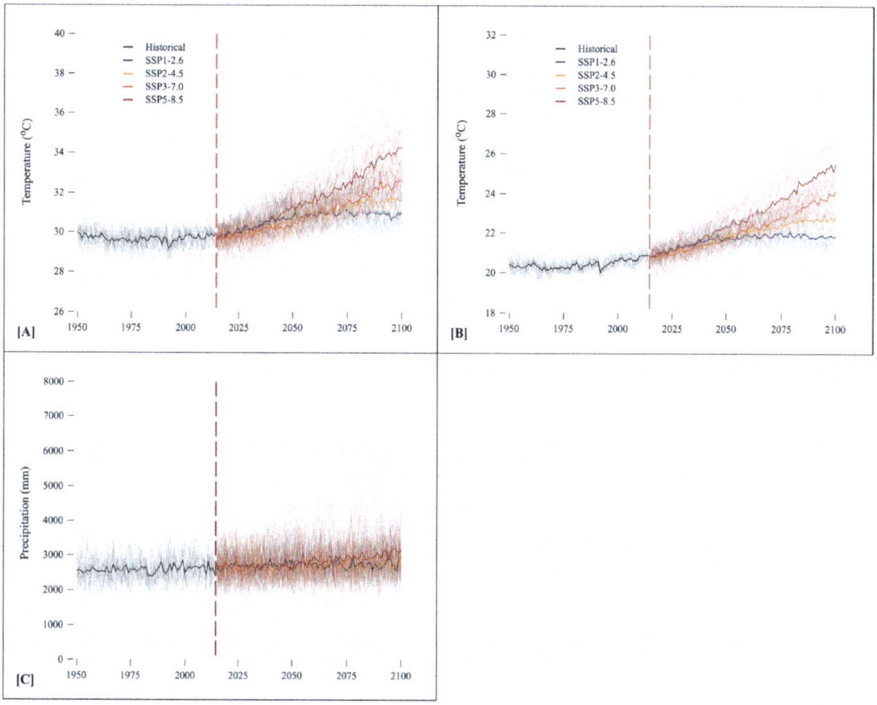

Fig. 16.2 The trend of variability in future climatic parameters: (A) maximum temperature, (B) minimum temperature, and (C) precipitation in Mymensingh.
Projections were based on open source multimodal Ensemble approach as depicted by Climate Change Knowledge Portal (CCKP) of the World Bank up to the year 2100 under four different Shared Socioeconomic Pathways (SSPs) (SSP1-2.6, SSP2-4.5, SSP3-7.0, and SSP5-8.5). Available at: https://climateknowledgeportal.worldbank.org/country/bangladesh (Accessed at: 5th April 2025).

Table 16.2 Level of significance in the studied parameters.

Parameter	*P* value	Significance
Plant height (cm)	0.0000	***
Branches per plant	0.0000	***
Siliquae per plant	0.0000	***
Seeds per siliqua	0.0015	**
1000 seed weight (g)	0.0036	**
Seed yield (t/ha)	0.0000	***

*** and ** indicate $P < 0.001$ and $P < 0.01$, respectively.

3.4 Effects of sowing dates on studied parameters

The results indicate that different sowing dates had a significant impact on the growth and yield attributes of mustard. In all tested parameters, seed sowing at P_2 (12th October) performed the best compared to sowing at C (1st November) (Fig. 16.3).

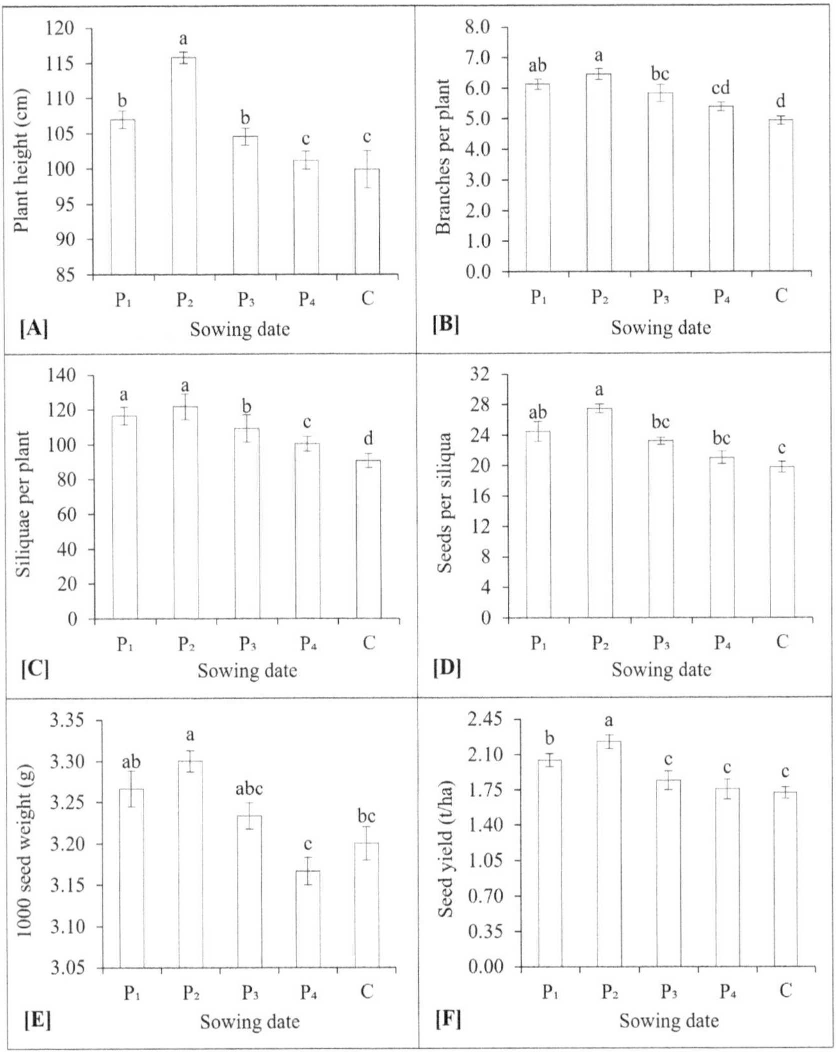

Fig. 16.3 Agronomic performance of mustard under differential sowing dates. (A) Plant height, (B) branches per plant, (C) siliquae per plant, (D) seeds per siliqua, (E) 1000-seed weight, and (F) seed yield. P_1: sowing on 5th October; P_2: sowing on 12th October; P_3: sowing on 19th October; P_4: sowing on 25th October; C: control (sowing on 1st November). Data are means of three replications. Vertical bars represent standard errors. Bar columns followed by different lowercase letters are significantly different based on Tukey's honest significant difference test at $P<0.05$.

Plant height was found to be significantly higher for P_2, followed by P_1, P_3, and P_4, compared to C. Although P_1 and P_3 had statistically similar plant heights, which were significantly higher than C, both were significantly lower than P_2 (Fig. 16.3A). Branches per plant were highest for P_2, but P_1 also showed statistically similar performance when compared with P_2. Sowing at P_3 showed moderate performance, which was statistically similar to P_1 but significantly lower than P_2 (Fig. 16.3B). Both siliquae per plant and seeds per siliqua were highest in P_2, but P_1 also performed statistically similarly in both parameters compared to P_2 (Fig. 16.3C and D). Sowing at P_3, on the other hand, performed significantly lower in both parameters compared to both P_1 and P_2 (Fig. 16.3C and D). The 1000-seed weight also demonstrated significant variation, with P_2 recording the highest weight, followed by P_1 and P_3, while P_4 and C had the lowest and second-lowest values, respectively (Fig. 16.3E). Seed yield was found to be maximum under the P_2 sowing date, followed by P_1. Compared to C, only P_1 and P_2 produced significantly higher seed yields, while sowing at P_3 or P_4 resulted in statistically similar seed yields compared to C (Fig. 16.3F). Seed yield under P_1 was found to be significantly lower than sowing at P_2 (Fig. 16.3F).

4 Discussion

Abnormalities in temperature and precipitation patterns are consequences of global climate change. Increasing GHG emissions and global warming have played significant roles in the existing alterations of climatic components such as temperature and precipitation in Bangladesh (Bhuyan et al., 2018; Basak et al., 2013; Hossain et al., 2020). In our current study, we observed an increasing trend in both annual maximum and minimum temperatures and a steady trend for precipitation within the timeframe of 1950–2020 (Fig. 16.1A, C, and E). However, the concerning fact was that, within 1991–2020, the rate of increase in both annual maximum and minimum temperatures was considerably higher (Fig. 16.1A and C). On the other hand, the precipitation trend within 1991–2020 showed a drastic decline despite maintaining a steady long-term trend (Fig. 16.1E). Monthly variability during the growing period of mustard (November to February) showed fewer changes in maximum and minimum temperatures. However, a decline in precipitation during 2011–2020 compared to the trend of 1991–2000 was clearly visible (Fig. 16.1B, D, and F). Hence, this suggests that the precipitation pattern is currently the dominant determining factor for planning mustard cultivation activities. Shaibur et al. (2018) reported reduced productivity in three cropping systems (T. Aman rice−Mustard−Boro rice/Sesame/Mungbean) in Jashore, Bangladesh, primarily due to reduced rainfall. Our results are also in agreement, indicating that, compared to temperature variability, the decline in precipitation is causing more damage to mustard productivity in Mymensingh. Recently, (Bazzaz et al., 2020) studied different mustard varieties and showed that late sowing is harmful to productivity. Relating this to our results, we can conclude that late-sown mustard was exposed to drought, which caused such responses. Several studies have shown that water scarcity during critical growth stages (vegetative, flowering, and grain filling) of mustard can lead to reduced photosynthesis, poor fertilization, seed setting, decreased yield, and lower oil content (Chauhan et al., 2007; Mostafaei et al., 2018; Raza et al., 2017).

To facilitate proper planning, we have utilized GCM-based predictions. This has enabled us to analyze and understand the current and future variability of climatic patterns and select an appropriate sowing date for mustard. We have considered GCMs archived in Phase 6 of the CMIP6. These models are the latest and most improved, as well as having higher spatial resolution than the CMIP5 models (Li et al., 2021; Stouffer et al., 2017; Zhu and Yang, 2020). The CMIP6 climate projections are based on a new set of emission and land-use scenarios generated using integrated assessment models aligned with different SSPs (Zhai et al., 2019). In the case of future climate projections, a multimodal ensemble approach, as depicted by CCKP of the World Bank, was adopted because of its higher accuracy and reliability (Zhi et al., 2012; Weigel et al., 2008; Barnes et al., 2019). The projection showed an increase in both maximum and minimum temperatures and a slight increase in precipitation (Fig. 16.2). This indicates that, in the future, temperature might pose a bigger threat to mustard productivity in Mymensingh than precipitation. However, the slightly increasing trend in the future precipitation pattern (Fig. 16.2C) suggests that the precipitation, which has already declined during November to February, might increase slightly or sustain its current trend. Therefore, mustard cropping during this time would possibly face water shortages throughout the season.

Water shortage during the early stages of crop growth is highly devastating (Passioura, 2006). During this period, crops remain comparatively sensitive. If a seedling ensures robust early growth, it can tolerate environmental extremes better at later stages (Jisha et al., 2013; Singh et al., 2010). Alternatively, poor early growth, especially due to water shortage, develops weak seedlings that fail to sustain yield and productivity (Anjum et al., 2017). From our climate analysis, we have observed good existing precipitation during October and considerable rainfall during November (Fig. 16.1F). Thus, in this current study, we conducted an experiment involving the early sowing of mustard to utilize the rainfall of both October and November. We planned to compensate for the water scarcity in later months and thereby maintain better productivity. Compared to the regular sowing practice (C: 1st November [control]), we found that sowing in P_1 (5th October) and P_2 (12th October) was significantly beneficial, improving plant height, branches per plant, siliquae per plant, seeds per siliqua, 1000-seed weight, and seed yield (Fig. 16.3). Although sowing in P_2 (12th October) was found to be better than P_1 (5th October), considering the less predictability of weather events, sowing between 5th and 12th October should be considered the best option (Fig. 16.3). Multiple previous studies have highlighted that water scarcity in mustard produces reactive oxygen species, which damage chlorophyll, membranes, and molecules (DNA, RNA, proteins, lipids) and reduce photosynthesis and yield (Sharma et al., 2025; Boomiraj et al., 2010; Sheoran et al., 2024; Bandeppa et al., 2019). However, the early planting of mustard could better utilize the existing soil moisture as well as benefit from the rainfall during October and November to sustain higher productivity.

5 Conclusion

Our study highlighted the critical role of climatic variability in determining mustard productivity in Mymensingh. While long-term climate trends from 1950 to 2020 showed an overall increase in both annual maximum and minimum temperatures with a stable precipitation trend, the recent period from 1991 to 2020 exhibited a distinctively steeper decline in precipitation during the mustard growing season. This precipitation deficit during the initial as well as crucial growth phases stands as a major limiting factor for mustard yield. Future climate projections based on CMIP6 models also indicated a continued temperature rise with frequent abrupt changes in precipitation. This suggests that precipitation is the current constraint, and in the future, temperature might likely become a significant constraint for mustard productivity. To mitigate the current moisture deficiency risk, our study explored the potential benefits of early sowing of mustard. Experimental results demonstrated that sowing mustard between October 5 and October 12 could effectively utilize the available soil moisture from early-season rainfall, leading to significant improvements in yield-contributing traits and overall productivity. Given the increasing climate variability, adjusting sowing dates to align with favorable moisture conditions can be an effective adaptation strategy. Our findings provide a strong basis for recommending early mustard sowing as a means to sustain productivity under changing climatic conditions.

CRediT author statement

Anamika: Conceptualization, Methodology, Formal analysis, Visualization, Writing original draft. **Md. Rahat Bari Rupok:** Data curation, Methodology, Investigation, Writing review & editing. **Ridwan Ahmed:** Conceptualization, Investigation, Writing original draft. **Md. Rashedur Rahman:** Conceptualization, Methodology, Supervision, Writing review & editing. **Mallappa Jadiyappa Madolli:** Formal analysis, Visualization, Writing original draft. **Sushil Kumar Himanshu:** Conceptualization, Writing review & editing. **Sheikh Faruk Ahmed:** Conceptualization, Formal analysis, Resources, Writing review & editing.

Acknowledgment

The authors would like to acknowledge the scholarly support of Climate Change Knowledge Portal (CCKP) of The World Bank in the successful conduction of the current study.

Conflict of interest

The authors declare that they have no known competing financial interests or personal relationships that could have appeared to influence the work reported in this paper.

Funding

This research did not receive any specific grant from funding agencies in the commercial, public, or not-for-profit sectors.

Data availability

The datasets used and/or analyzed during the current study are available from the corresponding author upon reasonable request.

References

Abel, D., Ziegler, K., Gbode, I.E., Weber, T., Ajayi, V.O., Traoré, S.B., Paeth, H., 2024. Robustness of climate indices relevant for agriculture in Africa deduced from GCMs and RCMs against reanalysis and gridded observations. Clim. Dyn. 14320894. 62 (2), 1077–1106. Available from: https://www.springer.com/journal/382. doi:10.1007/s00382-023-06956-8.

Ahmed, S.F., Ahmed, J.U., Hasan, M., Mohi-Ud-Din, M., 2023. Assessment of genetic variation among wheat genotypes for drought tolerance utilizing microsatellite markers and morpho-physiological characteristics. Heliyon. 24058440. 9 (11). Available from: http://www.journals.elsevier.com/heliyon/. doi:10.1016/j.heliyon.2023.e21629.

Alberti, P., Osborne, S., Mathew, F., Ali, S., Sieverding, H., Kumar, S., Nleya, T., 2019. Nitrogen requirements of ethiopian mustard for biofuel feedstock in South Dakota. Agron. J. 0002-1962. 111 (3), 1304–1311. doi:10.2134/agronj2018.06.0419.

Anjum, S.A., Ashraf, U., Zohaib, A., Tanveer, M., Naeem, M., Ali, I., Tabassum, T., Nazir, U., 2017. Growth and developmental responses of crop plants under drought stress: a review. Zemdirbyste-Agriculture. 1392-3196. 104 (3), 267–276. doi:10.13080/z-a.2017.104.034.

Bandara, J.S., Cai, Y., 2014. The impact of climate change on food crop productivity, food prices and food security in South Asia. Econ. Anal. Policy. 03135926. 44 (4), 451–465. Available from: http://www.journals.elsevier.com/case-studies-in-structural-engineering. doi:10.1016/j.eap.2014.09.005.

Bandeppa, S., Paul, S., Thakur, J.K., Chandrashekar, N., Umesh, D.K., Aggarwal, C., Asha, A.D., 2019. Antioxidant, physiological and biochemical responses of drought susceptible and drought tolerant mustard (*Brassica juncea* L.) genotypes to rhizobacterial inoculation under water deficit stress. Plant Physiol. Biochem. 09819428. 143, 19–28. Available from: http://www.journals.elsevier.com/plant-physiology-and-biochemistry/. doi:10.1016/j.plaphy.2019.08.018.

Barnes, C., Brierley, C.M., Chandler, R.E., 2019. New approaches to post processing of multi-model ensemble forecasts. Q. J. R. Meteorol. Soc 145, 3479–3498. doi:10.1002/qj.3632.

Basak, J.K., Titumir, R.A.M., Dey, N.C., 2013. Climate change in Bangladesh: a historical analysis of temperature and rainfall data. J. Environ 2, 41–46.

Bazzaz, M.M., Hossain, A., Farooq, M., Alharby, H., Bamagoos, A., Nuruzzaman, M., Khanum, M., Hossain, M.M., Kizilgeci, F., Öztürk, F., Çiğ, F., Sabagh, A.E., 2020. Phenology, growth and yield are strongly influenced by heat stress in late sown mustard (Brassica spp.) varieties. Pak. J. Bot. 05563321. 52 (4), 1189–1195. Available from: http://www.pakbs.org/pjbot/papers/1590018991.pdf. doi:10.30848/PJB2020-4(44).

BBS, 2024. Yearbook of Agricultural Statistics—2023 (35th Series). Bangladesh Bureau of Statistics (BBS), Ministry of Planning, Government of the People's Republic of Bangladesh, Dhaka. Available at: https://bbs.portal.gov.bd/sites/default/files/files/bbs.portal.gov.bd/page/1b1eb817_9325_4354_a756_3d18412203e2/2024-06-13-05-41-8d348db80ecf814b6f1876432643639e.pdf.

Bhandari, M.P., 2021. Climate change impacts on agriculture, a case study of Bangladesh, India, Nepal, and Pakistan. SocioEcon. Challeng. 2520-6214. 5 (2), 35–48. doi:10.21272/sec.5(2).35-48.2021.

Bhuyan, M.D.I., Islam, M.M., Bhuiyan, M.E.K., 2018. A trend analysis of temperature and rainfall to predict climate change for northwestern region of Bangladesh. Am. J. Clim. Change 7, 115–134. doi:10.4236/ajcc.2018.72009.

Boomiraj, K., Chakrabarti, B., Aggarwal, P.K., Choudhary, R., Chander, S., 2010. Assessing the vulnerability of Indian mustard to climate change. Agric. Ecosyst. Environ. 01678809. 138 (3–4), 265–273. doi:10.1016/j.agee.2010.05.010.

Chauhan, J.S., Tyagi, M.K., Kumar, A., Nashaat, N.I., Singh, M., Singh, N.B., Jakhar, M.L., Welham, S.J., 2007. Drought effects on yield and its components in Indian mustard (Brassica juncea L.). Plant Breed. 0179-9541. 126 (4), 399–402. doi:10.1111/j.1439-0523.2007.01394.x.

Chowhan, S., Islam, M., Rana, M.S., Khan, N.A., Ali, M.K.J., Ahmed, N.U., Rahman, M.M., 2023. Growth and development patterns in Mustard (Brassica spp.) as influenced by sowing time. J. Exp. Biol. Agric. Sci. 23208694. 11 (2), 325–338. Available from: https://jebas.org/ojs/index.php/jebas/article/download/1234/353/2743. doi:10.18006/2023.11(2).325.338.

Eckstein, D., Künzel, V., Schäfer, L., 2021. The Global Climate Risk Index 2021. Germanwatch, Bonn. Available at: https://bvearmb.do/handle/123456789/1306.

Ghaffar, A., Rahman, M.H.U., Ahmed, S., Haider, G., Ahmad, I., Khan, M.A., Afzaal, M., Ahmed, S., Fahad, S., Hussain, J., Ahmed, A., 2022. Adaptations in Cropping System and Pattern for Sustainable Crops Production under Climate Change Scenarios Improvement of Plant Production in the Era of Climate Change. CRC Press, Pakistan, pp. 1–34. Available from: http://www.tandfebooks.com/doi/book/9781000610826. doi:10.1201/9781003286417-1.

Gill, S.S., Gill, R., Kumar, G., Pareek, A., Sharma, P.C., Anjum, N.A., Tuteja, N., 2012. Mustard: approaches for crop improvement and abiotic stress tolerance. In: Tuteja, N., Gill, S.S., Tiburcio, A.F., Tuteja, R. (Eds.), Improving Crop Resistance to Abiotic Stress, John Wiley & Sons, pp. 1351–1368. doi:10.1002/9783527632930.ch52.

Hofmann, M., Volosciuk, C., Dubrovský, M., Maraun, D., Schultz, H.R., 2022. Downscaling of climate change scenarios for a high-resolution, site-specific assessment of drought stress risk for two viticultural regions with heterogeneous landscapes. Earth Syst. Dyn. 21904987. 13 (2), 911–934. Available from: http://www.earth-syst-dynam.net/volumes_and_issues.html. doi:10.5194/esd-13-911-2022.

Hossain, M.A., Reza, M.I., Rahman, S., Kayes, I., 2012. Climate change and its impacts on the livelihoods of the vulnerable people in the Southwestern Coastal Zone in Bangladesh. In: Filho, W.L. (Ed.), Climate Change and the Sustainable Use of Water Resources. Climate Change Management. Springer, Berlin, Heidelberg. pp. 237–259. doi:10.1007/978-3-642-22266-5_15.

Hossain, S.S., Mingying, M., Delin, H., 2020. Climate change in Bangladesh: evidence from temperature, precipitation, and rainfall. Int. J. Arts Humanit. Soc. Sci. Stud 5, 80–88.

Jisha, K.C., Vijayakumari, K., Puthur, J.T., 2013. Seed priming for abiotic stress tolerance: an overview. Acta Physiol. Plant 35, 1381–1396. doi:10.1007/s11738-012-1186-5.

Kapusta-Duch, J., Kopeć, A., Piatkowska, E., Borczak, B., Leszczyńska, T., 2012. The beneficial effects of Brassica vegetables on human health. Roczniki Państwowego Zakładu Higieny. 00357715. 63 (4), 389–395.

Langangmeilu, G., Verma, A.K., Sahu, M., Sarthi, D.P., Pusparani, K., 2023. Influence of different sowing date on the morphology, phenology and thermal indices of rapeseed: a review. Int. J. Environ. Clim. Change. 2581-8627. 13 (7), 707–715. doi:10.9734/ijecc/2023/v13i71923.

Lesk, C., Rowhani, P., Ramankutty, N., 2016. Influence of extreme weather disasters on global crop production. Nature 529, 84–87. doi:10.1038/nature16467.

Li, J., Huo, R., Chen, H., Zhao, Y., Zhao, T., 2021. Comparative assessment and future prediction using CMIP6 and CMIP5 for annual precipitation and extreme precipitation simulation. Front. Earth Sci. 2296-6463. 9. doi:10.3389/feart.2021.687976.

Man, R., Weir, R., 1988. Mustard, its taste and its role as a flavour enhancer. In: Jaine, T. (Ed), Taste: Proceedings of the Oxford Symposium on Food and Cookery, p. 136.

Masipa, T.S., 2017. The impact of climate change on food security in South Africa: current realities and challenges ahead. Jàmbá: J. Disaster Risk Stud. 2072-845X. 9 (1). doi:10.4102/jamba.v9i1.411.

Mostafaei, E., Zehtab-Salmasi, S., Salehi-Lisar, Y., Ghassemi-Golezani, K., 2018. Changes in photosynthetic pigments, osmolytes and antioxidants of Indian mustard by drought and exogenous polyamines. Acta Biol. Hung. 0236-5383. 69 (3), 313–324. doi:10.1556/018.68.2018.3.7.

Passioura, J., 2006. Increasing crop productivity when water is scarce—from breeding to field management. Agric. Water Manag. 03783774. 80 (1-3), 176–196. doi:10.1016/j.agwat.2005.07.012.

Rahman, M.R., Rahman, M.R., Fazal, M.J., Anwar, M.P., 2020. Phenological characterization and yield performance of hilly black rice cultivars under year-round cultivation in plain land ecosystem of Bangladesh. Trop. Agrobiodivers. 27167046. 1 (2), 66–71. doi:10.26480/trab.02.2020.66.71.

Raza, M.A.S., Shahid, A.M., Saleem, M.F., Khan, I.H., Ahmad, S., Ali, M., Iqbal, R., 2017. Effects and management strategies to mitigate drought stress in oilseed rape (*Brassica napus* L.): a review. Zymdirbyste Agric 104, 85–94. doi:10.13080/z-a.2017.104.012.

Shaibur, M.R., Nahar, N., Rahman, M.H., 2018. Probable relationship between climate change and cropping pattern in Jashore district, Bangladesh. J. Jessore Univ. Sci. Technol 2521, 5493.

Sharma, R., Lakra, N., Luhach, A., Islam, S., Zaid, A., 2025. Drought-induced differential biochemical and morphological responses in Indian mustard. J. Plant Biochem. Biotechnol. 0971-7811. 34 (1), 301–318. doi:10.1007/s13562-025-00963-y.

Sheoran, A.R., Lakra, N., Saharan, B.S., Luhach, A., Mandal, N.K., Seth, C.S., Sharma, D., Santal, A.R., Surekha, Sadh, P.K., Rajput, V.D., Duhan, J.S., 2024. Enduring drought: effects and strategies for Brassica crop resilience. Acad. Biol. 2837-4010. 2 (3). doi:10.20935/acadbiol6265.

Singh, R.K., Redoña, E., Refuerzo, L., 2010. Varietal improvement for abiotic stress tolerance in crop plants: special reference to salinity in rice. In: Pareek, A., Sopory, S., Bohnert, H. (Eds.), Abiotic Stress Adaptation in Plants: Physiological, Molecular and Genomic Foundation. Springer, Dordrecht, pp. 387–415. doi:10.1007/978-90-481-3112-9_18.

Singh, S.P., Mahapatra, B.S., Pramanick, B., Yadav, V.R., 2021. Effect of irrigation levels, planting methods and mulching on nutrient uptake, yield, quality, water and fertilizer productivity of field mustard (Brassica rapa L.) under sandy loam soil. Agric. Water

Manag. 18732283. 244. Available from: http://www.journals.elsevier.com/agricultural-water-management/. doi:10.1016/j.agwat.2020.106539.

Stouffer, R.J., Eyring, V., Meehl, G.A., Bony, S., Senior, C., Stevens, B., Taylor, K.E., 2017. CMIP5 scientific gaps and recommendations for CMIP6. Bull. Am. Meteorol. Soc. 0003-0007. 98 (1), 95–105. doi:10.1175/bams-d-15-00013.1.

Sylla, M.B., Elguindi, N., Giorgi, F., Wisser, D., 2016. Projected robust shift of climate zones over West Africa in response to anthropogenic climate change for the late 21st century. Clim. Change. 15731480. 134 (1-2), 241–253. Available from: www.wkap.nl/journalhome.htm/0165-0009. doi:10.1007/s10584-015-1522-z.

Tanimu, B., Danladi Bello, A.A., Abdullahi, S.A., Ajibike, M.A., bin Muhammad, M.K.I., Shahid, S., 2024. Enhancing reliability in climate projections: a novel approach for selecting global climate models. Phys. Chem. Earth. 14747065. 134. Available from: https://www.sciencedirect.com/science/journal/14747065. doi:10.1016/j.pce.2024.103598.

Vilavan, S., Das, D., Ullah, H., Gade, S.A., Ahmed, S.F., Cha-um, S., Praseartkul, P., Datta, A., Himanshu, S.K., 2024. Exploring the impacts of climate change and identifying potential adaptation strategies for sustainable rice production in Thailand's Lower Chao Phraya Basin through crop simulation modeling. Environ. Monit. Assess. 15732959. 196 (12). Available from: https://www.springer.com/journal/10661. doi:10.1007/s10661-024-13362-y.

Wang, X., Wang, S., Folberth, C., Skalsky, R., Hui, L., Liu, Y., Balkovic, J., 2024. Limiting global warming to 2°C benefits building climate resilience in rice-wheat systems in India through crop calendar management. Agric. Syst. 0308521X. 213, 103806. doi:10.1016/j.agsy.2023.103806.

Weigel, A.P., Liniger, M.A., Appenzeller, C., 2008. Can multi-model combination really enhance the prediction skill of probabilistic ensemble forecasts? Q. J. R. Meteorol. Soc 134, 241–260. doi:10.1002/qj.210.

Wheeler, T., Von Braun, J., 2013. Climate change impacts on global food security. Science 341, 508–513. doi:10.1126/science.1239402.

Yuan, X., Li, S., Chen, J., Yu, H., Yang, T., Wang, C., Huang, S., Chen, H., Xiang, A., 2024. Impacts of global climate change on agricultural production: a comprehensive review. Agronomy. 2073-4395. 14 (7), 1360. doi:10.3390/agronomy14071360.

Zhai, Y., Huang, G., Wang, X., Zhou, X., Lu, C., Li, Z., 2019. Future projections of temperature changes in Ottawa, Canada through stepwise clustered downscaling of multiple GCMs under RCPs. Clim. Dyn. 14320894. 52 (5-6), 3455–3470. Available from: http://link.springer.de/link/service/journals/00382/index.htm. doi:10.1007/s00382-018-4340-y.

Zhi, X., Qi, H., Bai, Y., Lin, C., 2012. A comparison of three kinds of multimodel ensemble forecast techniques based on the TIGGE data. Acta Meteorol. Sin. 0894-0525. 26 (1), 41–51. doi:10.1007/s13351-012-0104-5.

Zhu, Y.Y., Yang, S., 2020. Evaluation of CMIP6 for historical temperature and precipitation over the Tibetan Plateau and its comparison with CMIP5. Adv. Clim. Change Res. 16749278. 11 (3), 239–251. Available from: http://www.sciencedirect.com/science/journal/aip/16749278. doi:10.1016/j.accre.2020.08.001.

Assessment of the benefits, costs, and factors affecting the adoption of traceability systems in rice

17

Chawanrat Euafua[a], *Farhad Zulfiqar*[b], *and Muhammad Yaseen*[a]
[a]Faculty of Food, Agriculture, and Natural Resources, Asian Institute of Technology, Bangkok, Thailand
[b]College of Agricultural and Marine Sciences, Sultan Qaboos University, Muscat, Oman

1 Introduction

COVID-19 has indeed heightened consumer awareness regarding health and safety concerns, ethical sourcing, and supply chain transparency. The pandemic has led to a shift in consumer consumption patterns (Mbongwe and Qutieshat, 2023; Consumer Shift: Examining Female Generation y Consumers Activewear Consumption, 2023). Consumers are increasingly concerned about the presence of harmful chemicals in everyday products and their potential impact on health and the environment (Joshi et al., 2022). The COVID-19 pandemic has had a significant impact on farmers globally. The pandemic has disrupted agricultural activities, leading to unsold and rotten crops, financial crises, and acute hunger (Timilsina et al., 2020). The implementation of containment measures for the virus has caused significant disruptions to the provision of agri-food commodities and products, thereby impacting every component of the food system. Farmers in rural areas, who generally have minimal information compared to urban areas, face challenges in accessing health information about COVID-19, impacting their mindset and perception (Musviro et al., 2022).

In Thailand, farmers are still facing various problems from customers after the COVID-19 pandemic. One issue is the unstable rice prices and the middlemen benefiting more from selling rice, which has been a long-standing problem for farmers (Chareonwongsak, 2022). As the Thai government has developed a new economic model called Thailand 4.0 (a policy aimed at transforming the country's economic structure into a "value-based economy") to free the country from the middle-income cycle, the agricultural sector is one of the most affected. Consumers prioritize food safety, leading to increased demand for traceability systems in the food and agriculture industry. However, Thailand still faces challenges in agriculture due to labor shortages and a high proportion of older farmers. The government subsidizes agricultural

Precision Technologies for Digital Agriculture. https://doi.org/10.1016/B978-0-443-40513-6.00010-X

products to help small farmers, but increasing prices could lead to export problems or pressure to import or invest in domestic production at lower costs (Thana-dechophol et al., 2019). The purposes of this study are to compare the benefits and costs for users and nonusers, examine the factors that influence farmers' decisions to adopt or not adopt a traceability system, and identify the marketing channels used by users and nonusers.

2 Literature review

2.1 Innovations used in agriculture

Innovation in agriculture involves developing new technologies for raw material processing, packaging, food additives, and agricultural practices to reduce microorganisms, oxidation, and reactions; ensure safety; produce healthier, more nutritious food; prevent adverse changes in composition; and enhance food perception and enjoyment.

Definition of innovation

According to the Oslo Manual, "Innovation is the implementation of a new or significantly improved product (good or service), process, new marketing method, or a new organizational method in business practices, workplace organization or external relations" (Organisation for Economic Co-operation and Development & Statistical Office of the European Communities, 2005). Innovations are the introduction of new technologies, goods, services, organizational approaches, and business models that offer competitive benefits, as well as economic, scientific, technological, social, and environmental implications. Innovation is viewed as a creative means of developing fresh, improved, unique, and more efficient solutions to address market and customer demands.

Types of innovation

Oslo Manual mentions four types of innovation, which are shown in Table 17.1 below.

Innovations used in the agricultural sector

History of using innovation in the agricultural sector

The first innovation for agriculture can be traced back to the early days of human civilization, around 10,000 years ago (Alston and Pardey, 2016). However, during the 20th century, formal system research and development for agricultural innovation became common (Kalaitzandonakes et al., 2018). These innovations have played a significant role in transforming agriculture and contributing to the overall development of economies (Muralidharan, 2020).

Table 17.1 Types and definitions of innovation.

Types of innovation	Definition
Product innovation	Major enhancements to the capacity of goods or services, included both brand-new products and services, as well as major upgrades to already existing ones
Process innovation	Significant changes to the ways of manufacturing and distribution
Organisational innovation	The adoption of new organizational techniques. These adjustments might be made to the company's external interactions, workplace organization, or commercial procedures
Marketing innovation	The application of modern marketing techniques, changes in product packaging and design, advertising campaigns, and price strategies are a few examples of these

Organisation for Economic Co-operation and Development Statistical Office of the European Communities, 2005. Oslo Manual: Guidelines for Collecting and Interpreting Technological Innovation Data. Organisation for Economic Co-operation and Development.

The use of innovation in the agricultural sector nowadays

The need for ongoing innovation in agriculture is crucial for addressing global challenges such as food security, environmental sustainability, and economic development (Pisante et al., 2012). To meet the increasing demand for food due to population growth, research and development in agriculture are essential. Technological innovations in agriculture have a positive impact on food security, as they contribute to the modernization and sustainability of the sector (Dimitrova, 2022). Agricultural innovation ecosystems may be strengthened by encouraging farmer engagement, multidisciplinary training, and the growth of farmers with an entrepreneurial mindset. Agricultural innovation is mostly driven by farmers' ability to acquire and apply knowledge for innovation, as well as the larger network of connections and interactions they are part of (Yongabo, 2022).

Innovation use for farming

Innovations for agriculture include supporting farmer participation in deciding transition pathways, fostering interdisciplinary training and the development of entrepreneurial-minded farmers, and building an AgTech Pioneer ecosystem. The use of big data in agriculture can provide a scientific basis for innovative economic management and contribute to the construction of smart agriculture (Su and Wang, 2021). Innovations in the agricultural sector encompass technical or technological developments and changes in all areas of scientific and production activities (Saidova and Jurayev, 2021). Different forms of agricultural innovation have different impacts on the environment, with vertical and horizontal integration showing positive environmental contributions (Liu et al., 2021). The focus should be on converting the results of innovation into adoption by farmers and end-users, with lessons applicable to both the private and public sectors (Campos, 2020).

Innovations used for supply chain

Traceability is a key tool for ensuring the quality and safety of agricultural products and increasing consumer trust (Zhang et al., 2021). Blockchain technology provides a reliable solution for traceability in the agricultural product supply chain, offering transparency, decentralization, and a tamperproof nature. By using blockchain and InterPlanetary file system (IPFS), large quantities of transaction data can be stored securely, and consumers can quickly access information about the quality of agricultural products (Zarpelão et al., 2021). Using traceability codes, customers may access extensive data on the production, processing, logistics, and market circulation of agricultural goods, thanks to the deployment of agricultural IoT (Huang et al., 2021). Blockchain technology, with its ability to provide full information about a food product's journey from "farm to fork," has the potential to revolutionize food traceability and enhance reliability and safety (J. Wang et al., 2020). Implementing a traceability system based on blockchain and semantic segmentation can address the credibility and data storage issues traditional traceability systems face. Such a system can ensure real-time recording and monitoring of agricultural products, enabling immediate control of unqualified products and effective protection of consumer rights.

Innovation theory

Innovation theory refers to the study and understanding of the concept of innovation and its various aspects. It involves the development of new processes, products, or organizational improvements in an industry, with the aim of reducing costs and expanding market demand (Sengupta, 2014a). The meanings of innovation are constantly evolving and are influenced by contradictions and complexities. The practice of innovation is shaped by various factors, both internal and external to the organization (Sengupta, 2014b). There are multiple models and concepts that have been developed to explain and analyze the innovation process, ranging from simple linear models to more complex nonlinear ones (Pachura, 2012). The study of innovation theory has practical implications for creating favorable conditions for innovation and fostering a culture of innovation within organizations (Szymańska, 2020).

2.2 Rice in Thailand

Rice is a significant crop in Thailand, with the country being known for its high-quality rice varieties. The Thai government has implemented policies and regulations to ensure the quality and competitiveness of Thai rice in the global market (Napasintuwong, 2018). The International Rice Research Institute and various government agencies in Thailand have conducted research and disseminated knowledge to improve rice production and address environmental and health concerns (Kanjanamaneesathian, 2015). Archaeobotanical studies have shown that rice was a prevalent cereal in prehistoric Thailand, indicating its importance as a crop (Castillo, 2011). Stored rice in Thailand can be infested by several insect pests, depending on the stage of the milling process (Chanbang and Theanjumpol, n.d.). The Thung Kula Rong-Hai (TKR) area in Thailand is known for its high-quality Jasmine rice, which has been registered as a protected geographical indication (Napasintuwong, 2019). Rice in Thailand exhibits a large diversity of varieties due to its rich and diverse ecology.

A study conducted on 300 rice accessions from across Thailand revealed a population structure with five subpopulations based on location (Vejchasarn et al., 2021). Each subpopulation has unique genetic patterns, agronomic traits, and environmental conditions (Sae-Heng et al., 2021). The population structure was inferred through admixture and clustering analysis, supported by a phylogenetic tree, principal component analysis (PCA) analysis, and machine learning. In addition, the Sustainable Rice Platform Standard for Sustainable Rice Cultivation has been implemented in Thailand, with a focus on knowledge translation and behavior changes for sustainable rice cultivation (Fongfon et al., 2021). The rice root nematodes, *Hirschmanniella Oryzae* and *H. mucronata*, pose a threat to rice production in Thailand, particularly in Pathum Thani province. Purple rice landraces in northern Thailand, grown as wetland or upland rice, exhibit diversity in subspecies and genetic differentiation between ecotypes. These studies contribute to understanding the diversity and characteristics of rice in Thailand.

Types of geographic indicator rice in Thailand

There are 21 types of geographic indicator rice in Thailand (Table 17.2).

Table 17.2 Regions and types of geographic indicator rice in Thailand.

Region	Types of geographic indicator rice
North	• Chiangrai Khiaw Ngoo sticky rice • Lanna Kum rice • Phichit Kaw Gor Diew rice • Petchabun Rai Leum Pua rice • Phayao Hom Mali rice
Northeast	• Nakhon Ratchasima Thoungsamrit Hom Mali rice • Kalasin Khaowong sticky rice • Buriram Din Phu Khao Fai Hom Mali rice • Mueang Loei Siwgleang sticky rice • Mueang Loei Daeng sticky rice • Thung Kula Rong Hai Hom Mali rice • Surin Hom Mali rice • Sakon Nakhon Dhavapi Haang Golden Aromatic rice • KhaoHommali Ubon Ratchatani
Central	• Chainat Hom Khaojek rice • Saraburi Jek Chuey Sao Hai rice • Pathum Thani Hom Pathum rice

Continued

Table 17.2 Regions and types of geographic indicator rice in Thailand—cont'd

Region	Types of geographic indicator rice
Southern	• Chumphon Leuang Patew rice • Muang Phatthalung Sangyod rice • Narathiwat Hom Kradung-Nga rice • Phang Nga Rai Dawk Kha rice • Meuang Trang Bao Yod Rice

Department of Intellectual Property Publisher, 2016b. Thai Geographical Indication. Department of Intellectual Property, https://ipthailand.go.th/th/gi-001.html.

3 Methodology

3.1 Study site

The targeted group was rice farmers in Thailand's Surin province because Surin is very famous for producing rice, and it is one of the province's top economic crops. The province produces various types of rice, including geographical indication (GI) rice. Also, Surin's official website is the only one that provides the number of farmers who grow GI rice, along with the name of the coordinator to connect with the farmers.

3.2 Data collection instruments

The data collection instrument was a survey method using a structured questionnaire consisting of carefully crafted questions. The questionnaire included a mix of closed-ended and open-ended items to gather both quantitative and qualitative data. In addition, the survey method enabled efficient data collection, as it could be administered to a large number of participants simultaneously, ensuring a diverse range of responses. The total number of farmers in Surin was 186,295.

Sampling

The sample used in this research was calculated from the farmer population information in Surin province; the simple farmers were randomly selected from around the province to complete the questionnaire.

The calculation, based on the equation of Yamane (1967), will be used to determine the sample size of farmers.

$$n = \frac{N}{(1 + N(e)^{2)})}$$

Determined n = sample size
N = number of population (186,295 households)
e = significant level

Then, 10% of the margin of error applies to the equation.

$$n = \frac{186,295}{\left(1 + 186,295 * \left(\left((0.^2\right)\right)\right)} \approx 99\ Farmers$$

Since the population is heterogeneous, the sample size will be 100 for each group to prevent sampling errors. The sampling method will be chosen randomly.

3.3 Data collection

Both primary and secondary data are included in the data gathering process.

The primary data

The primary data collected by researchers (or assistants) was obtained through the operation of the questionnaire by asking questions orally. Various households were visited with the questionnaire to ask the farmers and gather the necessary data. They asked questions about the farmers' age, gender, education level, and years of farming experience to gain a better understanding of their demographic information. In addition, they inquired about the amount of profit earned from rice cultivation, the costs incurred in the process, the selling price of their rice, and their opinions on implementing a traceability system. This data collection process was specifically targeted toward farmers engaged in wet or in-season rice cultivation.

The data collection was separated into four parts, which were:

- General information
- Income and distribution channel
- Traceability system
- Recommendation

The secondary data

The secondary data were obtained from the Department of Intellectual Property of Thailand and the Geo-Informatics and Space Technology Development Agency of Thailand to collect the name list and addresses of farmers, and the additional data were gathered from books, articles, newspapers, websites, past studies, journals, and other sources.

3.4 Data analysis

After gathering primary and secondary data, the Technology-Organization and Environment (TOE) framework of innovation theory was used to analyze these data, which met the determined objectives and desired results.

- TOE framework

The uses of the TOE framework in this paper consist of Technological Factors, Organizational Factors, and Environmental Factors.

Data processing

Analytical model

The statistical analysis model was used to compare the costs and benefits of the users and nonusers. It was also used to analyze other related information. The variables used in the logit regression were tested for multicollinearity using the variance inflation factor, which quantified the strength and correlation between the predictor variables of a regression model in the statistical package for the social sciences (SPSS) program. After testing the variables, their variance inflation factor values were less than 5; therefore, all of the variables were used in the logit regression model.

Objective with its methods used to process and analyse, and variables.

Objectives	**Method used to process and analyze**	**Variables**
1. To compare the benefits and costs of users and nonusers	Mean, standard deviation, frequency analysis, profitability analysis, t-test, chi-square, propensity score matching method	Cost included the cost of production, such as seeds, fertilizer, land rent, equipment, transportation, and irrigation. Other profits, rather than selling rice, include selling rice straw, rice bran, or rice husk.
1. To examine the factors that influence farmers' decisions to adopt or not to adopt traceability systems.	Logit regression, mean, standard deviation, frequency analysis, *t-test,* chi-square	**Independent variables** • Social influence • Farmers' education • Size of farming operations • Gender • Age • Family member • Cooperative organisation membership • Proximity to a city, • Economics influences • Income from nonagricultural activities • Government subsidies • Attitudes of farmers toward traceability systems **Dependent variable** traceability adoption
1. To identify the marketing channels used by users and nonusers.	frequency analysis, Chi-square, logit regression	Effectiveness of the system and the selling channels used by farmers.

This research used quantitative research designs to analyze the data. The quantitative design involves collecting numerical data from both groups of farmers using surveys or questionnaires. The data can include information on farming practices, crop yields, input usage, income, expenses, and other relevant variables. The approach was used to explore farmers' attitudes, perceptions, motivations, and challenges regarding the adoption or nonadoption of traceability systems.

4 Data

Surin Province is situated in the southern part of the north-eastern area, between 103 and 105-degrees east longitude and 15 and 16 degrees north latitude. It is approximately 450 km from Bangkok by automobile and 420 km by train. It belongs to the southernmost group of north-eastern provinces, which includes Nakhon Ratchasima, Buriram, Chaiyaphum, and Surin provinces. Surin Province covers over 8,124.06 square kilometers or around 812,405.6 hectares, making it 1.58% of the entire country's size or 4.80% of the north-eastern region's area.

According to land statistics from the Surin Provincial Agricultural Economics Office in 2015, the area is 812,405.44 ha, with agricultural land accounting for 82.86% or 673,182.24 ha (of which 84.04% are rice fields and 60,075.2 ha or 8.92% are field crops). Characteristics of agricultural land ownership indicate that 57.42% is on their own property, with the remaining 42.58% farming on other people's land, such as renting or farming for free. Large regions are outside the irrigation area; therefore, most agricultural operations rely significantly on precipitation. Rice, cassava, sugarcane, and rubber are among the most important agricultural crops. Chickens, ducks, and cattle are among the most significant animal products. During the growing year, Surin Province's rice cultivation area in 2016/2017 was 47,8096.32 ha, down by 10,793.6 ha from the previous cropping year. The hectare's output volume is 194,302.24. The average yield is 2733.25 kg per hectare.

Surin jasmine rice is distinguished by its "fragrant, long, white, and soft" features, which refer to rice, brown rice, and white rice processed from light-sensitive fragrant rice types grown in Surin Province, namely Jasmine Rice 105 and 15 in the rice season. It has a natural aroma and is cultivated in four districts: Lamduan, Prasat, Chom Phra, and Khwao Sinrinthon District. TKR jasmine rice, brown rice, and white rice are made from photoperiod-sensitive rice types, namely Jasmine Rice 105 and RD 15, which are cultivated in the TKR area throughout the rice season (Office of the Permanent Secretary for Ministry of Agriculture and Cooperatives, 2023).

Surin Province has rice-growing fields in each district, consistently ranking at the top nationwide due to its status as an agricultural metropolis. Based on a poll on the population's working circumstances, almost half of the individuals are employed in the agricultural sector. Surin Province is the third-largest rice production area in the country, behind Ubon Ratchathani Province and Nakhon Ratchasima Province, and this area remains relatively stable each year. The trend in the amount of cultivated land is steady, ranging between 19.19 and 19.25 million hectares, leading to an annual production of at least 1.1 million tons. The mean production data for the last decade is around 1,120,673 tons annually. The average yield per unit of land has increased, leading to an improvement in overall production. Over the last five years, the average yield per hectare has ranged from 2,368.75 to 2,412.5 kg. Controllable and uncontrollable variables, such as weather and pests, play a role in the situation. A total of 94.88% of the province's agricultural fields are located beyond the irrigation zone. In 2017/2018, rainwater used for cultivating rice in Surin Province was delayed and did not follow the usual seasonal pattern. This resulted in damage to the rice crops, hurting the harvested area, quantity, yield, and quality of the rice. Natural catastrophes, which are uncontrollable, play a significant role in influencing the increase in rice production. Farmers have the opportunity to engage in cultivation and care practices to produce high-quality rice.

5 Results

In Table 17.3, the nonusers incurred significantly higher costs in pesticides and transportation, but there was no significant variation in the costs of rental, seed, and fertilizer. The total income of the users was significantly higher than that of the

Table 17.3 Costs and benefits compared between the users and nonusers.

Costs and benefits (Baht/hectare[2])	Users	Nonusers	*t*-test	*P*-value
Costs				
• Cost of rental	147.4	139.4	3.5	.174
• Cost of seed	1150	1566.7	5.7	.210
• Cost of fertiliser	9310.3	9400.1	3.8	.175
• Cost of labour and equipment	3150.2	6866.3	20.4	.834
• Cost of pesticide	0.00	260.2	7.8	<.00*
• Cost of transportation	21.31	141.2	3.7	0.056*
Total cost (TC)	13,779.3	18,373.9	5.6	.198
Benefits				
• Yield (kilo/hectare)	2000.7	5225.60	3.4	.203
• Price (Baht/kilo)	15.4	13.1	142.4	<.001***
• Income selling rice	31,820.1	18,671.9	3.18	.093*
• Income selling other rice	114.6	23.9		.148
Total income (TI)	31,934.7	18,695.8	3.2	.092*
Net income (TI-TC)	18,155.7	321.9	2.1	.067*
B/C ratio	1.70	1.35	9.478	<.001***

[2]36 baht = 1 USD*Significant at the level 0.1; **Significant at the level 0.05; ***Significant at the level 0.01. The significance values are for a two-group mean comparison *t*-test.

Table 17.4 Comparison of average incomes (Baht per hectare) from rice and other products between users and nonusers.

Comparison of income (Baht/hectare) between two groups	Users (n=106)		Nonusers (n=106)			
	$\bar{x}$	S.D.	$\bar{x}$	S.D.	t	P
Income from selling rice	31,820.1	707,842.7	18,671.9	23,104.6	3.18	.093*
Income from other products	114.6	3678	23.9	295.4	1.76	.148

*Significant at the level 0.1; ***Significant at the level 0.01.

nonusers; overall, the net income from both groups was significantly different from each other. The mean B/C ratio of the users was 1.70, while that of the nonusers was 1.35; these numbers demonstrated the financial advantage of the user group.

The t-test method was used to analyze the first hypothesis, which aimed to determine whether the income values of both groups were different from each other. Table 17.4 explained the results of data analysis comparing income differences from rice between the users and nonusers. The p-value indicated that the mean income from rice between both groups was significantly different at the 0.1 level. However, the mean income from other products for both groups was not significantly different at the 0.1 level.

The income from selling rice represented the income that the users and nonusers received from selling rice solely. The mean income of users was around 31,820.1 baht/hectare, while that of nonusers was 18,671.9 baht/hectare, with a significant value of 0.093, which was lower than the significance level of 0.1. Although the mean value of income from selling other products derived from rice was 114.6 baht, which was higher than that of the nonusers' group by around 90.7 baht, the results showed that the t-test between the users and nonusers had a significant value of 0.148, which was higher than the significance level.

Table 17.5 shows that the nonusers had a higher average of all costs than the users; the most outstanding components were pesticide cost and transportation cost. The pesticide cost shows a highly significant difference ($P<0.001$), and at the 0.1 significance level ($P=0.056$), there is a modest difference in transportation costs.

However, only the p-values for the cost of pesticide and transportation were significantly different at 0.01 and 0.1, respectively. The average costs of rental, rice seed, and fertilizer for the nonusers were just slightly higher, so the p-values for both groups were not significantly different.

Table 17.6 illustrates the results of the propensity score matching method, which was used to match the scores between the users and nonusers to reduce bias in this

Table 17.5 Comparison of average costs (Baht per hectare) for rice and other products between users and nonusers.

Comparison of cost (Baht/hectare) between two groups	User (n=106)		Nonuser (n=106)			
	$\bar{x}$	S.D.	$\bar{x}$	S.D.	t	P
Cost of rental	147.4	2607.91	139.4	1727.77	3.5	.174
Cost of rice seed	1150	17,162.67	1566.6	1995.27	5.7	.210
Cost of fertiliser	9310.3	186,863.89	9400.1	11,971.6	3.8	.175
Cost of labour and equipment	3150.2	9968.08	6866.3	12,873.45	20.4	0.834
Cost of pesticide	0.00	.000	260.1	687.28	7.8	<.000***
Cost of transportation	21.3	973.04	141.2	729.17	3.7	.056*

*Significant at the 0.1 level; ***Significant at the 0.01 level.

Table 17.6 Comparison of estimated net income (Baht/hectare) using propensity score matching methods.

	Nearest neighbour	Kernel	Radius	Stratified
Net income	19,976.3**	19,035.6**	18,068.5*	19,021.8**
S.E.	48,433.1	47,334.55	52,395.2	43,395.39
t-value	2.090	2.033	1.744	2.216

*Significant at the level 0.1; **Significant at the level 0.05; the analysis was done by using STATA program.

study's results. Logit regression was used as the first stage in this study before conducting the propensity score matching. By matching users and nonusers of the traceability system on important characteristics, the method in this paper eliminated the effect of producers' net income on the impact of traceability adoption. The region of common support for the sample was [0.11115039, 0.93375838], which means that there was a range of propensity scores between approximately 0.111 and 0.934 where both treated and control units exist.

Table 17.7 explained the results of data analysis comparing the means of factors affecting the use of the traceability system between the user group and the nonuser group. The p-value indicated that the means of gender, age, education, family members, size of the area growing rice, and income from nonagricultural activities were not significantly different at the 0.1 level. In contrast, the means of district, proximity to the city, type of enterprise participation, area of rice cultivation, and receipt of government incentives were significantly different at the 0.1 level. These findings suggested that these factors influenced the decision to use the traceability system among users and nonusers.

Table 17.7 Comparison of factors influencing traceability system adoption between users and nonusers.

Factors affecting the use of traceability Systems between Two Groups	Users (n=106)		Nonusers (n=106)			
	$\bar{x}$	S.D.	$\bar{x}$	S.D.	t	P
Proximity to a city	21.1	24.93	30.8	17.03	−15.605	<.001***
Gender	1.5	0.50	1.6	0.49	−.969	.334
Age	49.3	11.73	49.4	14.1	−.016	.987
Year in school	11.01	2.11	9.7	2.04	2.551	.011**
Family member	4.4	1.43	4.6	1.18	−1.364	.174
Factors affecting the use of traceability systems between two groups	**Users (n=106)**		**Nonusers (n=106)**			
	$\bar{x}$	S.D.	$\bar{x}$	S.D.	t	P
Type of enterprise participated	3	0.19	4.3	1.4	−9.156	<.001***
Size of area growing rice	31.7	124.52	14.8	7.88	1.394	.165
Receiving government incentive	0.4	0.5	0.3	0.45	2.055	.041**
Income from nonagricultural activities	94,074	170,890.39	112,696.8	209,820.32	−0.492	.623

6 Conclusion and recommendations

6.1 Conclusion

Overall, this research found that there was a difference in costs and benefits between the farmers who use the traceability system and those who do not. In this case, the traceability system might not be the direct factor that improves farmers' incomes. Still, despite that, the traceability system is part of making farmers' living conditions better, including receiving international standards, making products reliable, and being able to grow crops that can be sold, including exporting products abroad. The aim is to increase the international recognition of farmers' products. Growing GI rice, commonly known as Hom Mali 105 and RD15 in Surin province, is not the primary reason for the adoption of the traceability system. Not only do farmers who use the traceability system grow the rice, but farmers who do not use the system also grow this type of rice. Therefore, growing this type of rice is considered normal in this province.

6.2 Recommendation

This study showed that the use of the traceability method resulted in a significant difference in income between users and nonusers. Notably, the people who used it earned more money from selling rice than those who didn't. In addition, the net income of the users was higher than that of the nonusers. These results demonstrate that adopting the traceability method has a positive effect on farmers' bottom lines. The factors affecting adoption identified in this study include years of schooling, membership in cooperative organizations, proximity to urban areas, other sources of income, and the selling price of rice. Based on these factors, the government should establish cooperative groups to implement traceability systems for farmers in local areas and encourage their participation, not only in urban centers but also in districts farther from cities.

In addition to establishing groups, the government should arrange centers that provide knowledge and assistance to farmers in practicing organic farming in order to receive various certifications, including the use of traceability systems in different areas within the community. This includes setting up workshops in the community so that farmers can gain an overview of the work, as well as the advantages and disadvantages of the system, together with lectures, and establishing a fund to support the use of the system. The government should also establish distribution channels for products that utilize the traceability system. According to the survey, farmers who use the traceability system mostly export their products. Therefore, if the government can promote products that have both certifications and use the system to be sold domestically by organizing campaigns or events that feature these types of products, it will provide farmers with an incentive to adopt the system and make their products more widely known.

References

Alston, J.M., Pardey, P.G., 2016. Antipodean agricultural and resource economics at 60: agricultural innovation. Aust. J. Agric. Resour. Econ. 14678489. 60 (4), 554–568. Available from: http://www.blackwellpublishers.co.uk/journals/AJARE/descript.htm. doi:10.1111/1467-8489.12162.

Campos, H., 2020. The Innovation Revolution in Agriculture: A Roadmap to Value Creation the Innovation Revolution in Agriculture: A Roadmap to Value Creation. Springer International Publishing, Peru, pp. 1–234. Available from: https://link.springer.com/book/10.1007/978-3-030-50991-0. doi:10.1007/978-3-030-50991-0.

Castillo, C., 2011. Rice in Thailand: the archaeobotanical contribution. Rice. 1939-8425. 4 (3–4), 114–120. doi:10.1007/s12284-011-9070-2.

Chanbang, Y., Theanjumpol, P., n.d. Session 2: Biology, Ecology and Behavior Insect Pest Status in Stored Rice in a Rice Mill in Thailand. doi:10.14455/DOA.res.2014.12

Chareonwongsak, K., 2022. Self-Sustained Communities: Food Security in Times of Crisis. IntechOpen. doi:10.5772/intechopen.104425.

Department of Intellectual Property Publisher, 2016a. Thai Geographaical Indication. Department of Intellectual Property, https://www.ipthailand.go.th/th/gi-002.html.

Department of Intellectual Property Publisher, 2016b. Thai Geographaical Indication. Department of Intellectual Property, https://ipthailand.go.th/th/gi-001.html.

Dimitrova, A., 2022. Technological innovations in agriculture as a way to increase food security. Econ. Thought J. 00132993. 67 (6), 692–704. doi:10.56497/etj2267604.

Fongfon, S., Pusadee, T., Prom-u-thai, C., Rerkasem, B., Jamjod, S., 2021. Diversity of purple rice (*Oryza sativa* L.) landraces in Northern Thailand. Agronomy. 2073-4395. 11 (10), 2029. doi:10.3390/agronomy11102029.

Huang, Y., Xie, X., Ning, W., Hu, F., 2021. Design of agricultural product traceability platform based on 5G-IoT and blockchain. In: International Wireless Communications and Mobile Computing, IWCMC 2021. Institute of Electrical and Electronics Engineers Inc., China, pp. 323–326. 9781728186160 http://ieeexplore.ieee.org/xpl/mostRecentIssue.jsp?punumber=9497780. doi:10.1109/IWCMC51323.2021.9498577.

Joshi, R., Sahni, A., Malik, M.C., 2022. Consumer awareness regarding harmful chemicals in everyday products. Int. J. Bus. Forecast. Mark. Intell. 7 (4), 351. doi:10.1504/IJBFMI.2022.125793.

Kalaitzandonakes, N., Carayannis, E.G., Grigoroudis, E., Rozakis, S., 2018. Introduction: Innovation and Technology Transfer in Agriculture Innovation, Technology and Knowledge Management. Springer, United States, pp. 1–10. Available from: https://springer.com/series/8124. doi:10.1007/978-3-319-67958-7_1.

Kanjanamaneesathian, M., 2015. Bacillus for Rice Cultivation in Thailand. Springer Science and Business Media LLC, pp. 81–100. doi:10.1007/978-3-319-23183-9_5.

Liu, P., Qi, S., Li, D., Ravenscroft, N., 2021. Promoting agricultural innovation as a means of improving China's rural environment. J. Environ. Manag. 03014797. 280, 111675. doi:10.1016/j.jenvman.2020.111675.

Mbongwe, T.B., Qutieshat, A., 2023. A literature review on the impact of COVID-19 pandemic on consumer buying behavior towards private labels. Int. J. Soc. Sci. Res. Rev. 2700-2497. 6 (6), 138–144. doi:10.47814/ijssrr.v6i6.1182.

Muralidharan, C., 2020. Innovations in agriculture—a case of "Digital Agribusiness Models". Int. J. Curr. Microbiol. Appl. Sci. 23197692. 9 (8), 3030–3038. doi:10.20546/ijcmas.2020.908.342.

Musviro, Astuti, A., Rahmawati, P.M., Suhari, 2022. Farmers' problems in the era pandemic COVID-19. Nurse Health Jurnal Keperawatan. 2623-2448. 11 (1), 1–9. doi:10.36720/nhjk.v11i1.228.

Napasintuwong, O., 2018. Rice seed system in Thailand. Res. Pap. Econ. doi:10.22004/AG.ECON.284028.

Napasintuwong, O., 2019. PGI Hom Mali Thung Kula Rong-Hai Rice in Thailand. In: Sustainability of European Food Quality Schemes: Multi-Performance, Structure, and Governance of PDO, PGI, and Organic Agri-Food Systems. Springer International Publishing, Thailand, pp. 87–109. doi:10.1007/978-3-030-27508-2_5.

Office of the Permanent Secretary for Ministry of Agriculture and Cooperatives, 2023. *Basic Information about Surin Province.*

Organisation for Economic Co-operation and Development & Statistical Office of the European Communities, 2005. Oslo Manual: Guidelines for Collecting and Interpreting Technological Innovation Data. Organisation for Economic Co-operation and Development.

Pachura, A., 2012. Innovation theory: an epistemological aspects. Pol. J. Manag. Stud. 5, 122–129.

Pisante, M., Stagnari, F., Grant, C.A., 2012. Agricultural innovations for sustainable crop production intensification. Ital. J. Agron. 7 (4), 40. doi:10.4081/IJA.2012.E40.

Sae-Heng, P., Arunrat, N., Pumijumnong, N., Chareonwong, U., Stewart, T.N., Sereenonchai, S., 2021. Knowledge translation process of the sustainable rice platform (SRP) standard in Thailand. J. Community Dev. Res. (Humanit. Soc. Sci. 14 (2), 90–105. doi:10.14456/JCDR-HS.2021.18.

Saidova, D.N., Jurayev, F.M., 2021. Analysis of the state of innovative development of agriculture in the region. ACADEMICIA Int. Multidiscip. Res. J. 2249-7137. 11 (6), 529–537. doi:10.5958/2249-7137.2021.01654.2.

Sengupta, J., 2014a. Theory of innovation.In: Theory of Innovation: A New Paradigm of Growth. Springer International Publishing, pp. 1–25. doi:10.1007/978-3-319-02183-6_1.

Sengupta, J., 2014b. Theory of Innovation: A New Paradigm of Growth. Springer International Publishing, United States, pp. 1–154. 9783319021836. doi:10.1007/978-3-319-02183-6.

Su, Y., Wang, X., 2021. Innovation of agricultural economic management in the process of constructing smart agriculture by big data. Sustain. Comput. Inform. Syst. 22105379. 31, 100579. doi:10.1016/j.suscom.2021.100579.

Szymańska, E., 2020. Innovation theory: where is it going? Eurasian Stud. Bus. Econ. 23645075. 13 (1), 253–263. Available from: https://springer.com/series/13544. doi:10.1007/978-3-030-40375-1_18.

Thana-dechophol, T., Teamsuwan, V., Chamnankaew, U. (2019). Agricultural reform in Thailand. In: *Thaicid National Symposium, 11th.*

Timilsina, B., Adhikari, N., Kafle, S., Paudel, S., Poudel, S., Gautam, D., 2020. Addressing impact of COVID-19 post pandemic on farming and agricultural deeds. Asian J. Adv. Res. Rep. 2582-3248. 28–35. doi:10.9734/ajarr/2020/v11i430272.

Vejchasarn, P., Shearman, J.R., Chaiprom, U., Phansenee, Y., Suthanthangjai, A., Jairin, J., Chamarerk, V., Tulyananda, T., Amornbunchornvej, C., 2021. Population structure of nation-wide rice in Thailand. Rice. 19398433. 14 (1). Available from: http://www.springerlink.com/content/1939-8425. doi:10.1186/s12284-021-00528-2.

Wang, J., Jia, L., Zhou, Q., 2020. Agricultural product quality traceability system based on the hybrid mode. In: Advances in Intelligent Systems and Computing, vol. 1274. Springer Singapore, pp. 248–253. doi:10.1007/978-981-15-8462-6_29.

Yongabo, P., 2022. Technology and innovation trajectories in the Rwandan Agriculture sector: are value chains an option? Afr. J. Sci. Technol. Innov. Dev. 2042-1338. 14 (3), 697–707. doi:10.1080/20421338.2021.1889769.

Zarpelão, B.B., Junior, S.B., Badaró, A.T., Barbin, D.F., 2021. On the use of blockchain for agrifood traceability Food Authentication and Traceability, Academic Press, pp. 279–302.

Zhang, L., Zeng, W., Jin, Z., Yansen, S., Chen, H., Chen, X.-B., 2021. A research on traceability technology of agricultural products supply chain based on blockchain and IPFS. Secur. Commun. Netw. 1939-0122. 2021, 1–12. doi:10.1155/2021/3298514.

Smart agriculture technologies in Bangladesh: Current status, adoption, and policy implications

18

Md. Masudul Hassan and Md. Sadique Rahman
Department of Agricultural Finance and Management, Sher-e-Bangla Agricultural University, Dhaka, Bangladesh

1 Introduction

Smart agriculture technology (SAT) represents a transformative advancement in farming practices, employing digital innovations to improve production, sustainability, and resource management. This paradigm shift is facilitated by the integration of various technologies, including the Internet of Things (IoT), big data analytics, artificial intelligence (AI), and precision agriculture techniques. These advancements are essential for the global community to tackle the challenges posed by population growth, climate change, and the need for sustainable agricultural practices. SAT facilitates enhancements in farming operations. For example, IoT devices and big data analytics are used in precision agriculture to assess soil conditions, weather patterns, and crop health in real time, which is important for better agronomic management (Boahen and Choudhary, 2024; Bhat and Huang, 2021). This allows farmers to make informed decisions about irrigation, fertilization, and pest control, thus reducing resource waste (Cernisevs et al., 2024; Aliev et al., 2023).

Agriculture serves as a fundamental component of Bangladesh's economy, playing a crucial role in both gross domestic product (GDP) and employment contributions. Agriculture contributes around 11.02% to Bangladesh's GDP, employing approximately 45% of the workforce (BER, 2024; Metropolitan Chamber of Commerce and Industry, 2024). The Aspire to Innovate (a2i) Program (2023a) reports that there are 16.5 million farming families, of which 6.8 million cultivate land owned by others. Financial coverage in agriculture is 37.2%, highlighting deficiencies in access to credit and financial services. These statistics underscore the sector's significant contribution to the Bangladeshi economy and the necessity for policies aimed at improving financial inclusion and productivity. However, the agriculture sector encounters several constraints, such as climate change, diminishing soil fertility, postharvest losses, and suboptimal resource utilization. Therefore, SAT can present effective solutions through the integration of digital tools, precision farming methods, and automation aimed at enhancing productivity and sustainability.

Precision Technologies for Digital Agriculture. https://doi.org/10.1016/B978-0-443-40513-6.00001-9

Studies indicate that farmers employing precision agriculture techniques can realize yield increases of up to 20% while simultaneously decreasing input costs such as pesticides and fertilizers (Vatin et al., 2024). Robotics, drones, Global Positioning System (GPS), remote sensing, and computer imaging are all AI-powered technologies that can be used to predict disease and pest outbreaks, weather patterns, the best time and amount of fertilizers and pesticides to use, when to water plants, and when to harvest them, as well as predictions in agricultural analytics, market dynamics, and supply chain efficiency to improve economic, social, and environmental well-being. AI-based technologies serve as alternative methods to traditional practices for ensuring food security and environmental quality (Hossain and Islam, 2022). The incorporation of renewable energy sources, including solar-powered irrigation systems, supports sustainability objectives by decreasing dependence on fossil fuels and lessening environmental impact (Salehin et al., 2023). SATs are transforming the agricultural sector through the integration of advanced digital tools that improve productivity, sustainability, and resource management. The integration of SATs provides a strong framework for addressing the critical challenges faced by contemporary agriculture. As these technologies advance, they have the potential to significantly enhance global food security and environmental sustainability.

The relevance of SATs in improving yields, conserving resources, and reducing environmental consequences is highlighted by several studies (Boahen and Choudhary, 2024; Ioja et al., 2024; Fakhar and Khalid, 2023; Gopathoti, 2023; Peladarinos et al., 2023; Fatima et al., 2023; Sambas et al., 2023; Lohit and Mujahid, 2022; Kittichotsatsawat et al., 2021; Khanna, 2020). However, the adoption and challenges of SATs in Bangladesh have received little attention. To aid decision-making and guide the acceptance and implementation of SATs in developing countries like Bangladesh, this chapter attempts to offer a thorough review of the adoption status, challenges, and policy implications.

2 Methodology

This study utilizes a secondary data analysis strategy, leveraging existing literature, reports, and databases to investigate the subject issue. The process entails the systematic gathering, assessment, and integration of secondary sources to guarantee a thorough comprehension of the topic. The secondary data were obtained from peer-reviewed publications, books, governmental and nongovernmental reports, and publicly accessible databases. To guarantee integrity and reliability, only sources from esteemed publishers, academic institutions, and acknowledged organizations were utilized. A keyword-driven search method was employed to obtain pertinent papers, ensuring alignment of the data. The keywords are related to smart farming technologies, such as precision farming, IoT, smart farming, biotechnology, urban farming, Geographic Information Systems (GIS), etc. Priority was assigned to literature published recently to ensure the inclusion of the most recent information. The quality of the sources was evaluated according to authorship, publication venue, and the research methodology employed in the original studies.

Although secondary data analysis presents various benefits, including cost efficiency and access to extensive datasets, specific limitations have been recognized. These include insufficient oversight of data collection methodologies and probable information deficiencies. To address these difficulties, various sources were cross-referenced to improve validity and dependability. Since this study solely utilized publicly accessible secondary data, it eliminates any direct ethical issues concerning human subjects. Nevertheless, appropriate credits and attributions were provided to recognize the original authors and uphold academic integrity.

3 Findings

3.1 *Smart agriculture technology in Bangladesh: Current status and recent trends*

The key challenges in agriculture encompass reduced production stemming from insufficient innovation and mechanization, alongside postharvest losses nearing 40%, attributed to ineffective supply chains. Climate change effects, including river erosion, flooding, and rising sea levels, present further challenges. The absence of real-time data and integrated platforms complicates decision-making, while market and export constraints, such as inadequate standardization and weak global market connections, further hinder growth (Table 18.1). Addressing these challenges through SAT is essential for sustainable agricultural development.

However, the adoption of SAT in Bangladesh is still in its early stages, with relatively low adoption among the country's millions of small farmers. To date, the application of advanced precision agriculture tools has been primarily limited to research trials and pilot projects (Bhuiyan, 2021). For example, the application of drones in agriculture is minimal, with only a few drones being used for research purposes as of the mid-2020s (Bhuiyan, 2021). Overall, there has been almost no commercialized precision agriculture technology tailored to smallholder farms yet (Al-Amin et al., 2023). Krishoker Janala v.2, an AI-driven agricultural application, offers voice-activated crop disease identification and intelligent advice.

Table 18.1 Key challenges in the agriculture sector of Bangladesh.

Challenges	Description
Low production	Limited innovation in farming techniques and mechanization
Postharvest loss	~40% waste due to inefficient supply chains
Limited tech adoption	Poor integration of Fourth Industrial Revolution (4IR) technologies, IoT, and AI
Climate change impact	River erosion, flooding, and rising sea levels
Lack of real-time data	There are no integrated platforms for decision-making
Market & export limitations	Lack of standardization and poor linkage to global markets

Source: Aspire to Innovate (a2i) Program (2023a).

Table 18.2 Smart agriculture solutions in Bangladesh.

Smart solution	Expected benefits
Farmer-friendly smart data ecosystem	Improved access to finance, markets, and resources through integrated digital services
Krishoker Janala v.2 (AI-powered Farmer App)	Enables voice-based crop disease identification and provides smart recommendations
IoT-based smart fish farming	Automated monitoring of water quality and feed distribution
Digital agri-supply chain	Reduces middlemen dependency and improves direct farmer-market connections
Smart irrigation & fertilizer management	IoT-based sensors for precise irrigation and fertilizer application
e-Traceability of Agriculture & Livestock	Blockchain-based tracking for food safety and livestock monitoring

Source: Aspire to Innovate (a2i) Program (2023b).

IoT-enabled Smart Fish Farming facilitates automated monitoring of water quality and feed, whilst digital agri-supply chains diminish reliance on intermediaries and enhance direct linkages between farmers and markets. Smart Irrigation and Fertilizer Management use IoT sensors for accurate resource applications, while e-Traceability in agriculture and cattle employs blockchain technology to ensure food safety. These advances enhance efficiency, sustainability, and productivity within the agriculture sector (Table 18.2).

However, recent years have shown promising trends. Interest in "smart farming" is growing, driven by both government and private initiatives through SAT adoption. The use of IoT devices and drones has started to increase, helping improve crop monitoring and management on some farms (Mamun, 2024). A number of agri-tech startups have emerged to bring technology to agriculture—notable examples include iFarmer, Khamar-e, Integrated Precision Agriculture & Engineering (iPAGE), Krishi Shwapno, Bhalo Social Enterprises, and Fashol—which are at the forefront of applying data-driven farming solutions in Bangladesh (Future Startup Team, 2024; Bhuiyan, 2024; Kader and Amit, 2023; Shuvo, 2021; Ifaz, 2020). These early adopters and innovators indicate a shift toward more technology-driven agriculture. Market analysis reports also note that precision farming tools are *"gaining traction as farmers embrace technology-driven approaches,"* aided by government modernization efforts (6Wresearch, 2023a). In summary, while the current adoption rate is low, momentum is building, and SATs are gradually being introduced through pilots and early adopters. The smart agricultural programs in Bangladesh that are helping farmers with data-driven farming, financial resilience methods, and technology are summarized in Table 18.3.

Fig. 18.1 illustrates the productivity growth due to the adoption of IoT-based smart agriculture techniques in Bangladesh from 2015 to 2025. From 2015 to 2017, the growth was limited (<5%), but from 2018 to 2023, significant growth occurred due to successful pilot programs and increased awareness of its benefits. These

Table 18.3 Key smart agriculture initiatives in Bangladesh.

Smart agriculture initiative	Description	References
IoT-driven precision agriculture	A pilot project implemented IoT sensors across 500 farms, increasing crop productivity by 15% through better soil moisture and pest monitoring.	(Rahman, 2024)
Pest early-warning system (PRAMS)	PRAMS uses IoT sensors and data analytics to predict pest infestations. It helped protect 200,000 acres of crops in 2022.	(Rahman, 2024)
Agritech Startups (iFarmer, iPAGE, Khamar-e)	Startups are providing financing, precision farming tools, and smart hydroponics solutions, helping farmers reduce input costs and increase yields.	(Bhuiyan, 2024; Future Startup Team, 2024; Ifaz, 2020; Shuvo, 2021; Kader and Amit, 2023)
Research and institutional projects	Organizations like CIMMYT and BARI use drones for crop monitoring, and AWD irrigation techniques save 25%–30% of water in rice farming.	(Bhuiyan, 2021; Doody, 2021; IRRI, 2024; Paul and Rashid, 2013)
Weather-based crop insurance	Green Delta Insurance launched weather index-based insurance to financially protect farmers from floods and extreme rainfall. A total of 316 farmers benefited from the Tahirpur pilot project.	(Saini and Giriraj, 2020; Chowdhury, 2020; Green Delta Insurance, 2020; Ali, 2022; The Daily Star, 2019)

insights illustrate the continual evolution of Bangladesh's agricultural sector, highlighting the incorporation of intelligent technologies to improve production and sustainability (Ifty et al., 2023). Empirical evidence underscores that adoption levels vary by region and technology. For example, among the 19 evaluated technologies, low-cost irrigation solutions were the most adopted practices in coastal areas of Bangladesh (Mia and Roy, 2025). The adoption was positively associated with farmers' income, training exposure, knowledge and attitudes toward technology (Mia and Roy, 2025). Studies have indicated that the adoption of proven technologies and increased mechanization have already increased food productivity. Farmers are increasingly adopting eco-friendly and improved technologies that cause minimal environmental damage (Islam, 2025). These observations demonstrate that tangible progress is being made, even though adoption is still in its early stages.

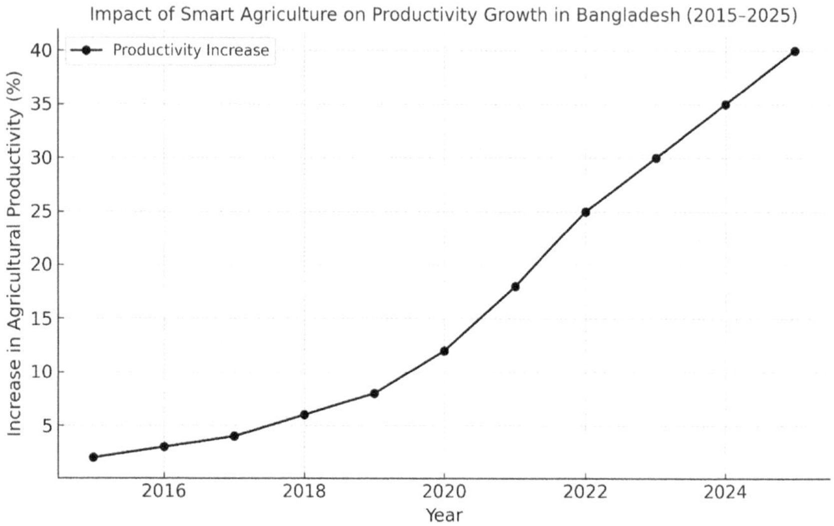

Fig. 18.1 Productivity growth of IoT-based smart agriculture practices in Bangladesh.

3.2 Suggested smart agricultural technologies (SATs) in Bangladesh

A range of technologies is being suggested to Bangladesh's farming sector as part of smart agriculture initiatives:

- **GIS and remote sensing:** GIS and satellite remote sensing are employed for field mapping and crop status monitoring. Bangladesh's agricultural officials are being trained to utilize satellite imagery for the detection of crop damage caused by floods or illnesses (Grain Feed & Milling Magazine, 2024). High-resolution photos facilitate the evaluation of plant health and the forecasting of yields across extensive regions.
- **Automated irrigation systems:** Automated irrigation systems, especially those utilizing IoT, facilitate accurate water management, which is essential for a nation like Bangladesh, where rice farming is prevalent. The alternate wetting and drying (AWD) technique demonstrates innovation by enabling fields to dry periodically prior to re-irrigation, thereby conserving water while maintaining yields (Sarker et al., 2009). This method has been advocated by entities such as the International Rice Research Institute (IRRI) and has been extensively embraced by Bangladeshi farmers to enhance water efficiency (IRRI, 2024). Studies demonstrate that AWD can decrease water consumption by 25%–30% while preserving or potentially enhancing rice yields in Bangladesh (Paul and Rashid, 2013).
- **Drones**: Drones fitted with sophisticated cameras and sensors are undergoing trials for airborne agricultural monitoring. Drones and sensors are revolutionizing agriculture by providing high-resolution aerial surveillance, enabling rapid identification of diseases, pests, and other factors impacting crop health. Research in Bangladesh indicates that drones can identify plant diseases with an accuracy of 70%–98%, facilitating early intervention (Bhuiyan, 2021). Over the past five years, around 15 distinct crop diseases have been effectively diagnosed with sensor-equipped drones (Bhuiyan, 2021). It is feasible to spray an entire hectare in about 10 minutes using a spraying drone because drones have large tanks

with a capacity of around 10 liters and a spraying rate of one liter per minute (Haque et al., 2021). Notwithstanding these characteristics, the utilization of drones in agriculture is currently restricted (partially due to financial and regulatory limitations), while pilot initiatives illustrate their promise. Studies indicate that drones and remote sensing technologies can identify insect outbreaks with over 85% accuracy, enabling farmers to implement timely preventive measures (Rahman, 2024). Nevertheless, obstacles such as high costs and the absence of regulatory frameworks hinder their extensive use in Bangladesh.

- **IoT sensors:** Smart farming makes use of a wide range of sensors (soil moisture sensor, leaf moisture sensor, soil salinity sensor, and soil temperature sensor) to guide agronomists and farmers in optimal crop planting, fertilization, and harvesting practices (Abdul-Salaam et al., 2017). IoT devices, including soil moisture sensors, meteorological stations, and intelligent insect traps, are being tested to gather real-time agricultural data. These sensors facilitate data-driven irrigation and pest control. In a recent pilot involving 500 farms, IoT sensors that monitored soil and pests enabled farmers to realize a 15% improvement in crop yields through optimized inputs, a benefit directly linked to precision agriculture practices (Rahman, 2024). These sensor networks can notify farmers of optimal watering times or increasing insect levels, facilitating prompt interventions.
- **AI and data analytics:** AI algorithms and decision-support software are progressively employed to evaluate agricultural data and inform farm management. In Bangladesh, AI tools are employed to predict pest infestations and suggest interventions; for instance, the Pest Risk Assessment and Management System utilizes sensor data to anticipate pest outbreaks, reportedly preserving over 200,000 acres of crops in 2022 through proactive pest control measures (Rahman, 2024). AI-driven precision agriculture technologies can ascertain optimal fertilizer application rates and irrigation schedules by analyzing field data in conjunction with weather forecasts (Hossain and Islam, 2022). Technologies such as robotics, drones, GPS mapping, computer vision, and predictive analytics are being investigated as "4IR" breakthroughs to enhance agricultural efficiency (Hossain and Islam, 2022). In addition to these research initiatives, AI-driven image recognition and predictive modeling tools embedded in mobile applications can scan and diagnose plant diseases from images with a high level of accuracy, thereby reducing pesticide use by up to 40% (DataInsightopia, 2025). Integrating such AI-enabled tools is vital for enhancing productivity, reducing input costs, and promoting sustainability. The use of advanced technologies such as deep learning for plant disease detection aligns with broader goals articulated in sustainable agricultural strategies that emphasize the role of information and communication technologies (ICTs) in promoting smart farming practices and information dissemination among farmers (Ahmed and Raha, 2023; Sharma et al., 2021).
- **Mobile apps and decision tools:** Although less advanced than drones or IoT, smartphone applications and digital platforms are integral to Bangladesh's precision agricultural environment. Applications such as Krishoker Janala and the Digital Crop Calendar offer site-specific guidance (e.g., planting schedules, input utilization) and assist farmers in making informed choices (Quader, 2023). These solutions enhance on-field hardware by providing tailored recommendations, frequently utilizing big data or expert systems in the background. They provide a digital conduit to deliver precise insights, such as fertilizer formulations or insect notifications for specific locales, directly to farmers' mobile devices.
- **Vertical and urban farming:** Vertical farming has emerged as a viable alternative to address escalating food demand, urbanization challenges, and resource constraints, while facilitating sustainable year-round urban agriculture (Kabir et al., 2023). Vertical farming, characterized by the cultivation of crops in stacked levels under regulated conditions, is gaining traction in Bangladesh. The vertical farming market in Bangladesh is anticipated to

grow substantially at a rate of 25.6% from 2025 to 2031 (6Wresearch, 2023b). Urban agriculture is increasingly prevalent in Bangladesh, especially in highly populated areas such as Dhaka and Chattogram. Methods such as rooftop gardening, container cultivation, and hydroponics are being implemented to maximize restricted urban areas for food production. These approaches facilitate access to fresh produce for low-income families, thereby improving food security and nutrition (Tarannum, 2023; FAO, 2021). Innovations such as the Integrated Floating Cage Aquageoponics System (IFCAS) have been developed to merge fish and vegetable cultivation in shaded ponds, thus improving resource utilization in constrained areas, as part of the EU-funded Agriculture and Nutrition Extension Project (Haque et al., 2015).

- **Biotechnology applications:** Bangladesh has achieved significant advancements in the development and adoption of genetically engineered crops (Foreign Agricultural Service, 2022). A notable instance is the development of Bt brinjal, genetically modified to withstand the fruit and shoot borer bug. Since its commercial introduction in 2013 (Choudhary et al., 2014), the cultivation of Bt brinjal has proliferated, with more than 31,000 farmers cultivating this variety (Shelton et al., 2020). This adoption has resulted in a 39% decrease in pesticide usage and a 43% increase in output, thereby improving farmers' profitability (Basu, 2019). Furthermore, salt-tolerant rice cultivars, engineered through biotechnological research, are being cultivated in the coastal regions of Bangladesh, assisting farmers in adapting to soil salinity caused by climate change. This innovation has fundamentally transformed rice breeding and adaptation in 19 coastal areas of Bangladesh, which account for 25% of the nation's rice production (Illius and Azad, 2022; Islam, 2019). Rice output in saline ecosystems may rise by 11.75% by 2050, contingent upon the distribution of salt-tolerant rice cultivars in 75% of salinity-affected areas with salinity levels up to 12 dS m^{-1} (Rahman et al., 2021).

To expedite agricultural transformation in Bangladesh, policymakers should prioritize cost-effective digital technologies that directly assist small farmers. Considering the socio-economic status of farmers, initially, mobile application-based platforms need to be expanded to provide farmers with real-time data regarding weather conditions, insect occurrences, and market trends. The use of affordable IoT sensors, alongside other technologies, can facilitate reduced consumption of water, fertilizer, and pesticides. Automated irrigation systems should be advocated in regions susceptible to drought and salinity to conserve water. Geographic Information Systems ought to be utilized at the institutional level to facilitate investments and extension services through crop zoning, disaster risk mapping, and land-use planning. Drones now offer limited utility for small farms; nevertheless, they are applicable in research, crop monitoring, and extensive commercial agriculture. To ensure the accessibility and adoption of digital innovations, policies must prioritize affordability, provide training, and collaborate with both public and private sectors.

3.3 Economic and environmental impact of smart agriculture technologies

Precision farming techniques offer significant economic advantages for farmers while enhancing environmental sustainability in Bangladesh. By optimizing input utilization and making data-driven decisions, farmers can enhance their yields and earnings while minimizing waste. The economic and environmental effects of adopting SATs and precision farming in Bangladesh are highly promising. The implementation of SATs enables farmers to "produce more with less," enhancing yields and profitability while minimizing resource usage and ecological impact. A policy research article highlighted that the advantages of smart farming for Bangladesh encompass "enhanced crop yield and quality, diminished production costs and environmental impact, improved access to finance and market intelligence, and increased resilience to climate change" (Tanvir and Bhattacharjee, 2024). Research demonstrates that implementing precision agriculture methods can markedly enhance crop yield. Research conducted in Bangladesh indicates that precision farming can enhance crop yields by around 22% or more in comparison to conventional approaches (Hossain and Islam, 2022). A separate study suggested that the widespread adoption of AI-driven precision agriculture might enhance crop yields in Bangladesh by 20%–30% for key crops (Kazi, 2025). These yield enhancements directly result in increased agricultural revenues and enhanced food security. Precision agriculture can optimize input use, applying the precise quantities of seeds, water, or fertilizer required instead of surplus. This signifies diminished production expenses over time, as farmers utilize fewer resources (Tanvir and Bhattacharjee, 2024).

The ecological advantages are of comparable significance. Precision agriculture primarily aims to optimize resource utilization, thereby reducing the excessive application of fertilizers, pesticides, and water. In Bangladesh, traditional methods frequently involve extensive fertilizer application (Fig. 18.2). The nation has long maintained one of the highest fertilizer usage rates per hectare in the region (Al-Amin et al., 2023). This not only increases expenses but also exacerbates soil deterioration and water contamination due to runoff. Precision agriculture methodologies

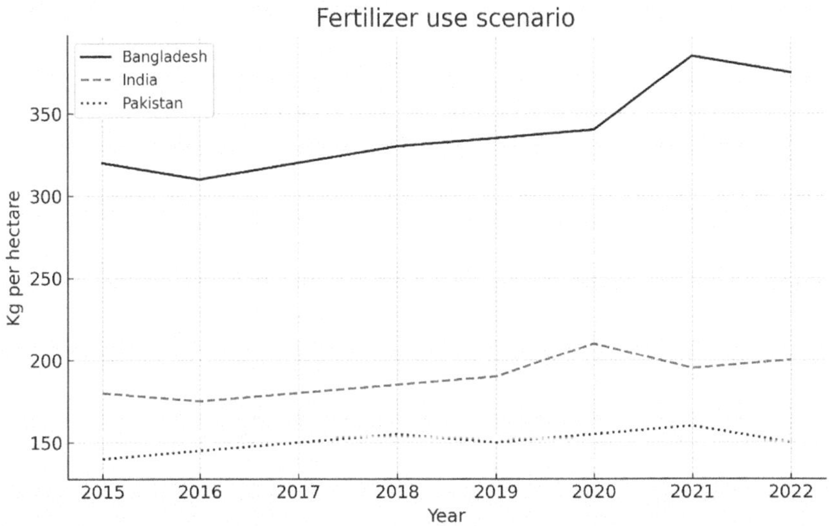

Fig. 18.2 Comparative fertilizer use by countries.
Source: https://www.theglobaleconomy.com/rankings/fertilizer_use/SAARC/.

mitigate this issue by administering inputs at customized rates and specific locations. For instance, employing GPS-guided nutrient management or site-specific fertilization enables farmers to sustain yields while utilizing smaller quantities of fertilizer, thereby diminishing the risk of chemical runoff into aquatic systems (UNDP, 2021). Enhanced pest monitoring ensures that pesticides are applied solely when and where necessary, thus reducing indiscriminate spraying. By minimizing input wastage, precision farming diminishes the environmental impact of agriculture (UNDP, 2021). A study in Bangladesh indicated that AI-driven solutions facilitate a 40% reduction in fertilizer usage, resulting in a decrease of CO_2 emissions by over 2 million kg. Consequently, farmers have experienced a rise of over 12% in yields and nearly 20% in sales for input and machinery partners (Seltier, 2023).

Water management represents a crucial domain in which precision methods significantly impact outcomes. Bangladesh encounters water limitations and climate-induced fluctuations, such as droughts and floods; thus, efficient irrigation is essential. Smart irrigation systems, informed by soil moisture sensors or meteorological data, can significantly enhance water conservation. Analysts estimate that AI-driven irrigation scheduling may decrease water consumption by approximately 30% while maintaining crop yields (Kazi, 2025). Techniques such as AWD in rice, a precision water management practice already implemented in certain regions of Bangladesh, illustrate these savings. Reducing water usage per crop contributes to the preservation of water tables and ecosystems while simultaneously decreasing farmers' pumping expenses.

3.4 Key challenges and barriers to adoption

Notwithstanding the promising uses, obstacles such as substantial initial investment costs, insufficient technical competence, and infrastructural limitations impede the extensive implementation of SATs in Bangladeshi agriculture. Some of the key barriers include:

- **Small and fragmented landholdings:** In Bangladesh, most farms are diminutive, with over 89% including less than 1 hectare; around 39% are categorized as marginal, measuring under 0.2 hectares, and roughly 50% are classified as small, under 1 hectare (Bhuiyan, 2021). The small, fragmented plots hinder the economic deployment of large machines or costly technologies such as drones. Precision agriculture technologies and techniques typically presume larger-scale operations; their adaptation to the diminutive farm sizes in Bangladesh presents a significant barrier. A drone may be neither cost-effective nor practicable for a 0.1-hectare area (Bhuiyan, 2021). Land fragmentation consequently restricts the quick implementation of certain precision techniques unless solutions are customized for smallholders.
- **High cost of technology:** Advanced technologies, including drones, sensors, and agricultural automation systems, are costly, rendering them inaccessible to the predominantly low-income farmers in Bangladesh (Bhuiyan, 2021). The initial expenditure on precision agriculture equipment and related software/services can be substantial. Agricultural drones and automated machinery necessitate significant capital investment, and even leasing them might impose a financial burden on a small farmer's budget. Widespread use is improbable

without affordable, scaled-down versions of precision technology or subsidy programs. The financial obstacle is further exacerbated by restricted access to funding for farmers to invest in innovative technology.

- **Limited farmer awareness and skills:** A significant technology literacy disparity exists in rural Bangladesh. A significant number of farmers lack familiarity with digital tools and contemporary agricultural data analytics. Insufficient education and exposure result in inadequate utilization of existing tools. Research indicates that inadequate digital literacy among farmers impedes the seamless integration of technologies (Tanvir and Bhattacharjee, 2024). The present deficiency of expertise and self-assurance in utilizing these instruments constitutes a significant obstacle.
- **Inadequate digital infrastructure:** The implementation of SATs depends on reliable connectivity (for IoT devices, GPS, and cloud services) and electricity. In certain rural regions, internet and electrical infrastructure remain inadequate. Insufficient digital infrastructure, exemplified by inconsistent internet access in remote areas, can hinder the implementation of smart farming solutions (UNDP, 2021). If sensors are unable to transmit data or farmers cannot use advisory applications due to connectivity problems, the technology becomes ineffective.
- **Regulatory and operational hurdles:** Emerging technologies occasionally encounter regulatory impediments. In Bangladesh, the utilization of drones for non-military purposes necessitates authorization from the Civil Aviation Authority (Bhuiyan, 2021). Although essential for safety, such laws increase administrative burdens and may hinder the deployment of agricultural drones.

3.5 Government policies and initiatives supporting SATs

The Government of Bangladesh has recognized the importance of SATs and has introduced several policies and programs to support their adoption:

- **National Agriculture Policy 2018:** The National Agriculture Policy clearly mandates the adoption of precision agriculture to enhance productivity. Clause 10.9 of the National Agriculture Policy 2018 emphasizes the necessity of precision farming techniques to enable farmers to enhance crop output sustainably (Ministry of Agriculture, 2018; Bhuiyan, 2021). This policy directive has prioritized precision agriculture in agricultural development initiatives. Since 2018, the Ministry of Agriculture has emphasized digitalization in agriculture, financing e-agricultural initiatives, and farmer education as components of this policy (Rahman, 2024).
- **National Agri-Mechanization Policy 2020:** Bangladesh's mechanization approach underscores precision agriculture as a pivotal future catalyst. The policy indicates that information technology-driven precision farming techniques will significantly contribute to the forthcoming phase of agricultural mechanization, enhancing crop yield through efficient resource utilization (Ministry of Agriculture, 2020; Bhuiyan, 2021). This corresponds with the government's overarching objective to shift from conventional practices to "smart" agriculture, which is more efficient and adaptable to climate change.
- **National ICT Policy 2018:** One of Bangladesh's most prominent regulatory structures for its digital agenda is the National ICT Policy. The policy vision can be realized through ten interconnected objectives, including but not limited to social fairness, integrity, universal access, education and research, job creation, bolstering exports, healthcare, the environment, climate and disaster management, and productivity. An important part of the policy goals has been to encourage the use of ICT in relation to environmental and agricultural

concerns. Community network development, agricultural information distribution, and innovations through community radio, telecenters, and web-based networks are some of the ways proposed to make the use of ICT easier. The government has accelerated the development of ICT infrastructure and public services under the Public-Private Partnership model, leading to a rise in the proportion of the country's GDP allocated to investments in ICT from approximately 2% to 6% (Aziz, 2020).

- **"Smart agriculture" initiatives (A2i program):** The government's a2i initiative has launched targeted projects to expedite agri-tech innovation. A Smart Agriculture Tech Lab was established to trial 4IR technologies, including agricultural robotics, drones, satellite data, and smartphone applications, with the objective of advancing precision agriculture solutions tailored for Bangladesh (Quader, 2023). Moreover, digital platforms such as Krishoker Janala (Farmer's Window) and an Agriculture Data Dashboard are government-endorsed resources for the dissemination of knowledge and consulting services for intelligent farming (Quader, 2023). These initiatives demonstrate a strong commitment to enhancing farming through increased intelligence and data utilization.
- **Training and capacity building initiatives:** Authorities are also endeavoring to enhance the capability for precision agriculture at the grassroots level. In April 2024, the Ministry of Agriculture, in collaboration with the Asian Development Bank, commenced training for Department of Agricultural Extension (DAE) staff in drone utilization and remote sensing for crop monitoring (Grain Feed & Milling Magazine, 2024). The DAE has established over 200 Digital Information Centers in rural regions to facilitate farmers' access to current agricultural knowledge, and it operates a call center (Krishi Call Center) to guide farmers, thereby indirectly promoting the adoption of precision practices through enhanced digital accessibility (Rahman, 2024).

The government considers "Smart Agriculture" a cornerstone of national advancement, establishing targets for extensive technological adoption. By 2041, the objective is for around 50% of farms to be outfitted with IoT devices and precision agriculture technologies as part of a technology-driven revolution (Mamun, 2024). This enduring vision is supported by continuous investments in rural connectivity and digital infrastructure to enable farmers nationwide to leverage advanced technology. The use of SATs is considered essential for realizing the Smart Bangladesh goals of a modern and efficient agricultural industry.

3.6 Success story: Digital advisory services (Krishoker Janala)—An AI-enhanced intelligent agricultural solution

The initiative for digital agriculture has produced favorable results regarding farmer expertise and crop output. As of 2022, these digital advising platforms have collectively engaged over 20 million farmers, resulting in yield enhancements of over 12% in certain regions through the optimization of agricultural methods (Rahman, 2024; Sultana et al., 2019). A farmer from Rajshahi observed that, by utilizing a smartphone application for accurate fertilization recommendations for his rice crop, he achieved a significantly higher yield per hectare compared to prior seasons.

Krishoker Janala is an AI-driven agricultural solution aimed at closing the knowledge gap between Bangladeshi farmers and agricultural extension agencies.

Originally conceived in 2014 by Upazila Agricultural Officer Abdul Malek in Mymensingh, Fulbaria Upazila, this program was subsequently enhanced and organized by a2i (Access to Information) with funding from the Service Innovation Fund. Throughout the years, the application has advanced to include state-of-the-art AI and machine learning technology to improve its efficacy in detecting crop illnesses, providing treatments, and minimizing inefficiencies in agricultural procedures (Aspire to Innovate (a2i) Program, 2023a).

Krishoker Janala served as an information store, comprising personally captured photos of 120 crop-related diseases and their corresponding solutions. Farmers can find and treat crop diseases without having to wait for an extension worker to visit them in person. The AI-generated suggestions ensure that the right amount of fertilizer is used, which cuts down on costs and reduces environmental damage. The service reduced agricultural service procurement time by 68% and saved associated expenditure by 96%. Farmer satisfaction with the application was elevated, with 66% indicating satisfaction and 30% demonstrating significant satisfaction (Aspire to Innovate (a2i) Program, 2023b).

4 Conclusion

SATs are gradually being adopted across Bangladesh. The current landscape of SATs in Bangladesh showcases a dynamic interplay of innovation, adoption, and policy implications that can significantly transform the agricultural sector. The findings of this study demonstrate that data-driven management using IoT sensors, AI algorithms, and precision tools can significantly enhance yields, reduce input use, and support adaptation to climate change. Pilot programs and agri-tech startups indicate that momentum is building, yet progress remains constrained by high initial costs, fragmented landholdings, limited digital infrastructure, and low technical literacy.

To accelerate adoption, coordinated policies and investments are essential. Government initiatives such as the National Agriculture Policy, mechanization policies, ICT, and smart agriculture initiatives already underscore the importance of precision farming, but these need to be complemented with targeted subsidies or credit schemes to make technologies affordable for smallholders. Extension services and community-based training should emphasize digital literacy and practical demonstrations. Strengthening digital infrastructure, developing locally adapted and affordable AI tools, and fostering public-private partnerships can further integrate AI-driven solutions into farming systems. It is imperative that the Bangladesh government and stakeholders collaboratively devise policies that foster innovation and support infrastructural developments to catalyze the transformation of agriculture into a more technologically advanced and sustainable sector. Such efforts are critical to ensuring that agricultural practices in Bangladesh can meet the challenges of modernity while also safeguarding food security for its growing population.

References

6Wresearch, 2023a. *Bangladesh precision farming software market (2025-2031) outlook.* Retrieved from https://www.6wresearch.com/industry-report/bangladesh-precision-farming-software-market

6Wresearch, 2023b. *Bangladesh vertical farming market (2025-2031) outlook.* Retrieved from https://www.6wresearch.com/industry-report/bangladesh-vertical-farming-market

Abdul-Salaam, G., Abdullah, A.H., Anisi, M.H., 2017. Energy-efficient data reporting for navigation in position-free hybrid wireless sensor networks. IEEE Sens. J. 17 (7), 2289–2297. doi:10.1109/jsen.2017.2665663.

Ahmed, M., Raha, S.M., 2023. ICTs and smart agriculture in Bangladesh: way of sustainable development. Asian J. Agric. Ext. Econ. Sociol. 41 (11), 74–83. doi:10.9734/ajaees/2023/v41i112263.

Al-Amin, A.K.M.A., Khondker, B.H., Lowenberg-DeBoer, J., 2023. *Precision agriculture could be smallholders' game changer.* Retrieved from https://policyinsightsonline.com/2023/06/precision-agriculture-could-be-smallholders-game-changer/

Ali, S., 2022. Agri Insurance Gains Ground as Farmers Get Climate Compensation. The Business Standard. Retrieved from https://www.tbsnews.net/economy/agri-insurance-gains-ground-farmers-get-climate-compensation-371881.

Aliev, R., Kurbanova, M., Samoylova, A., 2023. Transformative potential of digital agriculture for enhancing global food security. BIO Web Conf. 76, 05010. doi:10.1051/bioconf/20237605010.

Aspire to Innovate (a2i) Program, 2023a. Smart Agriculture. Aspire to Innovate (a2i) Program. Retrieved from https://a2i.gov.bd/wp-content/uploads/2023/06/Smart-Agriculture.pdf.

Aspire to Innovate (a2i) Program, 2023b. *Krishoker Janala v.2: An AI-based crop-disease suggestion, and solutions dissemination platform.* Retrieved from https://a2i.gov.bd/wp-content/uploads/2023/06/Krishoker-Janala-v.2.pdf

Aziz, A., 2020. Digital inclusion challenges in Bangladesh: the case of the National ICT Policy. Contemp. South Asia 28 (3), 304–319. doi:10.1080/09584935.2020.1793912.

Basu, M., 2019. *GM brinjals are helping Bangladesh farmers earn more, save more, study finds.* The Print. Retrieved from https://theprint.in/science/gm-brinjals-are-helping-bangladesh-farmers-earn-more-save-more-study-finds/296662/

BER, 2024. Bangladesh Economic Review, Ministry of Finance, Government of Bangladesh. Retrieved from https://mof.portal.gov.bd/sites/default/files/files/mof.portal.gov.bd/page/f2d8fabb_29c1_423a_9d37_cdb500260002/BER_2024_11.%20Chapter-2_Eng.pdf

Bhat, S.A., Huang, N., 2021. Big data and ai revolution in precision agriculture: survey and challenges. IEEE Access 9, 110209–110222. doi:10.1109/access.2021.3102227.

Bhuiyan, M.A.B., 2021. *Drones in precision agriculture in Bangladesh.* The Daily Observer. Retrieved from https://www.observerbd.com/news/343337

Bhuiyan, M.Z.H., 2024. *AI set to revolutionise agriculture in Bangladesh.* The Daily Star. Retrieved from https://www.thedailystar.net/business/news/ai-set-revolutionise-agriculture-bangladesh-3646046

Boahen, J.O., Choudhary, M., 2024. Advancements in precision agriculture: integrating computer vision for intelligent soil and crop monitoring in the era of artificial intelligence. Int. J. Sci. Res. Eng. Manag. 8 (3), 1–5. doi:10.55041/ijsrem29725.

Cernisevs, O., Surmach, A., Buka, S., 2024. Smart agriculture for urban regions: digital transformation strategies in the agro-industrial sector for enhanced compliance and economic growth. Sci. Horiz. 27 (4), 166–175. doi:10.48077/scihor4.2024.166.

Choudhary, B., Nasiruddin, K.M., Gaur, K., 2014. the status of commercialized bt brinjal in bangladesh. International Service for the Acquisition of Agri-biotech Applications, ISAAA Brief 47. Retrieved from https://www.isaaa.org/resources/publications/briefs/47/download/isaaa-brief-47-2014.pdf.

Chowdhury, F., 2020. *Climate-apt agriculture crop insurance: Reaching the unheeded margin through insurance*. Retrieved from https://www.fair1964.org/case_study/CaseStudy2020.pdf.

DataInsightopia, 2025. AI in agriculture revolution: the prospects of AI in agriculture industry of Bangladesh. Retrieve from: https://www.datainsightopia.com/2025/05/10/ai-agriculture-bangladesh/

Doody, A., 2021. International Maize and Wheat Improvement Center (CIMMYT). Retrieved from: https://www.cimmyt.org/news/a-view-from-above/

Fakhar, M., Khalid, M., 2023. Satellites to agricultural fields: the role of remote sensing in precision agriculture. Biol. Agric. Sci. Res. J. 2023 (1), 14. doi:10.54112/basrj.v2023i1.14.

FAO, 2021. Food and Agriculture Organization of the United Nations, *Urban gardening proving a success in Dhaka, helping hundreds of poor families*. Retrieved from https://www.fao.org/bangladesh/news/detail-events/es/c/1460861/

Fatima, S.A., Nasim, N.S.F., Ahmed, N.S., 2023. Enhancing agricultural operations: big data analytics using distributed and parallel computing. Int. J. Emerg. Eng. Technol. 2 (2), 1–7. doi:10.57041/v3jj9f69.

Foreign Agricultural Service, 2022. Bangladesh: Agricultural Biotechnology Annual. United States Department of Agriculture. Retrieved from https://www.fas.usda.gov/data/bangladesh-agricultural-biotechnology-annual-7.

Future Startup Team, 2024. *A list of Bangladesh's most fascinating agritech startups*. Retrieved from https://futurestartup.com/2024/10/23/a-list-of-bangladeshs-most-fascinating-agritech-startups/

Gopathoti, K., 2023. Precision agriculture and the role of machine learning. Int. J. Sci. Methods Intell. Eng. Netw. 1 (6), 01–08. doi:10.58599/ijsmien.2023.1601.

Grain Feed & Milling Magazine, 2024. *Bangladesh to introduce drone technology to assess crop losses*. Retrieved from https://gfmdhaka.com/news/bangladesh-to-introduce-drone-technology-to-assess-crop-losses

Green Delta Insurance, 2020. *Green Delta initiates index-based crop insurance in flood-prone Haor area as a tool for risk mitigation*. Retrieved from https://green-delta.com/green-delta-initiates-index-based-crop-insurance-in-flood-prone-haor-area-as-a-tool-for-risk-mitigation/

Haque, A., Islam, N., Samrat, N.H., Dey, S., Ray, B., 2021. Smart farming through responsible leadership in Bangladesh: possibilities, opportunities, and beyond. Sustainability 13, 4511. doi:10.3390/su13084511.

Haque, M.M., Alam, M.R., Alam, M.M., Basak, B., Sumi, K.R., Belton, B., Jahan, K.M., 2015. Integrated floating cage aquageoponics system (IFCAS): an innovation in fish and vegetable production for shaded ponds in Bangladesh. Aquac. Rep. 2, 1–9. doi:10.1016/j.aqrep.2015.04.002.

Hossain, M.B., Islam, M., 2022. Use of artificial intelligence for precision agriculture in Bangladesh. J. Agric. Rural Res. 6 (2), 81–96.

Ifaz, M.F., 2020. *Transforming agriculture finance in Bangladesh via technology*. Retrieved from https://ifarmer.asia/blogs/2?locale=en

Ifty, S.M.H., Ashakin, M.R., Hossain, B., Afrin, S., Sattar, A., Chowdhury, R., Tusher, M.I., Bhowmik, P.K., Mia, M.T., Islam, T., Tufael, Sunny, A.R., 2023. IOT-based smart agriculture in Bangladesh: an overview. Appl. Agric. Sci. 1 (1), 1–10. doi:10.25163/agriculture.119563.

Illius, S., Azad, A., 2022. Salt-tolerant Rice Varieties Bring Hope to Bangladesh Farmers. The Business Standard. Retrieved from https://www.tbsnews.net/bangladesh/salt-tolerant-rice-varieties-bring-hope-bangladesh-farmers-410202.

Ioja, I., Nedeff, V., Agop, M., Nedeff, F.M., 2024. Some possibilities of the aerial drones use in precision agriculture—a review. J. Eng. Stud. Res. 29 (4), 43–49. doi:10.29081/jesr.v29i4.006.

IRRI, 2024. *Prepaid irrigation cards and AWD may unlock a more sustainable rice farming future for Bangladesh*. Retrieved from https://www.irri.org/news-and-events/news/prepaid-irrigation-cards-and-awd-may-unlock-more-sustainable-rice-farming?

Islam, A., 2019. BRRI Develops Seven High-Yielding Aman Paddy Varieties. Dhaka Tribune. Retrieved from https://www.dhakatribune.com/bangladesh/agriculture/180500/brri-develops-seven-high-yielding-aman-paddy.

Islam, M. ., 2025. Bangladesh Advancing Towards Sustainable Agriculture through Climate Adaptation. Bangladesh Sangbad Sangstha. Retrieved from https://www.bssnews.net/agriculture-news/304985.

Kabir, M.S.N., Reza, M.N., Chowdhury, M., Ali, M., Samsuzzaman, Ali, M.R., Lee, K.Y., Chung, S.O., 2023. Technological Trends and Engineering Issues on Vertical Farms: a Review. Horticulturae 9 (11), 1229. doi:10.3390/horticulturae9111229.

Kader, M.R., Amit, S., 2023. The Future of AgriTech in Bangladesh: Awakening the Slumbering Giant. Center for Enterprise and Society, University of Liberal Arts Bangladesh. Retrieved from https://ulab.edu.bd/sites/default/files/CES%20Whitepaper_The%20Future%20of%20AgriTech%20in%20Bangladesh.pdf.

Kazi, Z., 2025. Bangladesh's AI Moment. Dhaka Tribune. Retrieved from https://www.dhakatribune.com/opinion/longform/372730/bangladesh-s-ai-moment.

Khanna, M., 2020. Digital transformation of the agricultural sector: pathways, drivers and policy implications. Appl. Econ. Perspect. Policy 43 (4), 1221–1242. doi:10.1002/aepp.13103.

Kittichotsatsawat, Y., Jangkrajarng, V., Tippayawong, K., 2021. Enhancing coffee supply chain towards sustainable growth with big data and modern agricultural technologies. Sustainability 13 (8), 4593. doi:10.3390/su13084593.

Lohit, V., Mujahid, M., 2022. Big data analytics in developing smart and sustainable solutions for the agricultural industry. Technoarete Trans. Adv. Data Sci. Anal. 1 (2). doi:10.36647/ttadsa/01.02.a001.

Mamun, S., 2024. From Digital Bangladesh to Smart Bangladesh: Charting the Path to an Advanced Nation. Dhaka Tribune. Retrieved from https://www.dhakatribune.com/others/special-supplements/353302/from-digital-bangladesh-to-smart-bangladesh.

Metropolitan Chamber of Commerce and Industry, 2024. Bangladesh economy FY 23-24. Metropolitan Chamber of Commerce and Industry. Retrieved from https://mccibd.org/wp-content/uploads/2024/10/Bangladesh-Economy-FY-23-24.pdf.

Mia, M.A.T., Roy, R., 2025. Adoption of climate-smart agriculture in the coastal area of Bangladesh. Int. J. Agric. Res. Innov. Technol. 15 (1), 39–47. doi:10.3329/ijarit.v15i1.82752.

Ministry of Agriculture, 2018. National Agricultural Policy 2018. Retrieve from: https://lpr.adb.org/resource/national-agriculture-policy-2018-bangladesh

Ministry of Agriculture, 2020. *National Agricultural Mechanization Policy 2020*. Retrieved from https://dae.portal.gov.bd/sites/default/files/files/dae.portal.gov.bd/policies/c94059

4d_3ca5_4665_b390_ace578255853/2021-08-25-06-05-7b2382934f7fa3773586d40840 61cd04.pdf

Paul, P.L.C., Rashid, M.A., 2013. Refinement of alternate wetting and drying irrigation method for rice cultivation. Bangladesh Rice J. 17 (1&2), 37–41.

Peladarinos, N., Piromalis, D., Cheimaras, V., Tserepas, E., Munteanu, R., Papageorgas, P., 2023. Enhancing smart agriculture by implementing digital twins: a comprehensive review. Sensors 23 (16), 7128. doi:10.3390/s23167128.

Quader, M.T., 2023. *Taking Bangladesh's agriculture to the next level*. Retrieved from: https://a2i.gov.bd/taking-bangladeshs-agriculture-to-the-next-level/

Rahman, T., 2024. *Fields to phones: Digital revolution in Bangladesh agriculture*. Retrieved from https://greenwatchbd.com/columns/62330/fields-to-phones-digital-revolution-in-bangladesh-agriculture

Rahman, Haque, M., Kabir, M.J., Islam, A., Sarkar, M., Mamun, M., Salam, M., Kabir, 2021. Enhancing rice productivity in the unfavourable ecosystems of Bangladesh. Bangladesh Rice J. 24 (2), 83–102. doi:10.3329/brj.v24i2.53450.

Saini, S., Giriraj, A., 2020. *Innovative index-based agriculture insurance: Where hope meets possibility*. Retrieved from https://ccafs.cgiar.org/news/innovative-index-based-agriculture-insurance-where-hope-meets-possibility

Salehin, M.U., Joy, J.I., Hasan, S.W., Babui, M.J., Khan, F., 2023. Utilization of solar energy in irrigation systems in Bangladesh. Energy Power Eng. 15 (12), 468–481. doi:10.4236/epe.2023.1512026.

Sambas, A., Mujiarto, M., Gundara, G., Refiadi, G., Mulyati, N., Sulaiman, I., 2023. Development of smart farming technology on ginger plants in padamulya ciamis village, West Java, Indonesia. Int. J. Res. Community Serv. 4 (3), 93–99. doi:10.46336/ijrcs.v4i3.483.

Sarker, M., Itohara, Y., Hoque, M., 2009. Determinants of adoption decisions: the case of organic farming in Bangladesh. Ext. Farm. Syst. J. 5, 39–40.

Seltier, E., 2023. *A winning idea: Agri-technology invigorates farms in Bangladesh. Trade Forum Features*. Retrieved from: https://www.intracen.org/news-and-events/news/a-winning-idea-agri-technology-invigorates-farms-in-bangladesh

Sharma, A., Jain, A., Gupta, P., Chowdary, V., 2021. Machine learning applications for precision agriculture: a comprehensive review. IEEE Access 9, 4843–4873. doi:10.1109/access.2020.3048415.

Shelton, A.M., Sarwer, S.H., Hossain, M.J., Brookes, G., Paranjape, V., 2020. Impact of BT Brinjal cultivation in the market value chain in five districts of Bangladesh. Front. Bioeng. Biotechnol. 8. doi:10.3389/fbioe.2020.00498.

Shuvo, A.H., 2021. iPage: An AI-based Start-up Helping Farmers Become Smarter, Richer. The Business Standard. Retrieved from https://www.tbsnews.net/features/panorama/ipage-ai-based-start-helping-farmers-become-smarter-richer-326683.

Sultana, M.S., Ali, M.S., Islam, M.R., Kabir, M.H., Hasnat, M.K., 2019. Effectiveness of Krisoker Janala in Disseminating Agricultural Information: an Innovative tool. Open J. Soc. Sci. 7 (3), 272–280. doi:10.4236/jss.2019.73023.

Tanvir, M.A.A., Bhattacharjee, R., 2024. Comparative analysis of 5G technology in the agriculture sector of Bangladesh. Int. J. Clin. Infect. Dis. 3 (5), 1–10.

Tarannum, T., 2023. *Urban farming practices*. New Age. Retrieved from: https://www.newagebd.net/post/opinion/247750/urban-farming-practices

The Daily Star, 2019. *Weather index-based insurance for mitigating agricultural risks in Bangladesh*. The Daily Star. Retrieved from https://www.thedailystar.net/round-tables/news/weather-index-based-insurance-mitigating-agricultural-risks-bangladesh-1745731

UNDP, 2021. *Precision agriculture for smallholder farmers*. Retrieved from https://www.undp.org/publications/precision-agriculture-smallholder-farmers

Vatin, N.I., Joshi, S.K., Acharya, P., Sharma, R., Rajasekhar, N., 2024. Precision agriculture and sustainable yields: insights from IOT-driven farming and the precision agriculture test. BIO Web Conf. 86, 01091. doi:10.1051/bioconf/20248601091.

Prioritization of erosion-prone areas for conservation treatments using the sediment yield index: A case study of the Lower Subansiri Dam project

19

Mriganka Kaustab Talukdar, Nitiwat Wongchanla, Aadarsh Rijal, Thin Thiri Aung, Shubham Anil Gade, and Sushil Kumar Himanshu
Agricultural Systems and Engineering, Faculty of Food, Agriculture and Natural Resources, Asian Institute of Technology, Pathum Thani, Thailand

1 Introduction

A geographical information system (GIS) is a powerful tool for the analysis, collection, storage, manipulation, and display of vector and raster data for specific applications. Therefore, utilizing GIS is vital in evaluating prospective locations for dams through hydrological analysis and modeling (Othman et al., 2020). Identifying critical areas within a basin for conservation efforts is essential but challenging due to the varying sedimentation rates across different hydrological units. Effectively managing natural resources requires prioritizing areas with the highest sediment yields for conservation measures, ensuring that regions contributing most to sedimentation receive immediate attention. However, this prioritization can be complex and costly, particularly in areas lacking sufficient data. Advanced remote sensing and GIS techniques can greatly aid the spatial analysis necessary to identify and prioritize these critical areas, thereby making the process more efficient and targeted (Palmate et al., 2022).

India ranks third globally, after the USA and China, in the number of large hydroelectric projects (HP) and is fifth in terms of production. To meet its soaring energy demands and rising per capita power consumption, the Government of India has focused on hydropower over the past few decades, particularly in the Himalayan and sub-Himalayan rivers of North and Northeast (NE) India (Menon et al., 2003). The construction of dams in India requires strict adherence to numerous rules and regulations, with approvals from various regulatory bodies, all while prioritizing safety. However, noncompliance often leads to controversy, exemplified by projects like the Subansiri Dam.

Precision Technologies for Digital Agriculture. https://doi.org/10.1016/B978-0-443-40513-6.00012-3

The construction of a dam in such a geologically dynamic area presents several challenges and risks. Firstly, the stability of the dam structure itself must be ensured amidst the threats of landslides and seismic events. Secondly, the reservoir behind the dam can exacerbate landslide risks in the surrounding areas by altering the natural drainage patterns and increasing the weight on the slopes. In addition, landslides can lead to sedimentation in the reservoir, reducing its storage capacity and impacting the efficiency of power generation.

The Sediment Yield Index (SYI) method, developed by the All-India Soil and Land Use Survey under the Government of India, offers a universally applicable approach for estimating soil erosion and sediment yield using actual spatial data (All India Soil and Land Use Survey, 1991). This method is recognized for its effectiveness in prioritizing basin areas affected by soil erosion and sedimentation. The SYI approach incorporates both the soil erosivity factor, which assigns a weightage value to the potential of the soil to erode, and the transportability of the detached soil, known as the delivery ratio, to conceptualize how sediment is delivered into river water (Arabkhedri et al., 2021).

The Subansiri Lower HP (SLHEP), with its proposed capacity of 2000 megawatts, aims to harness the immense hydropower potential of the Subansiri River (Baruah et al., 2009). The project involves the construction of a concrete gravity dam with a height of 116 m and the installation of 8 Francis turbines, each with a capacity of 250 megawatts. The catchment area of the dam spans 34,900 sq. km, with a storage capacity of 1365 million m^3. Once operational, the project is expected to generate approximately 7500 million kW/hr of power annually, contributing significantly to India's energy needs (Baruah et al., 2009).

Currently, the SLHEP is underway as a gravity dam on the Subansiri River, straddling the border between Assam and Arunachal Pradesh. Originating from the Tibetan Plateau, the Subansiri, also known as the "gold river," is the Brahmaputra River's primary tributary, flowing through Arunachal Pradesh's Miri Hills upon entering India.

Despite SLHEP's potential benefits, it has been mired in controversy since its inception. Concerns have been raised regarding its environmental impact, particularly on the fragile ecosystem of the Eastern Himalayas and the livelihoods of indigenous communities living in the region. Moreover, the project's location in a seismically active zone raises additional safety concerns, especially considering past incidents of dam failures and landslides in similar terrain. SLHEP is a significant yet contentious initiative aimed at harnessing the hydropower potential of the Subansiri River.

However, its construction in the geologically dynamic and ecologically sensitive region of the Eastern Himalayas presents numerous challenges and risks, including landslides, seismic activity, and environmental degradation. As the project progresses, it is crucial to carefully mitigate these risks and address the concerns of all stakeholders involved (Palmate et al., 2022). Water scarcity has been increasing over the years because of the inefficiency in receiving water resources from the dam. About 53% of the total geographical area of India endures extreme soil erosion and other forms of land degradation (MOEF, 2001). With 16% of the world's population living on 2% of the global land area, the population explosion scenario is serious.

Therefore, planning and management of land and water resources without deterioration is the main key for humankind. By applying GIS techniques, the project can save time, and it is more efficient and suitable for spatial planning. GIS can solve complex issues and handle large datasets for manipulation and retrieval. The work attempts to identify priority erosion-prone areas in the Subansiri watershed using the SYI model, which combines soil erosivity (related to rainfall characteristics) and erodibility (reflecting soil detachment and transportation potential) to assess sediment yield. The work attempts to identify priority erosion-prone areas in the Subansiri watershed using the SYI model, which combines soil erosivity (related to rainfall characteristics) and erodibility (reflecting soil detachment and transportation potential) to assess sediment yield (Rawat et al., 2013; Xu et al., 2023). By applying this model, the study aims to tackle the significant challenges of soil erosion and sedimentation, both of which threaten soil health and water quality (Dutta, 2016). The hypothesis suggests that the SYI model can effectively pinpoint high-risk areas, offering a strong foundation for prioritizing conservation efforts (Maqsoom et al., 2020). The SLHEP is being built in the challenging landscape of the Eastern Himalayas across Assam and Arunachal Pradesh. This area features a diverse geology of sedimentary, metamorphic, and igneous rocks and is susceptible to landslides due to its steep terrain and heavy monsoon rains. Factors contributing to these landslides include soil erosion, earthquakes, and human activities such as road building and deforestation.

In addition, the study explores the broader impact of these findings on watershed management. By incorporating additional factors like land use and climate change, the model's accuracy could be further enhanced (Xu et al., 2023). This integrated approach ensures that conservation resources are used efficiently, helping to reduce erosion and safeguard vital water resources (Rawat et al., 2013).

2 Study area

The study focuses on the area around the Subansiri Dam, situated at the confluence of Assam and Arunachal Pradesh in India. The region spans about 12,300 sq. km and is positioned at latitude 27°46'N and longitude 93°50'E. The landscape varies significantly, with the highest point reaching about 300 m above mean sea level (MSL). The local population primarily relies on agriculture, which dominates the economic activities in the area, making land use and agricultural practices critical for the community. The region receives an annual rainfall of 2000–2500 mm, which supports the cultivation of rice and mustard as the main agricultural products. The Subansiri Dam not only contributes to power generation but also supports a sophisticated irrigation network that boosts agricultural yields in both Assam and Arunachal Pradesh. The area is geologically composed of sedimentary rock structures.

The Subansiri River courses through the Himalayas within Indian borders for about 200 km in the NE and enters the plains of Assam through a gorge near Gerukamukh in the Dhemaji district. With a total river length of 520 km and a drainage basin encompassing 37,000 square km, the Subansiri flows about 130 km within Assam, moving through the districts of Dhemaji, Lakhimpur, and Jorhat. This segment of the river is positioned between 26°50'N and 27°35'N in latitude and 93°41'E to 94°23'E in longitude (Sarma et al., 2020).

3 Data used

The study uses geospatial datasets sourced from multiple online repositories, including digital elevation models (DEMs), land use land cover (LULC) data, soil data, and slope data. To extract relevant spatial data and perform statistical analyses, these datasets were further processed in the ArcMap 10.7 environment (Stefanidis et al., 2021).

3.1 *Digital elevation model data*

The Shuttle Radar Topography Mission (SRTM) DEM, with a spatial resolution of 30 m, was downloaded from the United States Geological Survey (USGS) website for the drainage network (Sarma et al., 2020). Based on the DEM, the following processes were performed in ArcMap software (Fig. 19.1).

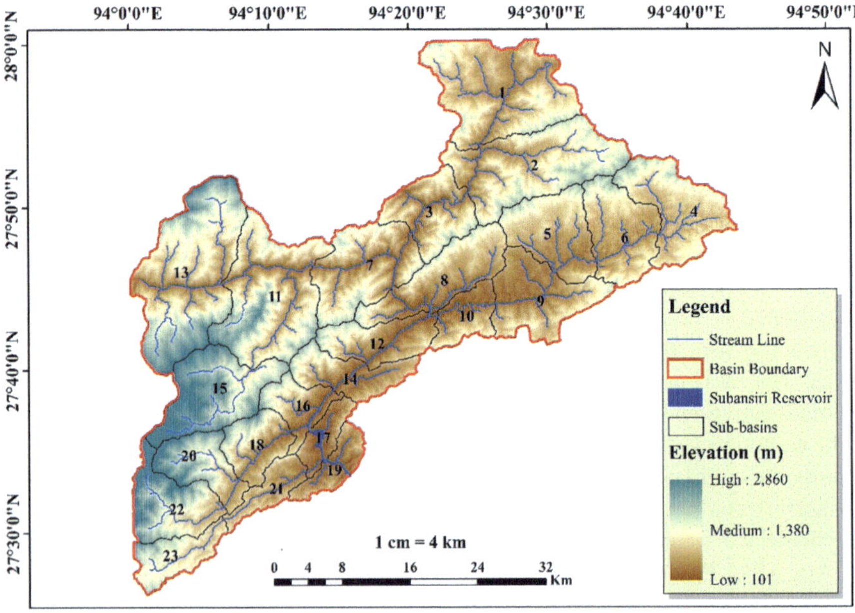

Fig. 19.1 Location map of the Subansiri Dam reservoir, with delineated subbasins and elevations obtained using SRTM DEM.

3.2 Land use land cover data

The LULC data of the study area were extracted from the USGS website using the Landsat-8 Operational Land Imager (OLI) satellite image (Nguyen et al., 2018). The processed map, containing six major LULC classes, was categorized as open forest, built-up area, crop land area, water body, dense forest, and bare land. It was observed that around 1631.975 sq. km of area, accounting for 89.16% of the total area of the watershed, is dense forest, while 114.409 sq. km, which is about 6.25% of the total area of the watershed, is open forest. The detailed data are shown in Fig. 19.2 and Table 19.1.

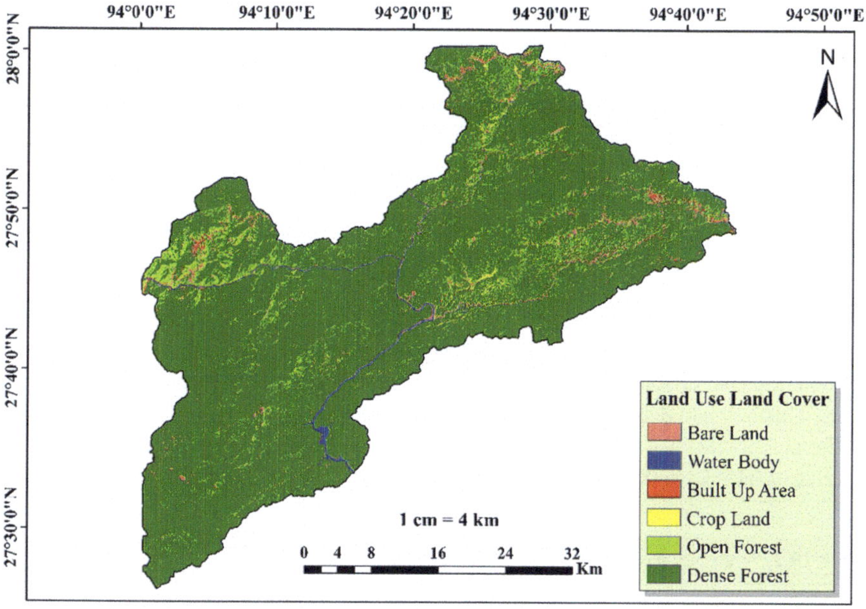

Fig. 19.2 Land use land cover (LULC) map of the study area.

Table 19.1 Different land use and land cover classes of the study area.

No.	LULC class	Area (km^2)	Area (%)
1	Open forest	114.409	6.25
2	Built-up area	3.58	0.20
3	Crop land	11.349	0.62
4	Water body	13.033	0.71
5	Dense forest	1631.975	89.16
6	Bare land	56.085	3.06
	Total	**1830.431**	**100**

Accuracy assessment

A confusion matrix was used to determine the accuracy of LULC classification. This matrix compares the land classifications from the map to the actual conditions on the ground, based on a sample of known data points. It lists different land types, such as open forest and built-up areas, and indicates the frequency with which the map correctly identified each type. The diagonal entries, from the top left to the bottom right, represent the correct classifications. To calculate the overall accuracy of the map, the sum of the correct classifications is divided by the total number of classifications. In this case, the map achieved an overall accuracy of 79.33%.

$$\text{Kappa coefficient (T)} = (\text{TS} \times \text{TCS}) - \Sigma(\text{Column Total} \times \text{Row Total})/\text{TS}^2 - \Sigma(\text{Column Total} \times \text{Row Total}) \times 100$$

Where TS = Total Samples, and TCS = Total Correctly Classified Samples.

Based on the formula, the Kappa score of 75.2% is better (Table 19.2). The confusion matrix allows researchers to assess not only the overall accuracy of the map but also how accurate it is for specific land types. This detailed assessment helps users make more informed decisions when utilizing the map (Rwanga and Ndambuki, 2017).

3.3 Slope

The slope percentage of the surface area is one of the major factors in estimating runoff and soil erosion in the watershed (Dutta et al., 2015). The slope percentage is divided into five different classes: gentle, moderate, moderately steep, steep, and very steep. This helps in determining the soil erosion intensity within the study area. Based on the categories, it was observed that 0.468% of the study area has a gentle slope, whereas 1.315%, 5.385%, and 15.289% of the area exhibit moderate, moderately

Table 19.2 Accuracy assessment with 150 samples.

Classes	Producer's accuracy (%)	User's accuracy (%)	Overall accuracy (%)	Kappa coefficient (%)
Bare land	78.26	72	79.33	75.2
Water body	100	96		
Built-up aera	82.35	56		
Crop land	86.36	76		
Open forest	63.33	76		
Dense forest	73.53	100		

Table 19.3 Slope classes and area percentage of the study area.

Slope category (%)	Classification	Area (km^2)	Area (%)
0–5	Gentle	8.87	0.468
5–10	Moderate	24.94	1.315
10–20	Moderate Steep	102.10	5.385
20–35	Steep	289.88	15.289
>35	Very Steep	1470.22	77.543
	Total	**1896.00**	**100**

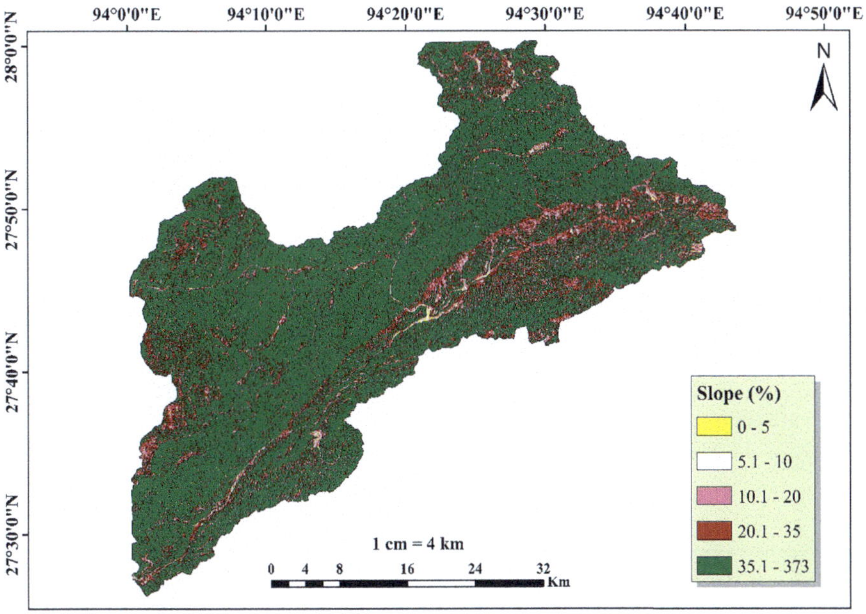

Fig. 19.3 Slope map of the border area of the study area.

steep, and steep slopes, respectively. The majority of the study area (77.54% of the total area) resembles a very steep slope, as observed in Table 19.3. These percent slopes were used for estimating the susceptibility to erosion intensity for the study area, as represented in Fig. 19.3.

3.4 Soil

The soil data were extracted from the soil portal of the Food and Agriculture Organization (FAO) of the United Nations. The soil texture map plays a significant role in determining soil erosion impacts (Pal et al., 2021). The study area was mostly covered by loamy-textured soil, including a well-balanced proportion of sand, silt, and clay particles (Fig. 19.4 and Table 19.4). Loamy soils are the

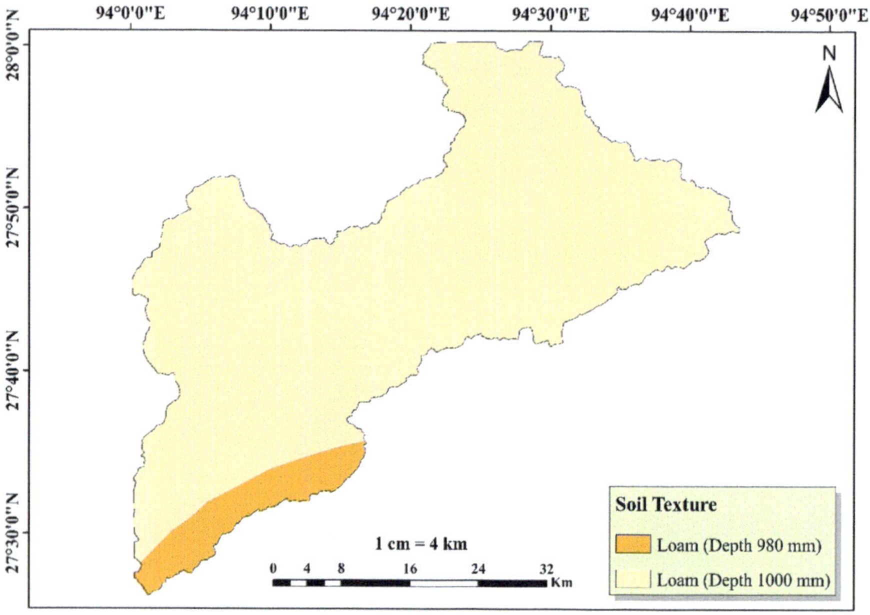

Fig. 19.4 Soil map of the border area of the study region.

Table 19.4 Soil depth and texture of the study area.

Soil depth (mm)	Depth category	Soil texture	Area (km^2)	Area (%)
980	Moderate	Loam	1707.517	93.280
1000	Deep	Loam	123.0084	6.720
		Total	**1830.5254**	**100.000**

dominant soil class in the watershed, mostly covering an area of 1707.5 sq. km and a total area of 93.28% at moderate soil depth.

4 Methodology

The dynamics of soil and sediment movement within an area are mainly influenced by the interaction between meteorological conditions and the surface features of the land. Intense and frequent rainfall plays a significant role in causing soil erosion and sediment flow. Other factors, such as land slope, land use (e.g., farming or construction), soil type, landform, and ongoing erosion or deposition processes, also play crucial roles in determining the movement of soil and sediment. The SYI method focuses on these land features to understand erosion and sedimentation in a river basin. Considering the above phenomena, maps and data about these land characteristics were used to calculate the SYI value. The flow chart below shows the steps

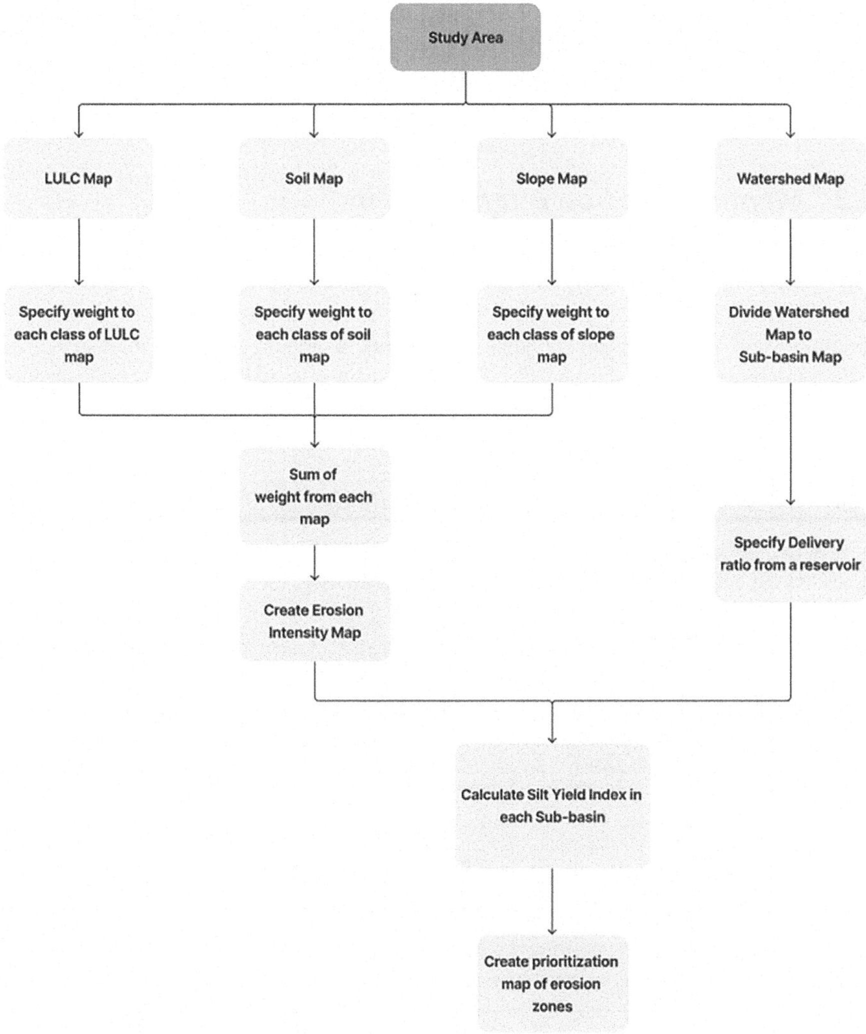

Fig. 19.5 Flow chart of overall methodology.

involved in creating the prioritization map of erosion zones by integrating data from each map of LULC, soil, slope, and subwatershed (Fig. 19.5).

4.1 Delivery ratio

The perception of the delivery ratio is an essential factor in transporting sediments through drainage networks. It is defined as the ratio of the total length of streams to the total area of the watershed, and it calculates the condition of stream flow, such as poor or good flow, in a watershed. In the context of the delivery ratio, several factors influence water delivery within a watershed, such as topography (e.g., land slope), the density and route of drainage, the shape and size of the subbasin, and the distance from the reservoir (Palmate et al., 2022). In this study, the delivery ratio of subwatersheds was assigned values ranging from 0.4 to 1 (Table 19.5).

4.2 Mapping erosion intensity unit

In the determination of erosion intensity, it requires extracting a thematic map, which is primarily based on the integrated information of soil texture, slope, and land use and land cover, as shown in Figs. 19.1–19.4. The composite map was overlaid onto the drainage map, including the subbasin boundaries, to identify erosion intensity units within the basin. These erosion units were then utilized to assign weights representing the relative severity of erosion.

4.3 Determining weights for basin units

Following the methodology of Palmate et al. (2022), weights for different LULC types, slope classes, and soil depth classes were assigned (Table 19.6). The lower weightage values correspond to areas with minimal erosion potential, while higher values indicate regions with significant erosion risk. These assigned weightages are

Table 19.5 Assigned delivery ratio to streams according to distance from the reservoir.

Nearest stream distance (km)	Delivery ratio
0–7	1
8–15	0.9
16–23	0.8
24–31	0.7
32–39	0.6
40–47	0.5
48–55	0.4

Table 19.6 Key factors and their weightage values used for the preparation of a composite erosion intensity map.

Land use land cover	Weightage	Slope class	Weightage	Soil depth class	Weightage
Dense forest	1	Gentle	1	Deep	1
Water body	2	Moderate	2	Moderate	2
Open forest	3	Moderate Steep	3	Moderate shallow	3
Built-up area	4	Steep	4	Shallow	4
Bare land	5	Very Steep	5		
Crop land	6				

used collectively to generate a composite erosion intensity map, which plays a critical role in calculating the SYI. The erosion intensity map, derived from these weighted factors, directly informs the SYI by quantifying the sediment yield expected from different basin units based on their erosion potential. Consequently, the calculated erosion risk scores, influenced by these weightages, are vital for effective land management and conservation planning, as they help identify areas that are more prone to erosion and require targeted intervention.

4.4 Calculation of sediment yield index

Accurate estimation of sediment yield at various temporal and geographical scales is crucial for addressing land degradation, establishing irrigation systems, and executing hydropower generation initiatives (Zarris et al., 2011). According to the All India Soil and Land Use Survey (1991), the SYI model is highly effective for the prioritization of subwatersheds. The SYI helps us understand how much sediment each part of a watershed contributes to water bodies, which is critical for planning how to manage soil erosion. The following equation was used to calculate SYI (Palmate et al., 2022):

$$\mathrm{SYI} = (\Sigma Ai \times Wi \times Di \times 100)/Aw, \quad I=1 \text{ to } N$$

Where Ai is the specific basin area, Wi is a number that represents the importance or weight of that area based on certain characteristics like its soil type or steepness, Di is the adjusted delivery ratio, and Aw is the total area of the watershed.

We calculated SYI for different areas (subbasins) of the watershed and then listed them from the highest to the lowest SYI values. Based on these values, we categorized areas into five priority classes for conservation efforts: very high, high, medium, low, and very low (Palmate et al., 2022). This categorization helps in focusing conservation efforts as required to effectively control erosion.

5 Results and discussion

In this study, the key findings are based on the erosion intensity mapping and the subsequent prioritization of treatment areas within the basin.

5.1 Erosion intensity map

Erosion intensity classes were categorized into five different groups (negligible, low, moderate, severe, and very severe). The results showed that around 3.05% (Table 19.7) of the area is under very severe erosion intensity, mainly in the northwest side of the study area, as shown in Fig. 19.6. Severe soil erosion occurs in almost three-fourths of the study area (72.67%), mainly found in dense forest, very steep slopes, and loam soil texture with a depth of 980 mm. Around 22% of the study area is under moderate erosion intensity, which is highly evident in the southwest part of the study area. The observed

Table 19.7 Erosion intensity classification with output weights and total areas.

No.	Sum of weightages	Erosion intensity class	Total area (km^2)	Total area (%)
1	0–3	Negligible	0.576	0.032
2	4–5	Low	29.3814	1.612
3	6–7	Moderate	412.517	22.633
4	8–10	Severe	1324.54	72.670
5	10–13	Very severe	55.6587	3.054
		Total	**1822.6731**	**100.000**

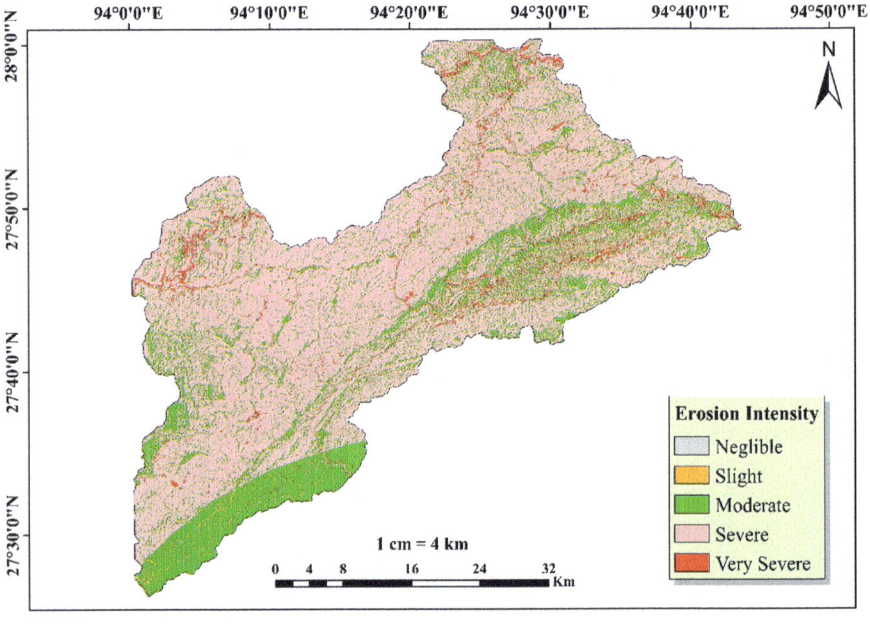

Fig. 19.6 Composite erosion intensity map of the study area.

spatial distribution, with higher erosion intensities occurring on steep slopes and specific soil types, aligns with findings from recent watershed analyses that integrate empirical erosion models with GIS. These studies highlight slope gradient and land cover as the dominant factors influencing variations in erosion across landscapes (Tilahun and Desta, 2023; Gebremichael et al., 2025). Recent studies conducted in the Subansiri basin and similar Himalayan foothill watersheds indicate that forested zones may still exhibit considerable erosion risk when characterized by steep gradients or soil mineralogy/grain size that makes the soils prone to detachment (Dutta et al., 2024).

Based on the data, management and planning should be more focused on the areas falling under the severe to very severe erosion intensity class. Comparable Revised Universal Soil Loss Equation, Sediment Delivery Ratio, and Sediment Transport Index (STI) assessments in Himalayan and adjacent basins have used similar classifications to identify hotspots for priority conservation interventions and recommend targeted, slope-specific measures like contouring, vegetative buffer strips, and check dams in those hotspot areas (Tilahun and Desta, 2023; Ghimire et al., 2024).

5.2 *Subbasin-wise treatment prioritization*

The SYI was calculated, and five priority classes were categorized based on the SYI, as shown in Table 19.8 (Palmate et al., 2022). The priority classes, along with their corresponding SYI values, are shown in Table 19.8. A total of 52.31% of the study area falls under high to very high priority ratings, with an SYI value of >751. Only 9.46% of the total area falls under a very low priority ranking, with SYI values less than 550. The area under medium priority ranking (651–750 SYI values) is 10.33%. In addition, 37.36% of the total area requires no treatment because it falls under very low to low priority rankings. High consideration should be given to management and planning in areas where the priority ranking is high to very high.

Moreover, the total area was divided into 23 subbasins, and SYI was calculated for each subbasin. Table 19.8 shows that, based on the SYI value, 6 subbasins are under very high priority, 7 subbasins are under high priority, 3 subbasins are under medium priority, 5 subbasins are under low priority, and 2 subbasins are under very low priority ranking. Therefore, focus should be directed toward those subbasins falling under high to very high priority ranking. This information is of paramount importance for informing soil and water management policy decisions, river basin planning, resource allocation, and the implementation of conservation strategies (Palmate et al., 2022). Recent applications of SYI and STI approaches suggest that identifying and treating the high-priority subbasins yield the most immediate benefit for reducing sediment loads reaching downstream infrastructure and reservoirs (Gebremichael et al., 2025; Ghimire et al., 2024). Prioritization maps, like Fig. 19.7, provide a practical basis for selecting subbasins for in-situ and ex-situ treatments. Such phased interventions can maximize sediment reduction while minimizing costs, making the present findings a useful tool for conservation planning in the Lower Subansiri Dam catchment.

6 Conclusion

To estimate the SYI for the Subansiri reservoir basin, various spatial information from different sources on percent slopes, LULC, and soils was used. Prioritization of critical areas and ranking were performed for the critical subbasin areas. The results revealed that a very high SYI value (>851) was obtained for subbasins SB 14, SB 16, SB 18, SB 15, SB 11, and SB 17. The subbasins were prioritized and ranked from 1 to 6 for conservation strategies and treatments, covering 22.98% of the total area. Subbasins SB 12, SB 20, SB 22, SB 7, SB 13, SB 10, and SB 19, with high SYI values (751–850), were ranked from 7 to 13 in the priority ranking, encompassing 29.33% of the total area. Subbasins SB 8, SB 21, and SB 3, with medium SYI values (651–750), were ranked from 14 to 16, covering 10.33% of the area. SYI-based

Table 19.8 SYI value, priority rating, and subbasin characteristics.

No	SYI range	Subbasins	SYI value	SYI rank	Area (km^2)	Area (%)	Priority rating
1	<550	SB 4, SB 5	469.486, 546.370	23, 22	172.376	9.46	Very low
2	551–650	SB 1, SB 2, SB 6, SB 9, SB 23	574.568, 578.068, 562.386, 637.095, 646.943	20, 19, 21, 18, 17	508.605	27.90	Low
3	651–750	SB TABLE3, SB 8, SB 21	681.722, 739.012, 728.144	16, 14, 15	188.285	10.33	Medium
4	751–850	SB 7, SB 10, SB 12, SB 13, SB 19, SB 20, SB 22	777.259, 758.100, 850.357, 773.284, 753.073, 837.050, 806.029	10, 12, 7, 11, 13, 8, 9	534.501	29.33	High
5	>851	SB 11, SB 14, SB 15, SB 16, SB 17, SB 18	873.272, 958.324, 922.400, 946.924, 871.541, 943.552	5, 1, 4, 2, 6, 3	418.902	22.98	Very High

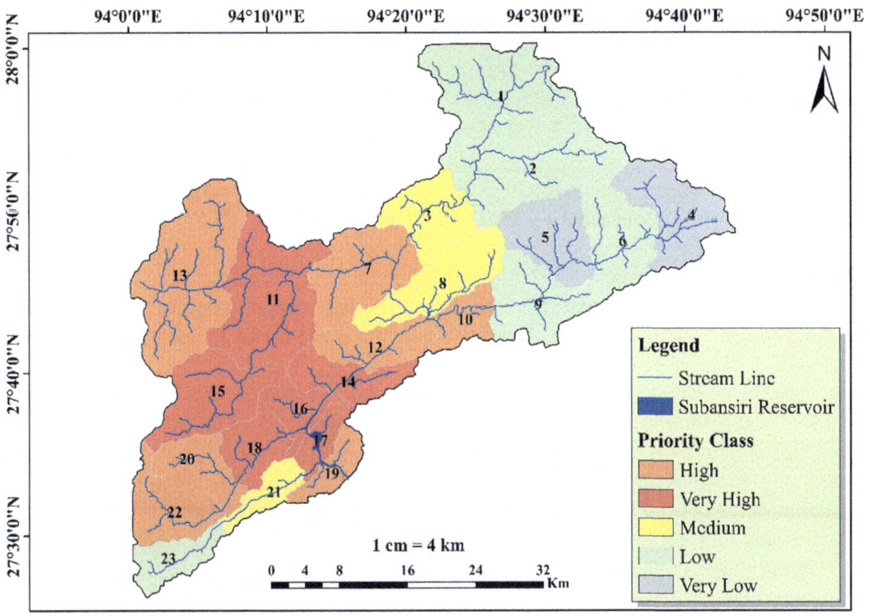

Fig. 19.7 Treatment prioritization for the study area.

prioritization offers a strategic approach to control erosion and sedimentation, crucial for both ecosystem sustainability and community livelihoods. Implementing targeted interventions such as vegetation planting, contouring, terracing, erosion control blankets, bioengineering techniques, reforestation, and conservation tillage is essential for protecting soil health and water quality.

Recent studies highlight that projected increases in rainfall intensity and altered runoff regimes in the Himalaya are likely to accelerate erosion and sediment transport (Baral et al., 2024; Huang & Sinclair, 2025). Integrating SYI-based prioritization into watershed management plans enables focused actions such as reforestation, contour trenching, and check-dam construction, which not only reduce sediment yield but also strengthen climate change adaptation by stabilizing slopes, mitigating siltation risks, and improving water security (Ghimire et al., 2024). These findings provide a scientific basis for spatially targeted conservation treatments and contribute to broader efforts toward climate-resilient watershed management across the Himalayas. Future work should integrate real-time datasets, advanced modeling, and socio-economic factors to improve precision and ensure long-term effectiveness.

References

All India Soil and Land Use Survey, 1991. Methodology of Priority Delineation Survey. AIS & LUP Technical Bulletin 9. All India Soil and Land Use Survey, Ministry of Agriculture, New Delhi.

Arabkhedri, M., Heidary, K., Parsamehr, M.R., 2021. Relationship of sediment yield to connectivity index in small watersheds with similar erosion potentials. J. Soils Sedim. 21 (7), 2699–2708. doi:10.1007/s11368-021-02978-z.

Baral, K., Pandey, V.P., Pradhan, A.M.S., 2024. Impact on hydrological alteration due to climate change in Seti watershed. J. UTEC Eng. Manag. 2 (1), 25–35. doi:10.3126/juem.v2i1.70406.

Baruah, D., Hazarika, L.P., Dutta, R., 2009. Probable downstream impacts of 2000 MW Lower Subansiri Dam in Arunachal Pradesh of Northeast India. Nat. Environ. Pollut. Technol. 8 (4), 835–844.

Dutta, B., Srivast, P., Boruah, A., 2024. Identification of Soil Erosion-Susceptible Areas Using Fuzzy Logic and Sediment Transport Index Aided by Mineralogical-Granulometric Analysis in Subansiri Basin. Research Square, Assam, India. doi:10.21203/rs.3.rs-4604071/v1.

Dutta, D., Das, S., Kundu, A., Taj, A., 2015. Soil erosion risk assessment in Sanjal watershed, Jharkhand (India) using geo-informatics, RUSLE model and TRMM data. Model. Earth Syst. Environ. 1 (4). doi:10.1007/s40808-015-0034-1.

Dutta, S., 2016. Soil erosion, sediment yield and sedimentation of reservoir: a review. Model. Earth Syst. Environ. 2 (3), 1–18. doi:10.1007/s40808-016-0182-y.

Gebremichael, A., Gebremariam, E., Desta, H., 2025. Assessment of soil erosion and sediment transport index in the Awash River Basin, Ethiopia: an application of the USLE model and GIS techniques. Discov. Sustain. 6 (1). doi:10.1007/s43621-025-01018-x.

Ghimire, S., Singh, U., Panthi, K.K., Bhattarai, P.K., 2024. Spatial sediment erosion and yield using RUSLE coupled with distributed SDR model. Water 16 (24), 3549. doi:10.3390/w16243549.

Huang, J., Sinclair, H.D., 2025. Sediment aggradation rates in Himalayan rivers revealed through the InSAR differential residual topographic phase. Earth Surf. Dynam. 13 (4), 531–547. doi:10.5194/esurf-13-531-2025.

Maqsoom, A., Aslam, B., Hassan, U., Kazmi, Z.A., Sodangi, M., Tufail, R.F., Farooq, D., 2020. Geospatial assessment of soil erosion intensity and sediment yield using the Revised Universal Soil Loss Equation (RUSLE) model. ISPRS Int. J. Geo-Inf. 9 (6). doi:10.3390/ijgi9060356.

Menon, M., Vagholikar, N., Kohli, K., Fernandes, A., 2003. Large dams in the Northeast: a bright future? Ecol. Asia 11 (1), 3–8.

MOEF, 2001. State of Environment Report. Government of India, New Delhi.

Nguyen, H.T.T., Doan, T.M., Radeloff, V., 2018. Applying random forest classification to map Land Use/Land Cover using Landsat 8 OLI. Int. Arch. Photogramm. Remote Sens. Spatial Inform. Sci. XLII-3/W4, 363–367. doi:10.5194/isprs-archives-xlii-3-w4-363-2018.

Othman, A.A., Al-Maamar, A.F., Al-Manmi, D.A.M.A., Liesenberg, V., Hasan, S.E., Obaid, A.K., Al-Quraishi, A.M.F., 2020. GIS-based modeling for selection of dam sites in the Kurdistan Region, Iraq. ISPRS Int. J. Geo-Inf. 9 (4). doi:10.3390/ijgi9040244.

Pal, S.C., Chakrabortty, R., Roy, P., Chowdhuri, I., Das, B., Saha, A., Shit, M., 2021. Changing climate and land use of 21st century influences soil erosion in India. Gondwana Res. 94, 164–185. doi:10.1016/j.gr.2021.02.021.

Palmate, S.S., Amrit, K., Jadhao, V.G., Dayal, D., Himanshu, S.K., 2022. Prioritization of erosion prone areas based on a sediment yield index for conservation treatments: a case study of the upper Tapi River basin. In: Advances in Remediation Techniques for Polluted Soils and Groundwater. Elsevier eBooks, pp. 291–307. doi:10.1016/b978-0-12-823830-1.00019-5.

Rawat, K.S., Tripathi, V.K., Mishra, A.K., 2013. Sediment yield index mapping and prioritization of Madia subwatershed, Sagar District of Madhya Pradesh (India). Arab. J. Geosci. 7 (8), 3131–3145. doi:10.1007/s12517-013-1007-1.

Rwanga, S.S., Ndambuki, J.M., 2017. Accuracy assessment of land use/land cover classification using remote sensing and GIS. Int. J. Geosci. 08 (04), 611–622. doi:10.4236/ijg.2017.84033.

Sarma, C.P., Dey, A., Krishna, A.M., 2020. Influence of digital elevation models on the simulation of rainfall-induced landslides in the hillslopes of Guwahati, India. Eng. Geol. 268, 105523. doi:10.1016/j.enggeo.2020.105523.

Stefanidis, S.P., Chatzichristaki, C.A., Stefanidis, P.S., 2021. An ArcGIS toolbox for estimation and mapping soil erosion. J. Environ. Prot. Ecol. 22 (2), 689–696.

Tilahun, A., Desta, H., 2023. Soil erosion modeling and sediment transport index analysis using USLE and GIS techniques in Ada'a watershed, Awash River Basin, Ethiopia. Geosci. Lett. 10 (1). doi:10.1186/s40562-023-00311-9.

Xu, D., He, Y., Tan, Q., 2023. Improvement of sediment yield index model through incorporating rainfall erosivity. Environ. Sci. Pollut. Res. 30 (13), 38141–38156. doi:10.1007/s11356-022-24923-4.

Zarris, D., Vlastara, M., Panagoulia, D., 2011. Sediment delivery assessment for a transboundary Mediterranean catchment: the example of Nestos River catchment. Water Resour. Manag. 25 (14), 3785–3803. doi:10.1007/s11269-011-9889-8.

Leveraging drone technology for agrichemical spraying and locust control: Policy insights from the Vidarbha region of Maharashtra, India

20

Vishal Katekar[a] *and Jeevan Kumar Cheruku*[b]
[a]National Institute of Disaster Management (NIDM), Ministry of Home Affairs, Government of India, New Delhi, India
[b]Maulana Azad National Urdu University, Hyderabad, India

1 Introduction

Sustainable agriculture is vital for food security and long-term economic growth (Hoummaidi et al., 2021). Agriculture currently occupies 34% of the world's land and consumes 70% of freshwater, putting immense strain on natural resources (Lee, 2019). Despite this, agricultural practices often lack sustainability (Tian et al., 2018). The rising demand for food, energy, and clean water (Hoff, 2011) necessitates the adoption of modern technologies to enhance agricultural productivity and efficiency.

The Vidarbha region of Maharashtra in India depends heavily on crops like cotton, soybean, and paddy, yet it faces significant challenges. These include low productivity, decreasing landholding sizes, inefficient resource use, frequent crop failures, and minimal market connections (Thota, 2021). Environmental factors, including pests and diseases, further impact agricultural output (Patil and Patil, 2020). Locust attacks in 2020 highlighted the inadequacies of traditional pest control methods, which rely heavily on labor and can be costly and environmentally harmful.

Drones have emerged as a promising solution for agricultural challenges. They have been used for various purposes in India, including during the COVID-19 pandemic for vaccine delivery. Notably, India was the first country to deploy drones for locust control (Business Standard, 2020). Drones can efficiently cover large areas, access rugged terrains, provide precise spraying, and reduce labor costs, thus addressing many issues in agricultural practices.

This chapter discusses an empirical study on implementing drone technology for agri-chemical spraying and locust control in the Vidarbha region. It explores the challenges faced by farmers and officials in adopting this technology and presents

Precision Technologies for Digital Agriculture. https://doi.org/10.1016/B978-0-443-40513-6.00002-0

strategies to overcome these obstacles. The chapter is structured as follows: Section 2 describes the data collection methods and reviews the literature on crop-pest issues and manual spraying, highlighting how drones can assist. Section 3 presents the results and discussion, while Section 4 offers future directions based on the study's findings. Section 5 concludes the study.

2 Methodology

This study employed both quantitative and qualitative approaches. The study area is the Vidarbha region, which is the northeastern part of Maharashtra in India. Data were collected through expert in-depth interviews with 10 farmers and four officials who have expertise in drone technology for crop spraying. The interview questions were open-ended, allowing the farmers and officials to share their experiences in more detail. The field study was conducted on the agricultural fields of respondent farmers to examine the effectiveness of drone technology in crop protection. This section includes a review of the literature on crop-pest diseases and agri-chemical spraying, the need for precision agri-chemical spraying using drone technology, a comparison between manual and drone-based spraying of agri-chemicals, use cases, and specialized applications. Finally, challenges in the adoption of drone technology in agriculture are discussed.

2.1 Crop pest disease and agri-chemical spraying

The constant increase in food demand due to worldwide population growth continues to put pressure on agriculture. However, pests and diseases are threatening productivity, causing annual losses of 20%–40% globally. The health risks posed by traditional spraying underscore the urgent need for safer, more efficient, and environmentally friendly spraying practices (Bharad et al., 2024).

India loses 15%–25% of its food production yearly—valued at Indian Rupee (INR) 1.48 lakh crore—due to pests, weeds, and crop diseases (Balakrishnan, 2021). The Food and Agriculture Organization warns that this could escalate to over 40% if not addressed. The Vidarbha region faced a major locust invasion in 2020, causing extensive crop damage (Chatterjee, 2020; FAO, 2021). Farmers traditionally combat this issue through manual spraying of chemicals, which endangers their health due to exposure to hazardous substances. The World Health Organization estimates 3 million pesticide poisoning cases annually, resulting in about 220,000 deaths globally—mainly in developing countries (Patil and Patil, 2020). In India, around 300,000 farmers suffer respiratory illnesses from chemical exposure, and 442 farmers died between 2013 and 2018, with 183 fatalities in Maharashtra. Surveys in Maharashtra indicate that 8–10 farm laborers per village experience health issues from ongoing pesticide exposure (Balakrishnan, 2021). Notably, insecticides like Monocrotophos—classified as highly hazardous—have contributed to farmer fatalities (Ghadyalpatil, 2017). In addition, snakebites account for about 58,000 farmer deaths annually during agricultural activities (Humane Society International, 2020).

2.2 The need for precision agri-chemical spraying using drone technology

As the basic resources required for food production, such as land and water, decrease day by day (Sylvester, 2018), the incorporation of precision agriculture practices is a possible solution to use available resources more optimally and rationally. Optimum utilization of resources in agriculture can reduce the cost of production, and their precise application based on crop requirements can enhance the yield from the same piece of land (FICCI, 2020). The precision agriculture approach is helpful to farmers in maximizing the efficiency of agricultural inputs such as water, fertilizers, and pesticides, thereby enhancing overall crop production (Shukla, 2021). The use of advanced technologies in agriculture has provided opportunities for precision agriculture, promoting the socio-economic and environmental dimensions of agricultural sustainability (Borkar, 2021).

One of the most promising uses of digital technologies in agriculture is the application of drone technology for crop management (Hovhannisyan et al., 2018). It is one of the cutting-edge technologies that is spreading quickly across the globe. In essence, it is an unmanned aircraft that can be flown and controlled remotely from the ground. Hence, there is minimal risk to human life. People are adopting this technology in a very short time because operating it does not require extensive expertise. Users can control it through a smartphone app, which enhances its user-friendliness (Dalai and Sabyasachi, 2019). Chen et al. (2019) examined the effectiveness of drone technology for crop spraying in a field study and found that drones provide more precise and accurate spraying compared to traditional methods. When operated by a single person from a safe location, the drone can assist in fertilizers fertilizers and crop protection chemicals. The device can be used for spraying fertilizers, herbicides, and crop protection chemicals by modifying the type of container used. This significantly reduces the time required and ensures a safe distance, providing precautions for the farmer while spraying fertilizers (Patil and Patil, 2020).

According to De Bellis et al. (2020), drones provide accurate and precise spraying, leading to better crop protection and reduced chemical waste. Drones can also aid in the timely detection and control of locust swarms, resulting in reduced crop damage. The use of drones in agriculture is not limited to spraying and pest control but also extends to crop monitoring, mapping, and yield estimation. Drone-based crop monitoring provides real-time information on crop health, water stress, nutrient deficiencies, topography, soil type, and vegetation cover of the land, enabling farmers to make informed decisions on land use, crop selection, nutrient and crop disease management, and to optimize harvest operations and marketing strategies. Drones are capable of performing activities over any geographical location, which is difficult for human beings (Pathak et al., 2020). Drones are being used to take pictures of pest-affected areas, providing an accurate and efficient evaluation of crop loss. This increases the speed and accuracy of compensation provision to farmers based on their crop damage (Iqbal, 2020).

Different pests in agriculture can be controlled by manual spraying, but pests like the desert locust, which are highly destructive and fly at greater heights, pose a significant challenge. Traditional manual spraying is ineffective against them. Drone technology provides a safer and more targeted approach for chemical application in combating desert locust attacks (Ochieng et al., 2023). In addition, drones can enhance pest detection, crop health monitoring, data collection, and crop loss assessment after extreme weather events. The integration of drones with artificial intelligence (AI) and machine learning further improves productivity and decision-making in agriculture (Singh et al., 2024).

2.3 Comparison between manual and drone-based spraying of agri-chemicals

Traditional manual spraying is labor-intensive and poses health risks due to exposure to toxic chemicals, particularly under challenging terrains (Employment News, 2022). Drones eliminate much of this exposure and ensure consistent dosing, making spraying tasks quicker and safer. These innovations not only save time and costs but also promote organic farming practices and reduce environmental pollution (Mrunalini and Deb, 2021). Drone technology provides an effective and efficient alternative to manual methods, as highlighted in Table 20.1.

In India, reliance on conventional methods contributes to inefficiencies in crop monitoring and pesticide application, negatively impacting productivity and farmer well-being (Bharad et al., 2024). Global warming and climate change further accelerate pest attacks on crops, highlighting the need for new technologies and methods for crop protection. Researchers have long focused on using remote sensing for pest

Table 20.1 Comparison between traditional chemical spraying and drone-based spraying.

Parameter	Manual spraying	Drones-based spraying
Convenience	Laborious, and time-consuming	Convenient, fast, and highly efficient
Exposure to chemicals	High	No/minimal exposure during spraying
Coverage of chemical application	Nonuniform coverage	Global Positioning System (GPS)-enabled precision spray with uniform coverage
Efficiency	Suboptimal crop protection.	Efficient coverage of land in a short period of time
Skilled	Unskilled and novice operators	Trained and certified operators
Return on investment (ROI)	Low	Better ROI offsets the high operational cost
Water consumption	More water consumption	Less water consumption

Source: *Adapted from FICCI (2020, p. 3)*

identification and management; however, its adoption is limited due to accessibility issues, high costs, and the expertise required to operate it (Novaković and Rikalović, 2025). Drone technology has demonstrated significant advantages in agricultural spraying, optimizing agri-chemical application, improving efficiency, and reducing labor intensity compared to traditional methods, such as knapsack sprayers and tractor-based spraying (Jayakumar et al., 2023).

Various studies show that drones can spray five times faster than the traditional method of agricultural spraying, thus saving time and reducing input costs. They eliminate exposure to toxic chemicals for humans. The lower use of input chemicals in agriculture will promote the organic approach in agriculture and prevent environmental pollution (Mrunalini and Deb, 2021). Spraying pesticides at the right time and in the right quantities is possible with drones. Thus, drones enable targeted and site-specific precision spraying (Majumdar, 2019). Table 20.2 summarizes the differences between manual chemical spraying and using drone technology.

Automation processes are believed to have reduced the input costs of agri-chemicals (pesticides and weedicides) by approximately 25%–30%. For example, using nano-urea at a cost of INR 240 per bottle for spraying an acre can result in savings of around 25%, thus lowering overall input costs. Furthermore, drones, with their small droplet sizes of about 50 μm compared to the 500 μm produced by hand spraying, can save up to 90% of the water used in agriculture (The Print, 2021). While traditional spraying methods require a whole day to cover 3–4 acres, drone spraying can be completed in just 5–7 minutes per acre (Beriya, 2022).

Drones utilize global positioning system (GPS) technology to precisely determine their starting point and automatically return to it upon completion. The data collected is then sent to users' mobile devices or laptops. Studies show drones can increase spraying efficiency up to 60-fold, greatly enhancing agricultural productivity (Meesaragandla

Table 20.2 Quantitative findings related to agri-chemical spraying activity.

Sr. No.	Indicators (per acre)	Manual spraying	Drone-based spraying	Difference	Difference (%)
1	Cost of chemical (in INR)	640	270	370	57.81
2	Cost of spraying (in INR)	800	600	200	25
3	Chemical requirement (in litres)	0.43	0.18	0.25	58.13
4	Water requirement (in litres)	90	10	80	88.88
5	Time requirement (in hours)	3	0.11	2.89	96.33

Legend: 1 Acre = 0.40468564 ha; INR = Indian Rupee
Source: *From field study, 2022.*

et al., 2024). Environmental sustainability is also improved, as drones reduce chemical usage and minimize waste (Nikam, 2024). Drone-based spraying does not cause soil compaction or crop damage, unlike ground-based machinery. Advanced technologies integrated into drones have enhanced spraying accuracy and efficiency, with potential reductions of up to 30% in pesticide use (García-Munguía et al., 2024).

The integration of AI and Internet of Things technologies has further expanded the capabilities of drones. These integrations enable precision targeting, real-time monitoring, early disease detection, and yield estimation (Guebsi et al., 2024). Drones also enable the precise, site-specific application of agro-chemicals, thereby reducing waste and environmental pollution (Rao et al., 2024). High-resolution imaging and advanced sensors improve crop analysis, soil assessment, and overall farm management efficiency (Toscano et al., 2024; Rishikesavan et al., 2024).

2.4 Use cases and specialized applications

Drone technology has several specialized applications in agriculture. The following are a few use cases of drone technology:

- *Weed detection and herbicide spraying:* Drones integrated with a high-resolution camera can significantly improve weed detection and the targeted application of herbicide. Furthermore, precision in weed management in the field is enhanced due to the rotor downwash effect (Meesaragandla et al., 2024).
- *AI-driven crop protection:* Drones integrated with AI systems analyze image and sensor data to detect pests, diseases, and nutrient deficiencies, enabling targeted application of pesticides and fertilizers. They can also operate autonomously across large fields, even in challenging terrains (Novaković and Rikalović, 2025).
- *High-capacity drones:* Hexacopter drones with advanced systems, such as obstacle avoidance, adaptive spraying, and mobile app integration, enable efficient pesticide application, real-time crop monitoring, and remote scheduling (Nikam, 2024).
- *Locust management:* Research shows that spraying biopesticides from a 10-meter flight height achieves optimal droplet density and high mortality rates in desert locusts, particularly in early instar stages. This highlights the potential of drones in effective and targeted locust management strategies (Ochieng et al., 2023).
- *Quadcopters for agro-chemical spraying:* Quadcopter drones can navigate rugged terrains and autonomously conduct precise spraying operations, reducing labor costs and enhancing operational efficiency (Rao et al., 2024).

2.5 Challenges in the adoption of drones

Drone technology offers numerous advantages in agriculture; however, several challenges limit its widespread adoption in this sector. Operating a drone and conducting agri-chemical spraying requires skilled professionals, a drone pilot license, and knowledge related to chemical formulation, safety labeling, and aerodynamic spray dynamics (Meesaragandla et al., 2024). Several operational challenges, including nozzle type, flight speed, battery life, and droplet deposition, directly impact spray efficiency. High initial costs, operational expenses, and training requirements also make drone-based spraying less accessible than conventional methods.

Regulatory, technological, and socio-economic barriers further constrain adoption. Financial limitations, high hardware and sensor costs, and evolving regulatory frameworks restrict large-scale deployment (Bharad et al., 2024). Technical constraints, such as battery life and communication range, remain critical for effective drone operation (Singh et al., 2024). Payload capacity and flight autonomy also significantly influence drone performance in agricultural fields (Toscano et al., 2024; Rishikesavan et al., 2024). Guebsi et al. (2024) emphasized that integrating drone technology with decision-support systems and AI solutions will further increase its benefits in agriculture by enhancing efficiency, accuracy, and sustainability.

The next section presents and discusses the findings of the study.

3 Results and discussion

This section presents findings and interpretations from open-ended interviews with farmers and officials. The findings are analyzed both qualitatively and quantitatively, focusing on direct comparisons of economic, environmental, and social indicators between traditional manual spraying and drone-based spraying.

3.1 Agrichemical spraying using drone technology

The study reveals that drone technology has enormous potential to transform agriculture from a labor-intensive to a technology-intensive profession. Tables 20.1 and 20.2 present the comparative results based on data collected from ten farmers. Figs 20.1 to 20.2 depict the drone-based agro-chemical spraying by farmers in the study area, which spans 84 acres.

Quantitative analysis revolves around issues such as the cost of chemicals and spraying, as well as the requirements for inputs like the amount of chemical, water, and time.

Cost of chemical and spraying

It is evident from Table 20.1 that the chemical cost per acre for manual spraying is INR 640, compared to INR 270 for drone spraying. The difference is INR 370, representing a 57.81% reduction. Usually, manual spraying lacks precision in terms of the quantity of chemical required in the field. Thus, drone-based spraying is 57.81% less expensive for chemicals than manual methods. Similarly, manual spraying costs INR 800 per acre, while drone-based spraying costs INR 600, resulting in a difference of INR 200, or 25%. Thus, drone-based spraying is 25% cheaper than manual spraying. The study's findings align with those of Beriya (2022).

Table 20.3 Qualitative findings related to agri-chemical spraying activity.

S. No.	Indicators	Manual spraying	Drone-based spraying
Agronomic Performance			
1	Quality of activity	Not uniform spraying	Uniform spraying
2	Fertiliser use efficiency	Less	More
3	Real-time information	No	Yes
4	Precise application of Agri-chemicals	Less	More
Environmental Impact			
5	Soil and water pollution	More	Less
Social Factors			
6	Chemical exposure	Yes	No
7	Human safety	Less	More
8	Mental stress during spraying	Yes	No
9	Remote operation	No	Yes
10	Labour-intensive work	Yes	No
11	Convenience	No	Yes
12	Social respect	No	Yes
13	Transparency in working	Less	More
14	Impact of geographical location	Yes	No

Source: *From field study, 2021–2022.*

Fig. 20.1 Farmer filling the drone tank with agri-chemical before spraying. Source: Image from field study, 2021–2022.

Fig. 20.2 Drone based agrichemical spraying in a rice crop field.
Source: Images from the field study, 2021–2022.

Chemical and water requirement

The chemical required per acre is 0.43 liters manually and 0.18 liters via drone, resulting in a difference of 0.25 liters, or 58.13% less, with drones. Similar findings are reported by Beriya (2022) and FICCI (2020). Manual spraying requires 90 liters of water per acre, while drone spraying uses only 10 liters—an 80-litre or 88.88% reduction. Pathak et al. (2020) and Shukla (2021) found similar results in their study.

Time requirement

Manual spraying per acre takes 3 h, while drone-based spraying takes 0.11 h—a difference of 2.89 h or 96.33% less. Beriya (2022), Thota (2021), and Majumdar (2019) reported similar findings in their studies.

It can be concluded that using drones for spraying chemicals is cost-effective and efficient in terms of the use of chemicals and water.

Table 20.3 provides a qualitative comparison between manual spraying and drone-based spraying activities related to agrochemicals. The indicators are divided into three categories: agronomic performance, environmental impact, and social factors. The analysis is as follows:

Agronomic performance

The indicators related to agronomic performance include the quality of activity, fertilizer use efficiency, real-time information, and the precise application of agri-chemicals. Indicator-wise benefits of drone-based spraying over manual spraying are discussed below:

1. *Quality of activity:* It was observed in the field that manual spraying was not uniform, whereas drone-based spraying was more uniform. Drone-based spraying of chemicals resulted in more efficient chemical use, better crop coverage, and increased yields—similar findings were reported by Shukla (2021) and FICCI (2020).
2. *Fertilizer use efficiency:* Manual spraying is reported to have lower fertilizer use efficiency, while drone-based spraying is reported to have higher efficiency. Drone-based spraying ensures precise application of chemicals and uniformity of coverage. This finding is consistent with the results of studies conducted by De Bellis et al. (2020).
3. *Real-time information:* Manual spraying reportedly lacks real-time updates, whereas drone-based spraying provides them. Real-time information helps farmers make informed decisions about when and where to apply pesticides, resulting in more efficient resource utilization. One farmer explained how real-time updates from the smartphone app supported his decision-making process:

   "The exact status of spraying is updated in the Smartphone app. This real-time information helped me to make an informed decision related to the spraying."

4. *Precise application of agri-chemicals:* Manual spraying is reported to be less precise, while drone-based spraying is more precise. Drone-based spraying led to more efficient chemical use and reduced negative impact on the environment. The finding aligns with studies by Chen et al. (2019), De Bellis et al. (2020), Majumdar (2019), Shukla (2021), and FICCI (2020). One of the farmers, who owns a drone and provides spraying services to other farmers on a rental basis, pointed out that:

   "Many farmers are contacting me for chemical spraying on their crops as they realised that drone-based spraying is efficient and cost-effective. Here, the expenditure is less on chemicals and labour compared to the traditional manual spraying."

Based on the findings from Table 20.2, it can be concluded that drone-based spraying is more efficient than manual spraying in terms of fertilizer use, precise application of agrochemicals, and uniform spraying.

Environmental impact

It was found that the use of drones had a positive impact on the environment in terms of energy consumption, soil pollution, and water pollution. Indicator-wise benefits of drone-based spraying over manual spraying are discussed below:

Soil and water pollution: It has been observed that the precision of drone-based spraying reduces the amount of chemicals that pollute both the soil and water. In drone-based spraying, contact between soil, water bodies, and chemicals is completely avoided. It uses an aerial mode of spraying from a distance of up to six meters within the inserted GPS coordinates by a drone pilot. One of the farmers, who is using drone technology for agricultural spraying, described the effects on soil and water pollution:

> *"When I used to do manual spraying on cultivated crops, toxic chemicals were easily mixed with the soil and water, polluting the surrounding water bodies and soil. Due to the drone-based aerial-spraying of agri-chemicals, there are fewer chances for soil and water pollution."*

The finding aligns directly with those of Mrunalini and Deb (2021) and Banerjee (2021). Based on the data in Table 20.3, it is evident that drone-based spraying is more effective than manual spraying in reducing soil and water pollution, as it uses less water and fewer chemicals compared to traditional manual spraying, thereby contributing to a more sustainable environment.

Social factors

Drone-based spraying of chemicals has social benefits. The indicators related to social factors include chemical exposure, human safety, mental stress during spraying, remote operation, labor-intensive work, convenience, social respect, transparency in working, and the impact of geographical location. The social benefits of drone-based spraying over manual spraying are discussed below, broken down by indicator.

1. *Chemical exposure:* Manual spraying resulted in exposure to chemicals, whereas drone-based spraying does not. The absence of exposure due to drone-based spraying led to better health outcomes for farmers and their families.
2. *Human safety:* Manual spraying presents a lower level of human safety compared to the higher safety of drone-based spraying. Reduced chemical exposure and remote operation lower the risk of snakebites and wild animal attacks, a point echoed by Patil and Patil (2020) and FICCI (2020).
3. *Mental stress during spraying:* Manual spraying caused mental stress, whereas drone-based spraying did not. Drone spraying required less physical work and is easier. It required less workforce, as a drone could spray at any time.
4. *Labor-intensive work:* Manual spraying required significantly more manual effort compared to drone-based spraying, which reduced labor requirements. As a result, farmers experienced improvements in working conditions and a reduction in labor costs. This study supports the findings by Patil and Patil (2020) and FICCI (2020).
5. *Convenience:* Manual spraying offered no convenience, while drone-based spraying did. This is due to remote operation, reduced physical demands, improved access to real-time information, and a shorter time required to complete spraying. The study's observations align with findings reported by Dalai and Sabyasachi (2019), Patil and Patil (2020), and FICCI (2020).
6. *Social respect:* Manual spraying is reportedly lacking in social respect, whereas drone-based spraying is seen as more modern and, thus, socially respected. One of the farmers using drones for agricultural spraying responded as follows:

> *"After the use of drone technology, I started getting recognition in society. Many farmers began seeking my advice and services related to spraying. District-level*

Agricultural Officers and assistants frequently visit me to know about the adoption of new technology in agriculture. The use of drone technology not only improved my income but also reduced the financial burden and mental stress. As it increased my living standard, I got more time to spend with my family, friends, and relatives. I am spending my time in different activities to increase my income."

7. *Transparency in working:* It has been observed that drone spraying is more transparent than manual spraying because it provides real-time information and tracks spraying progress.
8. *The impact of geographical location:* Manual spraying is more effective in flat areas and low-height crops, but it faces limitations in other geographical settings. Drone-based spraying, by contrast, can operate in any location due to its aerial capabilities—evidence supported by Pathak et al. (2020).

Based on the data in Table 20.3, it can be concluded that drone-based spraying is superior to manual spraying in terms of remote operation, convenience, social respect, transparency in working, and the impact of geographical location. Human safety is improved when performing drone-based spraying, as there are fewer chances of being bitten by a snake or attacked by wild animals. It also reduces the need for labor, chemical exposure, and adverse health effects.

3.2 Antilocust operation in the vidarbha region

The researchers interviewed officials from the Department of Agriculture involved in the antilocust operation in the Vidarbha region. Nagpur and Bhandara districts were highly affected by locust attacks. Chemical spraying was their only option, but manual methods were ineffective against flying locusts. Drones were used to spray over 500 liters of chemicals, breaking the chain of multiplication. Drones made it possible to spray at heights—including on tall trees—that are unreachable manually. Operations became faster, covering 2.5 acres in under 15 minutes, while drones also provided reliable information and efficient crop loss evaluation for financial relief. Interviews with government officials revealed the effectiveness of locust control and the disruption of their multiplication chain in Vidarbha, as drones provided more precise and efficient chemical application. Studies by Iqbal (2020) report similar results.

Field observations confirm that drone-based spraying is cheaper and requires less chemical, water, and time. It is more efficient and effective agronomically, environmentally, and socially than manual methods. The use of drones also proved to be an effective locust control solution in Vidarbha during the COVID-19 pandemic, highlighting the potential for adopting such technology in agriculture, as supported by earlier literature.

3.3 Challenges in the adoption of drone technology

Although drone technology offers socio-economic and environmental benefits, its adoption for chemical spraying is hindered by significant challenges. High initial investment costs, a shortage of skilled operators, and total reliance on imported drones and components create substantial barriers for small-scale farmers.

In the Vidarbha region, the lack of local drone pilot training institutes, high training fees, and elevated battery costs restricted access for smallholder farmers. Limited battery capacity necessitated multiple charged batteries and a separate generator for each operation. Stringent regulations by the Directorate General of Civil Aviation and the absence of local repair and spare parts centers further impeded the adoption of new technology. To reduce expenses, farmers often had to forgo advanced features, raising concerns about both affordability and ongoing maintenance. A farmer averred:

> *"I think the most important drawback of drone technology is its battery backup. It has very low battery backup. The drone and its related hardware are costly and not even available in the domestic markets. It is challenging for us to purchase spare parts for a drone. Operating and maintenance of a drone is difficult due to the limited resources and knowledge about such technology."*

Government officials also encountered challenges due to the lack of government-owned drones. They experienced difficulties in recruiting skilled personnel and drone pilots for antilocust operations. Officials had to rent drones from remote locations, which proved to be costly. A government official responded:

> *"We do not have our own drones and a permanent drone pilot with us, so we rented a few drones from a private player. However, we did not get sufficient drones in time for the anti-locust operation. This delay in spraying chemicals on locusts has hampered the operation's effectiveness."*

4 Future directions

The study concludes that drones have transformed agricultural practices in Vidarbha by enabling more effective and sustainable management of crops and resources. Addressing the challenges associated with adopting drone technology for agrichemical spraying requires targeted policy interventions in the Vidarbha region of Maharashtra. The following directions are proposed for the future course of action:

More liberalised regulatory framework

Drone operations for agricultural purposes beyond visual-line-of-sight (BVLOS, meaning flights where the drone operator cannot directly see the drone during flight) are now prohibited by rules, which makes it difficult to cover large agricultural fields in a single flight. There is a need to liberalize laws related to the use of cameras and advanced sensors on agricultural drones so that all farmers can benefit from the more advanced application of this technology.

Incentives and subsidies

The government should provide incentives for drone manufacturers to quickly speed up operations. Training facilities and operational tie-ups are needed to stimulate the use of drones in agriculture. Subsidies for the purchase of drones and related services should be given directly to producers. Currently, there is a provision for a "Kisan Drone Subsidy" of up to 50% of the drone's cost, based on social background and organizational type. However, there is a need for simplification in the process of obtaining the subsidy. The subsidy amount is also insufficient, as the cost of an agricultural drone is significantly higher. It is necessary to offer financing and leasing options to enable farmers and village-level entrepreneurs to rent or purchase drones. A government organization may purchase a drone and provide it to farmers on a rental basis. The success of drone technology in agriculture depends on the number of drones provided by the government to farmers, as they cannot afford to purchase the expensive drones.

Registration process

To ensure safer and easier drone adoption, a single-window ecosystem should be developed to facilitate the adoption of agricultural drones by farmers.

Digital Infrastructure and Training

There is a need to establish local drone manufacturing, assembling, and repairing centers so that farmers can benefit more from drones. Agricultural universities should standardize operational guidelines and ensure these are accessible in the regional language. Education and training institutions must adapt to provide affordable drone skills programs, as private options remain costly. Efforts should also be made to incorporate drone training into existing government skill development programs.

Fast approval and autonomous flight permission system

The government should permit fully autonomous AI-based drone flights, with restrictions on BVLOS operations, where the drone cannot be seen by the pilot, and on night missions beyond 400 feet. This will help increase the number of hours of agricultural drone operation, thereby lowering operational costs while maintaining a favorable benefit-cost ratio.

Custom hiring of drone services

Small and marginal farmers are hesitant to adopt drones due to the high price associated with purchasing them and the required expertise. The central and state governments should provide financial support for starting custom hiring centers at the local level. The custom hiring of drone application services will increase the use of drone technology for agri-chemical spraying and advanced applications by all types of farmers. The effective use of the existing local custom hiring network for other agricultural equipment and machinery will help connect the demand and supply chains for agricultural drone applications.

5 Conclusion

The agriculture sector in the Vidarbha region faces longstanding challenges due to locust attacks that have affected farmers for decades. Advanced digital technologies, particularly drone technology, offer significant potential to improve agricultural outcomes. This study demonstrates that drone-assisted crop spraying enhances efficiency and effectiveness in crop protection, leading to reduced costs and chemical waste. While drone-based spraying provides notable economic, environmental, and social benefits, these advantages can only be realized if the adoption challenges encountered by farmers and government officials are addressed. Policymakers and stakeholders must take decisive action to remove these barriers and enable the full potential of drone technology in Vidarbha's agricultural sector.

Acknowledgement

The authors are grateful to the National Institute of Disaster Management, Government of India, and the Department of Public Administration, Maulana Azad National Urdu University (MANUU), Hyderabad.

Conflict of interest

There is no conflict of interest.

Funding sources

The authors received no financial support for the research, authorship, and/or publication of this article.

References

Balakrishnan, R. (2021). *This woman entrepreneur's startup has developed an AI-based crop protection drone for climate-smart agriculture*. Retrieved March 8, 2023, from yourstory.com: https://yourstory.com/herstory/2021/10/woman-entrepreneurs-startup-ai-drone-crop-protection-agriculture/amp.

Beriya, A. (2022). *Application of Drones in Indian Agriculture, ICT India Working Paper, No. 73*. Center for Sustainable Development (CSD), Earth Institute, Columbia University, New York.

Bharad, N.B., Vagadia, V.R., Lakhani, A.L., Khanpara, B.M., Dulawat, M.S., 2024. A comprehensive review of unmanned aerial vehicle (UAV) sprayers used in modern agriculture. J. Sci. Res. Rep. 30 (10), 573–585. doi:10.9734/jsrr/2024/v30i102483.

Borkar, S. (2021). *How Agritech is Changing the Face of Sustainable and Industrial Agriculture in India*. Retrieved March 8, 2023, from inc42.com: https://inc42.com/resources/how-agritech-is-changing-the-face-of-sustainable-and-industrial-agriculture/

Business Standard. (2020). India is the first country to control locusts through drones: agriculture ministry. *Business Standard*. Retrieved March 8, 2023, from https://www.business-standard.com/article/economy-policy/india-first-country-to-control-locusts-through-drones-agriculture-ministry-120062800056_1.html

Chatterjee, B., 2020. First major locust attack in Maharashtra since 1993; 3 districts hit. Hindustan Times. Retrieved March 8, 2023, from https://www.hindustantimes.com/mumbai-news/first-major-locust-attack-in-maharashtra-since-1993-3-districts-hit/story-hll1tdD0Hty70JvAlKAGXL.html.

Chen, J., Zhang, J., Huang, J., 2019. The efficiency of unmanned aerial vehicle in pesticide spraying: a field study. J. Clean. Prod. 222, 647–656.

Dalai, P.R., Sabyasachi, S., 2019. Future of Drone Technology in India. Journal of Xi'an Shiyou University 15 (3).

De Bellis, L., Friedrich, K., Dotto, A.C., Torres, M.L., 2020. Unmanned aerial systems for agricultural applications: a review. Sustainability 12(24), 10410.

Employment News, 2022. Making India a Global Drone Hub, Ministry of Information and Broadcasting, New Delhi, India.

FAO. (2021). *Desert Locust*. Retrieved March 8, 2023, from www.fao.org: https://www.fao.org/locusts/en/

FICCI. (2020). *Technical Discussion Paper on Drone usage for agro-chemical spraying*. New Delhi: Croplife India and FICCI.

García-Munguía, A., Guerra-Ávila, P.L., Islas-Ojeda, E., Flores-Sánchez, J.L., Vázquez-Martínez, O., García-Munguía, A.M., García-Munguía, O., 2024. A review of drone technology and operation processes in agricultural crop spraying. Drones 8 (11), 674. doi:10.3390/drones8110674.

Ghadyalpatil, A., 2017. Vidarbha pesticide deaths: probe panel may file report this week. Livemint. Retrieved March 8, 2023, from www.livemint.com: https://www.livemint.com/Politics/YtxwY9FhCdap9ra8aBWx4J/Vidarbha-pesticide-deaths-Probe-panel-may-file-report-this.html. ().

Guebsi, R., Mami, S., Chokmani, K., 2024. Drones in precision agriculture: a comprehensive review of applications, technologies, and challenges. Drones 8 (11), 686. doi:10.3390/drones8110686.

Hoff, H. (2011). Understanding the nexus: background paper for the Bonn 2011 nexus conference: the water, energy and food security nexus. 19-18 Nov. 2011. Bonn.

Hoummaidi, L.E., Larabi, A., Alam, K., 2021. Using unmanned aerial systems and deep learning for agriculture mapping in Dubai. Heliyon Heliyon 7.

Hovhannisyan, T., Efendyan, P., Vardanyan, M., 2018. Creation of a digital model of fields with application of DJI phantom 3 drone and the opportunities of its utilisation in agriculture. Ann. Agrar. Sci. 16, 177–180.

Humane Society International. (2020). *NGOs' snake tracking project hopes to save lives as new data shows 58,000 human deaths in India from snakebites*. Retrieved March 8, 2023, from www.hsi.org: https://www.hsi.org/news-media/ngos-snake-tracking-project-hopes-to-save-lives/

Iqbal, M. (2020, June 02). Drones deployed for controlling locusts in Rajasthan. *The Hindu*. Retrieved March 8, 2023, from www.thehindu.com: https://www.thehindu.com/news/national/other-states/locust-attack-about-90000-hectares-hit-in-20-rajasthan-districts/article31694494

Jayakumar, V., Mohideen, A.B.K., Saeed, M.H., Alsulami, H., Hussain, A., Saeed, M., 2023. Development of complex linear diophantine fuzzy soft set in determining a suitable Agri-Drone for spraying fertilisers and pesticides. IEEE Access 11, 9031–9041. doi:10.1109/access.2023.3239675.

Lee, J. (2019). AgTech trends in 2019: precision agriculture and millennial farmers. *G2 Crowd Learning Hub*. 3 Dec. 2019

Majumdar, G. (2019). Drone-Aided Cotton Farming. *Cotton Statistics and News, 2019-20*(15). Retrieved March 8, 2023, from www.caionline.in

Meesaragandla, S., Jagtap, M.P., Khatri, N., Madan, H., Vadduri, A.A., 2024. Herbicide spraying and weed identification using drone technology in modern farms: a comprehensive review. Results Eng. 21, 101870. doi:10.1016/j.rineng.2024.101870.

Mrunalini, K., Deb, C.K., 2021. Chapter 16: drones in Agriculture. In: Rathore, S., B., V., Kumar, V.S., Digital Technologies in Agriculture. Biotech Books, New Delhi, pp. 171–182.

Nikam, K., 2024. Development of agriculture spraying drone. Int. J. Sci. Res. Eng. Manag. 08 (03), 1–5. doi:10.55041/ijsrem29644.

Novaković, N., Rikalović, A., 2025. The future of agricultural spraying: AI and UAVs for smart and sustainable practices. Forging the Future: Pioneering Approaches in Business 1, 681–690. doi:10.24867/future-bme-2024-077.

Ochieng, V., Rwomushana, I., Ong'amo, G., Ndegwa, P., Kamau, S., Makale, F., Chacha, D., Gadhia, K., Akiri, M., 2023. Optimum flight height for the control of desert locusts using unmanned aerial vehicles (UAV). Drones 7 (4), 233. doi:10.3390/drones7040233.

Pathak, H., Kumar, G., Mohapatra, S.D., Gaikwad, B.B., Rane, J. (2020). Use of Drones in Agriculture: Potentials, Problems and Policy Needs. ICAR-National Institute of Abiotic Stress Management, Baramati, Maharashtra, India.

Patil, K.M., Patil, M.M., 2020. Fertiliser spraying drone for agriculture application and controlling it over Wi-Fi. International Journal of Innovative Research in Science, Engineering and Technology (IJIRSET) 9 (9).

Raosaheb Gatkal, N., Madhukar Nalawade, S., Balasaheb Bhanage, G., Kumar Sahni, R., Ashok Walunj, A., Bhaskar Kadam, P., Musrrat Ali, N., 2024. Review of UAVs for efficient agro-chemical spray application. Int. J. Agric. Biol. Eng. 17 (6), 1–9. doi:10.25165/j.ijabe.20251801.8979.

Rishikesavan, S., Kannan, P., Pazhanivelan, S., Kumaraperumal, R., Sritharan, N., Muthumanickam, D., Firnass, M.M.R.A., Venkatesh, B., Teja, V.S., 2024. Prospects and challenges of drone technology in sustainable agriculture. Plant Sci. Today 11 (sp4). doi:10.14719/pst.5761.

Shukla, P. (2021). *How farmers can benefit using drones*. Retrieved March 8, 2023, from www.agritimes.co.in: https://agritimes.co.in/livestock/how-farmers-can-benefit-using-drones

Singh, N., Gupta, D., Joshi, M., Yadav, K., Nayak, S., Kumar, M., Nayak, K., Gulaiya, S., Rajpoot, A.S., 2024. Application of drones technology in agriculture: a modern approach. J. Sci. Res. Rep. 30 (7), 142–152. doi:10.9734/jsrr/2024/v30i72131.

Sylvester, G. (2018). E-Agriculture in action: drones for agriculture. FAO, Bangkok.

The Print. (2021). *Ministry of Agriculture-backed Startup IG Drones is helping farmers to boost income using Drone Geospatial Data*. Retrieved March 8, 2023, from the print.in: https://theprint.in/ani-press-releases/ministry-of-agriculture-backed-startup-ig-drones-is-helping-farmers-to-boost-income-using-drone-geospatial-data/767604/

Thota, P. (2021). *Precision farming and drones can help India make the big leap*. Retrieved March 8, 2023, from www.thehindubusinessline.com: https://www.thehindubusinessline.com/opinion/farming-the-potential-of-drones/article34573234.ece

Tian, H., Lu, C., Pan, C., Yang, J., Miao, R., Ren, W., Yu, Q., Fu, B., Jin, F., Lu, Y., Melillo, J., Ouyang, Z., Palm, C., Reilly, J., 2018. Optimising resource use efficiencies in the food-energy-water nexus for sustainable agriculture: from conceptual model to decision support system. Curr. Opin. Environ. Sustain. 33, 104–113. doi:10.1016/j.cosust.2018.04.003.

Toscano, F., Fiorentino, C., Capece, N., Erra, U., Travascia, D., Scopa, A., Drosos, M., D'Antonio, P., 2024. Unmanned aerial vehicle for precision Agriculture: a review. IEEE Access 12, 69188–69205. doi:10.1109/access.2024.3401018.

Index

Note: Page numbers followed by "*f*" indicate figures, "*t*" indicate tables.

D

E

K

L

M

Printed and bound by CPI Group (UK) Ltd, Croydon, CR0 4YY
25/09/2026
02223046-0001